# Meditation – Neuroscientific Approaches and Philosophical Implications

# Studies in Neuroscience, Consciousness and Spirituality

## Volume 2



For further volumes:
http://www.springer.com/series/10195

Stefan Schmidt • Harald Walach
Editors

# Meditation – Neuroscientific Approaches and Philosophical Implications

Springer

*Editors*
Stefan Schmidt
Department of Psychosomatic Medicine and Psychotherapy
University Medical Center Freiburg
Freiburg, Germany

Harald Walach
Institute for Transcultural Health Science
European Universtiy Viadrina, Frankfurt
Frankfurt, Germany

ISSN 2211-8918 ISSN 2211-8926 (electronic)
ISBN 978-3-319-01633-7 ISBN 978-3-319-01634-4 (eBook)
DOI 10.1007/978-3-319-01634-4
Springer Cham Heidelberg New York Dordrecht London

Library of Congress Control Number: 2013954596



Printed on acid-free paper

Springer is part of Springer Science+Business Media (www.springer.com)

# Contents

# Introduction: Laying Out the Field of Meditation Research

**Stefan Schmidt and Harald Walach**

Meditation is not just a simple research object. The rising and popular field of meditation research or contemplative science as it is called has a much larger impact on modern science and on our society as we assume at the first glance. This is because meditation is not only a fascinating research object but also quite a challenge for our current scientific practice. The content of this volume, which summarizes presentations and discussions of an expert meeting in Freiburg, Germany, documents the many facets and implications of meditation research. In this introduction we will touch on a few of them.

The first challenge meditation research brings to the established scientific paradigm is that it is pointing to the *neglect of experience*. Our personal first person experience, which, by the way, is our primary access to the world, cannot be shared with others directly. You cannot explain how chocolate tastes. In order to get the experience of the taste of chocolate others have to taste it themselves. Even worse, scientists are not even able to explain the phenomenal experience (qualia) using the current predominant approach in consciousness research (Chalmers 1995). The pragmatic solution taken so far was that personal experience was mainly devalued in comparison with third person approaches. And this scientific practice is also related to our cultural and social practice, where scientific beliefs often predominate personal experience. Classical examples are: Patients who have been told for many years that they cannot possibly have back pain because there is nothing to see on their x-ray; people having unwanted telepathic experiences being told that they are

S. Schmidt (✉)
European University Viadrina, Frankfurt (Oder), Germany

University Medical Center Freiburg, Freiburg, Germany
e-mail: stefan.schmidt@uniklinik-freiburg.de

H. Walach
European University Viadrina, Frankfurt (Oder), Germany

S. Schmidt and H. Walach (eds.), *Meditation – Neuroscientific Approaches and Philosophical Implications*, Studies in Neuroscience, Consciousness and Spirituality 2,
DOI 10.1007/978-3-319-01634-4_1, © Springer International Publishing Switzerland 2014

having illusions and being advised to seek psychological help because no scientific explanation is available; or people being shut away into psychiatric hospitals for no other reason than for hearing voices. The investigation into the practice of meditation has made a decisive difference here. The debate about including first person perspective into science had started already at the end of the 1990s (Shear and Varela 1999), and successively we were able to demonstrate that this is a feasible position which is also deeply rooted in the history of our western scientific paradigm (Walach and Runehov 2010; Walach 2011; see also the chapter by Walach in this volume).

Expanding on this first person approach we can now conceptualize the practice of meditation itself as a *scientific method*. Eastern cultures have investigated for millennia consciousness and the mind by introspective methods based on insights from meditation practice (Wallace 2007). Western science has abandoned introspective approaches since they were considered error-prone. What has been overlooked here is that any introspective approach needs the capability of stable attention regulation in order to arrive at an unwavering perspective for observing one's own inner processes. In the same way as you cannot play music on the piano without serious practice you cannot reliably observe your inner mental activities without having some stability in your attention. Meditation provides us with a technique to gain such stability, and thus trained mediators can be valuable participants in all kinds of psychological and neuroscientific experiments which rely on precise description of mental events (Lutz et al. 2007, see also Hasenkamp in this volume, Hinterberger in this volume for some examples).

Being an object and a method of research addresses another issue which is not usual for scientific practice. In order to study meditation most *researchers meditate themselves*. Some even argue that you cannot work in this area at all, if you do not have your own first person experience. This in turn means that the conceptualizations of most of our colleagues in this field (ourselves included) are among other factors also driven by our personal meditation experience and also the specific tradition we align ourselves to. In this sense meditation is no research object that one can remain separate from oneself in the way like for example a biologists can stay separate from a certain protein. Meditation science is mostly done by meditating scientists. For somebody familiar with different traditions it is often obvious when reading a publication whether the researchers in charge are coming from the Theravada Buddhist, Tibetan Buddhist or Transcendental Meditation tradition. But on the other hand, with science normally conceptualized as an objective method which should be independent of the individual conducting the research procedures, this fact is rarely mentioned. Jonathan Shear and Hoyt Edge, both in this volume, independently stress the importance and impact of this personal experience within the research context and make suggestions how to incorporate it into the research team or even into the process of forming teams including diverse perspectives on meditations by mixing researchers with background in different traditions.

But meditation is also a complex research object for yet more reasons. One is that meditation is almost *impossible to define* and we have to look for different approaches to conceptualize meditation within the scientific paradigm (see the chapter by Schmidt in this volume). But even the very idea of defining an object of research makes obvious where meditative practices and scientific approach diverge. Science needs definitions in order to create stable research objects. These well-defined objects, sometimes termed constructs, can then be empirically investigated by research teams all over the world. But let's take e.g. the practice of mindfulness. Mindfulness is entailing a radical reorientation towards one's own experience. One of the major insights gained by this approach is that there is no stability in experience; or in other words that objects and constructs will wax and wane and that they are thus essentially of a void nature, or at the very least quite fickle. This is, epistemologically speaking, a radically different approach to the world. Trying to find a definition for mindfulness seems to be a contradiction in itself.

Another issue where meditation is at odds with the standard scientific paradigm is the topic of the *individuality* of meditative states and experiences in meditators, despite similar formal traditions that are being followed. Here we are hardly at the beginning of even recognizing the problem. If you take a short glance at the colorful charts mapping brain activities of 50 different meditators in the chapter of Hinterberger (in this volume) you will get a first impression of the issue at stake. Standard empirical approaches normally imply that an intervention does the same to everybody and thus statistics across populations meditating or not meditating are computed and compared. The idea that this might be misleading came to our mind for the first time when we measured two very experienced nuns from the Theravada tradition in our neurophysiological laboratory in Freiburg, Germany. The two nuns, as well as the two sessions were quite alike. We spoke about techniques, meditation practices and different approaches and for us this all looked very similar in background, culturalization and practice of meditation. But the resulting EEGs of these two women were quite different. Thus, not only can we hardly compare inner experiences of meditators which each other, moreover, we even have to doubt that similar experiences result in similar third person data. Hinterberger in his chapter shows some data documenting this conundrum, and first approaches to deal with this variation in an adequate way.

With respect to this individuality of meditation we always have to recall the fact that meditation – although it is nowadays frequently applied in the clinical context – is *no passive drug*. Patients learning to meditate not only adopt a certain practice of attention regulation. Most of them will also experience a deep intellectual reorienting themselves and their lives. And it very likely is especially this part of finding new meaning and understanding which is part of the clinical effectiveness. In a qualitative study we have assessed the understanding of the concept of mindfulness by participants of a mindfulness-based stress reduction program (MBSR) and found huge differences, mainly regarding biographical aspects (see Schmidt, this volume). So while we still use the method for assessing a passive drug – i.e. the randomized controlled trial – to measure clinical effectiveness of mindfulness-based interventions, we need to remind ourselves that the model behind the changes observed does not really fit this method. This is another incentive for researchers within contemplative sciences to develop new methods which are more adequate to the topic. And this generally applies to most scientific approaches regarding meditation and its benefits. The goals of meditation are mostly levels of attainment within and of consciousness, and we have hardly any methods to assess these changes. The longstanding debate whether we can measure a potential 'level of mindfulness' (if such a thing even exists) by a self-report questionnaire (Belzer et al. 2013; Grossman 2011) may serve as a blue-print for this shortage and demonstrate that there are some tasks still on our way ahead.

If we continue to compare meditation with other topics of research, we can furthermore see that the investigation of meditation cannot be relegated to one scientific field alone. Contemplation research has an *interdisciplinary basis at its very roots* like hardly any other topic in science. This is also shown by the sociological aspects of the new field of meditation research which reveal some crucial differences to other disciplines. So far no major scientific societies for contemplative research have been founded. Only one specialized journal for the topic has been launched ("Mindfulness"). Initiatives regarding meditation research are coming from all kinds of different disciplines with psychology, neuroscience and maybe philosophy at the forefront, but also with impact from biology, cultural studies, Asian studies, or religious studies, not to mention organizations fostering the dialogue between science and spirituality, or between religion and society. Here we observe a new movement in the making, which in some aspects is quite different from others we have seen in science. The potential impact and benefit of meditation research to address and critically debate the basic assumptions of science themselves is obvious and, in our view, there is a lot to gain from this process.

This is especially true since science is currently the predominating paradigm for creating meaning in our modern societies. Any change within science will also result in societal changes, and this can already be seen with respects to the many effects scientists have reported in studies on meditation. These results inspire a cultural change which brings meditation-based and mindfulness-based approaches to many areas of our live. Meditation and mindfulness are now widely secularized and this transformation, due to a change in motivation, alters the practice itself (see Schmidt 2011 for a more detailed discussion). But this popularization of meditation is not only driven by science. It can also be seen from a different angle. With the current social acceleration of our culture (Rosa 2010), the ongoing individualization, as well as the resulting functionalization and capitalization of every tiny corner of our lives people are increasingly questioning our modern life style, and its consequences of continuing discontentment and stress. From a systemic perspective the popularity of meditation can be seen as a process of self-regulation of our society. And this social change in perspective requests, reciprocally, a confirmation from and by science. However, such a self-regulation will fail, if meditation based approaches are functionalized similar to current approaches of improving performance within the dominant model of submitting all human activities to the primary goal of economic productivity. And this can already be seen in many aspects. A mindfulness-based stress reduction course in the work place in order to raise production and profit of the company is a huge misunderstanding. The same program, with the primary motivation of improving quality of life by changing procedures, relationships and practices within the work-place has some potential for benefits. With respect to this potential social impact of contemplative practices it might be explainable why findings about meditation are often overrated in the public debate, or why the discourse about meditation changing the brain is so popular. The fact that meditation has the potential to change the way how we experience ourselves and the world might be of much more personal relevance.

In the end, we are coming back to the ever pending question what we need *to live a good life*. The philosopher Hoyt Edge (this volume) points out that neuroscience cannot find out what leads to a flourishing life. It is the classical realm of philosophy which addresses this question mostly in a rational and ethical fashion. Meditation with its spiritual background and its mainly Buddhist psychology, has the same aim, but uses a different approach. Rather than seeking overarching principles by rational considerations the impact is on staying in resonance with the world by grounding ourselves in experience and by inquiring and transforming the functioning of our mind. This is where we can benefit from the discourse on differences in traditional western and eastern cultural approaches to the world. It goes without saying that meditation research is not necessarily tied to a Buddhist world-view. It can also be approached from Western perspective and function to reinvigorate those spiritual traditions through experience.

Based on these considerations it is obvious that meditation and science share a special relationship. Meditation and the research on meditation will change science. The scientific findings about the effects of meditation change our cultures and societies. This, quite obviously, will lead to the fact that meditation based programs will become interwoven with many of our cultural processes (work, health, school), and this will also be changing the practice of meditation itself. It is with this background that we started the introduction of this volume on meditation research with the statement that meditation is not just a simple research object like any other. It is one of the most inspiring fields, science and culture can enter these days.

**Acknowledgments** Large parts of this volume are based on an interdisciplinary expert meeting taking place in Freiburg Germany in 2010. This meeting as well as the production of this book was sponsored by the Theophrastus Foundation, Germany and by the Samueli Institute, Alexandria; VA, USA.

## References

Belzer, F., S. Schmidt, G. Lucius-Hoene, J.F. Schneider, C.L. Orellana-Rios, and S. Sauer. 2013. Challenging the construct validity of mindfulness assessment – a cognitive interview study of the Freiburg mindfulness inventory. *Mindfulness* 4(1): 33–44. doi:10.1007/s12671-012-0165-7.

Chalmers, D.J. 1995. Facing up to the problem of consciousness. *Journal of Consciousness Studies* 2(3): 200–219.

Grossman, P. 2011. Defining mindfulness by how poorly I think I pay attention during everyday awareness and other intractable problems for psychology's (re)invention of mindfulness: Comment on Brown et al. (2011). *Psychological Assessment* 23(4): 1034–1040.

Lutz, A., J.D. Dunne, and R.J. Davidson. 2007. Meditation and the neuroscience of consciousness: an introduction. In *Cambridge handbook of consciousness*, ed. P. Zelazo, M. Moscovitch, and E. Thompson, 499–554. Cambridge: Cambridge University Press.

Rosa, H. 2010. *Alienation and acceleration: Towards a critical theory of late-modern temporality*. Malmö: NSU Press.

Schmidt, S. 2011. Mindfulness in east and west – is it the same? In *Neuroscience, consciousness and spirituality*, ed. H. Walach, S. Schmidt, and W.B. Jonas, 23–38. New York: Springer.

Shear, J., and F.J. Varela (eds.). 1999. *The view from within: First-person approaches to the study of consciousness*. Thorverton/Bowling Green: Imprint Academic.

Walach, H. 2011. Neuroscience, consciousness, spirituality – questions, problems and potential solutions: An introductory essay. In *Neuroscience, consciousness and spirituality*, ed. H. Walach, S. Schmidt, and W.B. Jonas, 1–21. New York: Springer.

Walach, H., and A.L.C. Runehov. 2010. The epistemological status of transpersonal psychology: The data-base argument revisited. *Journal of Consciousness Studies* 17(1–2): 145–165.

Wallace, B.A. 2007. *Contemplative science (where Buddhism and neuroscience converge)*. New York: Columbia University Press.

# Towards an Epistemology of Inner Experience

**Harald Walach**

**Abstract** Meditation research in the West has so far been a view onto the meditator from the outside, that is from a third-person perspective. Using inner experience, i.e. a first-person perspective, is uncommon, mostly because we do not have a reliable methodology, but also, because the prevalent mindset within science holds that such an enterprise is, ontologically speaking, of not much use. As long as consciousness is seen as purely derivative of matter and secondary to it, it cannot possibly have its own epistemological access to reality. I sketch here the historical conditions and systematic requirements that are necessary for an epistemology of inner experience to work. I hold that inner experience is not only a viable but also necessary mode of insight for a science that is more than natural science in the current sense. Many aspects of knowledge, such as values, creative insights into new theoretical models, intuition about new and fruitful avenues of research are strictly speaking only available to a first-person perspective and hence results of inner experience. The preconditions of such an epistemology are being discussed in this chapter.

## Inner Experience: Historical and Systematic Background

### *The Rise of the Empirical Method in Science and the Third-Person Viewpoint*

It is well known that the rise of science as we know it started in the late Middle Ages when the empirical method – observation and experiment, together with mathematical modeling of the results – became the method of choice to understand the regularities

H. Walach (✉)
European University Viadrina, Frankfurt (Oder), Germany

Samueli Institute, European Office, Frankfurt (Oder), Germany
e-mail: walach@europa-uni.de

S. Schmidt and H. Walach (eds.), *Meditation – Neuroscientific Approaches and Philosophical Implications*, Studies in Neuroscience, Consciousness and Spirituality 2,
DOI 10.1007/978-3-319-01634-4_2, 

in nature. Astronomy and physics were the first sciences to subscribe to these methods. In the West, the starting point of this development can be placed roughly in the twelfth century, when Adelard of Bath (1080–1152), a cleric who later became adviser to the king of England, came back from a long trip to Sicily and Palestine and brought books and instruments with him that allowed him to conduct optical experiments. This development was taken up by Robert Grosseteste (1175–1253), the later Bishop of Lincoln and one of the first university teachers to the newly founded Franciscan house in Oxford. He wrote various treatises on optical experiments and seemed to have conducted them. Here we find the noun "experimentum – experiment, experience" or the verb "experiri – to travel, to make an experience" for the first time used in connection with replicable scientific experience (McEvoy 1982; Crombie 1953). "Experience" here is beginning to mean a systematic approach to nature from the outside through the senses, using defined methods of observations that can be repeated at will. It is the beginning of the empirical method as we know it: looking at nature from outside, as an object, i.e. "objectively".

This method was taken up and propagated by a student of Grosseste's, by Roger Bacon (1214–1292). Bacon was important, because he became professor of philosophy in Paris with one of the longest regencies known in the history of the Middle Ages, from 1237 to 1247, commenting on Aristotle, and already advocating the empirical method as a foundation of science (Hackett 1995; Lindberg 1983; Clegg 2003; Power 2012). Bacon had entered the order of the Franciscan friars, and due to unrest within the order friars could only publish if the superiors had given permission. Due to these circumstances, Bacon was unable to reach a large audience after 1260. He wrote to the pope who asked him for a sketch of his teaching. This he wrote in 1267. It is known as his "Opus Majus – The Larger Work". He wrote, in quick succession, two shorter summaries, which were meant as short extracts or forwords, the "Opus Minus – The Smaller Work" and the "Opus Tertium – The Third Work". All these pieces contain his vision of a future science, albeit only as a sketch. These existed probably only in two copies, perhaps in three: one was Bacon's, and one was sent to the pope. Bacon's hope was that the pope would help him get the ban on writing lifted and allow him to install what was his vision of a universal science in earnest. This, he said, was based on experience, mathematics, and good knowledge of the old languages, especially Greek and Hebrew, apart from Latin which was the lingua Franca anyway. Experience, however, was the foundation of all knowledge, both of theological knowledge and knowledge of the natural sciences. He saw astronomy, optics, meteorology and knowledge about herbal remedies and dietary measures to prolong life as parts of his science, as well as farming, the art of forecasting historical events and a kind of technology, whereby the knowledge of optics would be applied, for instance by creating magnifying glasses to produce heat, and the like. It is interesting to see that Bacon had seen "*experimentum-experience*" already at the basis of *all* science, also of theology. This type of experience he called "inner experience"; I will come back to this.

Bacon's attempt to revision and reform the university teaching of his days and to install a universal science based on experience was unsuccessful. There are many historical reasons that can be found for it. The situation is altogether quite complicated. I have elaborated on this situation earlier (see Walach 2010) and the most solid historical analysis can be found in a book by Power (2012). Two reasons stand out: Bacon was not very prudent in dealing with his adversaries. He attacked the general of his order, Saint Bonaventure, and practically everyone who was powerful these days openly in his writings. Hence, it is unlikely that he found political support for his ideas. The pope, who was open and benevolent towards Bacon and his reform, died shortly after he had received the copies of Bacon's books, likely without having read them. The books found their way into the papal library, where they sat until, roughly 200 years later, Pico della Mirandola found them. From there his ideas found their way to the younger Bacon, Francis Bacon (1561–1626) (Goldstone and Goldstone 2005), who used a lot of these ideas for his "Novum Organum", the "New Instrument" that heralded the new scientific method.

The basic idea to found a science of nature on systematic experience of our senses and the analysis of the data derived therefrom has been the cornerstone of science ever since and has become the leverage for the power of the scientific method. Even though we have instruments that help us transcend the limitations of our senses – telescopes and microscopes, immunological and biochemical assays, genetic analytical methods, electric, magnetic and other resonance imaging, radiography and so forth – the basic set-up is all the same: an observation of systems from outside. With Roger Bacon and after him a process of objectifying nature has started that has not reached an end yet. We are observing nature from outside, and are trying to make sense of what we observe by using theoretical and mathematical tools to model the data. The power and beauty of this third-person perspective, or the objective view from outside, is that we seem to discover regularities that we can use to improve our condition. We have set ourselves to some degree free from the universal grip of nature and have fulfilled the vision that Francis Bacon in the seventeenth century has formulated: to develop ourselves from slaves to masters of nature.

In that same vein, we are now looking at meditation as a special state of mind and the brain. We are using observational methodology that is looking at the phenomenon of meditation from the outside. And we are starting to discover something extremely interesting: meditation is foremost a very private, a very personal exercise and experience, subjective in nature and bound to certain states of consciousness. Can those states be captured from outside *alone*? Will we be able to make sense of such observations without the inner flavor, the subjective experience? It is here that the first-person perspective of inner experience enters the scene. This has already been part of Roger Bacon's vision. So let us turn to this side of history once more.

## *Attempts to Introduce Inner Experience and First-Person Perspective into Science: Roger Bacon, Brentano, Husserl*

Roger Bacon, we saw, wanted to found the whole of what counted for science at that time on experience, not only the natural sciences proper, but also theology. What did he have in mind? In his "Opus Majus" he says, in Part VI (Bacon 1897, Vol. 2, p 167ff)[1]:

---

[1] Translation mine. To my knowledge there is as yet only this old edition of Bacon's Opus Majus, which I have used. I have reproduced the original Latin quote in full and discussed it in Walach (2010, p. 63 ff). Here one can also find a lengthier discussion of the issue with more textual proofs and arguments.

> Experience comes in two forms: One is through our outer senses, and this is how we experience what is in the heavens and below. This is human, scientific experience… but this experience does not suffice, because it does not give full evidence about material things on the basis of its difficulty, and about spiritual things it attains nothing. Therefore man's mind needs to be helped through other means, and this is, how the holy patriarchs and prophets, who were the first to give science to the world, received enlightenment, not only remaining in the senses…There are two ways of attaining knowledge. One is through natural experience, and one is through divine inspiration, which is much better… This inner science has seven steps… The seventh consists in ecstatic enlightenment and its modes according to which people receive understanding in various ways…

What Bacon does here is very interesting. He contrasts the outer, sense-bound experience of the material world with inner experience. What he calls "divine inspiration", "inner science", or "enlightenment" we can define as "inner experience", as first-person knowledge that is received through systematically training our mind and looking inwards. Bacon seems to have had in mind some science of inner experience. This can be gleaned from the fact that he quotes the old Franciscan tradition of inner mystical experience that dates back to Saint Giles of Assisi (1190–1262) (Théry 1934), a friend and follower of Saint Francis of Assisi, which comes in seven steps, the seventh of which is parallel to what we might call enlightenment. In the terminology of medieval mysticism it is called "raptus", which denotes an ecstatic state of complete absorption. In it, so Bacon holds following this tradition, people receive important insights that cannot be had through sense experiences.

Bacon was unsuccessful with his attempt at installing a comprehensive science based on – outer and inner – experience. Science progressed using outer experience alone. Inner experience was relegated henceforth to the realm of private religion and mysticism, but not part of the academic enterprise and science.

This remained the case until a second attempt to re-introduce inner experience into the enterprise of academic research was launched in modern times. This happened, as Franz Brentano (1838–1917) tried to start the new science of psychology in 1866 and 1874. Franz Brentano was a theologian and philosopher. In his habilitation defense at the university of Würzburg in 1866 he offered various theses, the most famous of which was: "The method of philosophy can only be the method of natural science" (Wehrle 1989, p. 45).[2] This was to be the beginning of his attempt to find a secure foundation for philosophy in the new methodology of psychology, which he started to found when he became professor of philosophy in Vienna in 1874 (Kraus 1919). He postulated that the new science of psychology was to be founded on what he called "descriptive psychology" which in his view was systematic observation of psychological acts – thinking, willing, remembering, experiencing – from within, inner experience in the terminology adopted here. This inner experience or "psychognosia", as he called it, would be, in Brentano's view, a better method to understand psychology and a starting point to gain new ground in philosophy as well (Brentano 1982). It would be through the same systematic enterprise of experience that science proper had used that we can found a science of inner experience, he thought.

---

[2] "Vera philosophiae methodus … nulla alia nisi scientiae naturalis est" The habilitation of Brentano is unpublished; reference and quote from (Wehrle 1989).

As we know, this program was not continued. Historically speaking, naïve introspectionism was unable to really unravel the mysteries and the working of the mind (Lyons 1986). Psychology adopted the methods of natural science, experiments and observation from the outside most of the time, and historically speaking, Wilhelm Wundt and his school have been more successful in influencing the new science of psychology than Brentano. But there began also a strong tradition of observing and hermeneutically understanding the content of one's own mind, namely in the methodology that was introduced by psychoanalysis through analyzsing the experience of transference and countertransference. Here, the inner experience of the analyst is used to understand what is going on in the mind of the patient. This method of recursively observing the changes in one's own mind, using this content to formulate a provisional understanding and communicating this understanding to the patient through interpretations or phantasies has become one of the cornerstones of narrative reconstruction used in psychoanalysis and clinical psychology. Interestingly, Freud very likely adopted this method from Franz Brentano's postulate of descriptive psychology through observing one's own mind, as Freud was one of Brentano's students in Vienna (Merlan 1945, 1949). Thus, at least in the methodology of psychoanalysis Brentano's ideas seem to have found fertile ground to grow on.

Another tradition that derived from Brentano's idea was phenomenology. Edmund Husserl had been a student of Brentano, and although Brentano sent him away to do his doctorate elsewhere the basic idea of his phenomenology is a direct sequel of Brentano's inner experience or psychognosis (Husserl 1919). Husserl's basic insight was that it is only through experience itself that the world is entering into our consciousness, and hence our consciousness and its experiences are the primary access points (Zahavi 2003). While Husserl tried to find a philosophical foundation for the direct experience of reality in how the world appeared to our consciousness in its experience, he also understood that there was need of a special state consciousness needs to be in, in order to grasp reality "as it is", and not "as it is filtered through the current state of our consciousness". Husserl called this special state by the Greek term "*epoché*". This word is derived from the Greek "ep-echein", which means, among others, to abstain, to refrain, to stop in activity. By his "epoché" Husserl wanted to signify a state of mind that abstains from the common judgments, that refrains from normal dogma and quick categorization and is, at the same time closely with the experience and observing it from above, as it were. In the terminology of Eastern psychology one might call it a witness type of consciousness. It would be extremely important and likely very fruitful to analyze to what extent Husserl's "epoché" can be paralleled to states of mind like mindfulness or witness consciousness in Eastern traditions. As far as I know this has not been done so far. In Husserl's words (Husserl 2009, orig. 1930, p. 39)[3]:

[3] Translation mine; original: "Die philosophische εποχη, die wir uns vornehmen, soll, ausdrücklich formuliert, darin bestehen, daß wir uns hinsichtlich des Lehrgehaltes aller vorgegebenen Philosophie vollkommen des Urteils enthalten und alle unsere Nachweisungen im Rahmen dieser Enthaltung vollziehen".

> The philosophical $\varepsilon\pi o\chi\eta$, which we are trying to achieve, should explicitly consist in abstaining completely from all judgment regarding previous philosophical teaching, and all our findings and arguments should happen within this abstention.

Thus, Husserl was trying to recover philosophy, i.e. knowledge about reality, from true, direct and immediate experience, unfiltered and unhindered by theoretical top-down processes that are conditioned by previous learning, theoretical concepts or philosophical higher order constructs. In the terminology adopted here he wanted to, among others, bring the mode of inner experience within the scope of providing knowledge in a philosophical-academic sense. In that interpretation, the thrust of an approach through an inner experience towards understanding the world had found a new voice in Brentano and Husserl.

Husserl's methodological requirement was "epoché", abstention of pure phenomenological consciousness in the face of experience. But how is this abstention, this "epoché" achieved? What exactly do we have to do? What is the proper method? How can we avoid the naïve introspectionism that was the grave of Brentano's methodology? Husserl seems to have seen this problem and hence introduced this new "state of consciousness" of abstaining from judgment, being with our experience without pre-determining anything through theory. But Husserl did not tell us how to actually achieve it. I suggest, this is where modern-day meditation research and old-day meditation-experience can offer each other a hand to recover what has been an epistemological impulse right from the beginning of the Western scientific enterprise: inner experience as access to reality from within, through consciousness experiencing itself.

## *The Situation and the Problem*

The situation, thus, is the following: The historical development of science in the West has favored the extravert type of experience, through the senses and their many aids gaining knowledge about the world from outside. The inner, more introvert type of experience, through consciousness itself, is the typical Eastern way, exemplified in the long traditions of Vedanta and Yoga in India or Buddhism in other parts of Asia (Barendregt 1996; Nisbett et al. 2001; Rao 2005). In the West, the method of empiricism, applied to the mind itself, has led to the rise of cognitive science and neuroscience. With it, and often surreptitiously, mind and consciousness, have received an interpretation and philosophical status of an entity that is completely derived of and dependent on its physiological substrate, the brain (Walach and Römer 2000, 2011; Römer and Walach 2011; Noë 2009; Walach 2007). However, if this is so, how can such a thing as in "inner experience", i.e. knowledge that is a result of a consciousness turned inwards and towards itself, be anything else than inconsequential play of a system idling? Is "inner experience" anything else than "default mode network" activity (Gusnard and Raichle 2001)?

Thus, in order to recover "inner experience" as a potential mode of knowledge, we need, at least and minimally, two things:

1. We need a proper ontology, at least a minimally acceptable ontology, that is not contrary to our scientific corpus of knowledge, yet allows for an access of consciousness to reality beyond the senses, through its own inner experiences.
2. We need an epistemology that tells us when such an inner experience might be really giving us some insight about reality.

Far from being able to offer a final solution, I will, in what follows, sketch some inroads.

## Complementarity Offers an Ontological Solution to Empower Inner Experience

### *Why a Reductionist Materialist-Monist Ontology is Insufficient*

We have offered a more extensive analysis of the situation in previous publications (Walach and Römer 2011; Römer and Walach 2011), and hence I will here only briefly recap the most important aspects. The current mainstream trend is easy to understand and beautiful in its simplicity. Within it the mind, conscious experience in general, is a result of the complex interactions of the organ that is biologically responsible for sustaining it, the brain. Similar to immunity being the result of the complex interactions of the immune system consciousness and the mind are the result of the interactions of the neuronal system. Mental properties are emergent of and dependent on this system, the brain. Without it, they will not be able to be sustained. Whether they have some causal consequences on the brain itself is currently debated. Through meditation research we find that by training our consciousness we also modulate brain structures (see the chapters by Austin, Esch, Farb, and Hasenkamp in this volume; Ott et al. 2011), which means, a forteriori, that our mind and our consciousness have causal influence on the physical substrate. If we train attention or mindfulness, the brain areas and connections engaged in these activities will grow and adapt, much as a muscle that will respond to training load.

But will a brain turned onto itself in meditation also be able to gain insights apart from its sense connections with the world? In other words, can consciousness also have some direct access to some other aspects of reality that sense experience does not provide and that we cannot gain by thinking about our experience? For this to be possible to happen one would have to postulate that the mainstream picture of consciousness and the mind being completely derivative of brain activity is incomplete. Although we have strong correlations between mental activities and brain activities – if brain areas are damaged, neuropsychology tells us that the sustaining mental capacities are damaged or gone as well – correlations are insufficient for proving causality.

We used the simple argument (Römer and Walach 2011) that our conscious activity – for instance observing our inner experiences, noticing our as yet unconscious impulses or tendencies, describing in words what has been only an unspecified feeling so far – actually change the very phenomena our conscious activity is directed towards. This is, phenomenologically speaking, the behavior of a non-classical system, i.e. a system that cannot be modeled using Newtonian thinking and causality. While in classical systems a behavior of the system is not changed through observation, it is the hallmark of a non-classical system that observation changes the observed. I can observe a flying ball and will not change its trajectory. In the same vein we can observe the EEG of another person without changing it. But as soon as we are observing our own EEG we are changing it, as all neuro-feedback research teaches. Else we could never have developed brain-computer interfaces (Hinterberger et al. 2004). Moreover, if I observe my own mental activity, I change it, and by changing it, I am affecting the physiological substrate, brain activity. In other words: a mental act, observing, changes the material basis for this very act, brain activity. However, this structure of an observation that changes the state of the observed entity, no matter whether the entity is of material or mental content, is non-classical, and for the description of such non-classical processes another type of theory is necessary than the normally applied classical theories. The minimal requirement is that this strange interaction of the influence of the act of observation can be modeled. The way to do this is to conceptualize the two systems as complementary.

## *Complementarity of Mental and Physical Systems as Minimal Requirement*

This is what we have proposed (Walach and Römer 2000, 2011; Römer and Walach 2011): We suggest viewing the mental system as complementary to the physical system of the brain. We do not need to make a strict ontological statement. That is to say, the duality arising from this proposal is purely phenomenological in nature; this is similar to Chalmer's analysis (Chalmers 1996). One could conceive of an ontological monism out of which the material and the mental systems co-arise (Atmanspacher 2003). We need not go into the full depth of ontology here. But minimally we would need to postulate a phenomenological dualism of mental and physical systems that are related to each other as complementary pairs.

The notion of "complementarity" signifies maximal incompatibility between two related pairs. While "opposites", such as "black" and "white" can be reduced to each other, as they are located on the same conceptual plane, as it were, complementary pairs cannot. They are orthogonal. That is, one cannot be expressed in terms of the other, or, in other words, we can never use the physical language of brain events to express the mental-psychological language of conscious events, and vice versa. This is at the basis of neuro-phenomenology, as advocated by Varela et al. (1991) and taken up by the notion of "Contemplative Science" by Farb in this volume. Rather the two domains need to be described separately but in a correlated fashion.

"Complementarity" also means that, philosophically and conceptually speaking, we cannot reduce a domain of first-person accounts to a domain of third-person accounts or vice-versa without making fatal category mistakes (Hoche 2008). Thus, there is no chance of simplifying complementary relationships without doing injustice to one domain. Although they are maximally incompatible, they are both necessary to describe a certain reality, in this case a living human being who has both a physical system, the brain, that sustains a mental system which is complementary to it, the mind.

Now, an interesting consequence of this situation is that, if this holds true, then our conscious system might have its own access to reality being a valid ontological expression of this reality just as our physical brain is. While our physical system needs the sense organs and physical contact with the world to understand and learn about the world, our consciousness, at least in this view, would not need direct sense contact. It can use itself as an instrument to gain knowledge about reality, at least about some aspects of it (see next section). And this is, I suggest, what happens in "inner experience" through meditation. Put differently, the minimal ontological requirement to make "inner experience" a viable epistemological option to gain knowledge is to postulate a phenomenological duality and complementarity between physical and mental systems. And I repeat: this does not mean ontological dualism. It might be simply a matter of phenomenology. But it would require a transcendental monism and not a materialist monist picture of the world. Granted that this is at least a plausible philosophical possibility, and a rational one, I hold, let us explore what happens, if we adopt such a stance.

## Towards an Epistemology of Inner Experience

### *Inner Experience as Potential Access to Reality*

If we concede, for the sake of argument, that physical and mental systems are not reducible to each other but different expressions of one reality related in a complementary way to each other, then we can perhaps see that, just as we can epistemologically access the world through outer experience, through our senses, we might be able to do this through our inner experience as well. While outer experience gives us access from a third-person perspective, through sense experience and through seeing the world and its events as objects, inner experience gives us subjective knowledge, a first-person perspective. It occurs when our consciousness is turned on itself, consciousness observing conscious events as they occur, thereby reducing background noise of thoughts, previous impressions and experiences, anticipation of future events. Neurobiologically speaking the activity of the default-mode network and mind-wandering are reduced (see the chapters by Esch and by Mrazek et al. in this volume; Pagnoni et al. 2008).

States of pure consciousness as such have been described by all contemplative traditions, often in different words. And those traditions have also held that in such states of pure consciousness insights about reality are possible. How and why this might happen can currently be only speculated about and I will not do this here. But phenomenologically speaking what can happen is that against the background of a blank and pure consciousness "ideas", "insights", "discoveries" fall into place or "come to mind". Such states of consciousness have been described as "witness consciousness" or "transcendental consciousness" (Travis and Orme-Johnson 1989; see also the chapter by van Wijk et al. in this volume). In the Zen tradition they are known as states of Zanmai (Kapleau 1969) during which one has access to what is called "big mind" (Suzuki 1970; Hakuin 1994). In the Vedanta tradition various states of "samadhi" are known, i.e. states of absorption in which consciousness turned on itself, and is at one with itself (Rao 2005; MacPhail 2013; Eliade 1975).

Granted, again for the sake of argument, that this is possible. What exactly is it about reality that is presenting itself here to the mind of the meditator? I suggest that, while outer experience, i.e. sense experience, tells us something about the form, the shape and the exact material make-up of the world, such inner experience tells us something about its structure. Perhaps the closest analogue to this process is scientific mathematical modeling and insight. Mathematics is a highly abstract science. It cannot be found, save in its very basic form through counting, in the outside world, but is derived from a kind of inner insight that has much phenomenological similarity to what I have in mind here. It is probably derived from the very same process. In order to mathematically model the relationship of observed data it is of no use to make more observations. Heinrich Kepler had a mass of observational data at hand which the astronomer Tycho Brahe, his teacher, had collected and published as "Tabula Rudolphina". But only his mathematical genius was able to find a model that could produce a mathematical structure which would then reproduce all the data, or at least approximately, that Brahe had collected. Kepler could account for the mercury perihel and for the obviously elliptic shapes of the planetary orbits by formulating his law, according to which the planets travel, in proportionate times, proportionally equal areas of elliptical trajectories. The insight into that structure is something that occurred to Kepler through long reasoning, absorption into the data and suddenly "came to his mind" (Holton 1973; Pauli 1955).

We can use mathematics and its role in scientific modeling as a close analogue to inner experience, as an epistemological road to insight that is independent of sense experience. It uses the material provided by sense experience, but its insights themselves are strictly abstract and come to mind, phenomenologically speaking, from inside. Other examples of such insights include the finding of the formalism of quantum mechanics by Schrödinger and Heisenberg, or the way how Einstein arrived at the formalism of general relativity, and in principle all examples of mathematical modeling that lie at the basis of the scientific enterprise. What happens, phenomenologically speaking, is that a scientist takes all the relevant material known, seeks an abstract structure, blacks out all previous and current discursive thinking, and focuses on the problem until the solution "pops up" out of the depth of his or her mind.

A good historical example is how Einstein arrived at his relativity theory. After a long and disturbing time he suddenly hit at the idea one morning, when he woke up. Brian, in his biography on Einstein writes, quoting an earlier biographer, Hoffmann (Brian 1996, p. 61):

> Banesh Hoffmann confirmed this account: "Einstein said his basic discovery came on waking up one morning, when he suddenly saw the idea. This had been going around and around at the back of his mind for years, and suddenly it wanted to thrust itself forward into his conscious mind... Einstein ... said: 'Ideas come from God.' Now he didn't believe in a personal God or anything like that. This was his metaphorical way of speaking. You cannot command the idea to come, it will come when it's good and ready. He put it in those terms: 'Ideas come from God'".

Here we have a good historical and phenomenological example. The condensation of a structure in a mathematical formalism "comes to mind". It is structurally equivalent with what I call here an inner experience. Thus, I submit, inner experience gives us access to reality like outer experience, only that, in contrast to outer experience which teaches us about content and material realities, inner experience teaches us about structures.

## *Inner Experience, Meaning, and the Structure of Reality*

If, what I have said so far, makes sense, then inner experience, insights that occur when consciousness is turned onto itself, gives us potential knowledge about the structure of the world. I have mentioned mathematics as an obvious example and mathematical modeling of scientific theories. Here we can see directly how such an insight can be relevant and how it is not derived from external empirical activity. Are there other areas that might be relevant? My suggestion is: Meaning making and values.

Human beings are meaning making semiotic animals, as it were. We cannot but make meaning out of all kinds of situations, sometimes for better and sometimes for worse. Meaning "making", however, is somewhat misleading, as meaning is rarely "made", but rather found, as Viktor E. Frankl used to say (Frankl 1964). This is not to say that meaning-making is not a constructive and creative process; it surely is. But it is a process that lies beyond our conscious and willful activity. It is rather like "finding" meaning in activities and occurrences. We find this meaning, paradoxically, if we let go of any intense effort, any conscious attempt or willful striving for meaning. Rather the meaning of a situation, an event, an occurrence, even of a piece of art or an encounter reveals itself in a kind of inner experience. The structure of this is rather similar to what happens in an insight that comes out of a meditative collection of our mind. Hence, so my interpretation, the renaissance of meditation in our days where activities abound and meaning is rare. It is rare, because it cannot be found in outer relations, through the senses and empirically relating to the world alone, but it needs to be found in the inner world of introspection and within the experience of turning our mind inwards. Here it suddenly occurs. This is the reason why, phenomenologically speaking, meditators often say that they meditate to find "meaning" in their lives (see also the motivation for meditation scale in the chapter by Schmidt in this volume). The meaning is the inner structure that holds the singular narratives of individual lives together and makes them whole, and the emotional-affective tone to it. To discover such an inner structure and its feeling is similar to discovering a structure that holds outer events together in the form of a mathematical formalism. It is the thread that runs through a life and that is woven by contemplating it, letting go of singular events and by looking at the whole picture. That this process might be quite constructive at the same time, or that it is about drawing and seeing connections that were always there is quite another truism. But in order to be able to do this one has to hold on, abstain from mental activity and allow the image to appear. And this is what happens when one meditates, very generally speaking.

Another potential area of relevance is values. Values do not occur in our outside world. We can count and document as many singular events and entities using our scientific methods, we will never find values. Values are abstract entities human societies subscribe to in order to make communal life possible and optimise (or "optimize"; either UK or American). But how do we arrive at them? Evolutionary psychology is trying to use evolution and what was beneficial for survival of the gene pool to understand how values operate. Another way of looking at them would be to see them as abstract structures that hold the fabric of human relationships together. In order to discover them, one would have to look, but not to the outside, but to the inside, where structures reveal themselves and the very special individual taste or emotional tone that goes with it. It is through religions that values were promoted and guaranteed previously. With the demise of formal religion in our days in Western and European societies it is difficult to universally uphold values. If we see religions as derived from inner experiences about reality and particular interpretations of these experiences, then values are direct consequences. If reality has been experienced in a certain way – for instance as benevolent or full of meaning -, then the value derived thereof is that of being benevolent and supportive of other human beings and the environment, to produce just a simple example. Thus, values are consequent on inner experiences of reality, I hold. If this is so, then a contemplative science of inner experience might be one way to recover, beyond societal agreements of a purely political and thus very fragile nature, values as basic structures of reality.

## *Requirements and Potential for an Epistemology of Inner Experience*

Now we have reached a decisive and interesting point in our discourse. If, by inner experience, we can discover something about reality, its inner structure as I postulate, how are we then to make sure that what we discover is in any sense approaching some truth? Science is, after all, about reducing error and increasing the truth about what we are discovering of the world. And the scientific enterprise has consisted in finding methods of improving the way how we reduce error and improve on the truth value of our observations and discoveries. This is difficult enough in outer experience of our senses and has taken some centuries to arrive at the sophistication we have today. What about inner experience? Is there a chance that we can find something similar?

Before we can answer this question we need to see the basic structural difference here: Outer experience is dealing with a referent in the outside world which most competent observers can observe and agree upon. Even though "competent observers" might mean a handful labs in the whole world, at least they have such a common referent they can observe, study, model and report on. The same is not true for inner experience. By definition the referent is subjective and hence an individual has privileged access that nobody else can potentially have. This is not a problem with subjective meaning, as this is supposed to be subjective and need not be shared by others. It is also not a problem for mathematical structures, as there are sufficiently well known procedures of deduction and formal operations that can be used to verify the results and its usefulness. But it is a problem for values and potential other insights about reality. Can there be a true epistemology of inner experience that will be scientific, i.e. potentially intersubjective, potentially disputable, potentially universal? Can, what is subjective by definition, become objective or can it be shared by others?

I think that there is a road, albeit not a simple one. Suggestions to model contemplative science along the lines of empirical science, to map inner experience onto the procedures to verify outer experience, do not hold water. This has to do with the subjective nature of inner experience as such. In order to build a bridge, we have to transform subjective, inner experience into something that can be shared, i.e. subjectivity that is intersubjectivity. This is a first-person plural account (Walach and Runehov 2010).

In order to achieve that we have to report experiences in a format that can be shared by others. Normally this will be language. But it might also be other cultural forms, such as art, music, dance, which all transmit and evoke certain experiences that can be shared. Let us take the simple route here and assume that experiences are shared in language. Normally this will be reports. They may be stories or metaphors. Suddenly we have reached another realm, the realm of hermeneutics, understanding and interpretation. This is equivalent to the process of interpreting data of outer experience and debating the relevance of it. But here the material is not quantitative in nature but qualitative. The material is expressed in language. And language can be understood. True, there is enough room for misunderstanding: The verbalization of the experience might be deficient. The experience might be too multifaceted to be able to be expressed in linear language. The emotional tone and the particular quality of the experience will likely remain unspeakable, except, perhaps in poetry. This is the reason, by the way, why most religious traditions seek refuge in paradoxes and seemingly irrational metaphors or metaphorical stories that are able to conserve the pluripotent meaning of inner experiences of reality. But by and large, language can be understood. It was one of the great insights of the German theologian Schleiermacher, one of the founding fathers of hermeneutics, to have seen the pluripotent meaning in the texts of the biblical stories and to demand a science of interpretation out of which hermeneutics grew (Gadamer 1975). So one direct route leads us to the art of understanding or the discipline of hermeneutics. Thus, part of the epistemology of inner experience would be to work on the hermeneutics of reports of inner experience (Louchakova 2005). We have numerous accounts of such experiences in the canonical texts of various traditions and it would be an important task for interdisciplinary religious studies to work out the core of the experiences and figure out whether there is such a thing as a common core.

Another way of promoting such an epistemology would be more psychological in nature and would be to develop behaviorally based measures of stages reached by such inner experiences. Piaget used behavioral tests to find out what stage of cognitive development a child operated at. Reich moved on from that, postulating a still different type of reasoning in older youths, "contextual and developmental reasoning" as he called it (Reich 2003). The basis for diagnosing such a state of functioning is the way how conundrums and dilemmas are solved. Using these approaches as a template, one could now move forward and think about behavioral or cognitive structural indicators of certain types of inner experiences. The medieval tradition held that wisdom is a result of such inner experiences. How would we document it? Certainly not by self-report. We could think about what indicators we would use to ascribe wisdom to a person. We could use conundrums to test how someone resolved it to figure out what kind of wisdom a person has acquired. The Eastern traditions see mindfulness as a result of meditative exercises and a continuous culture of consciousness. At the moment, mindfulness is measured using self-report questionnaires. But this is only a proxy measure and a problematic one at that (Belzer et al. 2013). One could think about behavioral indicators, and values have been mentioned as potential indicators (Grossman and van Dam 2011). But one would have to even go a step further and use not only values people cognitively subscribe to, but the behavior they display that document the values they are enacting. In that sense one could develop a whole science as to how to verify the results, the relevance and the intersubjective validity of inner experiences.

If my hunch is correct this will mean that the purview of what used to be theology, morality and ethics will be brought under the auspices of a science that has broadened out to encompass inner experience and has turned into Contemplative Science, at the same time broadening out the scope of science, making the current face of science more human and even more relevant to our lives, by taking in whole areas of human relevance that have been left out from the scientific enterprise for lack of methodology and theoretical foundation. Using inner experience as a potential road this one-sidedness could be remedied. Values, meaning and other constructs currently outside the scope of science would enter the stage. Taking first-person accounts, inner experience and Contemplative Science seriously would be the necessary steps.

## References


Atmanspacher, H. 2003. Mind and matter as asymptotically disjoint, inequivalent representations with broken time-reversal symmetry. *Biosystems* 68: 19–30.

Bacon, R. 1897. *The Opus Majus of Roger Bacon*, vol. 2. Oxford: Clarendon.

Barendregt, H. 1996. Mysticism and beyond: Buddhist phenomenology, part II. *The Eastern Buddhist* 29: 262–287.

Belzer, F., S. Schmidt, G. Lucius-Hoene, J.F. Schneider, C.L. Orellana-Rios, and S. Sauer. 2013. Challenging the construct validity of mindfulness-assessment – a cognitive interview study of the Freiburg mindfulness inventory. *Mindfulness* 4: 33–44. 10.007/s12671-012-0165-7.

Brentano, F. 1982. *Deskriptive Psychologie. Aus dem Nachlass herausgegeben und eingel. v. R.M. Chisholm und W. Baumgartner*. Hamburg: Meiner.

Brian, D. 1996. *Einstein – a life*. New York: Wiley.

Chalmers, D.J. 1996. *The conscious mind. In search of a fundamental theory*. New York/Oxford: Oxford University Press.

Clegg, B. 2003. *The first scientist. A life of Roger Bacon*. London: Constable.

Crombie, A.C. 1953. *Robert Grosseteste and the origins of experimental science 1100–1700*. Oxford: Clarendon.

Eliade, M. 1975. *Patanjali and yoga*. New York: Schocken Books.

Frankl, V.E. 1964. *Man's search for meaning: An introduction to logotherapy*. London: Hodder & Stoughton.

Gadamer, H.G. 1975. *Wahrheit und Methode. Grundzüge einer philosophischen Hermeneutik (4. Aufl.)*. Tübingen: Mohr.

Goldstone, L., and N. Goldstone. 2005. *The friar and the cipher: Roger Bacon and the unsolved mystery of the most unusual manuscript in the world*. New York: Doubleday.

Grossman, P., and N.T. van Dam. 2011. Mindfulness by any other name…: Trials and tribulations of sati in western psychology and science. *Contemporary Buddhism* 12: 219–229.

Gusnard, D.A., and M.E. Raichle. 2001. Searching for a baseline: Functional imaging and the resting human brain. *Nature Reviews Neuroscience* 2: 685–694.

Hackett, J. 1995. Scientia experimentalis: From Robert Grosseteste to Roger Bacon. In *Robert Grosseteste: New perspectives on his thought and scholarship*, ed. J. McEvoy, 89–119. Turnhout: Brepols.

Hakuin. 1994. *The essential teachings of Zen Master Hakuin*. Boston: Shambala.

Hinterberger, T., N. Neumann, M. Pham, A. Kübler, A. Grether, N. Hofmayer, et al. 2004. A multimodal brain-based feedback and communication system. *Experimental Brain Research* 154: 521–526.

Hoche, H.-U. 2008. *Anthropological complementarism. Linguistic, logical, and phenomenological studies in support of a third way beyond dualism and monism*. Paderborn: Mentis.

Holton, G. 1973. *Thematic origins of scientific thought: Kepler to Einstein*. Cambridge: Harvard University Press.

Husserl, E. 1919. Erinnerungen an Franz Brentano. In *Franz Brentano: zur Kenntnis seines Lebens uns seiner Lehre*, ed. O. Kraus, 151–167. München: Beck.

Husserl, E. 2009 orig. 1930. *Ideen zu einer reinen Phänomenologie und phänomenologischen Philosophie*, Philosophische Bibliothek Bd. 602. Humburg: Meiner

Kapleau, P. 1969. *The three pillars of Zen: Teaching, practice, enlightenment*. New York: Harper.

Kraus, O. 1919. *Franz Brentano: Zur Kenntnis seines Lebens und seiner Lehre*. München: Beck.

Lindberg, D. 1983. *Roger Bacon's philosophy of nature*. Oxford: Clarendon.

Louchakova, O. 2005. Ontopoiesis and union in the prayer of the heart: Contributions to psychotherapy and learning. In *Logos of phenomenology and phenomenology of the logos. Book four. The logos of scientific interrogation. Participating in nature-life-sharing in life*, Analecta Husserliana, ed. A.-T. Tymieniecka, 289–311. New York/Heidelberg: Springer.

Lyons, W. 1986. *The disappearance of introspection*. Cambridge, MA: MIT Press.

MacPhail, J. 2013. *Learning in depth: A case study in twin 5×5 matrices of consciousness*. Frankfurt (Oder): Europa-Universität Viadrina.

McEvoy, J. 1982. *The philosophy of Robert Grosseteste*. Oxford: Clarendon.
Merlan, P. 1945. Brentano und Freud. *Journal of the History of Ideas* 6: 375–377.
Merlan, P. 1949. Brentano and Freud – a sequel. *Journal of the History of Ideas* 10: 451–452.
Nisbett, R.E., I. Choi, K. Peng, and A. Norenzayan. 2001. Culture and systems of thought: Holistic versus analytic cognition. *Psychological Review* 108: 291–310.
Noë, A. 2009. *Out of our heads: Why you are not your brain, and other lessons from the biology of consciousness*. New York: Hill & Wang.
Ott, U., B.K. Hölzel, and D. Vaitl. 2011. Brain structure and meditation: How spiritual practice shapes the brain. In *Neuroscience, consciousness and spirituality*, ed. Walach Harald, S. Schmidt, and W.B. Jonas, 119–128. New York: Springer.
Pagnoni, G., M. Cekic, and Y. Guo. 2008. "Thinking about not-thinking": Neural correlates of conceptual processing during Zen meditation. *PLoS One* 3(9): e3083. doi:10.1371/journal.pone.0003083.
Pauli, W. 1955. The influence of archetypal ideas on the scientific theories of Kepler. In *The interpretation of nature and psyche*. New York: Pantheon Books.
Power, A. 2012. *Roger Bacon and the defence of Christendom*. Cambridge: Cambridge University Press.
Rao, K.R. 2005. Perception, cognition and consciousness in classical Hindu psychology. *Journal of Consciousness Studies* 12(3): 3–30.
Reich, K.H. 2003. *Developing the horizons of the mind: Relational and contextual reasoning and the resolution of cognitive conflict*. Cambridge: Cambridge University Press.
Römer, H., and H. Walach. 2011. Complementarity of phenomenal and physiological observables: A primer on generalised quantum theory and its scope for neuroscience and consciousness studies. In *Neuroscience, consciousness and spirituality*, ed. Walach Harald, S. Schmidt, and W.B. Jonas, 97–107. New York: Springer.
Suzuki, S. 1970. *Zen mind, beginners mind*. New York/Tokyo: Weatherhill.
Théry, G. 1934. Thomas Gallus et Egide d'Assise. Le Traité De septem gradibus contemplationis. *Revue néoscolastique de philosophie* 36, 2e Ser:180–190.
Travis, F.T., and D.W. Orme-Johnson. 1989. Field model of consciousness: EEG, coherence changes as indicators of field effects. *International Journal of Neuroscience* 49: 203–211.
Varela, F.J., E. Thompson, and E. Rosch. 1991. *The embodied mind. Cognitive science and human experience*. Cambridge, MA: MIT Press.
Walach, H. 2007. Mind – body – spirituality. *Mind and Matter* 5: 215–240.
Walach, H. 2010. *Notitia experimentalis Dei – experiential knowledge of God: Hugh of Balma's mystical epistemology of inner experience – a hermeneutic reconstruction*, Analecta Cartusiana, vol. 98:2. Salzburg: Institut für Anglistik.
Walach, H., and H. Römer. 2000. Complementarity is a useful concept for consciousness studies. A reminder. *Neuroendocrinology Letters* 21: 221–232.
Walach, H., and H. Römer. 2011. Generalized entanglement – A nonreductive option for a phenomenologically dualist and ontologically monist view of consciousness. In *Neuroscience, consciousness and spirituality*, ed. Walach Harald, S. Schmidt, and W.B. Jonas, 81–95. New York: Springer.
Walach, H., and A.L.C. Runehov. 2010. The epistemological status of transpersonal psychology: The data-base argument revisited. *Journal of Consciousness Studies* 17(1–2): 145–165.
Wehrle, J.M. 1989. *Franz Brentano und die Zukunft der Philosophie. Studien zur Wissenschaftsgeschichte und Wissenschaftssystematik im 19. Jahrhundert. Studien zur österreichischen Philosophie*. Amsterdam: Rodopi.
Zahavi, D. 2003. *Husserl's Phenomenology*. Stanford: Stanford University Press.

# The Meditative Approach to Awaken Selfless Insight-Wisdom

James H. Austin

**Abstract** This chapter reviews, first, the two complementary categories of meditation: concentrative meditation and receptive meditation. In this context, it then considers the representations and responses of the upper (dorsal) and lower (ventral) attention systems. The potentially reciprocal nature of the responses of the medial fronto-parietal regions vs. those of these two lateral attention systems is emphasized, because they are relevant to the mechanisms that could prompt an unexpected sensory stimulus to trigger the "peak" experience of awakening (J. *kensho*). In the model proposed, the reticular nucleus is assigned a key role in inhibiting the adjacent dorsal tier of thalamic nuclei. This disruption of the normal coherence of thalamo-cortical oscillations could, *simultaneously*, diminish Self-relational processing and liberate other-relational (allocentric) processing.

> The further I go, the better I see that it takes a great deal of work to succeed in rendering what I want to render: 'instaneity'.
>
> Claude Monet (1840–1926)

## Introduction

Vision happens spontaneously. How is it possible to see instantly, let alone paint instan*eity*? Only by integrating the functions of our multiple networks at multiple *levels*. Much of the brain's access to conscious associations hinges on messages integrated down in our thalamus. Its largest nucleus – the pulvinar – helps us "frame"

J.H. Austin, M.D. (✉)
Emeritus Professor of Neurology, University of Colorado Denver
School of Medicine, Columbia, Missouri, USA
e-mail: bklund@live.com

S. Schmidt and H. Walach (eds.), *Meditation – Neuroscientific Approaches and Philosophical Implications*, Studies in Neuroscience, Consciousness and Spirituality 2,
DOI 10.1007/978-3-319-01634-4_3, 

parts of a Monet painting, automatically selects which lily pad will dominate our visual foreground, or render a haystack as the salient focus of interest for our ongoing flow of aesthetic appreciation.

This chapter's aims are threefold: Part I reaffirms the fact that we use two complementary categories of meditation to train our two different forms of attention. Part II specifies several interactions between thalamus, basal ganglia and cortex that will contribute to much of the normal unification of soma and psyche we call our "Self." Part III discusses caveats for researchers. It then suggests some practical approaches that could help long-term meditators prolong a few more moments of clear, self*less* instaneity.

## Part I. Meditation Trains Attention

> I don't know anything about consciousness. I just try to teach my students how to hear the birds sing.
>
> Zen Master Shunryu Suzuki (1905–1971)

### *Two Different Systems of Attention; Two Different Categories of Meditation*

The opening sentences by Monet and Suzuki distill the essence of this chapter. Can their words help us understand the spontaneity of our consciousness? Yes, by inviting us to pause, to glance *up* toward that singing bird, and then re-experience that sharp point of attention which is leading our auditory and visual processing.

Consciousness is liberated to open up much wider during the extraordinary states of "awakening." These are called *kensho* and *satori* in Zen. What causes consciousness to be released to such a remarkable degree? (Skt: *moksha*) Has a heavy excess of personal baggage just been cast off? To understand how meditation could culminate in such self-less freedom, we need to go back to the very first steps in this long process. These steps begin with the two mutually reinforcing ways that meditators train attention. Table 1 condenses these two generic categories of concentrative and receptive meditation.

**Table 1** The attentive art of meditation; two complementary categories

| Concentrative meditation | Receptive meditation |
|---|---|
| A more effortful, sustained attention | A more effortless, sustained attention |
| Focused and exclusive | Unfocused and inclusive |
| A more deliberate, one-pointed attention. It requires voluntary top-down processing | A more open, universal, bare awareness. It expresses involuntary modes of bottom-up processing |
| More Self-referential | More other-referential |
| May evolve into absorptions | May shift into intuitive, insightful modes |
| Choosing to "pay attention" | A bare, choice-less awareness |

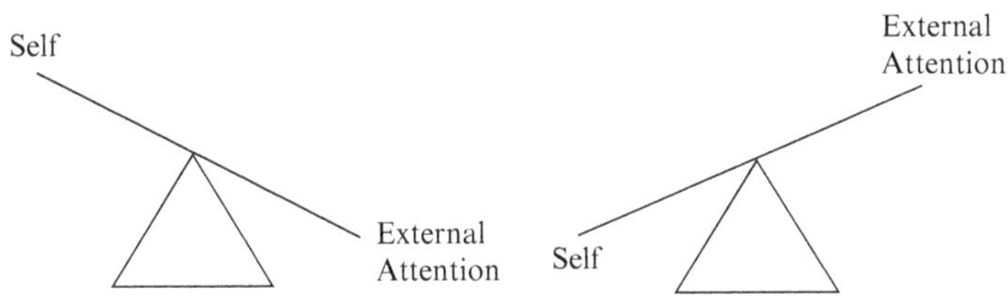


For centuries, Zen Buddhism has emphasized two major aspects of the long-term spiritual path. First, that it is crucial to train both focal and global forms of attention (Austin 1998, pp. 69–73, 274–276, 278–281; 2006, pp. 29–40, 179–187; 2009, pp. 1–47, 98–108, 139–141). Second, that it is essential to relinquish one's maladaptive Self. Yet, only recently has it been possible to appreciate the meditative implications of a striking set of observations: *attention and Self often relate normally in an almost reciprocal, see-saw manner* (Austin 2009, pp. 109–117, 191–196). Such an inverse relationship might be oversimplified visually as follows:

However, meditators can develop an imbalance on the long-term meditative path *if* they do not fully integrate their focal, concentrative practices with more open receptive styles of meditation. For example, a unilateral approach that over-intensifies the concentrative techniques can only reinforce Self-centeredness. This proves counterproductive to the meditator's primary long-term task: *reducing* egocentricity. In this chapter, we take a perspective that favors a more effective and balanced approach. Here the emphasis is more on the advantages of *letting go* into the *involuntary* forms of *bottom-up attentive* processing. It is these subtler, bottom-up kinds that can flourish when one's practice also opens up to include the more global techniques afforded by the receptive category of meditation.

Attention is an innate associative function. It expresses an underlying sophisticated intelligence. Experience teaches the brain where to scan for relevant information and how to point this incoming data toward an immediate goal. Attention's two systems – dorsal and ventral – draw their resources from diverse cortical, subcortical, and brainstem levels. In a highly complex field of research, we begin with the cortical levels of attentive functions, as recently reviewed and updated by Maurizio Corbetta and colleagues (Mantini et al. 2009; Shulman et al. 2009; Corbetta et al. 2008). Table 2 summarizes the findings in a field that is still evolving.

## *The Dorsal Attention System. Its Relation to the Voluntary, Concentrative Category of Meditation*

This dorsal attention network plays the decisive role when we harness and intensify our *voluntary* forms of *attentive processing*. In those final two words of that sentence, please notice which word comes first. *Attentive* leads; processing follows. Attention is the vanguard function of our so-called "cognitive load." Attention is the

**Table 2** Representations and responses of the dorsal and ventral attention systems[a]

| | The dorsal attention system | The ventral attention system |
|---|---|---|
| Major modules on the outside of the cortex | Intraparietal sulcus (IPS) Frontal eye field region (FEF) | Right temporoparietal junction (TPJ) Right inferior frontal cortex (IFC) |
| Functional anatomical representation | The right and left sides are symmetrical | Is predominantly right-sided |
| Is located closer to the course of: | The overlapping dorsal egocentric processing stream E) | The ventral allocentric processing stream (A) |
| Responds attentively to: | The opposite side of the environment | Both sides of the environment |
| Do intrinsic fMRI signals also fluctuate slowly and spontaneously? | Yes, more often in a reciprocal manner with the medial frontal and posterior parietal regions | Yes. On some occasions in a reciprocal manner with the medial frontal and posterior parietal regions |
| Basic modes of orienting, or of reorienting | An intentional shift. Orienting is more "top-down," voluntary, and goal-driven | A reflexive shift to stimuli outside the current focus of attention. Reorienting is more "bottom-up," and is strongly "stimulus-driven," especially when the stimulus is behaviorally relevant |
| What happens during deliberate searches designed to focus attention? | Activity is enhanced | Activity is suppressed. This reduces the possibility that some irrelevant stimulus might prompt an inappropriate shift toward reorientation |
| Responds most actively to | Prior cues denoting "what-where-when." It is also co-activated when fresh stimuli prompt reorienting | Salient task-relevant targets in unexpected locations. Fresh needs to disengage attention and to reorient it. Transitions between two events or behaviors |
| Potential small zones of "executive overlap" that could help integrate the two systems under varying contingencies | In the right prefrontal cortex; in the middle and inferior frontal gyrus. (Lesions in these frontal regions cause patients to be distractible and to perseverate) | |

[a]When the subjects are only resting passively, the regions representing these two attention systems are undergoing slow spontaneous, intrinsic fMRI activity. The subjects' eyes are then: either open and looking at a cross-hair; or are open in dim light; or are closed. No obvious task is superimposed; no stimuli are being added. The streams of consciousness flow in multiple directions under such conditions, as meditators know only too well from their "monkey-mind" ruminations. This table began with the earlier discussion in Fox et al. (2006). Its contents continue to evolve in keeping with the recent excellent reviews by Corbetta et al. (2008) and Shulman et al. (2009)

sharp point at the very tip of our otherwise blunt efforts at processing. Attention selects and impales its target in the earliest milliseconds. Then, during the next milliseconds and seconds, our brain mobilizes lesser degrees of attentiveness to sustain the momentum of its subsequent processing.

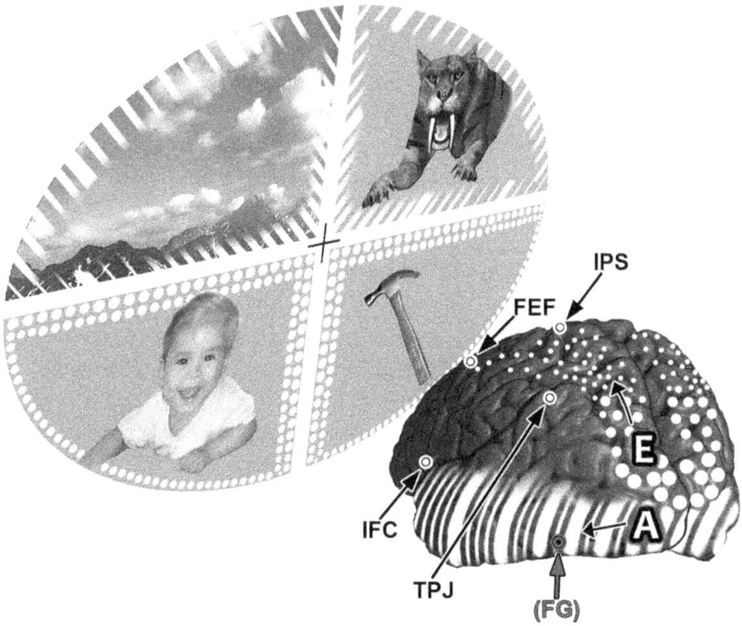


**Fig. 1 The inherent visual efficiencies of our dorsal and ventral attention system**. This view imagines the different kinds of scenery a normal brain might attend to with optimum visual efficiency. Our vantage point is from a position behind its left hemisphere. The end of the occipital lobe (*OCC*) is off to the bottom right. The brain is shown gazing up and off to the left into quadrants of scenery. The smaller black letters at the top of the brain identify two major *dorsal* regions: the intraparietal sulcus (*IPS*) and the frontal eye field (*FEF*). We activate both modules during our more volitional forms of "top-down" visual attention. Running through each of them, in turn, is the overlapping trajectory of the upper parietal → frontal processing system (shown as *white circles*). Why are there also rows of similar white circles in the scenery surrounding the *lower* visual quadrants containing the baby (at *left*) and the hammer (at *right*)? These circles suggest that this dorsal attention system can attend most efficiently to the way we handle tangible items when they are located down in the corresponding *lower* parts of the visual field and are *closer* to our body. In contrast, the two other modules identified reside at a lower level in the brain. They are the temporoparietal junction (*TPJ*) and the inferior frontal cortex (*IFC*). We activate them (chiefly on the *right* side) during involuntary forms of "bottom-up" attention. The diagonal white lines suggest the successive temporal and frontal processing networks whose functions can be directed by each lower module of this ventral attention system. Why do similar arrays of diagonal white lines also surround the *upper* visual quadrants? The purpose is to suggest that when this *ventral* temporal → frontal system becomes trained to stay more alert, it remains poised to detect items at a distance, especially when their stimuli enter the *upper* parts of the person's visual (and auditory) field. Consider the survival advantage. It would be helpful to quickly see a saber-tooth tiger and to detect its noises while this tiger is still located *a long distance away* from one's body. The *FG* in parentheses points toward the left fusiform gyrus, part of the color-sensitive cortex hidden on the undersurface of the temporal lobe. The large letters, **E** and **A**, illustrate the initial course of the related egocentric and allocentric processing streams, especially when their separate pathways partially overlap with, and their processing functions are governed by, their respective modules of the adjacent dorsal and ventral attention streams

Anatomically, we represent the higher levels of this voluntary attention system in our upper (more dorsal) cortical regions. Notice in Fig. 1 how these important regions are located *near the top* (more dorsally) and over the *outside* of the brain. Residing up here are the two major modules of our dorsal attention system: the intraparietal sulcus (IPS), and regions around the frontal eye field (FEF). Physiologically, each side of this right and left parieto→frontal attention system mobilizes our "top-down" attentional functions. Moreover, whenever we actively search for some target that lies off to one side of the environment, we activate those dorsal IPS and FEF modules that reside over on the *opposite* side of our brain.

When you pause to analyze how many actual steps are involved each time that you deliberately choose to intensify your meditative focus, it becomes clear that each step requires subtle degrees of intention. These are goal-driven, Self-chosen efforts. They are inherently "executive" in nature.

A key point: Please notice in Fig. 1 how this IPS and FEF attention circuit on the outside of the brain is overlapped by the general course of the dorsal *egocentric* processing stream (**E**). This Self-centered stream is described both in the caption of Fig. 1 and in further detail below. It turns out that our dorsal parieto→frontal, top-down attention functions can be *easily entangled with these nearby Self-centered frames of reference in ways that reinforce our egocentric networking functions.*

Yes, this overlapping is a more efficient system in its orientation toward personalized actions. However, suppose we always chose to meditate with some deliberate, explicit goal in mind. Wouldn't this choice seem to reinforce our already well-established Self-centeredness? It would, *because we would continue to be the active agency of how, where, when and why we've just chosen to "pay" attention.* Moreover, Table 2 indicates one consequence of such a deliberate attentive focus: it can *suppress* the activity of the ventral attention system. The message is: Don't overuse your already over-exercised top-down system of attention.

## The Ventral Attention System. Its Relation to the Involuntary, Receptive Category of Meditation

The ventral attention system also has two major modules. They too are represented over the *outside* of the cortex. However, both of these lateral modules are located much lower down (more ventrally). One major module is the temporo-parietal junction (TPJ). The other is the inferior frontal cortex (IFC) and its ventrolateral extension (Fig. 1). Down here, it is relatively easy for these lower attention modules to access the nearby ventral *allocentric* stream (**A**) and to service its dynamic processing needs. *Allo-* simply means other (from the Greek *allos*, other). This lower stream is also described both in the caption of Fig. 1 and in further detail below.

A key point: The functions of this ventral attention system are represented *a*symmetrically. *The right side dominates.* Its circuits cross over to co-opt those other ventral attentive mechanisms that are represented over on the left side of our brain. This means that the *right lower side* of our brain can assume responsibility for the instantaneity of attention. It assumes responsibility not only in the same instant that a bird calls from *anywhere*, but also when any other sudden, unexpected, *behaviorally-relevant* stimulus arises in the environment. Note the several directions from which such an unanticipated stimulus might enter. It can arrive from *either the right, or the left, or even from both sides of the environment.* At this moment, the right TPJ acts as a kind of "circuit-breaker," helping to switch attention from its prior direction toward this sudden new stimulus.

Physiologically, these reorienting functions of the ventral system are more involuntary, "bottom-up," and reflexive in nature. *We* don't direct them. *They* react instantly – and as choicelessly as does the knee jerk response – to whatever happens to arrive from the sensory world outside our skin surface.

Receptive practices of meditation become the wide-open setting within which such stimulus-dependent reactions can capture attention. The global awareness of the receptive brain remains poised to *receive* stimuli with essentially *no* Self-effort on our part. Moreover, when these temporo → frontal networks attend automatically to this global range of relevant stimuli, their impulses begin to flow forward along a lower trajectory. Figure 1 illustrates that this lower pathway is far removed from the impulses that contribute to our somatic Self as long as the latter messages arc high up to flow among our parieto → frontal regions. This lower course remains relatively free of the need to consult with our immediate physical axis of Self.

## *Neuroimaging During A "Resting" State; Activities, Activations, and Reactivities*

Normal human subjects can be encouraged to relax and to "rest" while their brains are being imaged in a scanner. Because the subjects have no obvious tasks to perform, you might think that their brains would be quiet. However, three highly active "hot spots" still stand out in the PET scans and functional MRI scans of these so-called "resting" brains (Austin 2009, pp. 70–76, 98–108, 287–288). The two larger active regions lie deep, next to the *midline* of the brain. Clearly, this medial position differs from the lateral locations over the outside of the brain that was just described for the two systems of attention. One large active region is on each side of the *medial* (innermost) *prefrontal* cortex. The other large active region is on each side of the *medial* (innermost) posterior *parietal* cortex. The third "hot" spot is smaller. It includes the angular gyrus on the outside (lateral aspect) of each inferior parietal lobule.

This particular fronto-parieto-triad of regions does more than stay especially active under resting conditions. It also *reacts* promptly when an event is superimposed. It can react in either of two directions. *How* it reacts depends precisely on which kind of event confronts the subjects and on how relevant this event is to them personally. For example, the activities in this triad can:

– *Increase* (and thus become further activated) when the subjects are assigned a task designed to generate their increasingly inturned degrees of personal *Self-involvement.* (In a highly introspective task, the whole context becomes intimately Self-referential.)

– Or *decrease* (and thus become *de*activated) each time the normal subjects try to perform a task that obviously requires them – *led by their attentive functions* – to incorporate more details from the external world into their behavioral response. Note: Any tasks assigned in a research experiment place a range of explicit and implicit demands on their human subjects. An important demand in either instance is that the subject must immediately start to engage some kind and degree of attention on this task. When the pioneering imaging researchers first assigned a task, and then saw that this "cognitive load" coincided with a *decrease* in their subject's prior neuroimaging "hot spots," it seemed natural to view this "cooling" response as a deactivation, one that seemed to have taken place in a "negative" direction. The choice made then (and perpetuated thereafter) was to describe such a decrease using this shorthand phrase, "task-negative."

In this chapter, we choose a different interpretation. We choose to probe much deeper. Herein, the deactivations that convention has attached to that word, "negative," are *not* rigidly yoked into one and the same concept with just any "cognitive load" and experimental task per se that might be assigned. No. Instead of stopping there, leaving us satisfied by this "task-negative" phrase, we need to search further.

We begin by linking such deactivations to the *initial pre-attention mechanisms that pointedly react to such a task*, not to the rest of the processing "load" of the "task" per se. These early mechanisms are the initial reactions any brain must mobilize in the first milliseconds (in ways not yet fully defined) *if it is to even start* to direct its subsequent processing toward whatever stimuli are involved in the task.

Once again, attention serves as the vanguard of our processing. Its sharp tip of *pre-attentive* subcortical mechanisms points the way, even though we still don't know precisely how they manage to accomplish this feat. For meditators, the over-simplified hypothesis now under consideration becomes straightforward:

- Deep mechanisms service the instaneity of our attentive responses.
- These *reflexive* mechanisms converge to *deactivate* our Self-centered regions.
- *They are trainable*. Therefore, it matters greatly how, where, and when we attend, and *what* we attend to.

## *Recent Developments*

The team of investigators based at Washington University in St. Louis continues to conduct pioneering research at the level of excellence. Word labels and semantic issues aside, the ongoing brain activities in a resting brain, their further activations, and their reactive de-activations constitute a crucial set of findings. They have stimulated a fertile field of international research that has vast implications.

New evidence each year provides new interpretations and new questions:

- Which allied regions share links into this mostly medial fronto-parietal network?
- How many dynamic functions does it seem to engage in, and which of these constitute the core of its Self-other continuum? (Spreng et al. 2009)

- Which underlying mechanisms, their checker-board patterns now acting reciprocally on *both* sides of the brain, cause their remarkable *de*activating responses to be so *inversely* related to those that activate our two lateral networks of attention?

Table 3 summarizes one sample of the recent data.

The left columns in Table 3 use the term RSN as an abbreviation for "resting state network." "RSN 6" refers to the cortical regions chiefly at the front end of the triad of regions that yield so-called "task-negative" (deactivating) responses. The category "RSN 1" includes two posterior parietal regions: the large one lies in the deep medial cortex, the smaller one lies out in the lateral cortex of the angular gyrus. In contrast, the column at far right ("RSN 2") refers solely to the *dorsal* attention network. It expresses the particular kind of *top-down* voluntary attention responses that have been discussed in the earlier paragraphs.

## Further Questions Related to Attention and the Self

Suppose a very substantial, goal-directed behavioral task fully commands our top-down network's attention. Table 3 illustrates what happens as the brain's dorsal attention network ("RSN 2") is activated. Instantly, our Self-referential ventral medial prefrontal network ("RSN 6") undergoes a sustained, *tonic deactivation*. In contrast, suppose a different task requires a lesser investment by our private Self-centeredness. Instead, let this new task call for a more open attentive scanning, the kind we require in order to access the many other circumstantial *environmental* details that we also need to solve the task. Now, the former resting "hot spots" become "cooler" back in the posterior parietal parts of the triad ("RSN 1"). They undergo a brief, *phasic deactivation* in the course of their now "selecting the wheat from the chaff," as it were.

Recent studies of the normal, contrasting responses of these same posterior regions confirm this interpretation. When new information is first being encoded, these regions show *lesser* degrees of fMRI activation (reflecting more of a deactivation component). In contrast, they show *greater* fMRI activity (more activation) when bottom-up attention is oriented into the act of retrieving relevant data from its prior repository in a memory storage compartment (Daselaar et al. 2009).

What about the earlier reported imaging activity that had been loosely ascribed to a relatively stable level of some so-called resting "default" condition? Could any of these numbers accurately represent only one particular, first-person *psychological* state, a single state that could actually represent that subject's *lowest possible* "baseline" consciousness? Formidable semantic and technical challenges would complicate any such earlier metabolic premise for the word, "default." We would not think that an earlier version of the original basal metabolic rate (BMR) test was "basal" if it were being performed on a patient who was also having a continuous Jacksonian focal seizure of his left leg. Each patient's mental state matters. Moreover, subsequent research has revealed that multiple networks also undergo

**Table 3** A sample of three different "Resting State Networks" and characteristics of their EEG and fMRI signatures[a]

| Overall functional category | The self-referential (Yet so-called "task-negative") network | | The dorsal attention network |
|---|---|---|---|
| "Resting state designation"[b] | "RSN 6" | "RSN 1" | "RSN 2" |
| Major modules | Ventro-medial prefrontal cortex, anterior cingulate (pre-genual), hypothalamus | Angular gyrus, posterior cingulate, precuneus, superior and medial frontal gyrus | Intraparietal sulcus (IPS), frontal eye field (FEF), ventral precentral gyrus, middle frontal gyrus |
| Chief EEG response to activation | Increased gamma power; also an increase at rest | Increased alpha and beta power; also an increase at rest | Decreased alpha and beta power (desynchronization) |
| Potentially reciprocal nature of fMRI responses | Is further activated during interoceptive, introspective, Self- referent functions. Even so, a sustained tonic de-activation occurs during attentive, goal-directed behaviors | Also responds during Self-referential internal processing, yet a brief, phasic de-activation is "time-locked" to the particular event that captures attention | Activation occurs during goal-directed, voluntary, attentive processing |

[a]Adapted from Mantini et. al. (2007). The subjects rested with their eyes closed. The article indicates that the retrosplenial region and hippocampus are included within RSN 1, and that the thalamus participates in its fluctuations. EEG frequencies above 50 cps are not included in the article, nor are data from the ventral attention network. The activity described in the dorsal attention network (RSN 2) reflects our fundamental, strong Self-centeredness. It also shows the strongest correlations (.77) with those of the other five networks

[b]During Zen meditation, what is a more "authentic baseline resting state?" One might describe it as a prolonged episode of clear awareness and *no* thinking (*mushin*). This discrete episode is not the usual state of the subjects in conventional "resting" neuroimaging experiments. Their discursive monkey-mind, driven by underlying emotions, wanders from one Self-preoccupied notion to the next

slow cyclic changes of similar amplitude in their functionings, in ways that reflect their own particular intrinsic, slow, spontaneous fluctuations (Mantini et al. 2007).

Because this chapter relates to Zen meditation, it seems desirable to introduce a new perspective. This calls for different words – words not rigidly bound to any potentially misleading "baseline" concept, nor to other interpretations of what the word "default" might seem to imply. The metaphor proposed seems useful – solely in this context – because it points toward the following questions and potential answers. Each of the next seven questions highlights the next themes in this discussion:

- Why are the normal, ongoing metabolic activities already so high up in this triad of (largely medial) regions of our fronto-parietal cortex? What underlying physiological source is driving them? Our own limbic system will be proposed as one major plausible source for the deeper, emotionalized impulses that keep rising up and driving this cortex.
- Do the different levels of this extended network share some vital biological properties? Both the earlier Self/attention diagrams, the discussions, and Table 3 summarize their potentials: they engage in *reciprocal*, see-saw behavior. In fact, the Self-other continuum of regions has the capacity to instantly switch directions. In which way? Into a different set of networking functions which are aligned in a direction *opposite* to those responses shown by our two *lateral* systems of attention.
- Are such inverse, see-saw shifts relevant for meditators? Yes. Why? One reason why the normal reciprocal reactions and fluctuations of this Self-other consortium are especially relevant is because *kensho-satori* is sometimes precipitated by a sensory trigger. Indeed, it is normally characteristic of this triad of Self-other regions to deactivate instantly when a more internalized task demands its full attention, or when the stimulus from an unexpected external situation captures attention (Table 3).
- Could such reactive deactivations constitute an adaptive response, have some practical, survival value? Speculation: Perhaps such brief deactivations might have helped ancestral brains to: (1) restructure a novel, urgent unexpected situation; (2) cut through their huge excess baggage of old irrelevant autobiographical-environmental trivia, and thus, (3) gain easier, *selective* access to only those few key circuits holding the working resources essential to solve the immediate task.
- If so, then which first steps would be most useful? Simply those selective steps that "cleared the decks for action" by stopping all current, ruminative, useless, mind-wandering.
- Where might such a relatively small, discrete deactivation block be most effective? At some bottleneck through which our usual maladaptive, over-conditioned limbic influences project up to over-stimulate our cortex. Notice that the partial nature of each deactivation implies a remarkable dual selectivity. It has the capacity to interrupt only those over-conditioned Self-centered memory traces that would prove maladaptive, yet to permit full access to other data relevant to the immediate situation, including those stored in the recesses of long-term memory.
- Where is such a bottleneck? One plausible region can easily be inhibited. It includes the limbic and other nuclei in the dorsal thalamus.

How does the foregoing cavalier survey of functional anatomy relate to meditation? Any well-designed training program of long-term meditation needs to cultivate a balanced portfolio of selective exclusions (restraints) and inclusions (reinforcements). As Table 3 (and later 4) will suggest, it falls to the more evolved association networks in the front of our brain to evaluate with prescience, and to prolong, only their most adaptive, quasi-intelligent, overview and behavioral responses. In such a realm of ideal behavior, the old Zen phrase, no-mind, (*mu-shin*) tends to sow confusion at first. However, it implies a similar, highly adaptive process, wide open to the full range of creative options. *Mu-shin* is not ignorance, nor is it the total absence of mentation. Rather does it imply a brain no longer governed by one's over-conditioned emotional responses.

## A SONARR Metaphor[1]

Along the meditative path, the author happened to drop into two very different, extraordinary, alternate states of consciousness. The first was an episode of internal absorption in 1974; the second was a taste of *kensho* in 1982. To a neurologist, it would become increasingly clear that thalamic mechanisms could be pivotal to the different phenomena of these two states. Yet, to introduce a loose metaphor here, as a way to point the general reader toward the levels of your own thalamus, requires both an apology and a brief explanation.

Even so, this acronym's basic working principle is reasonably familiar, because sonar systems are in such common use nowadays. ***So***und ***na***vigation systems help fishermen estimate the ***range*** of fish swimming at levels fathoms deep in the water. Ships of all kinds depend on sonar systems to detect subsurface objects, and to explore other topographical details of the depths. As the sound impulses oscillate back and forth, up and down, sonar technologies transform and amplify their energies. As a result, we can see useful visual images, hear meaningful auditory "pings" and "pongs."

Similarly, the acronym, SONARR, enters this chapter as a metaphor to point toward other basic facts: (1) The brain also has higher and lower levels; (2) These are also associated with oscillations, up and down, back and forth that resonate among layers of our own "***S***elf-***o***ther ***n***egative ***a***ttention ***r***eactive ***r***egions;" (3) Two of our special senses – vision and hearing – also enable us to perceive the results.

The discussions centered around Fig. 1 have already emphasized the very different upper and lower regions out on the cortical *surface* of the brain. However, these sites are not the only locations that increase and decrease their activities. What changes take place that cause such surface measurements to become "cooler?" Why do their fMRI signals decrease (become "negative") as hinted by the heading at the top of Table 3 that

[1] Please regard SONARR as serving solely as an acronym of temporary convenience. It is simply a metaphor pointing to different levels, useful solely in the context of these pages on selfless meditation.

borrows briefly from the early terminology? Hypothesis: Their patterns change because they are reacting to an earlier, deeper sequence of inhibitory events still farther down in the *thalamus*. The brain links these deep changes *reciprocally* with our earliest mechanisms of attention (governed by a sequence of switching mechanisms that still remain to be clarified). The second R in the SONARR system stands for ***regions***. It serves notice that mechanisms involving several *regions* at lower *levels* are involved in the deeper sequences, not just one change at one level in the thalamus per se.

So, regard this as a loose metaphor that can emphasize subsurface events. Let it serve to introduce our next topics in Part II: the brain's *lower levels* of *subcortical* responses that can shift the way our brain attends to our higher cortical processing functions.

# Part II. The Somatic and Psychic Aspects of Our Self; Their Interactive Origins in Thalamo-Cortical Oscillations

> When one goes into Zen meditation, one passes, as a usual process, through a psychic field, from the surface down to the depth, as if one were plummeting into a lake in a diving bell.
>
> Nanrei Kobori-Roshi (1914–1992)

> This emptiness is not something you can conceptualize; it is a state empty of ego, full of what can come through when that ego has been let go of.
>
> Shodo Harada-Roshi
> (cf p. 162 The Path to Bodhidharma)

## *Our Physical and Mental Sense of Self*[2]

A preliminary sounding will prove useful before we plummet into the deeper layers of thalamo-cortical interactions. Because we first need to highlight the two very different ingredients that blend into our individual sense of Self. This is not a new distinction. To the ancient Greeks, our soma was the tangible reality of our physical Self. Its intangible, mental aspect was our psyche.

When Kobori-Roshi spoke about the "psychic field," he was referring to the way the mental landscape of our psyche changes during progressively deeper levels of meditation.

When Zen master Shodo Harada-Roshi describes "a state empty of ego," what does he mean? He is describing an extraordinary alternate state of consciousness, one that is referred to in Zen as *kensho*. What does this word, "empty," mean in this context of *kensho*? To understand, let's start with our own physical body. Our soma is tangible. We can grasp its substance in our own two hands, feel our warm, moving flesh privately confirming its aliveness, notice that the same fingers transmit

[2] Self is capitalized throughout simply to suggest that its operations create many large problems.

**Table 4** Brain responses when "Risky" or "Safe" words are oriented either toward self or toward the outside environment[a]

| | Self-centered assessments | | Other-centered assessments | |
|---|---|---|---|---|
| | Risky | Safe | Risky | Safe |
| Word examples | "Bungee jumping" | "Reading" | "Tsunami" | "Light rainfall" |
| Speed of behavioral and evoked potentials | Slower (1,145 ms) | | Faster (1,111 ms) | |
| Transient response activity in the posterior cingulate cortex and the precuneus | Less | | More | |
| Response activity is more sustained | In the medial prefrontal cortex and the left supramarginal gyrus | | | |
| Comment | Note that a subtle, wordless "selfothering" assessment is implicit when a person judges that a tsunami is risky not just to others, but potentially to oneself as well (Austin 2009, pp. 58, 59, 73, 74) | | | |

[a]Adapted from Qin et al. (2009)

a very different perception when they grasp an apple. Two convincing sensory avenues instantly personalize such discriminations: our senses of touch and proprioception. Our *parietal* lobes refine these dual perceptions. They incorporate them into an elaborate body schema. In this way have we established the framework of our *somatic* Self, fleshed it out around a central physical axis.

Our psyche is different. It is an *in*tangible, amorphous creation. True, we can direct the psychic functions of our *I-Me-Mine* in personalized ways to achieve tangible external results. However, as children we soon discover that bad behavior can cause our soma much pain and our psyche much anguish.

Fear is the dominant emotion at the core of our psyche. Fear's powerful, pervasive influence is difficult to dissolve. If we can only set aside enough time, and introspect long enough to analyze the sources of our fear, we can start to identify which internal impulses within our psyche still drive some of our risky behaviors. Moreover, when we look out into that worrisome asphalt jungle *beyond* our own skin, our resulting apprehensions help identify which *external* risks we should avoid first. Researchers now have the tools to uncover many of these subtle, different emotional attitudes implicit in our psyche. How? By measuring the ways our brains' functional MRI signals respond not only to gross fearsome threats, but also to virtual risks, and even to events that tug on the sentimental levels of our softer emotions.

## The Power Inherent in Words; fMRI Correlates

A boundary separates *this* Self inside from *that* other world, outside. It's the surface of our skin. Neurobiology reflects this basic distinction between our intrinsic Self and that extrinsic other. Table 4 summarizes how normal brains respond with different EEG and fMRI patterns to this distinction between approach and avoidance behavior. The subjects are seeing very simple words. Yet each word hints how much risk is involved within the Self side – in contrast to the risk on the other side – of this Self/other boundary.

As might be expected, the deep midline cortex of the person's *medial prefrontal* region (MPFC) doesn't respond only when this subject makes an explicit egocentric assessment of risk. Its signals also tend to increase when the demands implicit in the task require this person inside to make subtle judgments about how much risk is inherent in "*their*" interactions with their environment outside their own body (Qin et al. 2009). Though we wear that robe on a layer just outside our skin, it still has woven in it the personal monogram of its owner.

On the other hand, it is chiefly our deep, *medial posterior* regions – those farther back in the SONARR system's parietal and posterior cingulate cortex – that help to assess a specific test word which clearly positions the acute risky situation as located in *that other world* outside our body. Other networks incline different people to avoid risks. Cautious women who avoid risks tend to have higher tonic baseline EEG activity over their right prefrontal cortex (Gianotti et al. 2009).

## Soma and Psyche Draw Upon Different Networks for Our Softer Emotional Sentiments

This time, suppose we observe a situation that directly involves someone *else's* body or psyche, not our own. How do we respond emotionally to a situation involving that other person's body? Now, we activate a part of our own deep medial posterior parietal cortex that is slightly different from the part activated when this situation involved the other person's *psyche*. Next, consider what happens when we empathize with someone else's *physical* (somatic) pain, or admire their unusual *physical* skill. Now, our medial parietal fMRI signals respond more in the (larger) superior and anterior part of our parietal cortex (Immordino-Yang et al. 2009). Suppose, on the other hand, that we develop empathy for the painful suffering involving someone else's *psyche*, or admire the *psychic* attributes of virtue that are expressed in that other person's behavior. Now, our medial parietal signals respond more in the (smaller) inferior and posterior part of our parietal cortex.

Zen Buddhism tends to use the word "empathy", to describe only the earlier, perceptive and imaginative side of our affective sensibilities. In contrast, authentic compassion (*karuna*) tends to ripen as a later development on the meditative path. This advanced habitual level of compassion *reaches out selflessly* to help, using only the most skillful and appropriate behaviors. With regard to this distinction, which other regions coactivated during the fMRI study of the normal witnessing subjects just cited above? Significantly, they were the particular regions in the observers' brains having the potential to adjust the somatic balance of their *body* with respect to gravity, not just the sites that might generate an empathic psychic tug, say from the limbic system or its thalamic relays. Such a basic somatic adjustment could serve as the requisite covert axial foundation for those subsequent motoric impulses that would enable the arms of each witness to reach out skillfully to correct an unfortunate situation.

As adults, we can hope that our more mature brains will enable us to express more compassionate attitudes for the body and psyche of another person than did those attitudes we had back when we were Self-centered children. In this regard, when children engage in Self-relational processing, they co-activate more of their anterior precuneus and posterior cingulate gyrus. In contrast, adults chiefly co-activate the posterior subregion of their precuneus (Pfeifer et al. 2007).

In the next section, we continue to plumb deeper into that posterior part of the spectrum of core functions represented among these *medial* parietal networks of the SONARR system. For the question remains: What subtle interactions could this extensive posterior physiology contribute to our larger sense of Self?

## *Basic Functions of the Deep Medial Parietal Cortex*

We register countless events in our lifetime, and consolidate them in our reference memories. Each such memory event is documented with circumstantial details from our environment at that very moment. Various lines of evidence suggest that many of these coded environmental scenic details are represented in "chunks" among contributions from our deep medial *parietal* cortex (Austin 1998, pp, 390–391; 2009, pp. 71–74). In what manner could we "represent" such incidental environmental details, and still tag such a hologram with our own personal initials? Perhaps we might do so (by analogy), in forms resembling the immediate, graphic, "entries" we make in a personal journal when we're traveling to a foreign country.

Here, the phrase, journal entries, implies only that after we first register and consolidate myriads of such coded entries in our private "memory bank," we can later draw on them when we need pertinent environmental details. At first, maybe we had barely noticed such events, then buried their incidental details at mostly subconscious levels. Even so, the sum of these subliminal coded entries still registers the circumstances of one person's decades-long *private history*: Our high school friends, our old bedroom, that street scene in the neighborhood where we once lived. When we access and retrieve these entries selectively, they also enable us to navigate our way anew in real life, or in dream worlds of our own invention.

Any such "journal" requires an underlying premise: a dynamic continuum of relationships centered on our *Self*, that receptive and executive agency back in the center. This Self-construct began in infancy with a physical axis that had its own 3-D spatial coordinates. Soon, the same centralized premise would superimpose a further, urgent *affective* life inherently its own. It developed an array of highly conditioned limbic associations. These gave rise to polarized feelings resonating with how this Self's psyche and soma *felt subjectively* at the time. Thereafter, many journal entries were infused by positive and negative emotions. This meant that they no longer registered the simpler objective facts about *where-when-and-how* each personal incident had once occurred. On the other hand, they did have the capacity to register, consolidate, and narrate a full impressionistic range of our fictional life experiences.

Recent research illustrates that complex narrative potentials are implicit in this same deep medial parietal region. The evidence suggests how its association links could enable us normally to draw together a mosaic of detailed journal entries, then use them to navigate through a Self-other "storyline" that could go on to recreate such a private version of our own long life history.

The study specifies two cortical regions in the deep medial posterior SONARR system (Whitney et al. 2009). They are the *right* precuneus and the posterior/middle cingulate gyrus. Both are activated when subjects listen with great care to the way a novella unfolds. The cognition task assigned to these experimental subjects is daunting. They must detect the slender thread of coherent meaning that links each of the 15 major twists and turns of the narrative!

How can these listeners pass so rigorous a test for coherence when a complex story shifts its settings and characters back and forth so many times? Only if they construct an intricate web of visual/auditory/spatial images, then assemble them into an *imagined sequence* of relevant events. Soon, their new mental structure develops a logical plot. It evolves into a plausible storyline, one that enables each character to react to his or her environment in specific ways during that particular moment. To accomplish such a major task, instant parallel communications are essential at *multiple levels* all along the Self-other continuum of core functions affiliated with the psyche.

## The Dorsal Thalamus Interacts with the Dorsal Cortex

As Part I closed, the notion of a SONARR metaphor was introduced for a straightforward reason. It was to alert the reader to the fact that when any event instantly catches our attention, *several layers of Self-centered regions* will be drawn into a sequence of interactions at levels high and low. Suppose we now descend further to inquire: which nuclei farther down at the level of the thalamus interact most influentially with their cortical counterparts up in the SONARR regions just discussed?

Surely the large *dorsal pulvinar nucleus* is the thalamic co-partner of first interest. Its interactions resonate not only medially with the precuneus but also up and out with the laterally-placed angular gyrus in the inferior parietal region. Moreover, the *anterior nucleus* of the thalamus interacts with the posterior cingulate gyrus; and the *lateral dorsal nucleus* of the thalamus interacts with the neighboring retrosplenial cortex just behind it.

In this same dorsal tier of thalamic nuclei is a fourth nucleus, the *lateral posterior nucleus*. It makes a vital contribution to the somatic matrix of our physical Self. How so? By virtue of its having dedicated interconnections with our *superior parietal lobule*. This most dorsal part of the parietal lobe is our major somatosensory association cortex. It contributes an elaborate 3-D schema to our *physical* sense of Self.

Finally, a fifth nucleus in the dorsal thalamus completes the interactions between this upper level of the thalamus and its co-partners high up in the dorsal cortex. This is the *medial dorsal nucleus*. It maintains an intimate, to-and-fro communication with the Self-referential ventro-medial prefrontal cortex, as well as with the rest of the large prefrontal cortex (Table 3).

What makes these five interactions between the dorsal thalamus and dorsal cortex so important? What do they normally accomplish? Subliminally, they inform us which way our own head is facing. They blend this emergent forward impression with that of our own body schema, and jointly infuse these coded functions into the activities of our Self-centered processing stream as it courses up through the superior parietal lobule (**E** signifies this egocentric stream in Fig. 1). Moreover, three of these five thalamic nuclei become especially intriguing. Why? Because these three sponsor *the affective resonances and influences of our psychic Self.*

These anterior, lateral dorsal, and medial dorsal nuclei are called the *limbic* nuclei of the thalamus, because each is strongly interconnected with the limbic system. Indeed, they are the major route through which our limbic system relays its over-conditioned emotional and instinctual messages up to excite its cortical partners. So, it is not surprising to find that the two larger "hot spots" of the *medial* cortical SONARR system – one medial frontal, the other medial parietal – share vital connections with their limbic partners down in the dorsal thalamus. Nor that the other hot spot out in the angular gyrus is a special target of their neighboring nucleus, the dorsal pulvinar. The attentive reader may begin to wonder: could this whole dorsal tier of nuclei down in the thalamus – led by its three limbic nuclei – represent the potential "bottleneck" referred to back in Part I? It could.

What other words help describe how these five dorsal thalamo-cortical partners normally interact? They share in fast, to-and-fro oscillations. As these oscillations resonate up and down, to and fro, they blend the higher-level functions of our dorsal cortex with those of their five corresponding thalamic nuclei down at the next lower level (Austin 2009, p. 88, Fig. 6, color plate 4 after 168).

Our next challenge is to explain how the normal physiological functions of these upper cortical and lower thalamic levels correlate with the two categories of relationships outlined earlier. Specifically, how do these shared thalamo-cortical functions clarify the relationships: (1) between attention and meditation; and (2) between meditative attention and the dynamics of Self-other SONARR relationships?

Let us begin by returning to the largest thalamic nucleus, the pulvinar. The way this pulvinar helps us instantly frame one salient focus of interest in a large Monet painting recommends it an obvious choice for helping to understand the psychophysiology of meditation. The more so because the pulvinar can draw instantly on our polymodal auditory and visual impulses as they speed up from the midbrain colliculi, and almost instantly on a host of our other relevant associations as this information descends to it from most of the posterior cerebral cortex.

## *The Pulvinar; Proposals for Future Pulvinar Research in Meditators*

Recent functional MRI studies reveal how silently and pre-attentively the dorsal pulvinar manages such covert operations. It stays actively engaged even though we may not be consciously aware that we are devoting attention instantly and automatically to some portions of our external environment (Smith et al. 2009). Moreover, the right and left pulvinar each register these subliminal involuntary responses when stimuli enter their visual response field from chiefly the *opposite* side of the environment. In this respect, the pulvinar already attends and responds the same way as do its cortical partners high up in the dorsal cortex. The ventral pulvinar appears more responsive to visual stimuli. Suppose, however, that we make an executive, top-down choice to concentrate and attempt to "pay" conscious attention. Now the pulvinar's responses increase only slightly. (The increase averages the same 21 % in both the dorsal and ventral subdivisions of the pulvinar.)

When the eyes are closed, functional connectivity is reduced between the right pulvinar nucleus and the visual cortex (Zou et al. 2009). This reduction occurs both in the medial regions of the primary visual cortex and in its lateral association cortex. The right pulvinar nucleus is the larger, and has more nerve cells, in keeping with the right hemisphere's predominate role in being more activated during attention.

When researchers apply similar fMRI procedures to study meditators in the future, their data could then help answer the following crucial questions:

- Do prior years of a training program designed to include *receptive* meditative techniques further enhance the speed and scope of this subtle, automatic, ongoing tonic visual drive? Does this happen even at times when the meditators are not overtly making an explicit, conscious choice to superimpose their top-down attention on obvious visual stimuli?
- Are the concentrative and receptive styles of meditative training really complementary? When we combine their assets, do they mutually enhance and stabilize several aspects of the reactive kinds of attention – those triggered reflexly by an abrupt stimulus that occurs unexpectedly? Does this dual approach to training also enhance those instantaneous, *pre*-attentive varieties of flexible "pop-out" attention that normally involve the pulvinar? Does it also enable such a span of pre-attention functions to last longer?
- In the course of such future longitudinal investigations, which discrete thalamic and cortical regions become deactivated and/or activated simultaneously?

At present, few well documented case reports describe patients who have one small lesion limited just to their pulvinar. Recent evidence from two patients suggests that the pulvinar's spatial maps are coded similarly to those of the visual cortex that are illustrated schematically in Fig. 1. Namely, the dorsal pulvinar also represents the *lower fields* of visual space; the pulvinar's ventral regions also represent the *upper fields* of visual space (Ward and Arend 2007).

These authors further suggest that our pulvinar has the capacity to represent 3-dimensional, coordinate systems in visual space. Moreover, in doing so, they add that the pulvinar can also use frames of reference that are "object-based." What does "object-based" mean? An "object-based" (allocentric) frame of reference is sensitive to more than the form of a particular object, "out there." Its frame of reference can relate the spatial position of that first object to the 3-dimensional coordinates of some adjacent object "out there." (Austin 2009, pp. 54–64, 133–135)

By definition, such object-based frames of reference are independent of the human observer's 3-D coordinates in space. Therefore, *no personal subjectivities* of the egocentric Self are intruding into the field of vision for as long as such a frame of reference remains exclusively "object-based." Accordingly, a substantial impression of detached objectivity can accompany a long-sustained moment of this other-referential perspective.

Why is it so crucial to understand the basic ego/allo distinction? Because when meditative training enables longer moments of other-referential (allocentric) processing to become free – suddenly or incrementally – and to flow spontaneously through ventral pathways, their networks can then serve as the matrix for selfless forms of intuitive processing that become creative, insightful, and adaptive.

## *A Pivotal Thalamic Role in Our Sudden Response to a Triggering Stimulus*

Subtle shifts occur deep in the thalamus at times when we become relaxed and are seemingly "at rest." These help adjust and integrate the normal range of our usefully balanced, rapidly oscillating functions (as suggested in Table 3). Because the thalamus is the major "gateway" on the sensory-motor-limbic paths to the cerebral cortex, it is poised to play a key role (directly and/or indirectly) in the genesis and patterning of those bilateral, *slow*, spontaneous, reciprocal, fluctuating cycles of the brain's *intrinsic* activation and deactivation, as cited in the fifth layer of Table 2.

Other thalamic nuclei create decisive shifts in the net balance of these cortico-basalganglia-thalamo-cortical looping circuits. For example, let an abrupt sensory stimulus activate the requisite triggering mechanisms, and several layers of sequences within the thalamus could suddenly coalesce into a configuration that precipitates the resulting state of *kensho*. Thus, the *reticular nucleus* of the thalamus also becomes acutely informed at almost the same instant that this triggering stimulus enters and sets off the initial sensory impulses. It releases a wave of inhibition capable of briefly *deactivating all five of the Self-centered nuclei in the dorsal tier of the thalamus*. How can a nucleus do this? By what mechanism? The reticular nucleus releases a potent transmitter: GABA (gamma-aminobutyric acid). This can disorganize the normal in-phase relationships of the brain's usual synchronized oscillations and firing rhythms. These oscillations would otherwise still be resonating – up and down – in the looping circuits that normally connect this dorsal thalamus with its dorsal cortical partners (Austin 2008; 2009, pp. 87–94). As a result of this de-synchronization, the usual Self-referential spectrum of functions previously conferred by the mostly medial *cortical* level of the SONARR system would share in the resulting deactivation.

The resulting change in the state of consciousness would be distinctive. It would be rendered "empty of ego," just as Harada-Roshi describes. Those intrusive veils of the old subjective Self would finally have parted: No over-conditioned limbic networks would remain to rile up and muddy the field of consciousness. Instead – in the long-term meditator whose sensitized modes of attention had become balanced and stabilized – bottom-up attention would now serve as the sole vanguard of a liberated, clarified, receptive field of allocentric consciousness. And, thus as

Harada-Roshi also beautifully describes, does the resulting state finally become "full of what can come through" when no ego stands in the way.

*What* signals can now come through in all of their fullness? Simply the way that these other-referential pathways normally perceive all other things in the environment – clearly, "as THEY really are." During this long-sustained present moment, the brain is finally free from all the language and other burdensome obstructions imposed by the Self. Throughout this instant of *kensho*, "emptiness" of self coincides with "suchness" in *one* single field of consciousness (Austin 2009, pp. 549–553). No duality remains.

But how can a whole dorsal tier of our thalamo-cortical functions be deactivated with such exquisite selectivity? The reticular nucleus "cap" on the thalamus can cast this inhibitory net to envelop our entire thalamus over its 3.3 cm anterior-posterior extent (Mai et al. 2004). Moreover, its inhibitory actions are strengthened by those of its two smaller GABA allies, the zona incerta, and the anterior pretectal nucleus (Austin 1998, p. 502; 2010, pp. 373–407).

An active, noise-cancelling electronic headphone provides a convenient analogy. It generates a sound wave profile of oscillations that are 180° out-of-phase with those of undesirable background noises. This interference pattern tends to cancel unwanted sound energies, yet – at the same time – it allows desired signals to come through and be heard.

## *Neuroimaging Caveats*

Many contemporary reports refer data to the "thalamus," as if it were some homogeneous anatomical structure. Single neuroimaging methods do not yet have sufficient resolving power to measure the activations and deactivations within each discrete individual thalamic nucleus. In the interim, it cannot be assumed that some preliminary *gross* estimate of activity down on only one whole side of the *entire* "thalamus" must correlate only with the gross activity of its several cortical partners high up in the hemisphere on that same side. In fact, a recent fMRI study shows that the thalamic activity on one side during an ordinary "resting state" can tend also to correlate with that of the *opposite* cortex higher up in the *other* hemisphere (Zhang et al. 2008).

A variety of integrative roles is apportioned among very different thalamic nuclei. We don't yet know how they all interact to pattern our activations and deactivations into such a delicate balance. We don't know precisely which bioelectric and metabolic mechanisms converge to switch their functions instantly, or help generate the similar shifts of their slower, intrinsic fluctuations. "Not knowing" is not confined to the neurosciences. It is an age-old phenomenon in Zen training, one that Zen masters cultivate in their trainees and exploit unexpectedly at every opportunity (Austin 1998, pp. 119–125).

An earlier generation PET scan becomes of interest in regard to thalamo-cortical interactions (Austin 1998, pp. 281–284; 2009, pp. 12–87). The subject had been meditating for 2 h, had let go of thoughts, and was barely attending to the faint movements of abdominal breathing. Vision and hearing were masked. Cortical activities were relatively greater lower down over the right hemisphere. Thus, the lower occipito-temporal pathway on the undersurface of the right side was substantially more active, and clearly included the region of the right fusiform gyrus (FG). These findings are interpretable as consistent with an asymmetrical trend toward a more "bottom-up," ventral mode of attentive, allocentric processing. In contrast, both medial prefrontal regions showed minimal activity, and the lower portions of the right thalamus were less active than on the left.

## *Looping Circuits Engage Deeper Levels: The Cortical-Striato-Pallido-Thalamo-Cortical Circuit*

The basal ganglia are a paired collection of deep nuclei on either side of the brain. They contribute to our learned patterns of sensori-motor responses that – through repetition – gradually enable our behaviors to become more habitual and automatic. During these complex integrations, several levels of our basal ganglia nuclei are interposed in large looping circuits that interconnect them with both the cortex and the thalamus (Austin 1998, pp. 155–156, 197–201, 392–397, 671) .

Starting at the top level of these nuclei, the *dorsal* striatum includes the large caudate nucleus and putamen. The *ventral* striatum is a paralimbic structure that includes the nucleus accumbens. Dopamine energizes these striatal nuclei. It is supplied by nerve fibers rising up from two other nuclei farther down in the midbrain: the substantia nigra and the ventral tegmental area. In contrast, the nerve cells of the globus pallidus release GABA into the ventral anterior thalamic nucleus. The first action of this GABA is inhibitory (Kopell and Greenberg 2008; Austin 2010, Fig. 15.5).

By repeatedly engaging in meditative practices, can we transform the functions of these looping circuits? Can we do so in ways that help us modify our prior dysfunctional habits of thought as well as other habitual behaviors? These are crucial questions. Their significance becomes clear in the context of the following sentence from Graybiel's recent review. The words provide a succinct definition of the nature of our habits (Graybiel 2008).

"Habits are sequential, repetitive, motor, or cognitive behaviors elicited by external or internal triggers that, once released, can go on to completion without constant conscious oversight."

The last four words are operative words that summarize a major theme of this chapter: "without constant conscious oversight." So let it be clear: (1) We are considering a program of endlessly *repeated long-term* meditative practices; (2) The majority of their results unfold *unconsciously*, during a very gradual trait-change in basic attitudes. How is this possible? In the previous sections, we cited a sequence of interactive processes that serve to *decondition* an array of our maladaptive,

Self-centered *emotional* responses. In broad brush strokes, that earlier model assigned a key role to the thalamus as one region having the capacity to liberate an enhanced flow of other-referential attentive processing from its prior bondage to an over-emotional, egocentric frame of reference.

There exists a more subtle way that meditative practices can cultivate and enhance the normal "plasticity" of the brain. It is by encouraging the *inherent normal resources* of our slowly maturing brain to evolve *at a greater rate* – decade by decade – an increasingly mature variety of attitudes and behaviors at widely distributed levels. In one sense, we're referring to an *acceleration* of that same basic gradually self-correcting "universal intelligence" – our "horse-sense," if you will – which is every person's biological heritage (Austin 1998, pp. 660–663; 2006, pp. 399–401; 2009, pp. 237–244).

The broad brush strokes used to outline the incremental mechanisms of this second model are much more theoretical at present. These subtle mechanisms, enabling our habitual behaviors to be transformed, focus mostly on the levels of function within the dorsal and ventral striatum. Here, the glutamate → nitric oxide → cyclic GMP metabolic pathway is a plausible candidate for gradually re-sculpturing many intricately interactive looping circuitries. These changes could also serve to gentle our habitual attitudes and to channel the energies of our habitual sub-conscious behavioral expressions in more compassionate directions (Austin 1998, pp. 412–413, 653–659, 675–676; 2006, pp. 279–288; 2009, pp. 260–261).

## Part III. Practical Issues Relating to Zen Meditation

> The Buddhist Way flourishes in a mind utterly free from conceptual thought processes, whereas discrimination between this and that gives birth to a legion of demons!
>
> Zen Master Huang-po (died c. 850)

> Keep your mind clear like space, but let it function like the tip of a needle.
>
> Zen Master Seung Sahn (1927–2004)

### *Introspections for Future fMRI Research on Zen Meditators*

Zen values our normal thoughts, feelings, words, and objective introspections. However, it emphasizes the assets of their arriving skillfully, and of their being applied efficiently at the most appropriate times. While current fMRI research likewise exhibits numerous assets, it too is burdened with innate technical and conceptual liabilities (Van Horn and Poldrack 2009). Future researchers need to design studies that strike an appropriate balance between: (1) *their* legitimate needs for hard numerical data, and (2) the no less valid – but much softer – needs that meditators must also have on *their* Buddhist Way to enter into *clear, silent, emotionally-thought-free* mental space. This requisite balance – between the sharp needle tip of top-down attention and the clarity of wide-open objective receptivity – crucially influences how the data can be interpreted.

Both the Zen Buddhist schools and the neurosciences maintain rigorous scholarly traditions. Yet all too often, a few members in each camp underestimate the absolute importance of what Masters Huang-po and Seung Sahn are each saying. They're pointing toward that same openly receptive advanced attitude *and* state of *clear, wordless, emotionally-thought-free consciousness which responds habitually with instant discernment to whatever happens next*. Its attributes include:

- Openess.
- Receptivity.
- Clarity.
- Wordlessness.
- No-thought.
- Instaneity.
- Discernment.

Lest Huang-po's words had not driven the point deeply enough, Master Huang-lung added this comment two centuries later: "Zen doesn't need to be studied. The important thing is to stop the mind. Then, once you've become free of ruminations, with each step you'll be walking the Way." The basic Zen message seems clear: Our usual mental field is too preoccupied with its Self-centered subjective thought stream of "monkey-mind" concerns. As a result, we can't open up our innate receptivity to perceive clearly the simple fresh objective realities of the real world right in front of our nose.

Meanwhile, researchers are burdened by academic demands. Soon, they are led to impose this same top-heavy burden on their experimental subjects: words, concepts, multiple assigned stimulus → response tasks. Conscientious subjects become overly preoccupied with top-down multitasking. Emotional responses warp their fMRI signals and their performances as well.

Beginning meditators do need formal instruction in traditional techniques during their early months and years of practice. Concentrative approaches do assist top-down attention to focus like the sharp tip of a needle. Then, both brain and body begin to discover how to cultivate the subtler, more advanced, receptive attentive practices. These also allow consciousness to become increasingly clear, spacious, effortless, selfless (*wu-wei*), intuitive, and improvisational (Austin 1998, p. 207; 2009, pp. 43, 144, 146). In the interim, researchers must remember: practitioners need years to cultivate and fully express such skillful spontaneity – the bottom-up kinds that tend to "happen" only when well-trained silent illuminative meditation practices have also *let go of conceptual thoughts* (let alone of artificial tasks which force their Self to discriminate between "this" and "that.")

It was never the stated purpose of one recent fMRI report to serve as a study of meditators (Christoff et al. 2009). However, a variety of readers might be mislead into thinking along meditative lines when they encounter terms like "meta-awareness," "mind-wandering," and phrases like "parallel recruitment of executive and default network regions." Accordingly, it is instructive to review the recent premises and experimental conditions during which 15 student subjects were required to perform a basic "sustained attention to response task."

This task demanded attention. Its intent was to introduce major errors in the subjects' performance. Indeed, the fact that the task involved a new number every 2 s, and an appropriate button press did encourage the subjects' minds to wander. Thereafter, why did the experimenters also superimpose two specific visual questions, unpredictably, every minute or so? Because their next immediate goal was to elicit samples of self reports from the subjects whose minds (by then) were usually wandering. Notably, these probing questions then required each subject (*96* times during one experiment) to select a single answer from a scale that offered 7 potentially different self reports.

In themselves, these experimental conditions would seem likely to generate multiple subconscious and conscious emotional reactions. Under even remotely similar circumstances, were a Zen meditator also to be required repeatedly to introspect (e.g. "Am *I* completely on task?"), one might anticipate that some of this subject's Self-referential regions would become activated and also expect that increased fMRI signals could develop within the so-called "executive network."

## *The Power of Silence*

To date, fMRI research has not yet taken full advantage of those longer, deeper quiet intervals of clarity, spaciousness, and everpresent awareness. These develop only when advanced meditative practitioners – in keeping with Master Huang-po's advice – *let go of all their mind-wandering impulses and conceptual thoughts, yet still retain a background of keen awareness*. Meanwhile, why can the casual application to meditative research of loose terminologies in the literature (such as "default state" and/or "resting state") continue to sow confusion? Because neither word has an exact psychological mooring. Neither word accurately describes that same subject's precise mental state at that very moment. Table 3 suggests that one source of such heterogeneity is represented among those different networks which infuse their blend of distributed functions along a continuum. This extends from our chiefly Self-referential regions in the front of the brain (RSN 6) towards what constitutes, in a sense, the more "Self-in-other" mixture of representations distributed much farther back (RSN 1). The second footnote to this table indicates a second source of heterogeneity: each subject's ruminitive monkey mind.

## *Other Practical Issues Relating to the Functional Anatomy of Selfless Meditation*

Even the single-celled amoeba needs to distinguish between itself and its outside environment. Human beings employ similar pairs of words – not only Self/other, but inside/outside, this/that. The word pairs suggest our general tendency to identify things in terms of their opposite aspects. Many such polar differences become *complementary* when viewed from the perspective of the basic, underlying Taoist principle of yin/yang organization. Conceptual pairings that might first seem simple tend to blur even further when researchers probe complex human equations, attempting to specify what their correlates mean in more precise neuroimaging terms.

One problem is that each human subject is unique. Each of us is still constructing a long detailed personal history. How can our episodic memory system integrate external landmarks into some of its earlier memory links? One way is by connecting the retrosplenial cortex into links with its close neighbors – the hippocampus and amygdala – nearby in the medial temporal lobe (Greicius et al. 2009). In contrast, our posterior cingulate cortex, (cited in Table 2), lies immediately in front of this same retrosplenial region. Yet, *its* strongest interconnections are with our anterior cingulate and medial prefrontal cortex. These messages travel as "long distance calls" when routed through the cingulum, a long nerve-fiber bundle.

The way we interpret other important psychophysiological issues also hinges on similar fine anatomical distinctions. For example:

- Is it correct to attribute our inturned, Self-referential functions exclusively to the front of the brain, say just to our ventromedial prefrontal cortex? No. As Table 2 suggests, the Self-centered functions of our omni-Self are too widely distributed. Indeed, some speculate that whenever we generate and project an empathic process that "reads" another person's mind, we first refer back to our own private mental networks – making the way they function the source of a relevant proxy (Mitchell 2009).
- Do connections between identical cortical nerve cells also subserve such separate functions as mind-reading, mirroring, and the spectrum of Self-other activities? Examples already discussed in previous pages suggest how some of the specific views ventured by network "sub-dividers" might prevail over other views held by those prone to be "lumpers." The peak activities recently identified in separate local subregions of the SONARR network provide a further example of the former approach (Summerfield et al. 2009). Each separate local site can be correlated, respectively, either with an actual (real) event that was directly Self-experienced or with a second-hand event that had only risen into experience when the subject had merely passively viewed a movie or TV.

In the interim, two groups of French investigators, representing two different schools of thought, have each drawn interesting conclusions that become complementary, yet divergent. The contrasts happen to illuminate some age-old distinctions between: (1) the meditators who enter the early, shallow, states of meditative absorption; and (2) those advanced practitioners who undergo *kensho* and develop the characteristic deep insights and trait changes of the much later stages on the long spiritual path. For example, the first school would favor restricting the concept of Self to just the most elementary steps that lie off at the *physical* end of a whole spectrum of brain activities (LeGrand and Ruby 2009).

This view has one advantage: it does highlight those simplest sensori-motor integrations – functions also enhanced in early Zen training – that might be attributable chiefly to the brisk, reflexive responses of one's *somatic* Self. A disadvantage lies in what such a restriction might seem to exclude. Consider how many attributes could be further cultivated by repetition during a program of advanced Zen training. These are referable to the vast association networks and intuitive resources of an evolving human *psyche*. Would not such narrow interpretations exclude Homo sapiens from the potential to develop mature degrees of wisdom and compassionate social behaviors, limiting a meditator to the most primitive reflex level accomplishments of some earliest hominid progenitor?

The second school interprets the existing data as favoring a view weighted more toward social and environmental outcomes (Duval et al. 2009). It emphasizes the mature repertoire of our *psyche*. It leaves room for this psyche to become transformed toward a smaller self, and then to engage actively in the dynamic interpersonal welfare of a larger cultural community. From such a Mahayana-like perspective, how might this kind of *living Zen* manifest itself in contemporary society? After decades of practice, it would have developed the potential to evolve toward sage levels of wisdom that deliver selfless compassion skillfully. These evolved traits are exemplified in the bodhisattva ideal, an outlier on the spectrum of altruism far beyond the usual human expressions of self-othering affirmative functions.

- Does recent neuroimaging research suggest other beneficial ways that meditative training could reshape such a continuum of Self-other interactions? It points toward a practical approach. This simplified approach draws on the resources from various levels in ways that simultaneously: (1) relieve the Self-inflicted suffering caused by our maladaptive limbic ruminations; and (2) redirect these wasted emotional energies toward affirmative ends. What is a most basic source for these inappropriate limbic influences? Often our most troublesome emotional reactions stem from over-conditioned networks that share connections with our amygdala, the right more so than the left (Austin 1998, pp. 85–94; 2009, pp. 228–232, 251–252).

A disturbing *negative* emotion begins as soon as a normal subject glimpses the kinds of unpleasant visual image that are used in the International Affective Picture System (Schmitz et al. 2009). This disturbing event activates the right amygdala. It also effectively *narrows* the scope of what that person's visual fields can take in and remember. In contrast, a *positive* emotional image activates different brain regions and coincides with a positive emotional state. Among these positive response regions are the *right* medial orbito-frontal region and the *right* parahippocampal place area. An important correlate at this point is the way the subject's visual field now appears to literally "open up" and to include more details out at the periphery.

- How can a program of meditative training minimize such maladaptive emotional responses?

One way is by superimposing higher level cognitive resources. These often reside among the interactions that link the dorsal medial frontal gyrus with the dorsal cingulate gyrus (Urry et al. 2009). Noteworthy also is the recent report by Hölzel and colleagues (Hölzel et al. 2010). After only 8 weeks of mindfulness-based stress reduction training, their subjects *reduced* both the amount of reported subjective stress *and* reduced the concentration of the gray matter in their right baso-lateral amygdala. Other reports link intriguing positive attributes with the functions of the medial orbital region cited above. Among them are exploratory and improvisational behaviors, and the development of pro-social attachment responses. These are the kinds of adaptive responses that can favor our using insight to solve problems (Austin 2009, p. 256).

Two phrases from a song of the last century seem now to be taking on simultaneous functional anatomical correlates. They are: "Eliminate the negative," and "latch on to the affirmative."

- Do other patterns of brain activation and deactivation correlate with the different ways we actually feel our emotions? We can feel big differences between merely liking something, and being overpowered by a strong motivational and instinctual drive that makes us urgently want it (Austin 1998, p. 264). Countless millions of patients are suffering from the somatic complications of pathological obesity and from the psychic distress caused by other overdriven emotional impulses. Understanding how we regulate our appetitive drives clearly assumes worldwide importance.

Model experiments demonstrate which sites in the brain are responsible for that first thin gloss of preference which enables a rat to merely "*like*" a palatable food (Berridge 2009). This mere *liking* correlates with a series of tiny, active "hot spots." Each site stands out on a long chain that links into much larger "islands." Such an archipelago of excitations begins with the parabrachial nucleus down in the pons. This is a major source for the neurotransmitter, acetylcholine (Austin 1998, pp. 164–169). Its course then leads forward into the larger, habit-forming nuclei of the ventral pallidum and the outer shell of the nucleus accumbens. Thereafter, what do the animal's really urgent *wantings* look like? They seem to be driven by invasive, almost lava-like expansions among these local excitations. First, the individual "hot spots" enlarge during a pattern of flow that soon overruns an entire nuclear island. Next, similar activations spread outside these initial sites, going on to consume the higher-level nuclei up in the amygdala and the dorsal striatum. The suggestion is that in this manner we, too, can be consumed by the fires of longings that soon seem to rise up and "get the better of us."

Anxiety was recently studied in 24 human subjects. When they were only slightly anxious about the possibility that they *might* receive an electrical shock, fMRI signals increased in their subgenual anterior cingulate, hippocampus, and right amygdala (Mobbs et al. 2009). In contrast, when they knew that this electrical shock was imminent, their fMRI signals increased in networks that linked the mid-dorsal anterior cingulate and medial dorsal thalamus with their lateral midbrain. Moreover, the subjects made more panic-induced locomotor errors when signals increased in their left midbrain periaqueductal gray, dorsal anterior cingulate, and right insula.

Recent structural MRI studies in 44 long-term meditators have detected larger gray matter volumes in certain regions along those pathways which normally help us learn how to regulate our appetites and fears. These larger regions include the right orbito-frontal cortex, right thalamus, left inferior temporal gyrus and right hippocampus (Luders et al. 2009).

## *Looking Up into the Upper Fields of Vision; Looking Down into the Lower Fields*

What kinds of triggering events have shifted a person's attentive processing into states of selfless insight? In an intriguing number of instances, the sensori-motor mechanisms that converge in the act of *looking up* seem to have precipitated these moments of awakening (Austin 2008, pp. 211–230; 2009, pp. 113–116). Which sensory and motor processes that combine when we normally look *up* might help trigger the arrival of these rare states? Is the act of looking *up* into our upper visual fields so different from looking *down* into our lower visual fields? It is. (Fig. 1).

Some skills involved in such visuo-spatial discriminations are innate. Others we adopt simply as cultural conventions. Still other skills we learn and refine only through practice. Our visual behaviors are reflected in the different styles of meditation discussed in Tables 1 and 2. At the University of Florida, researchers in Heilman's team have investigated the ways the attentional bias of our right cerebral hemisphere differs from that of the left. Certain visual performance skills of our *right* hemisphere show biases that are highly relevant to the allocentric theme of the present pages. These normal physiological biases appear to combine their operations at both the sensory-receptive *and* motor-activity levels (Drago et al. 2007).

For example, when do these sensori-motor mechanisms jointly enable our normal right hemisphere to *attend*, to *process*, *and* to *perform* with greater efficiency?

1. When the stimulus from the visual target enters our *left upper visual field*;
2. When this left upper visual target is located at distances increasingly *farther away* from our own reaching hand.

## *The Innate, Prior Impression of "Light-from-Above"*

Other research indicates that when we look normally at an object with both eyes open, we show a curious, innate visual bias. How is this physiological bias manifested? It leads us to assume that this object is already being subtly illuminated by an *external* source of diffuse light. Where does this "prior" source of diffuse illumination seem to come from? It too appears to originate (a priori) from up in that *left upper quadrant of our vision* (Mamassian and Goutcher 2001).

In this regard, it was intriguing to find that paintings by European artists gave evidence of a strong similar tendency. Often, they represented their light source as having originated from over on the *left* side (Mamassian 2008). This analysis was based on 659 paintings hanging in the Louvre in Paris. They represented the seven centuries from the thirteenth to the nineteenth century. The artists' scenes during this interval were sorted into two categories: portraits (194), and non-portraits (465) (Most of these non-portraits were landscapes).

Huge differences emerged. The artists in these earlier centuries were 8.6 *times* more likely to have painted the illumination in their portraits as arising from a source off on the *left* side as opposed to a source off on the right side. Their non-portrait paintings were also more likely (some 2.9 *times* more likely) to have been illuminated from the *left* side than from the right. However, this left-sided bias among European artists for their chosen source of illumination peaked between the sixteenth and the eighteenth centuries. After that, it fell off. The evidence suggests that painters then began to question their earlier cultural assumptions, felt free to drop some long-held artistic conventions, and began to express their individual preferences.

## *Help From "Above?" Gazing Up Toward Distant Hills and Mountains*

Sometimes by elevating one's angle of vision and gazing up toward distant lofty peaks, the whole psyche seems to become more inspired. Psalm 21 brings an affirmative Judeo-Christian perspective to this issue: "I will lift up mine eyes unto the hills, from whence cometh my help."

Might several of these inherent physiological tendencies combine? Could they render a person especially responsive, at certain times, when looking up toward distant scenery in the visual fields above the horizon? Recent functional MRI research tracks the responses of our *lower* visual pathways as they extend forward into the parahippocampal gyrus of the medial temporal lobe (Arcaro et al. 2009). Residing here are extra large visual response fields. Moreover, the visual associations of the *upper* visual fields are coded preferentially. They represent *whole scenes*. This means that details are in focus simultaneously throughout *both the center and the periphery* of this scenery when it registers in the upper fields above the visual horizon.

Paul Cézanne (1839–1906), born in Aix-en-Provence, was impressed by the lofty scenery of nearby Mont Sainte-Victoire. In painting this mountain some *87 times* (Collester 2009), he exemplified our instinctive need to "read nature" and to *see into it*, as he said, "beneath the veil of interpretation."

As we now return to Monet's theme of instaneity (and marvel at any brain's capacity to grasp a whole landscape scene) it helps to recall the intriguing properties linked with other associations in the large visual brain (Austin 1998, pp. 240–249). Nerve cells in the temporal lobe and farther forward along the allocentric pathway also have large visual fields. These properties enable them to infuse subtle shades of meaning, atmospheres, and values throughout an entire scene. It is in this manner that the allocentric pathway has innate potentials to unify and transfigure the appearance of what is seen into that of "the Big Picture."

## In Conclusion

Selflessness is not an abstract concept. It is not some word confined to a printed page. The arrival of selflessness coincides with the insight that sees into all things as *THEY* really are. When can such a state occur? Only when *You* finally get out of your own way. Only when your brain happens to *let go* of all of your former impressions of Self and drops its old veils of egocentric interpretation. Within this cavernous vacancy of Self, consciousness simultaneously opens up to the direct experience of the underlying objective reality of the world outside. The term, *kensho*, refers to just such a sustained moment of selfless insight-wisdom: *Just This*. At this point, direct experience comprehends the incomprehensible paradox: the simultaneous interpenetrations of emptiness and suchness.

*Kensho* is "nothing special" in Zen – only a brief state of consciousness. Yet this experience amply confirms the value of drawing sharp distinctions between the two different physiologies of our ego- and allo-centric processing. The distinctions are not academic. Indeed, Self/other distinctions are of fundamental biological importance. They also help clarify what can actually develop when we cultivate balanced modes of attention on the long-term Path of authentic spiritual training (Austin 1998, pp. 186, 491–492, 498, 533; 2006, pp. 183, 202–203, 324–325, 370–371).

Meanwhile, consider carefully: what do you approach, reach out for, and grasp hold of? What do you avoid and retreat from? Advanced meditators, in particular, can get attached to long-held conventional allegiances that become counterproductive. Remain open to examine the potential pros and cons of habitual techniques. Some techniques might be subtly reinforcing your Self-centered, wordy, concept-burdened forms of top-down attentive processing. Become free to experiment and improvise, to gravitate instead toward those silent other-referential alternatives. Their receptivities can become more open, alert, and increasingly thought-free (Austin 2011). Effortless, wordless, openly receptive attention remains the primary setting in which selfless insight is likely to develop incrementally and (on rare occasions) be triggered incidentally.

This chapter emphasizes chiefly these more bottom-up receptive approaches. The intent is to encourage meditators to cultivate more of their often neglected, non-linguistic, intuitive allocentric resources. Consider adopting only those techniques that survive the rigorous testing of your own experience. A few may turn out to coincide in practical ways with your own innate physiology and with how far along *you* are as an individual meditator on the long-term trajectory of your particular path of training.

## References

Arcaro, M., S. McMains, B. Singer, and S. Kastner. 2009. Retinotopic organization of human ventral visual cortex. *Journal of Neuroscience* 29: 10638–10652.

Austin, J. 1998. *Zen and the brain. Toward an understanding of meditation and consciousness.* Cambridge, MA: MIT Press.

Austin, J. 2006. *Zen-brain reflections. Reviewing recent developments in meditation and states of consciousness.* Cambridge, MA: MIT Press.

Austin, J. 2008. Selfless insight-wisdom. A thalamic gateway. In *Measuring the immeasurable: The scientific case for spirituality*, 211–230. Boulder: Sounds True.

Austin, J. 2009. *Selfless insight. Zen and the meditative transformations of consciousness.* Cambridge, MA: MIT Press.

Austin, J. 2010. The thalamic gateway: How the meditative training of attention evolves toward selfless states of consciousness. In *Effortless attention: A new perspective in the cognitive science of attention and action*, ed. B. Bruya, 373–407. Cambridge, MA: MIT Press.

Austin, J. 2011. *Meditating selflessly; A practical, neural-Zen perspective*. Cambridge, MA: MIT Press.

Berridge, K.C. 2009. 'Liking' and 'wanting' food rewards: Brain substrates and roles in eating disorders. *Physiology and Behavior* 97: 537–550.

Christoff, K., A. Gordon, J. Smallwood, R. Smith, and J. Schooler. 2009. Experience sampling during fMRI reveals default network and executive system contributions to mind wandering. *Proceedings of the National Academy of Sciences* 106: 8719–8724.

Collester, J. (June 14, 2009). Cézanne's legacy confronts high-speed rail in France. *The Christian Science Monitor* 28

Corbetta, M., G. Patel, and G. Shulman. 2008. The reorienting system of the human brain: From environment to theory of mind. *Neuron* 58: 306–324.

Daselaar, S.M., S.E. Prince, N.A. Dennis, S.M. Hayes, H. Kim, and R. Cabeza. 2009. Posterior midline and ventral parietal activity is associated with retrieval success and encoding failure. *Frontiers in Human Neuroscience* 3: 13. doi:10.3389/neuro.09.013.2009.

Drago, V., P. Foster, D. Webster, et al. 2007. Lateral and vertical attentional biases in normal individuals. *International Journal of Neuroscience* 117: 1415–1424.

Duval, C., B. Desgranges, F. Eustache, and P. Piolino. 2009. Looking at the self under the microscope of cognitive neurosciences: From self-consciousness to consciousness of others. *Psychologie Neuropsychiatrie Vieil* 7: 7–19.

Fox, M., M. Corbetta, A.Z. Snyder, J.L. Vincent, and M.E. Raichle. 2006. Spontaneous neuronal activity distinguishes human dorsal and ventral attention systems. *Proceedings of the National Academy of Sciences USA* 103: 10046–10051.

Gianotti, L., D. Knoch, P.L. Faber, D. Lehmann, R.D. Pascual-Marqui, C. Diezi, C. Schoch, C. Eisenegger, and E. Fehr. 2009. Tonic activity level in the right prefrontal cortex predicts individuals' risk taking. *Psychological Science* 1: 33–38.

Graybiel, A. 2008. Habits, rituals, and the evaluative brain. *Annual Review of Neuroscience* 31: 359–387.

Greicius, M.D., K. Supekar, V. Menon, and R.F. Dougherty. 2009. Resting-state functional connectivity reflects structural connectivity in the default mode network. *Cerebral Cortex* 19(1): 72–78. doi:10.1093/cercor/bhn059.

Hölzel, B.K., J. Carmody, K.C. Evans, E.A. Hoge, J.A. Dusek, L. Morgan, et al. 2010. Stress reduction correlates with structural changes in the amygdala. *Social and Cognitive Affective Neuroscience* 5(1): 11–17.

Immordino-Yang, M.H., A. McColl, H. Damasio, and A. Damasio. 2009. Neural correlates of admiration and compassion. *Proceedings of the National Academy of Science USA* 106: 8021–8026.

Kopell, B., and B. Greenberg. 2008. Anatomy and physiology of the basal ganglia: Implication for DBS in psychiatry. *Neuroscience and Biobehavioral Reviews* 32: 408–422.

Legrand, D., and P. Ruby. 2009. What is self-specific? Theoretical investigation and critical review of neuroimaging results. *Psychological Review* 116: 252–282.

Luders, E., A. Toga, N. Lepore, and C. Gaser. 2009. The underlying anatomical correlates of long-term meditation: Larger hippocampal and frontal volumes of gray matter. *NeuroImage* 45: 672–678.

Mai, J., J. Assheuer, and G. Paxinos. 2004. *Atlas of the human brain*, 2nd ed. San Diego: Elsevier. The zona incerta extends two-thirds of the way immediately below the thalamus.

Mamassian, P. 2008. Ambiguities and conventions in the perception of visual art. *Vision Research* 48: 2143–2153.

Mamassian, P., and R. Goutcher. 2001. Prior knowledge on the illumination position. *Cognition* 81: B1–B9.

Mantini, D., M. Corbetta, M. Perrucci, G. Romani, and C. Del Gratta. 2009. Large-scale brain networks account for sustained and transient activity during target detection. *NeuroImage* 44: 265–274.

Mantini, D., M.G. Perrucci, C. Del Gratta, G.L. Romani, and M. Corbetta. 2007. Electrophysical signatures of resting state networks in the human brain. *Proceedings of the National Academy of Science USA* 32: 13170–13175.

Mitchell, J.P. 2009. Inferences about mental states. *Philosophical Transactions of the Royal Society* 364: 1309–1316.

Mobbs, M., J. Marchant, J.D. Hassabis, B. Seymour, G. Tan, M. Gray, et al. 2009. From threat to fear: The neural organization of defensive fear systems in humans. *Journal of Neuroscience* 29: 12236–12243.

Pfeifer, J., M. Lieberman, and M. Dapretto. 2007. "I know you are but what am I?" Neural bases of self- and social knowledge retrieval in children and adults. *Journal of Cognitive Neuroscience* 8: 1323–1337.

Qin, J., T.M. Lee, F. Wang, L. Mao, and S. Han. 2009. Neural activities underlying environmental and personal risk identification tasks. *Neuroscience Letters* 455: 110–115.

Schmitz, T.W., E. De Rosa, and A.K. Anderson. 2009. Opposing influences of affective state valence on visual cortical encoding. *Journal of Neuroscience* 29: 7199–7207.

Shulman, G., S. Astafiev, D. Franke, D. Pope, A. Snyder, M. McAvoy, and M. Corbetta. 2009. Interaction of stimulus-driven reorienting and expectation in ventral and dorsal frontoparietal and basal ganglia-cortical networks. *Journal of Neuroscience* 29: 4392–4407.

Smith, A.T., P.L. Cotton, A. Bruno, and C. Moutsiana. 2009. Dissociating vision and visual attention in the human pulvinar. *Journal of Neuropsychology* 101: 917–925.

Spreng, R., R. Mar, and A. Kim. 2009. The common neural basis of autobiographical memory, prospection, navigation, theory of mind, and the default mode: A quantitative meta-analysis. *Journal of Cognitive Neuroscience* 21: 489–510.

Summerfield, J.J., D. Hassabis, and E.A. Maguire. 2009. Cortical midline involvement in autobiographical memory. *NeuroImage* 44: 1188–1200.

Urry, H., C. van Reekum, T. Johnstone, and R. Davidson. 2009. Individual differences in some (but not all) medial prefrontal regions reflect cognitive demand while regulating unpleasant emotion. *NeuroImage* 47: 852–863.

Van Horn, J.D., and R.A. Poldrack. 2009. Functional MRI at the crossroads. *International Journal of Psychophysiology* 73: 3–9.

Ward, R., and I. Arend. 2007. An object-based frame of reference within the human pulvinar. *Brain* 130: 2462–2469.

Whitney, C., W. Huber, J. Klann, S. Weis, S. Krach, and T. Kircher. 2009. Neural correlates of narrative shifts during auditory story comprehension. *NeuroImage*. doi:10.1016/j.neuroimage.

Zhang, D., A.Z. Snyder, M.D. Fox, M.W. Sansbury, J.S. Shimony, and M.E. Raichle. 2008. Intrinsic functional relations between human cerebral cortex and thalamus. *Journal of Neurophysiology* 100: 1740–1748.

Zou, Q., X. Long, X. Zuo, et al. 2009. Functional connectivity between the thalamus and visual cortex under eyes closed and eyes open conditions: A resting state fMRI study. *Human Brain Mapping* 30: 3066–3078.

# Meditation as First-Person Methodology: Real Promise—and Problems

**Jonathan Shear**

**Abstract** Meditation as a scientific first-person investigative tool has been discussed for decades, but remains largely a mere idea. One reason may be lack of relevant theory. Maps of mind developed by meditation traditions could prove helpful. The map used by orthodox Indian traditions, for example, identifies six phenomenologically distinct levels (senses, discursive thinking, discriminative intellect, pure individuality and pure consciousness/emptiness). This map, if valid, would have implications for many fields. It would indicate, for example, that the introspective awareness of major philosophers such as Descartes, Hume and Kant was open to particular levels and not others, and suggest why each favored particular theories and found particular problems unresolvable. Identification of physiological correlates of the levels could provide evidence for the map's validity. Significant correlates of the deepest level already appear to be identified. Research relevant to other levels has been conducted. Identifying correlates of all the levels would provide an objective way to evaluate many mind-related questions. It would also provide an objective, tradition-invariant way to identify individuals capable of sustaining attention at specific levels and using meditation to investigate diverse levels-related topics. Meditation research faces strong questions of appearances of bias.

A consortium, including researchers associated with competing traditions and non-associated researchers, overseeing replications and meta-analyses could respond to these questions directly.

---

J. Shear (✉)
Department of Philosophy, Virginia Commonwealth University, Richmond, USA
e-mail: jcs@infionline.net

S. Schmidt and H. Walach (eds.), *Meditation – Neuroscientific Approaches and Philosophical Implications*, Studies in Neuroscience, Consciousness and Spirituality 2,
DOI 10.1007/978-3-319-01634-4_4, 

## Section I

Some years ago, Francisco Varela and I co-edited a book on "First-Person Methodologies" entitled *The View from within: First-Person Approaches to the Study of Consciousness*. My own particular interest was in the use of meditation as a methodology for scientific exploration of consciousness. Meditation traditions had indicated that meditation could provide types of inner awareness that could prove very useful here. They reported that it is possible to experience a pure, unfluctuating level of inner awareness devoid of all content and activity, and suggested that this "pure" level of awareness should provide a platform for exploring all of consciousness with minimum internal distortion and maximum signal-to-noise ratio. They also indicated that particular types of meditation practices could train people to maintain unfluctuating awareness of diverse contents of consciousness for extended periods of time, as would be necessary for careful inner exploration. All of this suggested the possibility of using meditation to explore the internal domain in scientifically productive ways. And the need for effective first-person methodologies in consciousness studies was conspicuous.

It has now been 14 years since *The View From Within* first came out. Several authors of this book were contributors, and Alan Wallace and I wrote chapters on some of the above ideas. The fact that modern Westerners were reporting relevant experiences in meditation was particularly encouraging. But while over the last 14 years there has been a great deal of research *on* meditation in terms of the effects of different practices on such things as physiological correlates, psychological development and behavioral effects and social implications, not much progress appears to have been made in the *use* of meditation as a *methodology to explore* consciousness. The promise of meditation as exploratory methodology thus still appears to remain largely just that—mere promise.

This perhaps should not be so surprising. Good research requires more than good investigative methodologies; it requires relevant theory to guide their use. But what kind of theories are likely to be most relevant here? Within our traditional Western intellectual frameworks, developed independently of knowledge of what meditation might and might not be able to do, the answer might not be obvious. Eastern meditation traditions on the other hand have developed a variety of maps of consciousness organizing the states, structures and contents of consciousness their meditative methodologies have displayed over the centuries, and integrating them with the features of consciousness we are ordinarily aware of. These maps obviously reflect the kinds of things that meditation has traditionally been thought to be capable of displaying. And some of them appear highly relevant to modern questions about consciousness. It is thus natural to think that such traditional might suggest ways to use meditation that are both well-suited to their capabilities and of contemporary scientific interest.

## Section II

Consider for example the traditional map of levels of mind, developed over many centuries and widely accepted, by Yoga, Vedanta, and other "orthodox" Indian systems (*darshanas*). The map discriminates six levels. Each level is defined phenomenologically. All of the levels are said to be experienceable. And the mental content at each more "superficial" level is said to depend on and develop through the "deeper" ones. From surface to depths, the "levels" are

i. Senses
ii. Discursive thinking
iii. Discriminative intellect
iv. Pure individuality or ego
v. Pure bliss (pure positive affect)
vi. Pure consciousness (pure emptiness)

The first three levels are easy to understand. Every normal person experiences levels (i) and (ii). We are all familiar with level (i), the senses. Level (ii) amounts to thinking in words, as occurs in ordinary internal discourse. Level (iii) involves the discriminative mental functioning that underlies and is presupposed by coherent use of words. It is phenomenologically independent of (i) and (ii), as observable in ordinary experiences of preverbal thought.

The deeper layers are less likely to be familiar. They all lie outside the ordinary range of experience, they are all highly abstract, and, for most people at least, they are all usually first experienced clearly only as a result of meditation. Different traditions use different terms to refer to them. I will generally use terminology deriving from Yoga and Vedanta, the traditions I am most familiar with. But nothing should be taken to depend on this. Ordinary language should, in my opinion, be able to indicate the phenomenological nature of each level well enough that anyone who has had the relevant experience should easily be able to identify what is being referred to.

Level (iv), "pure individuality" or "ego," is highly abstract. It contains no specific phenomenal objects such as the sensations, thoughts and images of levels (i) – (iii). It is in effect experience of one's internal phenomenological manifold itself, devoid of all specific objects. In a traditional analogy, it is likened to being *a disembodied observer in the midst of vast emptiness*. Yet phenomenologically speaking, it is not entirely empty. For despite the absence of objects, the phenomenological I-it "intentional" relationship remains. *One* is *there* as disembodied pure observer, a bare "mind's eye" so to speak, aware *of* the vast empty phenomenological space surrounding in all directions. In traditional terms one is said here to "hold one's individuality" in a "void of abstract fullness" surrounding it, steady like "a lamp in a windless place."[1] Thus, as empty as this experience seems, it still retains the *structure* of ordinary experience. After one emerges from the experience, it is also often remembered as permeated by abstract well-being or bliss.

In level (v), "pure bliss (positive affect)," even the abstract subject-object duality of level (iv) is no longer present. Here, in other words, there is no longer any sense of being an observer having a vantage point at all. All that remains is utter abstract bliss or well-being itself.

Level (vi), "pure consciousness (pure emptiness)," is even more abstract. For it contains no phenomenal content at all. Even bare subject-object orientation and abstract, objectless bliss are absent. It is thus completely *unimaginable*, for anything one can imagine has phenomenal content, and the identifying characteristic of this experience is the absence of all such content. It is, nevertheless, experienceable. What is it experienced as? Not *as* anything. Just itself. But unlike unconsciousness, one can look back and *remember* it. What does one remember it *as*? Not *as* anything. Just itself.

Different traditions interpret the experience differently. Advaita Vedanta refers to it as "pure Being," since all that one can say directly about the experience is that it *was*. Vedanta and Yoga both also refer to it as "pure consciousness," for one can hardly be thought to remember something where one's consciousness was not present. Buddhist traditions generally refer to it as "non-being" and/or "emptiness," emphasizing the absence of every conceivable thing. Taoism often refers to it in terms of "non-being" and "nothingness." Zen Buddhism refers to it in all of these ways, depending on context, in order to separate the experience as it is in itself from attachment to all conceptual and philosophical perspectives. All such terminological questions aside, however, there is widespread agreement that an internal level completely devoid of phenomenal content lies at the ground of everyone's awareness, and can be experienced and remembered.[2]

## Section III

This intelligibility of this map appears independent of whatever positions we may individually hold with regard to such things as "self," "being," "emptiness," "pure consciousness" and other metaphysically and/or philosophically charged terms and

---

[1] Compare, for example, *The Bhagavad Gita*, Maharishi Mahesh Yogi (1974), p. 412.

[2] Compare, for example, Zen (Austin 1998, p. 474):

> [inner] space becomes the object of consciousness, followed by an awareness of objectless infinity, and then by absorption into a void which has 'nothingness' as its object. Finally… there evolves 'neither perception, nor nonperception.

and Tibetan Buddhism(Wallace 1999, pp. 182 and 186):

> One's consciousness is now left in an absence of appearances … as if one's mind has become indivisible with space … [and, later the more advanced state of] the ultimate nature of awareness, free of all conceptual meditation and structuring, transcending even the concepts of existence and non-existence.

concepts. I was happy to find this view emphatically seconded by Samdong Rinpoche, noted scholar and head of the Dalai Lama's government in exile, who made it clear that in his opinion all the major traditions share the same phenomenologically defined experiences, despite the traditions' often strenuous metaphysical differences (Shear 2006, pp. 360, 368). He also made it clear that experiences fitting all the map's levels are readily recognized within his Tibetan Buddhist tradition. The show of hands at this seminar also indicates that perhaps a fourth of its participants feel they have had some of the deeper experiences described above, including in some cases the deepest. These results exemplify the ready intelligibility of the map, and the ease with which experiences that appear to correspond to its levels can be identified in a rough and ready way, independently of significant differences of techniques, traditions and metaphysical perspectives.

Evaluating claims of sameness of experience across different traditions raises many important questions, and we will return this topic later. For the moment however, it should be apparent that the map itself can be understood purely phenomenologically, independently of the metaphysical contexts in which it developed. This, I think, is an essential aspect of its contemporary usefulness. In this regard, a comment is in order about the terms we have been using for the fourth level of the map, namely "pure ego" and "pure individuality." Questions about the existence of anything corresponding to our psychological sense of ego have been sharply debated for centuries in both Eastern and Western thought. So we need to be careful here to distinguish (a) the readily identifiable experience the terms are intended to refer to here from (b) the many metaphysical assertions that they have often been associated with. In particular we should note that neither the experience nor its status in the phenomenological map by itself implies anything about the truth or falsity of ontological claims about the "substantial" existence of individual "ego" or "self." The move from phenomenology to ontology here is not so straightforward. The use of "ego" and "individuality" for level (iv) type experience is found in diverse traditions with very different theories about the ultimate ontological nature of self, and appears to reflect a natural psychological response to the experience itself.[3] My own use of terms such as "pure ego" and "pure individuality" is intended to reflect this apparent psychological fact, rather than any of these conflicting ontological claims.

[3] Yoga takes the experience we are calling pure "ego" or "individuality" to display the existence of a substantial, continuing, individual self. Buddhist texts also describe this experience of individual awareness in the midst of vast emptiness. But in contrast to Yoga, they emphatically reject the idea that this is evidence of a continuing individual substantial self, an idea which they take to be an illusion akin to, but much deeper (and more problematic) than that of physical things as abiding substances, rather than collections of phenomena. Advaita Vedanta embraces Yoga on the whole. But, like Buddhism and unlike Yoga, it holds that the idea of individual substantial self is ultimately illusory. To make the situation even more complex, both Advaita and Far-Eastern Mahayana, but not Theravada, go on to assert the existence of absolute, eternal "Self" (distinguished from the "illusory" commonsense "self") experienceable in enlightenment. Such examples make it clear that recognition of the level (iv) experience naturally identified as the core of individual awareness is independent of, and should not be identified with, any of these traditions' competing ontological theories about the "reality" of "self."

## Section IV

It is easy to see that this map of phenomenologically identifiable levels, if valid, could prove very useful both for reevaluating existing mind-related theories and for new types of objective research into consciousness. In my own field of philosophy for example, both the introspective accounts and the theories of major philosophers clearly indicate that the philosophers' introspective awareness of each was often open to less than the full range of the map's levels. This in turn suggests why each philosopher favored particular theories and was particularly interested in specific seemingly unresolvable problems. Consider, for example, Descartes, Hume and Kant, three of the most influential philosophers in modern Western philosophy.

Descartes, often referred to as the "father" of modern (Western) philosophy, describes his own introspective procedures and experiences in first-person language in his seminal *Meditations*. His most famous philosophical conclusion, of course, was his cogito *ergo sum*. The meaning of this expression and the validity of its conclusion have been much debated from Descartes' own time to the present. Traditionally the cogito has generally been regarded as a piece of (probably flawed) reasoning. But Descartes himself emphatically denied that his cogito was a piece of reasoning at all, and insisted instead that it was "a simple mental intuition" that his mind "sees, feels, handles" (Descartes 1978, pp. 299 and 301). His experiential language could hardly be clearer. What was it that he "intuited"? It was, as he put it, that "there is this consciousness … of this and this only I cannot be deprived;" or, in other words, "I am certain that I am a conscious being" (*ibid.*, pp. 69, 76).

Descartes then lets us know that he regarded this "consciousness" as not being confined to thinking at all, but as underlying all of his mental activities (thinking, doubting, asserting, understanding, willing, imagining, etc.) (*Ibid.*, p. 76). This, of course, is just how level (iv) of the map, pure "ego" or "individuality," is regarded in the relevant meditation traditions. Descartes then reported that he came to recognize this "finite" consciousness as existing only in the context of infinite consciousness, which he calls "God" in an "intuition" as clear and indubitable as his original "intuition" of himself. His discussion of these claims is embedded in scholastic terminology, is often very difficult, and few people (including myself) today find it at all convincing. Nevertheless, his terminology suggests a kinship with the only other "indubitable intuition" he describes, and is so famous for, namely the cogito. And two explicitly experiential passages in the *Meditations* suggest strongly that here, too, his claims, however well or poorly articulated they may be, reflect Descartes' own experience. Both of these passages occur in the third "Meditation." The meditation begins with his first-person assertion

> I will now shut my eyes, stop my ears, withdraw all my senses; I will even blot out the images of corporeal objects from my consciousness; or at least (since this is barely possible) I will ignore them as vain illusions…. (*ibid.*, p. 76)

And it ends with the equally first-person,

> I wish to stay a little in the contemplation of God … to behold. wonder at, adore the beauty of this immeasurable Light, so far as the eye of my darkened understanding can bear it …

> even now the ... contemplation ... makes us aware that we can get from it the greatest joy of which we are capable in this life. (*ibid.*, p. 91)

These passages are easily overlooked today, since contemporary philosophers are generally more interested in arguments than experiential reports. But in the context of our discussion of Eastern maps of consciousness they are especially interesting. For the first passage describes a very common type of meditation, practiced in many traditions. We have already seen how otherwise puzzling aspects of Descartes' account of the cogito appear readily intelligible in light of level (iv) experiences described by meditation traditions. The second passage above describes more advanced experiences that can develop from this one. For they often report that when someone at level (iv) (ego), turns attention away from level (iii) (intellect) and "looks" inward with "darkened understanding" (having left intellect behind) through level (v) (bliss) "toward" level (vi) (unbounded consciousness, or "God," as Descartes puts it).[4]

In other words, if we take these first-person reports at face value, they appear to reflect just what one would expect if his or her inner awareness were "stationed," at least momentarily, at level (iv), looking "inward" to level (v), seeing it with some clarity, and through it toward (vi), still apparently seen unclearly (inasmuch as there is no suggestion, here or elsewhere in the *Meditations*, of unbounded contentless consciousness by itself, as contrasted with mere background for other experiences). This hypothesis, if correct, would help us understand why Descartes might have been so interested in establishing that the "self" exists as independent from the world of the senses and all sensory content, and also had to be single, simple and unimaginable. It would also help us understand the disconnect between Descartes and other philosophers, ancient and modern, who might not either have had experience of these deeper levels, or even known such experience is possible.

I should emphasize, however, that what we have said should not be taken to imply that the conclusions Descartes drew from his experiences (about self as substance, mind-body dualism, continuation of the self after death, the existence of God as creator, "clear and distinct" ideas, etc.) are correct. Phenomenological observations are one thing, and conclusions drawn from them are quite another. We will return to this question later. It should also be noted that the above analysis is little more than a rough beginning. The full body of Descartes' work would have to be taken into account to be to be really confident of claims about the scope of his introspective awareness.[5] Nevertheless, what we have noted here should be enough to suggest how knowledge of the map can in principle help us understand major, otherwise often very puzzling elements of both his introspective reports and mind-related theories.

---

[4] Compare, for example, Maharishi Mahesh Yogi (1974, p. 312).

[5] For a fuller discussion of Descartes in this context, see Shear (1990, Chap. 4).

## Section V

Let us now turn to Hume. Locating his perspective in terms of our levels map is rather straightforward. One of the most conspicuous features of Hume's work is its attention to how the intellect integrates mental content into concepts. His analyses of principles of association, contiguity, resemblance and logico-temporal relations basic to causality are all well known and highly influential.[6] These analyses are filled with accounts of verbal meanings as collections of associated internal images or "impressions" that correspond well with the chains of associated preverbal images reported by creative thinkers, experienced meditators and others familiar with ordinary preverbal thought. He also paid especially close attention to relationships between internal experiences and sensory inputs. Hume's work thus appears to reflect deep awareness of, and concern for, the content of level (iii) intellect, and its relation to levels (ii) discursive thought and (i) sensory awareness.

Hume moreover makes it very clear that he neither has, nor can even conceive of, anything like experience of the deeper levels (iv) pure individuality and (vi) pure consciousness. For in support of his rejection of Descartes' account of self, he reports that

> Whenever I enter most intimately into what I call *myself*, I always stumble on some particular perception or other, of heat or cold, light or shade, love or hatred, pain or pleasure, I never can catch myself at any time without a perception, and never can observe anything but the perception. So long as my perceptions are removed for any time, as by sound sleep, so long am I insensible of *myself*, and may truly be said not to exist.
>
> If anyone upon serious and unprejudic'd reflection, thinks he has a different notion of *himself*, I confess that I can no longer reason with him. All that I can allow him is, that he may be in the right as well as I, and that we are essentially different in this particular. (Hume 1980, p. 252)

It would thus appear that Hume's introspective awareness, at its deepest, was stationed firmly at level (iii) looking "outward" towards levels (ii) and (i), with no experience of the deeper levels at all. Thus it is quite natural that he should reject Descartes' account of self, inasmuch as it requires such experiences even to be empirically intelligible, much less supported—all questions of whether it actually arose in response to these experiences notwithstanding. Nevertheless, staunch empiricist that he is, Hume explicitly allowed the possibility of someone else's being "essentially different" from him here. As it turns out however, the real difference appears only to be one of knowledge of an appropriate, effective meditation technique

## Section VI

Kant, too, appears readily locatable on the levels map. He accepted many of Hume's analyses, saying that they woke him from his "dogmatic slumbers." But, unlike Hume, Kant carefully distinguished the internal phenomenal manifold, separated it

---

[6] They led, for example, directly to the empirical notion of causality as empirically identifiable patterns of events, now generally taken for granted by philosophers of scientists and scientists alike as freed from traditional (and commonsensical) "metaphysical" connotations.

from its contents, and argued that for it to function in the way it does, it has to be internally unified and presented to a unified ego. For, he argued, every experience has to be extended in space and/or time (otherwise, we can add, it would be too tiny and/or quick to be experienced at all), and to be *that* experience, all of its parts have to be presented to one and the same experiencer (otherwise, it would be a different experience). Furthermore, Kant argued, this unifying ego has to be "transcendental" to the phenomenal manifold. It cannot, in other words, be characterized by any "empirical data," be "accompanied by any further representation," or have any "special designation" at all (Kant 1964, pp. 131, 153, 329, etc.). Kant's arguments here are notoriously difficult. The underlying logic can however perhaps be conveyed simply by noting that if the ego had any special designation or empirical qualities of its own, it would be incompatible with otherwise possible experiences where that "special designation" was absent.[7] All questions of the difficulty of Kant's arguments aside, however, his conclusion is clear: the self can only be understood as "pure consciousness," a "bare consciousness" that accompanies all of one's experiences (Kant 1964, pp. 331, 382, etc.).

The congruences between Kant's position and Yoga, Vedanta and other orthodox *darshanas* have often been noted on by Indian scholars. But there is also a singular difference. Yoga and Vedanta developed in the context of centuries of reports of experiences of pure consciousness. Kant made abundantly clear that he, like Hume, never had such experiences. Indeed Kant, not being so empirically-minded as Hume, went much further than Hume did, and insisted that it is actually *impossible* for any human being to have such an experience. Nevertheless, Kant was careful to add that this impossibility is not a *logical* one. For he held that it is not *logically* impossible that some other type of being might be able to have such an inconceivable experience, even though we ourselves "cannot form the least conception" of what it might be (*ibid.*, p. 157).

Unlike Descartes and Hume, Kant does not refer explicitly to his own experience here. But it is apparent that he himself had no knowledge of experiences of pure consciousness, either as localized in the midst of the empty phenomenal manifold (level iv) or by itself, independent even of this empty manifold (level vi). Nevertheless he could hardly have emphasized its importance more, reasoning that as the nature of the "I," it *had* to exist as ground of all knowledge and experience, "the supreme principle of all employment of the understanding." Kant wrestled with the problem

---

[7] One can unpack the logic involved here a little further as follows: Logically speaking, any empirical quality or "special designation" (by the usual definitions of "empirical" and "special") can in principle (if not in actual fact) be either present or absent from a person's experience. So no matter what empirical quality or "special designation" one might choose, it is logically possible for one to have an experience where that quality or "special designation" is absent. So since one has to be present at all of one's experiences, no such quality can be an adequate marker for the presence of one's ego or self. More can be said, of course, to unpack the relationships of this somewhat formal argument both to the distinction between actual and possible experiences and to our commonsense intuitions about the ego as experiencer. See, for example, Shear (1990, Chap. 4, pp. 93–99 and 104–106).

at length,[8] and finally concluded unhappily that the (a) necessary existence and (b) impossibility of experiencing the qualityless, transcendental level of self constitute a profound "paradox" that inescapably "mocks and torments" even the wisest of men. If the levels model is correct, such sustained conviction that pure consciousness exists at the core of self, concern with its experienceability, and inability to experience it, are just what we might expect of someone whose introspective awareness was often stationed deep within level (iii), intellect, intuiting the deeper level (iv) of ego, but not knowing how to access it.[9]

## Section VII

Some important caveats should be noted. The above examples were not intended to imply that any of the philosophers' respective metaphysical theories about self are correct. They were instead offered to illustrate how different types of introspective access can be expected to significantly affect the theories and conclusions philosophers may develop. When philosophers such as Descartes, Hume and Kant report that they did not and/or could not have experiences characteristic of specific levels of the map, I think it is reasonable to accept their assertions at face value. This would imply that their introspective awareness had not reached (or at least had not been remembered as reaching) beyond particular levels. It can also, I think, often be reasonable to conclude that specific texts can provide significant evidence of experience of particular levels. Concluding that the awareness of a particular philosopher was typically, or even often, stationed at a particular given level however is a very different matter. Even a single experience of a given level could be enough to influence a person's thinking significantly.[10]

It might also seem natural to think that if the map accurately displays the structure of people's inner awareness in general, familiarity with more of its levels should, all other things being equal, be likely to lead to more accurate theories about consciousness. But philosophers with putatively equivalent phenomenological access have often differed on important issues, both within and between major Eastern traditions, as we saw earlier with regard to the "self." The relation between good data and good theories is not so simple.

Nevertheless, new empirical knowledge can inform our theories directly by falsifying significant claims. The levels map describes classes of phenomenologically identifiable experiences new to modern Western discourse that appear capable in

[8] We can note that the sections on the "transcendental unity of apperception"—that is the "I"—were only sections of the *Critique of Pure Reason* that Kant felt constrained to rewrite substantially between the first and second editions.

[9] This surmise, reflecting passages from Kant's *Critique of Pure Reason*, would appear to be reinforced by paradoxes of "will" and accounts of sublime positive affect central to Kant's later ethics and aesthetics as well.

[10] My thanks to philosopher Neil Sims for noting the importance of taking this point into consideration.

principle of doing this. The level (vi) experience of pure consciousness, as reported and discussed in the scientific and philosophical literature, for example, already appears to have falsified both (a) Hume and Kant's claims that experiences without phenomenal content are impossible and (b) modern phenomenologists' assumption that experience must always have "intentional" subject-object (I-it) structure. The map, in other words, draws attention to types of experience and cognitive functioning which, if the map is correct, need to be taken into account by philosophers and others studying consciousness if they want their mind-related theories to be correct and comprehensive.

## Section VIII

The levels map does more than merely draw our attention to types of experience not ordinarily attended to by modern Western thinkers however. For it maintains not only that the types of experiences it describes are identifiable phenomenologically, but that they are stratified, with more superficial levels developing from deeper ones. This is supposed to be directly experienceable by people sufficiently adept with various types of meditation procedures. The mental contents of untrained people are also supposed to develop in this way, even though they may not be aware of this fact themselves. These claims, if correct, have implications for a wide variety of theories in fields ranging from philosophy to psychology, neurophysiology and consciousness studies as a whole. As soon as we attempt to analyze them objectively, however, serious problems arise.

Consider first the reports of unusual experiences made by people adept with particular meditation practices. When people spend considerable time practicing mental techniques that are taught and interpreted in the context of well-defined conceptual systems, it is not unreasonable to wonder to what degree the resulting experiences might be shaped by these systems, rather than purportedly universal features of human awareness. It is also not unreasonable to wonder to what extent either the experiences and/or the reports might be distorted by wishful thinking and the desire to justify time spent in the practice. Such questions easily raise doubts about the objective significance of reports of experiences of deep levels of inner awareness. The significance of the map's implicit claim about ordinary people is even more problematic. For what objective sense can be made of the claim that the content of people's subjective awareness develops through unperceived stages? Senior teachers sometimes respond that meditation traditions have developed a variety of time-tested ways to ensure that certified adepts are not fooling themselves and actually have the kinds of experiences reported. This is certainly true. But the nature of these tests is often kept private, so it is not easy to see how unknown tests, derived within a tradition and applied only to people who have gained experiences through that same tradition, could escape questions about tradition-specific metaphysical and other factors that might significantly color not only the tests, but also the experiences themselves. And in the absence of knowledge of their details, claims of objectivity associated with such tests are hardly likely to be satisfying to people outside the relevant tradition, and scientists in general—all questions of their accuracy aside.

## Section IX

It may well be possible, however, to evaluate the validity of the map objectively. Identification of physiological correlates of experiences of the different levels, both individually and in appropriate sequences, could provide evidence for the map's culture-invariance and objective validity.[11] Identification of correlates of meditation practices and experiences is now a lively field, and some of the results already appear applicable to levels-related questions. Meditation traditions such as Yoga and Zen for centuries have reported respiratory suspension as a correlate of the level (vi) experience of pure consciousness (pure emptiness). Chinese Zen ("Ch'an") sometimes even uses the expression "breath stops" as a name for this experience. And as noted earlier, studies of people practicing Transcendental Meditation (TM) show strong correlations between reports of level (vi) experiences and suspension of perceptible respiration, with no change of $CO_2$ and $O_2$ levels in the blood or compensatory breathing afterward, and high levels of frontal EEG alpha coherence (see, for example, Travis and Wallace 1997; Farrow and Hebert 1982, Badawi et. al. 1984).[12] So far as I know it remains to be determined whether these physiological measures might also be correlates of other deep "transcendental" experiences, including in particular the objectless, uniquely simple level (v) and level (iv) experiences of pure bliss and pure individuality. Nevertheless, real progress in identifying tradition-independent correlates of level (vi) experience seems to be being made.

The correlation of reports of level (vi) experience with these kinds of physiological markers in particular already addresses questions of whether the experiential reports should be understood as products of such things as tradition-generated beliefs and wishful thinking. For it would hardly seem plausible that different types of procedures, practiced in different cultures with competing belief systems, would produce identical parings of unusual experiential reports and unconscious (and in the past generally both unknown and undetectable) physiological states,[13] unless the

[11] The present analysis will focus exclusively on the potential usefulness of physiological correlates. The reasons for this are purely methodological, and should not be taken to reflect a preference for a physicalist metaphysics. Behavioral markers can also be scientifically useful in levels-related research. See also note 14 below.

[12] The point under discussion here is only the identification of these correlates of the level (vi) experience, not their putative relationship to other features of awareness. We can note nevertheless that low frequency rhythms such as alpha (and theta) appear to reflect top-down information processing, in contrast to high frequency rhythms (beta2 and gamma) associated with bottom-up processing of the contents of experience (Razumnikova 2007). If we take pure contentless consciousness as the "top" locus of "top-down" processing—as meditation traditions regularly do—then it should be no surprise that alpha might turn out to be associated with this experience. Frontal alpha coherence is also associated with cortical excitability, and has been postulated to underlie inner wakefulness (Travis et al. 2010).

[13] Any bodily awareness would, of course, be incompatible with empirically contentless level (vi) experience.

reports reflected the natural subjective correlates of the states themselves. Such markers, in other words, give us good reason to conclude that the descriptions of the experiences as empty of all identifiable phenomenological content reflect the nature of the experiences themselves, rather than culture-dependent variables, and, where the correlates have actually been found, to take people's reports seriously.[14]

## Section X

Significant work thus already appears to have been done concerning level (vi). So far as I know no comparable work has yet been published regarding levels (iv) and (v). These two levels, like level (vi), however are defined in terms of single, phenomenologically extremely simple experiences, and finding physiological correlates could in principle be expected to proceed in the same rather straightforward way, however difficult it might turn out to be in practice. The other levels however are filled with all sorts of different kinds of experiences, and locating relevant correlates might be expected to be more complex. In this regard I would like to cite a well-known study led by Antoine Lutz, and use some of its results to suggest how more complex level-related research might progress. This study discussed (among other things) research on correlates of meditation experiences reported by advanced Tibetan Buddhist monks, and described as having an unconditioned, objectless feeling of compassion as their "sole content." The study reported robust gamma-band oscillation and long-distance phase-synchrony, including in some instances the highest levels ever reported in a nonpathological context, as correlates of this highly abstract, intensely blissful, "nonreferential" meditative state (Lutz et. al. 2004).

Identifying and distinguishing different "objectless" experiences can be phenomenologically tricky. Nevertheless if, by way of example, we simply take the reports of the experience at face value, (a) the absence of phenomenal objects indicate clearly that it is deeper than level (ii) discursive thought, and (b) its phenomenological content, the feeling of compassion, distinguishes it from the "emptiness" characteristic of levels (iv), (v) and (vi). At the same time its bliss, simplicity, and absence of phenomenal objects (c) suggest close kinship to level (iv), pure individuality. The experience, in sum, would thus appear to be located somewhere deep in level (iii) preverbal discrimination, close to level (iv) pure individuality. Interpreting the potential level-related significance of the observed physiological correlate will be much more complex, even in principle, however, as might be expected with experiences having explicit phenomenological content, even where the content is abstract and objectless. The extremely high levels of gamma synchrony might turn out to be uniquely associated with this specific nonreferential experience. It might on the

[14] This analysis is purely methodological, and depends only on the hypothesis of the existence of strong natural correlations between subjective experiences and objective nervous system states, taken for granted by modern science and many Eastern meditation traditions alike. As such it is intended to be independent of metaphysical theories and preferences about such things as materialism, idealism, non-dualism, the directionality and/or existence of mind-body causality, etc.

other hand reflect the way the experience was attended to, above and beyond its specific content of compassion. It might also perhaps turn out to be a feature of a general class of abstract, bliss-permeated experiences locatable at the depths of level (iii), independently of whether they were gained by Tibetan monks, adepts from other traditions, or people with no connection with any meditation tradition at all. If the latter, high gamma synchrony might also turn out be part of an effective marker of deep level (iii) experiences in general. Whether gamma synchrony might eventually play such a role remains to be seen. But the example should be enough to suggest how studies of deep, meditation-related experiences could potentially lead to discovery of tradition-independent physiological markers of the different levels and/or sublevels identifiable by standard scientific protocols, if such markers actually exist.[15]

## Section XI

All of this at present is only a suggestion. We are only just beginning to identify putative markers of even a very few deep meditative experiences, and reliable markers of experiences putatively located at all the levels, and many types of experiences within some of the levels, would need to be identified. Such research ideally would involve first-person investigators capable of sustaining the type of experience putatively characteristic of each level long enough for appropriate measurements to be conducted. This however is a level of first-person expertise quite beyond what has generally been available to scientists exploring the subjective domain, and locating people who might rationally even be suspected to be capable of this might seem to be very difficult, if not impossible. However when I asked Samdong Rinpoche[16] about this at a seminar in Kolkata not too long ago, he was confident that Tibetan monks capable of doing this could be identified for the purpose of such research, and added that this is the kind of thing that the Dalai Lama would be likely to be interested in.

With the help of such Buddhist adepts, a basic research program to evaluate the overall significance of the map might then ideally unfold in the following way. Finding consistent physiological markers of experiences of each level as reported by these adepts would be an important first step. Finding the same correlates of comparable reports by adepts from other traditions would indicate that the correlates were tradition-independent. This in turn would provide evidence for the tradition-independence of the map's phenomenological categories. Finding the same correlates of experiences as described by people in general would provide objective evidence, independent of any connection with training in meditation, of the general significance of the map's phenomenological categories.

---

[15] I would like to thank both David Orme-Johnson and Fred Travis for their critical feedback on this and the preceding section, and their help in selecting research references.

[16] Noted scholar and head of the Dalai Lama's government in exile.

The sequential aspect of the map would begin to be supported if adepts from different traditions reported in common that they experienced all the levels in the appropriate sequence as their attention moved in meditation from surface to depth and back, and the reports were corroborated by the appropriate physiological correlates. Another step would be to see whether experiences of such sequential development where reported by non-meditators (as, for example, described in many accounts of creative thinking[17]) were accompanied by the same correlates. A further, more difficult step would be to examine whether people's experiences in general are immediately preceded by the markers of the deeper levels as the map would lead us to expect, even when, as ordinarily happens, they unfold too quickly for such putative prior stages to be noticed by the subjects themselves. Success at all of these points would provide significant empirical corroboration for the map's sequential aspect as well.

Such a research program might of course fail. The map could be incorrect in various ways. Even if the map is correct, finding the hypothesized correlates is not likely to be easy. Given the speed with which mental contents ordinarily develop, identifying correlates of hypothesized earlier stages might be expected to require more time-sensitive instrumentation than currently available, in addition to identification of presently unknown parameters. Nevertheless if the appropriate markers were eventually observed in the proper sequences as described above, the sequential aspect of the map would appear to be corroborated as well. The implications of this for mainstream philosophy, psychology, psychophysiology and other disciplines would be far reaching. It would also, among other things, draw valuable critical attention to currently less than mainstream theories of human development that emphasize the existence and importance of deep levels of inner awareness such as those described by the map.

It would also be very significant methodologically. For it would give us an objective way to identify individuals able to maintain awareness at any given level for the purpose of identifying and observing its contents and processes systematically. This, when complimented by correlating results with investigators trained in different systems to filter out idiosyncrasies of perception, meditation procedure and conceptual context, could allow us to identify people capable of serving as much more reliable first-person information-gathering "instruments" than anything previously available to Western science.

All of this is at least in principle possible.

The levels map in short suggests a major research program. And this is just one map, from one group of traditions. Many other maps, and many other traditions with their own maps, exist, each with its own potential research programs.[18] The potential is huge.

---

[17] Cp, for example, Shear (1990), Chap. 5.

[18] For related research suggested by a Buddhist map, for example, see Shear (2007).

## Section XII

Let us now to turn from this optimistic picture to a major problem facing meditation research in general. The mere appearance of potential bias and/or vested interest is a serious matter in modern science. Such appearances are particularly widespread in the field of meditation research, where researchers often meditate themselves, express support for meditation traditions (and their beliefs), and/or are employed by meditation-related institutions. Skepticism by researchers outside the field is widespread, and even researchers in the field often express concerns about research on types of meditation other than their own.

A first, major line of defense against bias in modern science is publication in established, refereed journals. Editors and referees take their responsibility to evaluate the objectivity and potential significance of studies they publish very seriously. But scientific protocols also require independent replications. All too often however seemingly significant, published replications and meta-analyses have been conducted only by researchers associated with the techniques and traditions studied. This leaves the appearance of possible bias, and the scientific status of results, unresolved.

One way to address this problem would be to establish a meditation research consortium composed of both (a) researchers associated with different, putatively competing traditions and (b) non-affiliated researchers to (c) oversee—and sign off collectively on (i) replications and or falsifications of existing, potentially significant studies, (ii) meta-analyses of already published results, and (iii) new comparative studies. Such a consortium could, I think, go a long way towards directly addressing questions of bias and vested interest, while building on existing work. Less ambitiously, a cross-tradition, virtual network of researchers willing to collaborate with researchers studying traditions other than their own and to act as expert consultants and/or co-authors could also help address these problems as well. At the very minimum, such collaboration is necessary for research on meditation to become a real scientific field in the full sense of the term, rather than merely a collection of more or less related work.

## Section XIII

I would like to conclude with an ethically-related remark. It is becoming increasingly apparent that the kinds of benefits often traditionally claimed for meditation are especially important in today's world. As standards of living have risen throughout the world, so have levels of depression and other serious mental problems. Development of mental potential is increasingly becoming recognized as a need, rather than a mere luxury, in modern societies. So it is only natural that many millions of people have become actively interested in meditation and the developmental claims traditionally associated with it. In today's world, however, people are bombarded with all sorts of claims, accurate and inaccurate, including claims about meditation. So it should not be surprising that, if when people learn meditation techniques on the basis of inappropriate expectations, they can become disillusioned, stop practicing, and even reject the idea of meditation in general. Polls indicate already that this has happened for millions of people in the US alone. This is unfortunate, since the many benefits meditation can bring are then not gained, misinformation spreads, and the real value of meditation, which of course goes far beyond the benefits generally advertised, becomes obscured.

The remedy here is, I think, solid research determining which procedures produce what results, on what subpopulations, over what time frames. To be most effective this research will need to be collaborative, with experts in the different tradition contributing their knowledge of variables most relevant in their own areas, and helping to disseminate solid results about all the techniques studied, regardless of their own attachments. Such research will have to emerge eventually. But whether we are concerned with enlightenment, or simply reduction of stress and more harmonious life in society, it seems to me that we have an ethical obligation to support this sort of collaboration.

I hope that the present volume, bringing together researchers from different traditions to share their unique perspectives and results, can contribute to this larger project.

## References

Austin, J.H. 1998. *Zen and the brain*. Cambridge, MA: MIT Press.

Badawi, K., K. Wallace, D. Orme-Johnson, and A.M. Rouzere. 1984. Electrophysiologic characteristics of respiratory suspension periods occurring during the practice of the Transcendental Meditation program. *Psychosomatic Medicine* 46(3): 267–276.

Descartes, R. 1978. Meditations on first philosophy In *Descartes philosophical writings*. Trans. Elizabeth Anscombe and Peter Thomas Geach. Indianapolis: The Library of Liberal Arts, Bobbs-Merrill Education Publishing.

Farrow, J.T., and J.R. Hebert. 1982. Breath suspension during the Transcendental Meditation technique. *Psychosomatic Medicine* 44(2): 133–153.

Hume, D. 1980. *A treatise of human nature, edited by L.A. Selby Bigge*. Oxford: Oxford University Press.

Kant, I. 1964. Critique of pure reason, Trans. Norman Kemp Smith. New York: Macmillan & Co.

Lutz, A., L.L. Greischar, N.B. Rawlings, M. Ricard, and R.J. Davidson. 2004. Long-term meditators self-induce high-amplitude gamma synchrony during mental practice. *Proceedings of the National Academy of Science* 101(46): 16369–16373.

Razumnikova, O.M. 2007. Creativity related cortex activity in the remote associates task. *Brain Research Bulletin* 73(1–3): 96–102.

Shear, J. 1990. *The inner dimension: Philosophy and the experience of consciousness*. New York: Peter Lang Publishing.

Shear, J. 2007. Eastern methods for investigating mind and consciousness. In *The Blackwell companion to consciousness*, ed. M. Velmans and S. Schneider, 697–710. Malden: Blackwell Publishing.

Shear, J., and S.P. Mukherjee, eds. 2006. "Discussion." In *Consciousness: A deeper scientific search*, 353–373. Kolkata: The Ramakrishna Mission Institute of Culture.

Travis, F.T., and R.K. Wallace. 1997. Autonomic patterns during respiratory suspensions: Possible markers of transcendental consciousness. *Psychophysiology* 34: 39–46.

Travis, F., D. Haaga, J. Hagelin, M. Tanner, A. Arenander, and S. Nidich. 2010. A self-referential default brain state: Patterns of coherence, power, and eLORETA sources during eyes-closed rest and the Transcendental Meditation practice. *Cognitive Processing* 11: 21–30.
Wallace, A. 1999. The Buddhist tradition of Samatha: Methods for refining and examining consciousness. In *The view from within*, ed. F. Varela and J. Shear, 175–187. Thorverton: Imprint Academic.
Yogi, Maharishi Mahesh. 1974. *The Bhagavad-Gita: A new translation and commentary, chapters 1–6*. Maryland: Penguin.

# Using First-Person Reports During Meditation to Investigate Basic Cognitive Experience

**Wendy Hasenkamp**

**Abstract** This chapter describes a line of research that seeks to incorporate first-person subjective input into the analysis of meditation-related brain activity and connectivity, as a way to better define and understand everyday mental functions. I present a basic model of naturalistic cognitive fluctuations between mind wandering and attentional states derived from the practice of focused attention meditation. This model proposes four phases in a cognitive cycle: mind wandering, awareness of mind wandering, shifting of attention, and sustained attention. We developed a paradigm to leverage the common experience of awareness of mind wandering during this style of meditation, using subjective reports to drive the analysis of brain imaging data. Results revealed activity in specific brain networks associated with each cognitive phase. Further, participants with more meditation experience exhibited altered patterns of neural activity and resting state functional connectivity compared to participants with less experience. These neural patterns may be involved in the development of cognitive skills such as maintaining attention and disengaging from distraction that are often reported with meditation practice, and suggest mechanisms for how benefits may transfer "off the cushion." Implications for neurophenomenological investigations are discussed, as well as future directions and possible extensions of the model.

## Introduction

As evidenced by the present volume, as well as an exponentially increasing number of scholarly and lay publications, research on the mechanisms and effects of meditation is burgeoning. These investigations are undertaken with a variety of goals, but

W. Hasenkamp, Ph.D. (✉)
Mind & Life Institute, Hadley, MA 01035, USA
e-mail: wendy@mindandlife.org

S. Schmidt and H. Walach (eds.), *Meditation – Neuroscientific Approaches and Philosophical Implications*, Studies in Neuroscience, Consciousness and Spirituality 2,
DOI 10.1007/978-3-319-01634-4_5, 

most often in an attempt to understand how meditation "works" – that is, how does meditation practice lead to its frequently-reported benefits, including reduced stress, increased clarity, and a general enhancement of well-being? These pursuits are of great value, particularly for clinical applications, and the results of such investigations have been artfully summarized elsewhere (Chen et al. 2012; Chiesa and Serretti 2011; Chiesa and Serretti 2009; Rubia 2009). The goal of this chapter is to outline another use of contemplative practice in research – one that leverages the cognitive experiences that occur during meditation, coupled with the finely honed ability of practitioners to provide subjective report on these experiences, to gain a better understanding of the neural correlates of basic mental processes. This approach has great potential to advance cognitive neuroscience, and deepen our understanding of the human mind and the possibilities for its transformation. Facilitated in part by the work of Francisco Varela and colleagues (Varela et al. 1991), there has been a recent upsurge of interest in incorporating subjective information regarding mental states into neuroscientific explorations of the human mind. In this chapter, I describe one method for studying cognition based on this neurophenomenological perspective, and discuss implications and possible extensions of the approach.

Most forms of meditation employ fundamental cognitive processes that are also involved in many everyday experiences. For example, processes such as focused attention, detection of distraction, disengagement from ongoing thoughts, logical analysis, emotional engagement, cognitive re-framing, and meta-awareness are all used in various forms of meditation currently being taught in western contexts. Indeed, the goal of any given meditation is often to train one or more of these particular capacities, with the assumption that these skills will transfer "off the cushion" to be available in daily life. In addition, repeated meditation practice results in an increased familiarity with and ability to report on subtle changes in mental state in the practitioner (Lutz et al. 2008; Lutz and Thompson 2003). Thus, contemplative practice, when carefully understood and studied, can yield insights into these other processes in a more refined way.

The advent of non-invasive technologies such as functional magnetic resonance imaging (fMRI) has allowed for investigation of brain activity associated with human cognition in real time. Many studies have used such neuroimaging techniques to investigate brain activity during meditation, and while much has been learned and some consistencies are emerging (Chiesa and Serretti 2010; Chiesa 2009, 2010; Fell et al. 2010; Green and Turner 2010; Rubia 2009), theoretical and methodological limitations have often made results difficult to interpret. For example, the theoretical assumption that meditation is a single mental state that is achieved and maintained has supported a methodological approach of averaging brain activity over a block of time spent "in meditation" (often several minutes) and interpreting the static picture that results as a representation of brain activity during meditation. However, this underlying assumption is likely only accurate in the case of advanced practitioners, and only when the style of meditation has a goal of a single-pointed stable mental state. Rather, novice and intermediate practitioners often experience dynamic fluctuating cognitive states, even when trying to remain stably engaged in focused attention practices, oscillating between mind wandering and focused states. Thus, the common statistical practice of averaging brain activity over extended periods of time introduces caveats not commensurate with subjective experience.

It would seem advisable, given the aforementioned limitations, for researchers to find ways of leveraging the ability of meditators to report on their cognitive experience and adjust their methodological approaches to include this information. Cognitive models of meditation that incorporate subjective report can provide a framework from which to design and analyze experiments with finer sophistication, yielding deeper insights into the mental processes involved. Herein, I describe a simple model that begins to examine the cognitive processes involved in the practice of focused attention (FA) meditation in greater detail.

## A Basic Cognitive Model

FA meditation is intended to help the practitioner enhance awareness of his/her cognitive states while developing attentional control (Lutz et al. 2008). Indeed, research has demonstrated that FA meditation improves attentional skill in several domains (Jha et al. 2007; Lutz et al. 2009; MacLean et al. 2010; van Leeuwen et al. 2012; Zeidan et al. 2010). During FA practice, an individual attempts to maintain focus on a single object (e.g., the sensation of breathing), bringing attention back to the object whenever the mind wanders (Gunaratana 2002; Wallace 2006).

In line with many traditional accounts, our model proposes that during FA meditation, particularly for novices, one's subjective experience follows the general structure outlined in Fig. 1. When attempting to sustain focus on an object, an individual inevitably loses this focus and experiences mind wandering. At some time during mind wandering, the practitioner becomes aware that his/her mind is not on the object, at which point he/she disengages from the current train of thought and shifts attention back to the object, where it stays focused again for some period of time. In our original analysis, we termed these states MIND WANDERING (representing loss of focus), AWARE (representing the awareness of mind wandering), SHIFT (representing disengaging and shifting of focus back to the breath) and FOCUS (representing maintenance of attentional focus on the breath). The subjective experience of these states is a cycle that iterates repeatedly throughout a session of FA meditation. Thus, the practice of FA meditation is not a single cognitive state, except perhaps in very advanced practitioners. Instead, it involves a dynamic fluctuation between states of FOCUS and MIND WANDERING, incorporating the more transitory states of AWARE and SHIFT.

As a practitioner gains fluency and expertise, the processes of vigilance and adjustment become more automated, allowing one to rest in an "effortless" state of concentration or awareness (Lutz et al. 2008). In contrast, the present model is concerned with early stages of practice, which are more closely related to everyday experiences of attention and distraction.

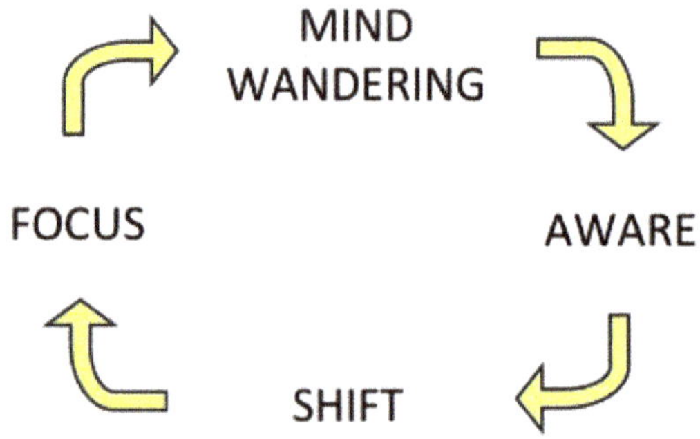


**Fig. 1 Basic cognitive model of FA meditation.** A theoretical model of dynamic cognitive states experienced by a non-expert practitioner during a session of FA meditation. When attempting to sustain focus on an object (*FOCUS*), an individual inevitably loses this focus and experiences wandering of attention (*MIND WANDERING*). At some time during mind wandering, the practitioner becomes aware that his/her mind is not on the object (*AWARE*), at which point he/she disengages from the current train of thought and shifts attention back to the object (*SHIFT*), where it stays focused again for some period of time (*FOCUS*). The cycle iterates repeatedly over a session of FA meditation

## Two Large-Scale Neural Networks

The mental processes of mind wandering and attention are increasingly becoming associated with activity in different distributed brain networks (Fig. 2). A task-negative, or default mode network (DMN) has been associated with task-independent, spontaneous thought processes, also known as mind wandering (Buckner et al. 2008). Mind wandering processes are directed away from a primary task and toward personal goals, and cover a broad range of mental functions, including memory, planning, and theory of mind (Smallwood and Schooler 2006). The DMN consists of hubs in the medial prefrontal cortex and posterior cingulate cortex, and also includes inferior parietal and lateral temporal regions (Buckner et al. 2008). The medial prefrontal cortex has been specifically implicated in self-related cognitive processing (Northoff and Bermpohl 2004), which is arguably a central feature of much mind wandering experience.

Conversely, a task-positive network has been associated with various attentional, present moment, and task-related processes, and is active during rest in an anticorrelated manner with the DMN (Fox et al. 2005; Fransson 2005). This large task-positive attention network can be subdivided in several ways to yield smaller and more distinct subnetworks. One such division distinguishes the salience and executive networks (Seeley et al. 2007). The salience network is thought to be involved in the immediate, present moment processing or detection of relevant stimuli, and involves the dorsal anterior cingulate cortex and bilateral anterior insula (Craig 2009; Seeley et al. 2007). The executive network, also referred to as the frontoparietal attention network, consists of dorsolateral prefrontal cortex (dlPFC) and posterolateral parietal regions, and is involved in controlling attentional

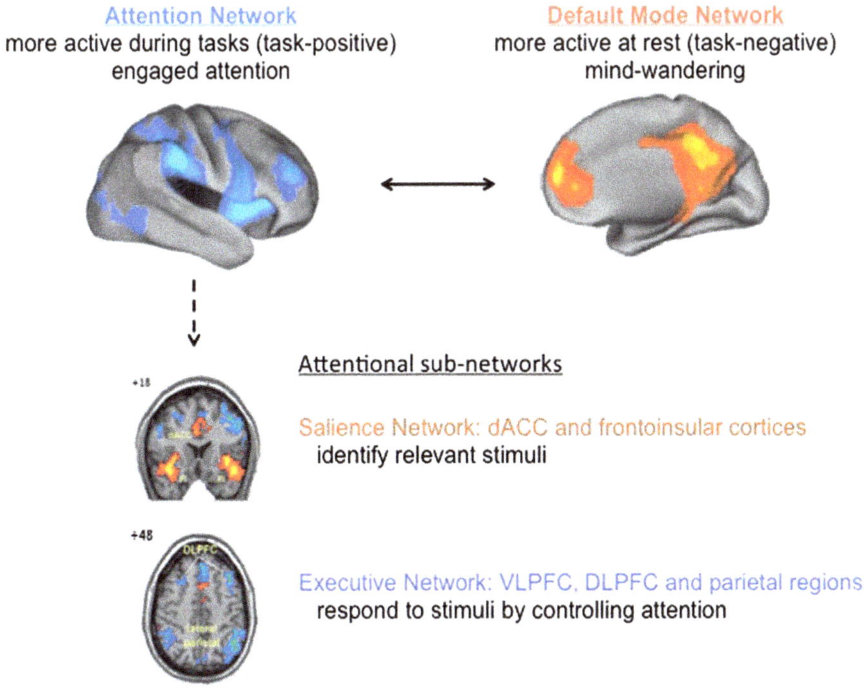


**Fig. 2 Brain networks associated with mind wandering and attention.** The brain can be divided into two large-scale distributed networks – the *default mode network* (*top right*), activity in which is associated with mind wandering states, and the *attention network* (*top left*), activity in which is associated with focused states demanding attentional resources. Within the larger attention network, multiple sub-networks can be identified. Two shown here are the *salience network*, comprised of the dorsal cortex (*dACC*) and bilateral insula, and the *executive network*, comprised of the dorsal and ventral prefrontal cortex and parietal regions. The salience network is thought to be involved in identifying relevant and salient stimuli in one's environment that require attention; the executive network is involved in disengaging and reorienting attention to these identified stimuli (Images modified from Buckner et al. 2008 and Seeley et al. 2007)

resources to deal with immediate or future demands (Corbetta et al. 2008; Corbetta and Shulman 2002; Seeley et al. 2007).[1]

## Attentional Sub-Processes in FA Meditation

The processes of detecting salience and controlling attention are highly relevant in the context of FA meditation (Fig. 3). Broadly speaking, the goal state, set by the intention of the practitioner via executive management systems in the brain, is to

[1] See also the chapter by Austin in this volume.

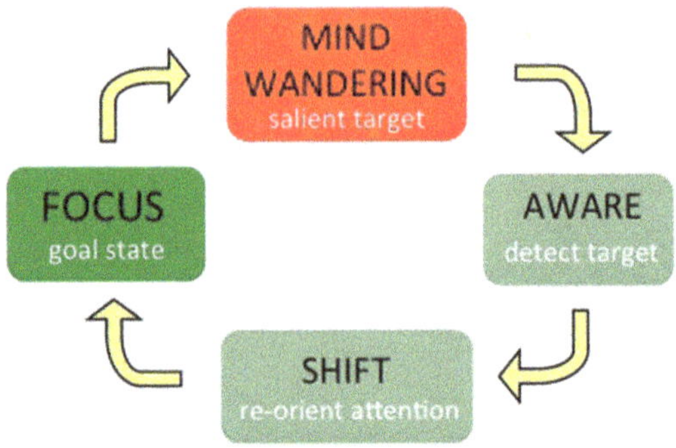


**Fig. 3 Attentional sub-processes in FA meditation.** During the practice of FA meditation, the goal state is one in which attention rests stably on the object of choice (*FOCUS*). In order to maintain this state, one must monitor one's own mental state and detect any attentional deviation from the object. In this case, the state of MIND WANDERING – one in which the object of attentional focus has been lost – becomes a salient target to be detected by monitoring systems. Once detected (at the *AWARE* moment), one must then disengage and re-orient attention (*SHIFT*) back to the chosen object. *Green* coloring represents top-down regulatory processes (subserved by the attention network) and *red* coloring represents bottom-up processes (subserved by the default mode network)

keep attention stably placed on an object. Ongoing regulation of this attention proceeds as a thermostat or feedback system, monitoring the environment – in this case, the internal, mental environment – for variation from the goal state (i.e., error detection). This could also be described as the process of conflict-monitoring, where the current state is in conflict with the goal state (Alexander and Brown 2010; Posner and Rothbart 2009). In this case, the primary relevant "target" that is to be detected becomes any mental state in which the object has been lost to attention. In the model here, the mental process of MIND WANDERING represents such a salient target. The meta-cognitive ability to monitor the contents of one's mind allows for detection of the MIND WANDERING target, and strengthening this ability is one of the main endpoints of FA training (Lutz et al. 2008). With continued practice, it would be hypothesized that detection of such target states would become increasingly rapid and sensitive. Indeed, this aligns well with subjective report of many practitioners, as well as traditional practice manuals (Wallace 2006). Once the target is detected, attentional disengagement and re-orienting must be employed in order to re-engage with the object (Posner and Petersen 1990; Posner et al. 1984). Again, it is expected that these processes of attention management are trained and improved by FA practice. The result of training in these attentional processes allows for longer and more stable retention of focus on the object, and thus, reduced object loss and faster recognition and re-orienting when object loss does occur.

This cycle also represents an interesting dynamic fluctuation between top-down and bottom-up mental processes, where attention systems are responsible for top-down management (e.g., setting the goal, sustaining attention, conflict monitoring, detection of salient targets/error detection, disengagement and re-orienting), and the natural tendency towards mind wandering or distraction is a bottom-up process

(Corbetta and Shulman 2002). For example, as you read this chapter, your top-down brain systems (e.g., attention regulation networks) set the goal of reading and processing the words on these pages. However, bottom-up processes related to your immediate surrounding environment (e.g., a colleague knocking at your door), internal bodily state (e.g., the strong desire for coffee), or other mental demands (e.g., the need to plan for a lecture tomorrow) will likely create distractions, sensory or cognitive in nature, intermittently pulling your focus away from the content of these pages. Top-down systems then regain the upper hand; meta-cognitive monitoring allows you to realize you have lost focus, and then you must disengage from the distraction, re-orient your attention towards the chapter and re-engage with the information content. When considered in the larger context of daily life, this endless dance between focus and distraction forms the undercurrent of much of our subjective experience.

## Incorporating Subjective Input into Neuroimaging Research

In order to incorporate subjective input into analytical cognitive models, we must first identify reportable cognitive states, that is, those that are accessible to conscious awareness. Throughout the process of FA meditation, perhaps the most consistently reportable moment is the experience of becoming aware of mind wandering (i.e., AWARE). Subjectively, this is a highly salient moment of meta-awareness, and also a necessary step in order to transition back to FOCUS.

In the experiment described here, intermediate level meditation practitioners performed 20 min of breath-focused FA meditation while undergoing whole-brain fMRI scanning, pressing a button each time they realized their mind had wandered (signifying the AWARE moment), and then returning their focus to the breath. This temporal information given by the button press allowed for separation of the data into subjectively meaningful epochs around AWARE, based on the model above. We used this information to construct the four intervals of AWARE, SHIFT, FOCUS and MIND WANDERING around the button presses, in 3-s epochs (for a detailed explanation of the analysis, see Hasenkamp et al. (2012).[2] Results show activity in brain regions associated with the task-positive attention network during AWARE, SHIFT, and FOCUS phases, and activity in default mode regions during MIND WANDERING phases (Fig. 4). Below, we briefly summarize these results; for a full discussion and deeper consideration of the findings, please see the original publication (Hasenkamp et al. 2012).

[2] Briefly, the TR containing the button press, as well as the preceding TR, constituted the AWARE phase, corresponding to awareness of mind wandering (3 s total). The two TRs (3 s) before the AWARE phase were cognitively defined as MIND WANDERING, representing loss of focus. The two TRs (3 s) following the AWARE phase made up the SHIFT phase, representing the shifting of attention back to the breath. Finally, the two TRs (3 s) following the SHIFT phase made up the FOCUS phase, representing maintenance of FA on the breath.

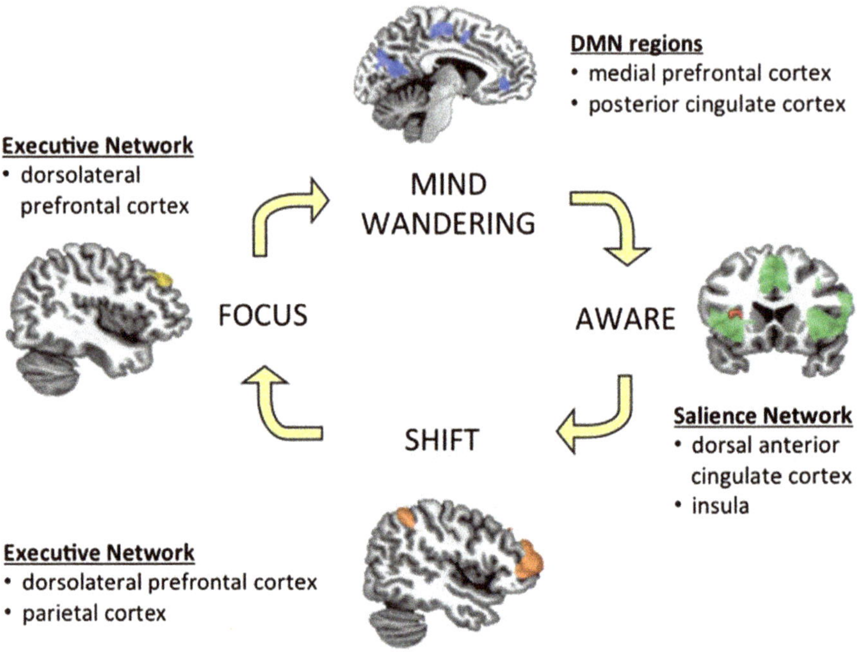


**Fig. 4 Summary of brain activations during the four cognitive phases of FA meditation.** Determined by subjective input (a button press at the moment of *AWARE*), we constructed temporal windows to represent the four phases of FA meditation (3 s each), and calculated brain activity for each phase across participants. During MIND WANDERING, we saw activations in default mode regions, including the medial prefrontal cortex and the posterior cingulate cortex. At the moment of awareness of mind wandering (*AWARE*), we observed highly robust activations in the salience network, specifically in the dorsal anterior cingulate cortex and bilateral insula (*red* areas indicate activity during a simple button-pressing task, as control for purely motor effects). Following AWARE, during the SHIFT phase when participants were disengaging and re-orienting their attention, we saw activations in the executive network, consisting of right lateralized dorsolateral prefrontal cortex and parietal cortex. Finally, during sustained attention in the FOCUS phase, activation persisted in the right dorsolateral prefrontal cortex. The known function of these brain areas corresponds well with the mental processes associated with each of the phases (see text for description)

*AWARE.* Analysis of the AWARE phase revealed robust activations in bilateral anterior insula and dorsal anterior cingulate cortex (Fig. 4, green). These regions are consistent with the subdivision of the attention network known as the salience network, described above (Seeley et al. 2007). Anterior insula and dorsal anterior cingulate have been implicated in a diverse range of cognitive processes, including conflict monitoring and error detection, interoceptive-autonomic arousal, the moment of perceptual recognition, self-regulation, emotional aspects of pain, empathy, musical chills, pleasurable touch, and present moment awareness (reviewed in Craig 2009; Seeley et al. 2007; Singer et al. 2009). Detection of relevant or salient events is important in each of these processes, which has led to the suggestion that these brain regions act together to comprise a general salience network (Seeley et al. 2007). Interestingly, while most paradigms have implicated this network in the detection of external salient events such as visual targets, the detected event in this paradigm—a state of mind wandering—was internally generated and purely cognitive in nature. This extends the scope of the salience network and supports recent suggestions that it may indeed function to detect general salience, regardless of environment or modality (Corbetta et al. 2008; Craig 2009; Seeley et al. 2007).

*SHIFT.* During the SHIFT phase, we observed significant activation in lateral PFC (dorsal and ventral) and lateral inferior parietal cortex, with larger clusters and more robust activation in the right hemisphere (Fig. 4, orange). These frontoparietal regions are consistent with the subdivision of the task-positive attention network known as the executive network, which acts on relevant stimuli (thought to be identified by the salience network) by re-orienting or directing attention while maintaining a goal (Corbetta et al. 2008; Corbetta and Shulman 2002; Seeley et al. 2007). Thus, what is known about the function of this network corresponds well with the hypothesized cognitive processing occurring in this phase: shifting or re-orienting attention from mind wandering back to the breath.

*FOCUS.* During maintenance of attention in the FOCUS phase, a cluster in the dorsolateral prefrontal region of the executive network remained active from the SHIFT phase (Fig. 4, yellow-orange). This may represent persistent neural activity underlying working memory, or "keeping a goal in mind," to maintain sustained attention on the focal object (Curtis and D'Esposito 2003; D'Esposito 2007). The DLPFC has been specifically implicated in active rehearsal, which consists of "the repetitive selection of relevant representations or recurrent direction of attention to those items" (D'Esposito 2007). Active rehearsal would be central to the sustained attention we hypothesize is occurring in the FOCUS phase, providing repetitive selection of, or attention to, the object (e.g., the sensations of breathing).

*Mind wandering.* During the MIND WANDERING phase, we detected activity in posterior cingulate cortex, medial PFC, posterior parietal/temporal cortex and parahippocampal gyrus (Fig. 4, blue); these regions have been repeatedly associated with the DMN in prior studies (Buckner et al. 2008). This pattern supports recent work associating the DMN with mind wandering processes (Buckner et al. 2008; Christoff et al. 2009; Mason et al. 2007).

Overall, activations in these phases were consistent with results from previous research showing that the respective attentional brain areas are associated with awareness (salience), disengagement and re-orienting (executive control), and maintenance (sustained attention). We also detected activity during MIND WANDERING in brain regions frequently associated with the DMN, mentalizing and self-related processing. This pattern of network activations is consistent with cyclic alternation between default mode and task-positive networks, in which DMN activity is associated with mind wandering, and attentional subnetworks are associated with awareness, shifting attention, and maintaining attention.

## Effects of Meditation Experience

We also sought to investigate whether there was an effect of practice time on these activation patterns—that is, does brain activity look different during these epochs dependent on the amount of time a person has meditated over their lifetime? By correlating estimated lifetime practice hours (see Hasenkamp and Barsalou 2012 for algorithm) with each of these activation maps, we identified several brain regions where activity varied according to meditation experience, particularly during the SHIFT phase. One region we chose to examine further was the ventromedial prefrontal cortex (VMPFC), a region in the DMN, which has been particularly implicated in self-related processing. In this area, activity returned to baseline more quickly following the button press in participants with more meditation experience than in those with less experience (Fig. 5). This suggested a possible increased ability in individuals with greater meditation experience to disengage from self-related or DMN processing upon awareness of MIND WANDERING. Research is increasingly suggesting that neuroplastic changes occur following repeated meditation practice (Baron Short et al. 2010; Brefczynski-Lewis et al. 2007; Farb et al. 2007; Farb et al. 2012; Hölzel et al. 2011; Tang et al. 2010; Xue et al. 2011; Zeidan et al. 2011), in a similar manner to other forms of experience-dependent plasticity. In this case, perhaps disengaging from ongoing thought content (as is required during FA meditation during the SHIFT interval) intentionally and repeatedly, induces neural

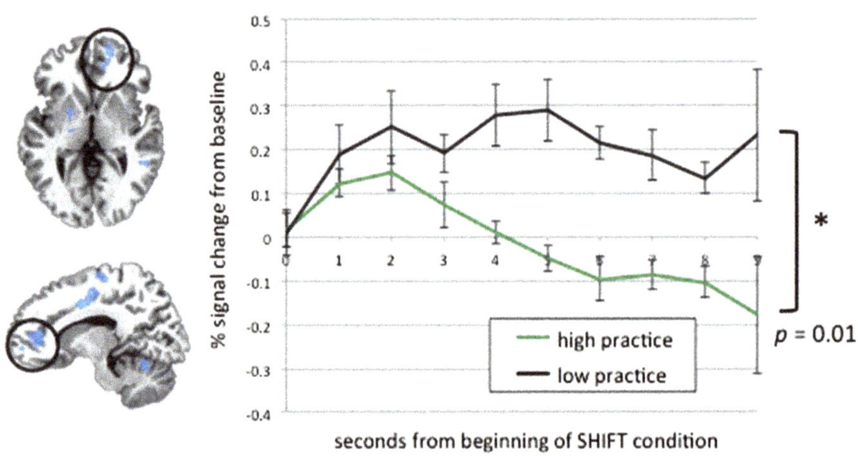


**Fig. 5 Correlations of brain activity with lifetime meditation experience.** *Left*: Activity during the SHIFT phase was negatively correlated with practice time in several brain regions (shown in *blue*). The VMPFC cluster that was examined at right is *circled*. *Right*: Time courses from the VMPFC cluster were extracted and hemodynamic response functions were calculated from the onset of the SHIFT phase for each subject. Percent signal change (from MIND WANDERING, mean ± s.e.m.) over time is plotted for high (n = 5) and low (n = 9) practice participants. Activation in this cluster was significantly reduced in high practice compared to low practice participants across the modeled time series. *Main effect of group over time by repeated-measures ANOVA, $p = 0.010$

changes that facilitate this process. This would correspond with the subjective experience of many practitioners and descriptions in practice manuals.

## Implications

The study described above reveals several important findings, not only for understanding neural correlates of cognitive fluctuations during FA meditation, but also for research methodology. By incorporating subjective input into the analytical design, we were able to obtain a fine-grained picture of cognitive states that were occurring in real time during FA meditation. The neural activations shown in Fig. 4 occurred over a total period of only 12 s. Thus, the conventional method of averaging brain activity over several minutes would surely have obscured the richness of these subjective experiences and associated neural activations. As described above, these mental states also occur in daily life during other tasks where focused attention is required. This means that similar methodologies could be implemented to examine general processes of mind wandering and attention monitoring and regulation.

Another implication with regard to neuroimaging study design is the somewhat paradoxical outcome that by reducing the time window of analysis (3 s in this study), but basing the precise temporal location of that window on relevant subjective input, the researcher likely *gains* statistical power. Indeed, the activations at the moment of the button press (AWARE) in the above study were so robust that the *p*-value threshold had to be lowered to $5.0 \times 10^{-6}$ in order to obtain a meaningful activation map. By comparison, standard block-design neuroimaging paradigms utilize periods of 30 s or more. For tasks where the mental processes involved are consistent (although this is always an assumption, even in standard tasks), using longer time periods increases power tremendously, making block designs preferable. However, for tasks such as FA meditation with non-advanced participants, where mental processes are inherently variable, incorporating subjective input and reducing the time window around self-reported events may impart a statistical advantage. Event-related designs increase power in a similar way, examining narrow temporal windows around experimenter-defined events. The difference here is that the *participant* determines the events based on their own experience. An initial concern is the necessarily variable number of events across subjects, with irregular and unpredictable temporal spacing. Researchers traditionally take great pains to avoid this kind of experimental variability; however, the above study suggests that such variability is not a hindrance when the experimenter can be more certain about conserved mental states.

## Examining Brain Activity "Off the Cushion"

Detailed temporal analyses such as those described above are clearly useful in understanding the moment-to-moment cognitive shifts underlying the subjective experience of focused attention. However, to understand how cognitive skills gained during such meditation practice may be transferred "off the cushion," we must also investigate brain states more reflective of daily life. One way to assess non-task related neural activations is to examine "resting state" brain activity. The nature of the brain's resting state has generated much interest among neuroimaging researchers in recent years (Lee et al. 2012; Snyder and Raichle 2012). These investigations focus on brain activity and functional connectivity when people are asked to rest quietly and not engage in any particular mental activity. Functional connectivity in particular has been examined as an indicator of brain areas that oscillate in a highly correlated pattern over time. Strong functional connectivity is often taken to suggest that brain regions are acting together, or have increased ability to communicate as compared to regions with weak functional connectivity. Changes in resting state functional connectivity have been identified after learning or practice paradigms (Leavitt et al. 2012; Schultz et al. 2012; Wang et al. 2012), and can be interpreted as an outcome of re-wiring or neuroplastic processes.

We reasoned that the networks identified as being involved in FA meditation (Fig. 4) are likely candidates for experience-dependent plasticity resulting from repeated practice. Thus, these regions may express differential resting state functional connectivity depending on the amount of meditation experience a person has accumulated. To investigate this possibility, we created seeds[3] to represent the activation patterns seen in each of the four conditions above. We then used these seeds to perform functional connectivity analysis on resting state data from the same participants. Participants were dichotomized into groups with high and low levels of lifetime meditation practice, and results were compared between groups to evaluate the effect of meditation experience on the functional connectivity of these networks. We hypothesized that participants with more meditation experience would exhibit increased functional connectivity of attentional networks, possibly reflecting plasticity induced by repeated engagement of these networks during contemplative practice.

Figure 6 shows representative findings from this analysis (for full results, see Hasenkamp and Barsalou 2012). Functional connectivity to the DLPFC, which was active during sustained attention in the FOCUS phase, was increased to several regions in the right insula in the high practice group. The insula has been implicated in a vast array of tasks, including present moment awareness and sensation of internal states (Craig 2009; Hasenkamp et al. 2012). Increased coherence of signal between right DLPFC and right insula at rest suggests that individuals with more meditation experience may have an enhanced awareness of present moment experience and more access to internal bodily states when employing executive processes in daily life. In addition, the VMPFC region found to have reduced activity during the SHIFT phase in high practice participants (Fig. 5) showed increased functional connectivity to bilateral regions of the inferior parietal lobule in these same participants. These parietal regions are part of the executive network, and have been

[3] In functional connectivity analysis, the term "seed" is used to refer to the distinct brain area to which all other brain activity is compared. Brain activity over time is plotted in the seed region, and temporal data from all other points in the brain is correlated with this pattern. If the correlation is strong, a given area is said to have strong functional connectivity with the seed region.

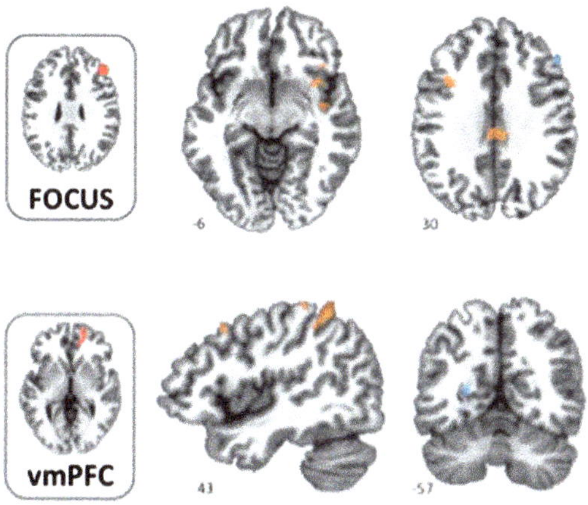


**Fig. 6 Functional connectivity differences between participants with high and low levels of meditation experience.** *Upper*: Increased functional connectivity in high practice participants between the right DL*PFC* (active during the FOCUS phase, *red*) and several attentional brain regions, including right insula, left D*LPFC* and mid-cingulate. *Lower*: Increased functional connectivity in high practice participants between the VMPFC (correlated with practice time, *red*, see Fig. 5) and left inferior parietal lobule. *Orange* indicates increased functional connectivity with more meditation experience; *blue* indicates reduced functional connectivity with more meditation experience

specifically implicated in attentional disengagement (Posner et al. 1984). This finding supports conclusions from our previous analysis (Fig. 5); with increased coherence between these regions, experienced meditators may have improved capacity for disengagement of thought content mediated by the medial PFC region.

These results add to growing evidence that the amount of time an individual spends practicing meditation is associated with activity and connectivity changes in the brain, particularly in attentional regions (Baron Short et al. 2010; Brefczynski-Lewis et al. 2007; Hasenkamp et al. 2012; Saggar et al. 2012). In this analysis, we utilized seed regions that were identified through the use of subjective input provided during meditation, and examined connectivity during the resting state in the same participants. This approach arguably allowed us to examine specific regions that would be most likely to have undergone plasticity from repeated meditation practice. Compared to selecting brain regions based solely on the literature, the leveraging of subjective input enabled a directed approach that likely increased the probability of detecting meditation-related changes.

## Extending the Model

While the basic model we propose above is useful as a starting point for this research, it is clearly an over-simplification of experience, and could be extended in many ways. I outline some possibilities here in the hopes of stimulating future

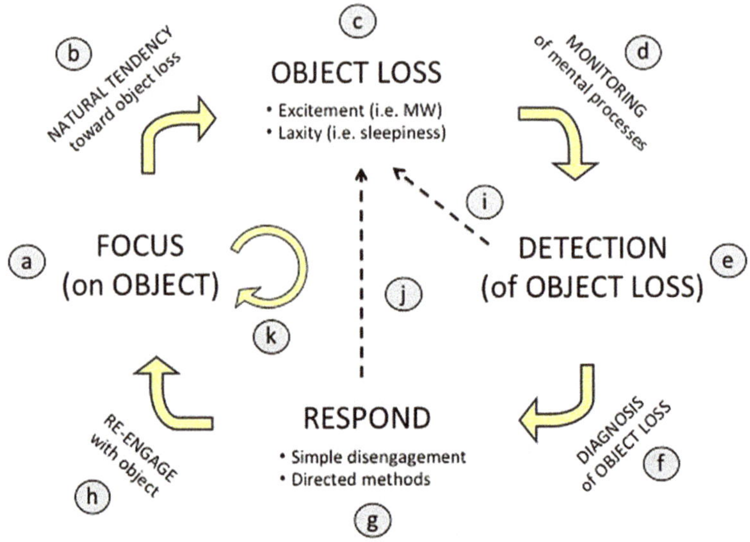


**Fig. 7 Possible extensions to the cognitive model of FA meditation.** In this expanded model of focused attention practice, the main phases have been re-labeled (**a**, **c**, **e**, **g**) to reflect a more general fluctuation between retention and loss of the object. In addition, the transitions between states have been labeled as processes (**b**, **d**, **f**, **h**) that enable each successive state. Finally, multiple exceptions to the serial nature of the four-phase model have been noted, including derailment (**i**, **j**) and parallel awareness/passing thoughts (**k**). For a full description of each point, see the text

research that incorporates a neurophenomenological approach (see Fig. 7). In doing so, I reconsider the labels of the four cognitive states as well as the processes that enable them (denoted by arrows). It should be noted that other useful models of this process have been proposed (Tang et al. 2012; Vago and Silbersweig 2012), and the present re-formulation is offered as only one possible next step.

First, it may be more generalizable in future research to broaden the MIND WANDERING state so that it encompasses all possibilities in which attention is "off" the chosen object of focus. This is denoted in Fig. 7 as OBJECT LOSS. This classification allows for more possibilities within the OBJECT LOSS state than just mind wandering; indeed, attention can also drift from the object into a state of dullness or sleepiness (Fig. 7c). This is a different subjective experience than the more engaged and energized state of mind wandering. In traditional Buddhist accounts, these mental states are referred to as "excitement" (meaning active thinking, roughly equivalent to mind wandering) and "laxity" (meaning a state of dullness or sleepiness; Wallace 2006). Furthermore, the arrow indicating the transition from FOCUS to OBJECT LOSS can be labeled to reflect that this process seems to be a natural tendency of the untrained mind (Fig. 7b).

Moving forward in the cycle, the previously named AWARE moment could be more generally described as DETECTION of object loss (Fig. 7e), to more clearly reflect the target detection function of the salience network that performs this process. Moreover, the process that enables this detection can be described as monitoring of attention, or meta-awareness (Fig. 7d). Once the DETECTION has been made, there may or may not be a process of diagnosis—that is, determining how the attention was off the object (e.g., via excitement or laxity; Fig. 7f). This process can also be referred to as "labeling," or "noting." Whether this process is engaged in will depend on the intention of the practitioner. It is also debatable whether diagnosis occurs simultaneously with detection or sequentially. It may be that in the early stages of learning, the processes are quite distinct, and with greater experience they become more indistinguishable such that the act of DETECTION automatically incorporates some level of diagnosis.

Once DETECTION occurs and a diagnosis is made (or even if none is made), one can then RESPOND to the state of OBJECT LOSS (Fig. 7g). This can be done in many ways, and will also depend on the manner of OBJECT LOSS. For example, in the case of excitement, one may simply disengage with the thought content that has arisen, "dropping" the thoughts in order to return to FOCUS. This is a common instruction in western styles of practice (Gunaratana 2002; Kabat-Zinn 1990; Wallace 2006). Another approach that is encouraged in some practices is to counteract the nature of the OBJECT LOSS through physical and/or mental strategies; for example, if laxity has occurred, one can open the eyes, look up, take a deep breath, perform certain visualizations, or find another way to increase energy and support vigilance (Gunaratana 2002; Wallace 2006). Whatever strategy is used, be it mere disengagement or a more directed approach, this is the beginning of the attentional response. Following this initial adjustment, one must then re-engage with the chosen object by directing attention through the process of re-orienting (Fig. 7h). This re-engagement enables the FOCUS state to once again be obtained (Fig. 7a). Maintenance of this state will require working memory and executive systems.

Another area where this model could be extended is in the serial nature of the cycle that is implied by the arrows. In reality, of course, one's subjective experience does not always proceed in such a clean and step-wise fashion. Rather, there are times when the cycle can "short circuit" after OBJECT LOSS before a successful return to FOCUS is achieved. For example, upon DETECTION, it is possible to simply return to the distracted or dull state of OBJECT LOSS without fully disengaging or re-engaging (Fig. 7i). Similarly, one can begin to RESPOND with disengagement and re-engagement, but not return to FOCUS before another state of OBJECT LOSS emerges (Fig. 7j). Lastly, based on the subjective report of many practitioners, there can be a sense of parallel mental states,[4] whereby one retains some amount of FOCUS on the object, but passing thoughts or brief states of dullness can co-arise, and be experienced without full OBJECT LOSS (Fig. 7k).

---

[4]This proposal of parallel states of processing may or may not agree with Buddhist theory, depending on tradition. It is proposed here as a result of anecdotal reports collected during the course of the research described herein, as opposed to an attempt to align with any particular textual account.

It is important to note that in neither the original model nor this revised version, do we mean to suggest that each of these cognitive states has a consistent duration. Indeed, significant individual variability undoubtedly exists in the precise temporal nature of the cognitive fluctuations discussed here. In our previous analysis, we used 3-s windows surrounding the button press – an assumption based partly on participant feedback and partly on methodological restrictions. It may be possible in future research to more accurately determine the temporal patterns of these states by incorporating additional subjective input. For example, one approach could be to include an additional button press when the state of FOCUS has been re-achieved. This resulting window between the two button presses would then denote the entirety of the RESPOND process. Similarly, one could envision adding a separate, distinguishable button-press to indicate detection of excitement (button 1) vs. laxity (button 2). While these approaches could provide additional accuracy and would certainly be interesting to investigate, they may also complicate analysis in several ways. For example, the addition of a second button press upon return to FOCUS would result in differing time windows for each cycle, necessitating a more complex analytical model for neuroimaging data. Moreover, the requirement of additional button presses could further interfere with the naturalistic experience of FA meditation, thereby disrupting the very process that is to be investigated. Careful consideration and experimentation is needed to navigate the balance between an approach that reduces complex mental processes to distinct, reportable moments, and one that embraces the totality of phenomenological experience.

## Conclusions

In this chapter, I review a recent attempt to incorporate subjective report into the neuroscientific study of cognitive processes associated with FA meditation and the shifts between attentional states. While still an uncommon approach, other work using subjective report is showing similar success (Christoff et al. 2009; Fox et al. 2012). Moving forward, it is hoped that cognitive science will begin to broaden its view of what constitutes valid evidence, and strive to develop rigorous methods of utilizing first-person information. In addition, the study of meditation in its various forms can be considered not only as a means to understand "mechanism of action," but to more clearly elucidate basic cognitive functions that are involved in everyday mental processes. It is becoming ever clearer that simple, third-person "objective" measurements will not suffice if we are to understand a topic as inherently subjective as the human mind. The marriage of first- and third-person investigations offers great potential for those wishing to more clearly elucidate this deeply important area of our shared experience.

**Acknowledgments** The author would like to extend warm gratitude to Lawrence Barsalou, John Dunne, Christy Wilson-Mendenhall and Arthur Zajonc for assistance in developing the original research as well as the proposed theoretical extensions described here.

## References

Alexander, W.H., and J.W. Brown. 2010. Computational models of performance monitoring and cognitive control. *Topics in Cognitive Science* 2(4): 658–677. doi:10.1111/j.1756-8765.2010.01085.x.

Baron Short, E., S. Kose, Q. Mu, J. Borckardt, A. Newberg, M.S. George, et al. 2010. Regional brain activation during meditation shows time and practice effects: An exploratory fMRI study. *Evidence-based Complementary and Alternative Medicine* 7: 121–127. doi:10.1093/ecam/nem163.

Brefczynski-Lewis, J.A., A. Lutz, H.S. Schaefer, D.B. Levinson, and R.J. Davidson. 2007. Neural correlates of attentional expertise in long-term meditation practitioners. *Proceedings of the National Academy of Sciences USA* 104: 11483–11488. doi:10.1073/pnas.0606552104.

Buckner, R.L., J.R. Andrews-Hanna, and D.L. Schacter. 2008. The brain's default network: Anatomy, function, and relevance to disease. *Annals of the New York Academy of Sciences* 1124: 1–38. doi:10.1196/annals.1440.011.

Chen, K.W., C.C. Berger, E. Manheimer, D. Forde, J. Magidson, L. Dachman, et al. 2012. Meditative the rapies for reducing anxiety: A systematic review and meta-analysis of randomized controlled trials. *Depression and Anxiety* 29: 545–562. doi:10.1002/da.21964.

Chiesa, A., and A. Serretti. 2010. A systematic review of neurobiological and clinical features of mindfulness meditations. *Psychological Medicine* 40: 1239–1252. doi:10.1017/S0033291709991747.

Chiesa, A., and A. Serretti. 2011. Mindfulness-based interventions for chronic pain: A systematic review of the evidence. *Journal of Alternative and Complementary Medicine* 17: 83–93. doi:10.1089/acm.2009.0546.

Chiesa, A. 2009. Zen meditation: An integration of current evidence. *Journal of Alternative and Complementary Medicine* 15(5): 585–592. doi:10.1089/acm.2008.0416.

Chiesa, A. 2010. Vipassana meditation: Systematic review of current evidence. *Journal of Alternative and Complementary Medicine* 16(1): 37–46. doi:10.1089/acm.2009.0362.

Chiesa, A., and A. Serretti. 2009. Mindfulness-based stress reduction for stress management in healthy people: A review and meta-analysis. *Journal of Alternative and Complementary Medicine* 15: 593–600. doi:10.1089/acm.2008.0495.

Christoff, K., A.M. Gordon, J. Smallwood, R. Smith, and J.W. Schooler. 2009. Experience sampling during fMRI reveals default network and executive system contributions to mind wandering. *Proceedings of the National Academy of Sciences USA* 106: 8719–8724. doi:10.1073/pnas.0900234106.

Corbetta, M., G. Patel, and G.L. Shulman. 2008. The reorienting system of the human brain: From environment to theory of mind. *Neuron* 58: 306–324. doi:10.1016/j.neuron.2008.04.017.

Corbetta, M., and G.L. Shulman. 2002. Control of goal-directed and stimulus-driven attention in the brain. *Nature Reviews Neuroscience* 3: 201–215. doi:10.1038/nrn755.

Craig, A.B. 2009. How do you feel–now? The anterior insula and human awareness. *Nature Reviews Neuroscience* 10: 59–70. doi:10.1038/nrn2555.

Curtis, C.E., and M. D'Esposito. 2003. Persistent activity in the prefrontal cortex during working memory. *Trends in Cognitive Science* 7: 415–423.

D'Esposito, M. 2007. From cognitive to neural models of working memory. *Philosophical Transactions of the Royal Society B: Biological Sciences* 362: 761–772.

Farb, N.S., Z.V. Segal, and A.K. Anderson. 2012. Mindfulness meditation training alters cortical representations of interoceptive attention. *Social Cognitive and Affective Neuroscience*. doi:10.1093/scan/nss066.

Farb, N.S., Z.V. Segal, H. Mayberg, J. Bean, D. McKeon, Z. Fatima, et al. 2007. Attending to the present: Mindfulness meditation reveals distinct neural modes of self-reference. *Social Cognitive and Affective Neuroscience* 2: 313–322. doi:10.1093/scan/nsm030.

Fell, J., N. Axmacher, and S. Haupt. 2010. From alpha to gamma: Electrophysiological correlates of meditation-related states of consciousness. *Medical Hypotheses* 75(2): 218–224. doi:10.1016/j.mehy.2010.02.025.

Fox, K.R., P. Zakarauskas, M. Dixon, M. Ellamil, E. Thompson, and K. Christoff. 2012. Meditation experience predicts introspective accuracy. *PLOS One* 7(9). doi:10.1371/journal.pone.0045370.

Fox, M.D., A.Z. Snyder, J.L. Vincent, M. Corbetta, D.C. Van Essen, and M.E. Raichle. 2005. The human brain is intrinsically organized into dynamic, anticorrelated functional networks. *Proceedings of the National Academy of Sciences USA* 102: 9673–9678. doi:10.1073/pnas.0504136102.

Fransson, P. 2005. Spontaneous low-frequency BOLD signal fluctuations: An fMRI investigation of the resting-state default mode of brain function hypothesis. *Human Brain Mapping* 26: 15–29. doi:10.1002/hbm.20113.

Green, R., and G. Turner. 2010. Growing evidence for the influence of meditation on brain and behaviour. *Neuropsychological Rehabilitation* 20(2): 306–311. doi:10.1080/09602010903172239.

Gunaratana, B. 2002. *Mindfulness in plain English*. Somerville: Wisdom Publications.

Hasenkamp, W., and L.W. Barsalou. 2012. Effects of meditation experience on functional connectivity of distributed brain networks. *Frontiers in Human Neuroscience* 6. doi:10.3389/fnhum.2012.00038.

Hasenkamp, W., C.D. Wilson-Mendenhall, E. Duncan, and L.W. Barsalou. 2012. Mind wandering and attention during focused meditation: A fine-grained temporal analysis of fluctuating cognitive states. *NeuroImage* 59(1): 750–760. doi:10.1016/j.neuroimage.2011.07.008.

Holzel, B.K., J. Carmody, M. Vangel, C. Congleton, S.M. Yerramsetti, T. Gard, et al. 2011. Mindfulness practice leads to increases in regional brain gray matter density. *Psychiatry Research* 191: 36–43. doi:10.1016/j.pscychresns.2010.08.006.

Jha, A.P., J. Krompinger, and M.J. Baime. 2007. Mindfulness training modifies subsystems of attention. *Cognitive and Affective Behavioral Neuroscience* 7: 109–119.

Kabat-Zinn, J. 1990. *Full catastrophe living: Using the wisdom of your body and mind to face stress, pain, and illness*. New York: Delta.

Leavitt, V.M., G.R. Wylie, P.A. Girgis, J. Deluca, and N.D. Chiaravalloti. 2012. Increased functional connectivity within memory networks following memory rehabilitation in multiple sclerosis. *Brain Imaging and Behavior*. doi:10.1007/s11682-012-9183-2.

Lee, M.H., C.D. Smyser, and J.S. Shimony. 2012. Resting-state fMRI: A review of methods and clinical applications. *American Journal of Neuroradiology*. doi:10.3174/ajnr.A3263.

Lutz, A., H.A. Slagter, J.D. Dunne, and R.J. Davidson. 2008. Attention regulation and monitoring in meditation. *Trends in Cognitive Science* 12: 163–169. doi:10.1016/j.tics.2008.01.005.

Lutz, A., H.A. Slagter, N.B. Rawlings, A.D. Francis, L.L. Greischar, and R.J. Davidson. 2009. Mental training enhances attentional stability: Neural and behavioral evidence. *Journal of Neuroscience* 29: 13418–13427. doi:10.1523/JNEUROSCI.1614-09.2009.

Lutz, A., and E. Thompson. 2003. Neurophenomenology: Integrating subjective experience and brain dynamics in the neuroscience of consciousness. *Journal of Consciousness Studies* 9: 31–52.

MacLean, K.A., E. Ferrer, S.R. Aichele, D.A. Bridwell, A.P. Zanesco, T.L. Jacobs, et al. 2010. Intensive meditation training improves perceptual discrimination and sustained attention. *Psychological Science* 21: 829–839. doi:10.1177/0956797610371339.

Mason, M.F., M.I. Norton, J.D. Van Horn, D.M. Wegner, S.T. Grafton, and C.N. Macrae. 2007. Wandering minds: The default network and stimulus-independent thought. *Science* 315: 393–395. doi:10.1126/science.1131295.

Northoff, G., and F. Bermpohl. 2004. Cortical midline structures and the self. *Trends in Cognitive Science* 8: 102–107. doi:10.1016/j.tics.2004.01.004.

Posner, M.I., and S.E. Petersen. 1990. The attention system of the human brain. *Annual Reviews Neuroscience* 13: 25–42. doi:10.1146/annurev.ne.13.030190.000325.

Posner, M.I., and M.K. Rothbart. 2009. Toward a physical basis of attention and self regulation. *Physics of Life Reviews* 6: 103–120. doi:10.1016/j.plrev.2009.02.001.

Posner, M.I., J.A. Walker, F.J. Friedrich, and R.D. Rafal. 1984. Effects of parietal injury on covert orienting of attention. *Journal of Neuroscience* 4: 1863–1874.

Rubia, K. 2009. The neurobiology of meditation and its clinical effectiveness in psychiatric disorders. *Biological Psychology* 82: 1–11. doi:10.1016/j.biopsycho.2009.04.003.

Saggar, M., B.G. King, A.P. Zanesco, K.A. Maclean, S.R. Aichele, T.L. Jacobs, et al. 2012. Intensive training induces longitudinal changes in meditation state-related EEG oscillatory activity. *Frontiers in Human Neuroscience* 6: 256. doi:10.3389/fnhum.2012.00256.

Schultz, D.H., N.L. Balderston, and F.J. Helmstetter. 2012. Resting-state connectivity of the amygdala is altered following Pavlovian fear conditioning. *Frontiers in Human Neuroscience* 6. doi:10.3389/fnhum.2012.00242

Seeley, W.W., V. Menon, A.F. Schatzberg, J. Keller, G.H. Glover, H. Kenna, et al. 2007. Dissociable intrinsic connectivity networks for salience processing and executive control. *Journal of Neuroscience* 27: 2349–2356. doi:10.1523/JNEUROSCI.5587-06.2007.

Singer, T., H.D. Critchley, and K. Preuschoff. 2009. A common role of insula in feelings, empathy and uncertainty. *Trends in Cognitive Science* 13: 334–340. doi:10.1016/j.tics.2009.05.001.

Smallwood, J., and J.W. Schooler. 2006. The restless mind. *Psychological Bulletin* 132: 946–958. doi:10.1037/0033-2909.132.6.946.

Snyder, A.Z., and M.E. Raichle. 2012. A brief history of the resting state: The Washington University perspective. *NeuroImage* 62(2): 902–910. doi:10.1016/j.neuroimage.2012.01.044.

Tang, Y.Y., Q. Lu, X. Geng, E.A. Stein, Y. Yang, and M.I. Posner. 2010. Short-term meditation induces white matter changes in the anterior cingulate. *Proceedings of the National Academy of Sciences USA* 107: 15649–15652. doi:10.1073/pnas.1011043107.

Tang, Y.Y., M.K. Rothbart, and M.I. Posner. 2012. Neural correlates of establishing, maintaining, and switching brain states. *Trends in Cognitive Science* 16(6): 330–337. doi:10.1016/j.tics.2012.05.001.

Vago, D.R., and D.A. Silbersweig. 2012. Self-awareness, self-regulation, and self-transcendence (S-ART): A framework for understanding the neurobiological mechanisms of mindfulness. *Frontiers in Human Neuroscience* 6. doi:10.3389/fnhum.2012.00296

Van Leeuwen, S., W. Singer, and L. Melloni. 2012. Meditation increases the depth of information processing and improves the allocation of attention in space. *Frontiers in Human Neuroscience* 6: 133. doi:10.3389/fnhum.2012.00133.

Varela, F., E. Thompson, and E. Rosch. 1991. *The embodied mind: Cognitive science and human experience*. Cambridge, MA: The MIT Press.

Wallace, B.A. 2006. *The attention revolution: Unlocking the power of the focused mind*. Somerville: Wisdom Publications.

Wang, D.Y., D.Q. Liu, S.F. Li, and Y.F. Zang. 2012. Increased local synchronization of resting-state fMRI signal after episodic memory encoding reflects off-line memory consolidation. *NeuroReport* 23(15): 873–878. doi:10.1097/WNR.0b013e3283587c96.

Xue, S., Y.Y. Tang, and M.I. Posner. 2011. Short-term meditation increases network efficiency of the anterior cingulate cortex. *Neuroreport* 22(12): 570–574. doi:10.1097/WNR.0b013e328348c750.

Zeidan, F., S.K. Johnson, B.J. Diamond, Z. David, and P. Goolkasian. 2010. Mindfulness meditation improves cognition: Evidence of brief mental training. *Consciousness and Cognition* 19: 597–605. doi:10.1016/j.concog.2010.03.014.

Zeidan, F., K.T. Martucci, R.A. Kraft, N.S. Gordon, J.G. McHaffie, and R.C. Coghill. 2011. Brain mechanisms supporting the modulation of pain by mindfulness meditation. *Journal of Neuroscience* 31(14): 5540–5548. doi:10.1523/JNEUROSCI.5791-10.2011.

# I Am I From Moment to Moment: Methods and Results of Grasping Intersubjective and Intertemporal Neurophysiological Differences During Meditation States

**Thilo Hinterberger**

**Abstract** An essential aspect of counsciousness is rooted in the subjectivity of the human experience. Such subjectivity implies that we are all individual beings, and each moment is unique. This uniqueness of each moment is a fundamental wisdom taught in many meditation schools. However, most research focusing on uncovering the psychophysiological correlates of meditation practice reports on statistical averages across subjects and across the time of the meditation task. Such averaging cannot show individual states and unique peak experiences within an individual. On the other hand, such statistics seem to be necessary for generalization. This chapter illustrates the inter-individual differences among 50 meditation participants with various meditation proficiencies. The inter-individual differences were measured with 64 channels of EEG plus peripheral measures during meditative practice. The results show that meditation brings about highly individual brain patterns. Furthermore, they illustrate how meditative states vary over time. Finally, methodological suggestions are given as to how an intra-individual and inter-individual analysis could be presented. Therefore, this chapter is dedicated to present details of the data which are invisible in common statistical analysis.

## Introduction

This chapter reports a study with 50 meditators, of various proficiencies, who were measured with 64 channels of EEG and several peripheral physiological measures during a meditation session. These meditation sessions were analyzed extensively with statistical methods (Hinterberger et al. 2011a, b). Spectral band power as well as

T. Hinterberger (✉)
Applied Consciousness Sciences, Section of Psychosomatic Medicine,
University Medical Center Regensburg, Regensburg, Germany
e-mail: thilo.hinterberger@klinik.uni-regensburg.de

S. Schmidt and H. Walach (eds.), *Meditation – Neuroscientific Approaches and Philosophical Implications*, Studies in Neuroscience, Consciousness and Spirituality 2,
DOI 10.1007/978-3-319-01634-4_6, 

coherence patterns have been correlated with meditation experience and self-reported mindfulness. Additionally, physiological data from different measurement phases and meditation tasks were compared. Generally, these statistical methods obtain their power from a huge collection of data samples from a sufficient number of participants in various states of consciousness. Following statistical analysis, the uniqueness of individual patterns and single events occurring during these meditation sessions is washed out through averaging processes within the statistical procedure. Most psychophysiological studies in the domain of meditation research require commonalities in participants in order to achieve generalizable statements from their measurements. A long history of EEG research in various types of meditative practice exists (see e.g. Cahn and Polich 2006; West 1979, 1980; Kasamatsu and Hirai 1969). In contrast, single subject studies do not allow for any generalization and are therefore rather unpopular, even though they sometimes report remarkable findings (e.g., Lehmann et al. 2001).

From this perspective, there seems to be a gap between science as a method for retrieving generalizable knowledge, and the individual experience, which can be of great individual meaning not only for the meditator as a participant, but also for the scientist as a researcher who wants to learn from a participant. Often, an experience that provokes great individual meaning in a person can significantly alter his or her behavioral traits. From this perspective, individual and unique experiences become collectively relevant, are no longer private, and should be studied more intensively. Therefore, in spiritual research all individual experiences which are associated with meaning should be treated as potentially relevant, and should not be excluded from the scientific process. As a consequence, we have to face the question: What are the possible scientific approaches for researching unique individual experiences? This chapter focuses on methods and results of individual and event-specific data evaluation. The scientific value which could be extracted from such analysis is discussed.

## Methods

Before focusing on the methods and results of individual data analysis it is important to describe the data sets, the experimental design, and the standard processing of physiological events in detail. This information creates the basis for further research questions.

### *Experimental Design*

All physiological data were recorded with a 72 channel QuickAmp amplifier system (BrainProducts GmbH, Munich, Germany). EEG was measured using a 64 channel ANT electrode cap with active shielding and Ag/AgCl electrodes, which were arranged according to the international 10/10 system. The system was grounded at the participants' shoulder. Data were recorded with a common average reference and filtered in a range from DC to 70 Hz at a sampling rate of 250 Hz and 22 bit resolution. For correction of eye movement and blink artefacts, the vertical EOG (electrooculogram) was measured by placing an electrode above and below one eye. Additionally, respiration was measured both with a respiration belt and via the skin conductance at the second and third finger of the non-dominant hand. Heart rate variability measures were extracted through ECG (electrocardiogram) measurement using two electrodes on the chest.

Prior to the measurement, participants answered a short initial questionnaire soliciting meditation experience in terms of duration and type of practice. In addition to the frequency of meditation, descriptions of the posture and method of their meditative practice, in as much detail as possible, were obtained. Measurements started with an initial 15 min baseline session in which they sat in their meditation posture for 5 min with eyes opened, 5 min with eyes closed, and 5 min reading a text from either a book or a computer screen. After a short break, a meditation session of 20–30 min duration was carried out in which they were asked to meditate in their routine way. During this time they held a button in one hand which they could press to indicate extraordinary events. After the meditation session, a report was written mentioning all events, feelings, emotions, thoughts and properties of the session. During recording, the experimenter observed the raw data stream. He tapped on the table whenever highly significant changes in the EEG waves, such as Gamma oscillations, spikes, or slow waves, were noted. Information on any subjective experience felt by the meditator during the time immediately before the tapping was requested and marked in the report. Finally, a 10 min guided meditation was carried out. This short meditation provided another four tasks: resting in the state of presence, resting in the highest possible emptiness or thoughtlessness, concentration on the 3rd eye, and concentration on the body axis and its projection down to earth and up to the sky. The EEG and physiological data from these eight epochs formed the data set for each participant. After the EEG and physiology recordings, the participants were asked to answer two questionnaires. At first self-report measure for mindfulness was assessed using the 30-question Freiburg Mindfulness Inventory (FMI) (Walach et al. 2006). The second questionnaire focused on the assessment of exceptional experiences and dreams (QEE), comprising 25 questions (Kohls and Walach 2006). The QEE measures two aspects, frequency and individual evaluation, regarding positive mystical experiences, psychological deconstruction and ego loss, psychopathological factors, and visionary experiences and dreams. These questionnaires were validated in their original German version. For the FMI an English version was available. We have translated the QEE into English and the FMI and QEE into Japanese to make them available in a convenient language of the participants. The results of those questionnaires and their physiological correlates were reported elsewhere (Hinterberger et al. 2011b). The entire session lasted between 2 ½ and 3 h.

For further analysis of the EEG-data reported here, artefact-free epochs of the eight conditions were selected according to Table 1. In addition to the analysis of the eight standardized conditions, a reference condition (No. 4 in Table 1) was

**Table 1** Definitions for eight standard conditions plus one reference condition. The reference condition (4) is either the eyes open or eyes closed condition or a mixture of both, depending on the eye state during meditation

| Condition type | No | Task | Duration/min |
|---|---|---|---|
| Reference conditions | 1 | Eyes open | 5 |
| | 2 | Eyes closed | 5 |
| | 3 | Reading | 5 |
| | 4 | Reference (formed by eyes open and/or closed) | 5 |
| Meditation | 5 | Self-selected meditation style | 20–30 |
| Guided meditation | 6 | Presence | 2 |
| | 7 | Emptiness | 2 |
| | 8 | 3rd eye | 2 |
| | 9 | Body axis | 2 |

introduced as either the eyes open or eyes closed condition, or, when meditators had their eyes half open during meditation, the mean between eyes open and closed. For this report, only a few selected comparisons were used.

## *Participants*

Altogether, the data base comprises 50 participants, non-meditators and meditators of various meditation proficiency, aged from 22 to 68 years (mean 45, 17 female, 33 male). All participants carried out the same experimental procedure as described above. The inclusion criteria were that they carry out a meditative spiritual practice on a regular basis and/or should be used to the practice of meditation, or, for the eight non-meditators, that they should not be familiar with meditation. The total meditation proficiency was calculated by multiplying the years of practice by their weekly practice sessions and the time per session. The results were between 12 and 21,185 h (mean 3,357 h) for the meditators. While 12/50 participants were non-meditators or had less than 40 h of meditation practice, 18/50 had between 1,000 and 5,000 h, 6 between 5 and 10,000 h and 7 participants had more than 10,000 h of estimated meditation practice throughout their life. The mean was exactly 4,000 h of meditation and 14 years of meditation practice on average. As this report focuses on individual data, the total meditation proficiency is shown in Fig. 1 for all participants. The participants also were inhomogeneously associated with different kinds of spiritual traditions such as Zen-Buddhism (11), Qi-Gong (4), Tibetan Buddhism (4), Sahaja Yoga (8), western contemplative methods (7), spiritualists or mediumistic practitioners (5), and eight non-meditators. Some of the meditators were also involved in spiritual healing and shamanism. Six participants were Buddhist monks living in monasteries in Japan. The measurements were carried out at various locations, predominantly rooms normally used for meditation or the participants' homes. The majority of the Buddhist practitioners were Japanese or Chinese and were measured in Japan.

 

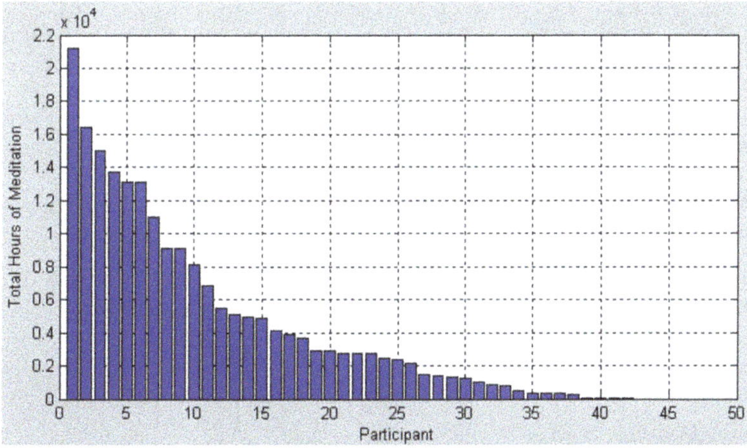


**Fig. 1** Figure: Sorted distribution of the meditation proficiency for all 50 participants. The most proficient meditator revealed an estimated total practice of more than 21,000 h in his life

All meditators participated voluntarily and gave their informed consent. The study was approved by the school ethics committee of the University of Northampton/UK and the ethics committee of the University Medical Center Freiburg i.Br./Germany.

## *EEG Data Processing*

### Data Preprocessing

Data analysis was performed using Matlab version 7.3. All EEG data were cleaned from high amplitude artefacts. After detrending the DC recorded EEG data sets, all EEG channels were corrected for eye movements using a linear correction algorithm similar to the one suggested by Gratton et al. (1983). The peripheral physiological data (EDA, ECG, respiration) are not reported here.

### Power Spectral Density (PSD)

A power spectrum time series was calculated using the Fast Fourier Transform (FFT). FFT was applied to the windowed EEG time series which was convolved using a Nutall window and shifted in steps of 0.5 s. A window size of 2 s was chosen for calculation of the FFT frequency coefficients. Their squared values resulted in the power spectral density (PSD). The following seven frequency bands were calculated by merging the FFT coefficients: Delta (1–3.5 Hz), Theta (4–7.5 Hz), Alpha (8–11.5 Hz), Beta1 (12–16 Hz), Beta2 (16.5–25 Hz), Gamma1 (25.5–47 Hz), and Gamma2 (54–70 Hz). The gap between 47 and 54 Hz was necessary to eliminate possible 50 Hz contamination of the electricity supply. EEG PSD was calculated for each participant, recording condition, electrode, frequency band, and time window.

### Statistical Considerations

For statistical comparisons of recording conditions, effect sizes and t-tests using PSD measures were calculated based on the assumption of normally distributed data. Therefore, it was necessary to take a closer look at the distribution of the data sets. For spectral EEG measures, a normal distribution is obtained after log-transformation values into log(PSD).

## *Approaches for Individual Data Analysis*

Usually, statistical data analysis is completed after averaging the data to a degree at which the dimensionality of a complex data set (e.g. participants x recording conditions x electrodes x data samples or spectral bands) has been reduced sufficiently. Results can then be easily overseen or a hypothesis answered simply with yes or no. Such a reduction usually masks many properties of a data set which would be detectable in a less reduced but more complex presentation style. In order to provide deeper insight into individual data structure we are attempting the extraction of individual data properties in several ways:

1. Rather than reporting group means, e.g. from gamma activities during meditation, topographic gamma mappings can be shown for each individual meditator. This provides an overview of inter-individual variances.
2. Rather than treating a meditation session as one single state, it can be highly informative to view the variations within a session for each individual. This is possible with the so called *State Monitoring* approach described below.
3. If this is realized it could be valuable to match physiological changes observed at certain times with subjective experiences, and also to match subjective events such as peak experiences with physiological changes. Therefore, events of potential relevance have to be extracted and time stamped to the physiological data stream. This requires a procedure that allows for marking specific events occurring during a meditation session with minimal interference with the meditation that would distract the meditator. In our experiment, three approaches have been implemented: (a) the meditator could report an event by pressing a button during the session. At the end of the session s/he was prompted to verbally report the meaning of each button press. (b) The researcher scanned the recording data stream and also paid attention to the meditator and the environment. Whenever he identified an unusual pattern he tapped lightly on the table. After the session, the meditator was asked to report his/her subjective state before each tapping. (c) A detailed time-resolved off-line data analysis was applied to uncover events of especially deviant activation patterns. Those events could be compared with post-session reports.

The results derived from these approaches are described and displayed in the following paragraphs.

## Individuality in Meditation-Related Gamma Activity

An often debated question in EEG research during meditation is the relevance of gamma band activity and whether it indicates high levels of meditation proficiency. For example, Lutz et al. (2008) report huge increase in gamma activity in meditators during meditation, which evokes the question of the reliability of such results. Data from our study of 50 meditating participants of varying proficiency levels and from various traditions allow a closer look into individual gamma activations during meditation. These data were analysed for gamma changes in comparison with both a resting state and a meditation session by comparing the final 10 min with the first 5 min of meditation as reported below.

### *Gamma Band Power Topography*

Generally, gamma band analysis in the EEG has to be treated cautiously due to the various potential sources of artefacts. An averaged gamma PSD mapping shows that gamma is strongest in lateral areas which are also sensitive to muscular artefacts. Gamma power in prefrontal and temporal areas can be about 10 times higher than in centroparietal regions. Such a distribution can be very likely explained by muscular artefacts originating from facial and jaw muscles. Therefore, it is important to identify those participants with high muscular tension which dominates those averaged mappings. This suggests that researchers should be careful in the interpretation of statistically significant results in frontal and temporal regions. Gamma effects in central areas more likely originate in the brain.

Before analyzing the data on an individual level, the average *gamma1* and *gamma2* band power activity over all 50 participants was illustrated during different task conditions by calculating a *t*-test over participants. Therefore, standardized mean differences of PSD values between two task conditions were calculated first for each participant separately. A one sample *t*-test (N=50) was calculated separately for each electrode and each frequency band using these effect sizes.

According to Fig. 2, the most proficient meditators in particular show a significant decrease in *gamma1* band power during meditation compared to the resting state. These changes are even more prominent in the beta band which is not displayed. It is a broad band effect which is not visible in the high *gamma2* band. As stated previously, gamma band results should always be considered with care as it is extremely important to pay attention to possible artefacts that could cause the result. Generally, the most prominent source of gamma artefacts is of

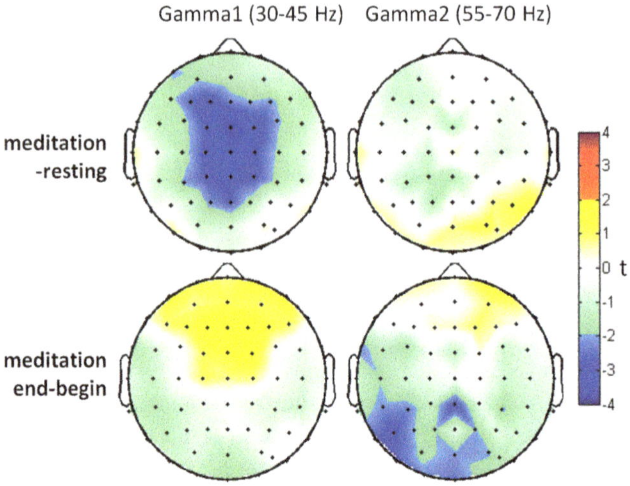


**Fig. 2** Significance maps resulting from a *t*-test over all 50 participants between two task conditions for the low and high gamma band are displayed. The *upper* graphs show the difference between meditation and resting state while the *lower* graphs show the difference between end and beginning of the meditation session

muscular origin. These EMG (electromyographic) artefacts are visible as more or less broad-band noise in the gamma range with frequencies up to a few hundred Hz. Therefore, if EMG signals contaminate our results they should show up in both the *gamma1* and *gamma2* band. This is not the case for the significant *gamma1* decrease in highly proficient meditators. Further, muscles such as facial muscles, e.g. clenched teeth or tension in the forehead, predominantly disturb the EEG in prefrontal, frontal and temporal regions. The findings in Fig. 2 do not seem to be heavily influenced by such artefacts. The significant *gamma1* decrease however is located mainly in central, frontocentral and centroparietal areas, and not in temporal and prefrontal regions. That means that the significant gamma 1 decrease is not a contaminant by EMGs but can be considered as a serious result.

In the following step we asked whether this gamma1 decrease in meditation can be seen similarly in the inexperienced and highly proficient meditators. The upper third proficient meditators were analyzed and compared to the lower third proficient group. According to Fig. 3, those meditators with no or little meditation proficiency only tend to show a weak but global and broad band decrease in gamma power in both bands, while the effect could be clearly associated with the high proficiency meditators.

Such overall significance measures were calculated from the averages of the contrasted conditions individually determined for each participant. Unfortunately, group statistics completely mask the individual differences which can be highly significant within a participant. Therefore, individual brain surface mappings are displayed in the next paragraph.

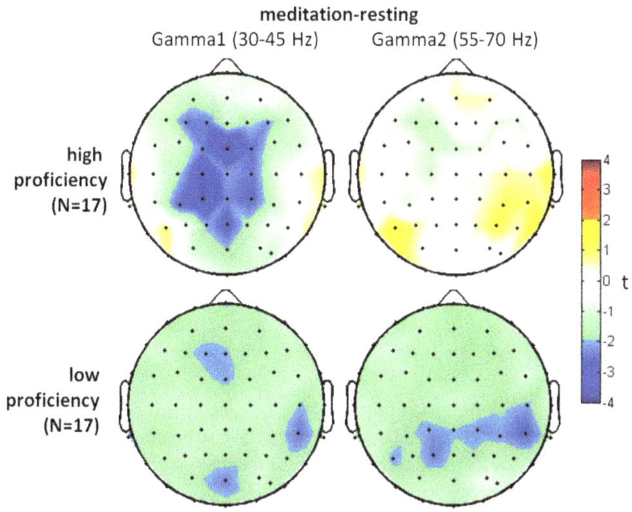


**Fig. 3** The contrasted conditions 'meditation versus resting' were analysed separately for the 17 highest proficient meditators (*top*) as well for the 17 with the lowest proficiency (*bottom*)

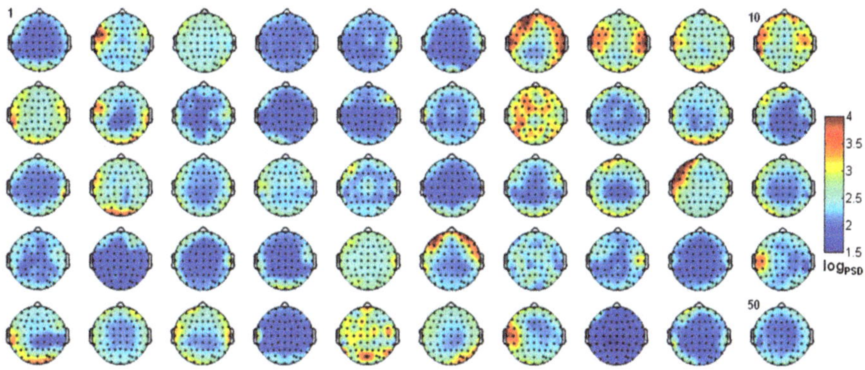


**Fig. 4** Resting state gamma during 5 min sitting relaxed with closed eyes. All 50 participants are shown sorted by their meditation proficiency (*top row* are the least experienced). Averaging over time was done using the median of FFT band power values

## *Inter-Individual Differences in Gamma Power*

In order to obtain a clearer picture of the distribution of gamma resting state power in the EEG, Fig. 4 illustrates the topographic gamma activity during rest in each participant separately. All graphs in the style of Fig. 4 show mappings of the total gamma power (*gamma1* and *gamma2*) in a sequence sorted by the meditation proficiency in total hours of practice, as displayed in Fig. 1.

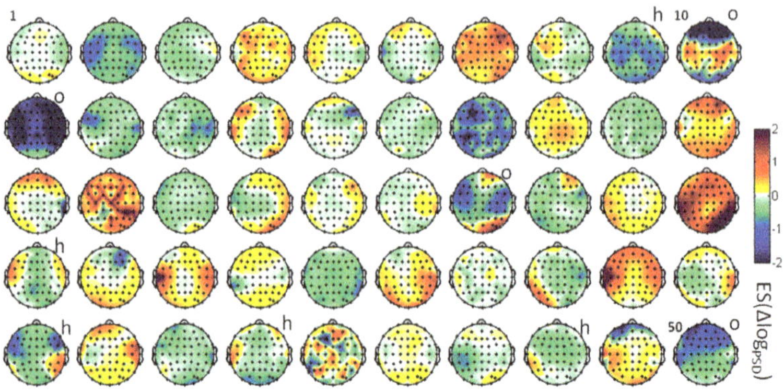


**Fig. 5** Effect sizes of the difference in gamma power between meditation and resting. *Yellow-red* indicated an increase while *green-blue* indicates a decrease of gamma in the meditation session. Legend: *h* stands for meditation with half open eyes: *o* means that the eyes were kept open during the meditation

Taking an overview of these 50 time-averaged surface mappings, it was easy to identify some with extremely high frontal and temporal activities. These activities were likely to be contaminated by muscular artefacts from facial muscles. Results from these participants should be selected with care in further analysis.

The individual changes of gamma PSD from resting to meditation were displayed as effect sizes in order to demonstrate their relevance. In Fig. 5 the differences of means for each condition were divided by the merged standard deviation of the FFT windows calculated each second. Here we also can see differences not only in the distal regions, but also in the central areas.

Meditators keeping their eyes open during meditation tend to show reduced frontal and temporal gamma activity compared to a resting period with open eyes. This becomes plausible with the assumption that in normal resting they were looking around, leading to a steady cognitive load and updating of the visual system, which involves frontal structures connected to, e.g. memory updating processes. In contrast, during meditation they looked constantly at one neutral spot, so no cognitive load due to visual input was required. This should result in a reduction of gamma activity as can be seen in the four participants with open eyes.

Interestingly, these results do not seem to match the subgroups which could be defined by considering the meditation styles reported by the meditators. Also, meditation experience does not seem to reflect a correlation with gamma differences, as can be easily seen in Fig. 5.

Who were these people with extraordinarily high gamma activities? In the high gamma group was a non-meditator (7), a spiritual healer and trance medium (20), a Zen-monk (22), a Sahaja Yoga practitioner (30), and a person who refers himself to both Buddhism and Daoism, practicing various forms of spiritual techniques in his meditation (39). This indicates that the high meditation gamma can neither be attributed to a specific tradition nor to a specific form of meditation, as these subjects all described their meditation contents differently.

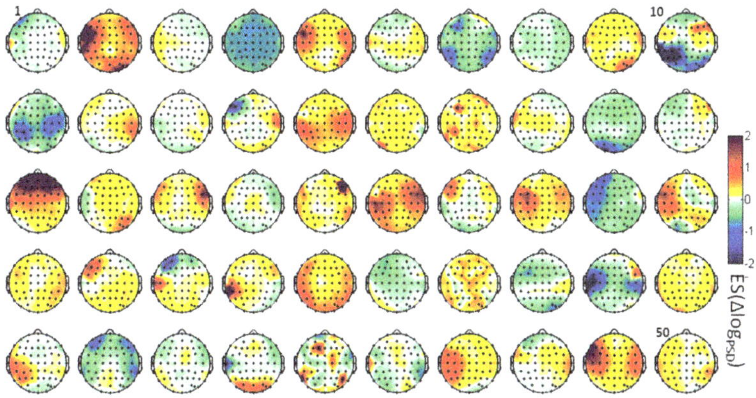


**Fig. 6** Effect sizes of the difference in gamma power between presence and resting, in both conditions the eyes were closed. *Yellow-red* indicated an increase while *green-blue* indicates a decrease of gamma in the state of presence

### *Individuality in a Standardized Condition*

One might argue that these idiosyncratic meditations of 50 participants cannot be compared as they had no common meditation task. Therefore, all meditators were asked to fulfil a number of standardized tasks as described previously. One quite well defined task was to sit relaxed with closed eyes, maintaining a state of highest possible presence in the here and now. This state was contrasted with the eyes closed resting state (Fig. 6).

## Event-Related Meditation State Analysis

### *Extraction and Categorization of Events*

The task-related states as described in Table 1 were available for all meditators. They can be regarded as well-defined except for the fifth task, i.e. the meditation session. The meditation task was highly individual and often contained a higher variability than the other eight conditions. One reason might be that it is more likely that a specific state of mind or physiological state can be maintained over 2–5 min than for a period of 20–30 min. Another reason might be that meditation is, for some meditators, more than sitting quietly maintaining a mindful state during the whole session. Some meditators, either voluntarily or involuntarily, access different states of consciousness, or their meditation technique leads them into certain different states during the time. Therefore, meditators were asked to press a button or give a signal whenever they experienced a specific state of consciousness or something that they thought could be of interest. Some meditators did not press the button at all because they did not ascribe much meaning to their experiences or they did not want to get distracted by such activity. In the following, button pressing or notification were called a ‘meditator initiated marker’. In the discussion between meditator and experimenter after the session, most of the button presses and some of the tapping could be associated with a specific subjective experience.

Another type of event was marked by the experimenter during the session (experimenter initiated marker). These event markers were indicated by lightly tapping with a pen on the table, a glass or a gong whenever the experimenter noticed significant changes in the ongoing raw EEG data stream, or other significant behavior in the meditator. In the discussion after the session, the experimenter asked for the subjective experience during the few seconds before the tapping.

The following tables list the two event types. The descriptions of the various subjective experiences were highly individual and it is hard to find groups of experiences with a similar content. Here, as a first categorization, we distinguished three types of events:

1. *Events of distraction* either from external, i.e. environmental sources such as noise or from internal sources such as pain
2. Events of *pure body sensation* either neutrally or positively evaluated by the participant, however without spiritual aspects
3. Events with imaginative or spiritual content

This grouping seemed to be necessary in order to distinguish the spiritual experiences from common experiences such as distraction and usual body sensations. For a more detailed characterization of an arbitrary event, four major categories of distinction have been chosen.

1. The first category describes the *sensory channel* involved in the experience. This can be a body sensation, a vision, or audition, e.g. when hearing inner voices. As several participants reported experiencing some kind of “energies”, a report of such energy was treated as a sensory channel in this study.
2. A further characteristic is the *origin of the experience*. Events such as noise in the environment are termed as coming from *outside*. Sensation or feelings originate from *inside the body*. Mental images, for example, are usually treated as originating from the *mind*. However, there are also mental images, feelings, or sensations that were attributed to an external source such as a spirit. Such events are categorized as *extrasensory*.
3. The content of a specific event is often mentioned by the meditator as a *specific interpretation*. The most prominent content categories are pain, happiness, relaxation, and being in connection with someone or something. “Connection” refers to a transcendent or transpersonal connection to an energy or spirit entity. It should be noted that these contents only refer to the time before the event and not to the entire meditation session. Thus, a meditator could be in a state of happiness during the whole session and suddenly feel some distractive pain which he marked with a button press. In such a case the table would only report the pain as a special event.
4. Finally, the *marking of an event* could be initiated either by a meditators button press or by the experimenter.

**Table 2** Events that were reported as subjective experiences by the meditators after the meditation session

| | | Events of | | |
|---|---|---|---|---|
| Category | Specification | Distraction % (No/Total No) | Pure body sensation % (No/Total No) | Imaginative or spiritual content % (No/Total No) |
| Sensory channel | Sensation | 33 (5/15) | 100 (6/6) | 12 (6/52) |
| | Vision | – | 50 (3/6) | 38 (20/52) |
| | Audition | 40 (6/15) | – | 2 (1/52) |
| | "Energy" | – | – | 23 (12/52) |
| Attributed origin of event | Outside | 40 (6/15) | – | 4 (2/52) |
| | Own body | 33 (5/15) | 100 (6/6) | 13 (7/52) |
| | Mind | 27 (4/15) | 17 (1/6) | 77 (40/52) |
| | Extrasensory | – | – | 56 (29/52) |
| Specific interpretation | Pain | 20 (3/15) | – | – |
| | Happiness | – | – | 6 (3/52) |
| | Relaxation | – | 33 (2/6) | 15 (8/52) |
| | "Connection" | – | – | 37 (19/52) |
| Event marked by | Meditator | 27 (4/15) | 50 (3/6) | 48 (25/52) |
| | Experimenter | 73 (11/15) | 50 (3/6) | 52 (27/52) |

Table 2 provides an overview of the specific events that occurred during a 20–30 min meditation session. It gives the total number of events accumulated over all 50 participants. However, events were only marked in 16 of 50 participants. The remainder did not mark specific events with a button press and the experimenter also did not notice events of interest.

Distractive perceptions were mainly due to external noise (40 %) or uncomfortable or painful body sensations (33 %). The remaining 27 % can be allotted to distractive thoughts. The contents of the pure body sensations were described as relaxation (3/6), warmth (2/6) and shivering (1/6). The major category of events referred to spiritual and imaginative contents (52/73). Here, 38 % were perceived as visions, 23 % as a kind of energy, 12 % as sensations and only one event was perceived as speech. 77 % of the events can be regarded as mental processes without physical origin and 56 % of these could be attributed to the extrasensory domain. 37 % of the events were experienced as in connection with a form of energy, a spirit or an imagined person. 15 % were associated with relaxation, and 6 % with a feeling of happiness.

### *Neurophysiological Characteristics Associated to Events*

The following table shows a comprehensive summary of the major events that caused the meditators to press the event button during the meditation session. The categories were defined according to the verbal report which was given mainly after the meditation session to explain the meaning of the event (Table 3).

**Table 3** Overview of the most prominent changes in activity during a maximum of 60 s before the meditators pressed the event button. P. refers to the participant number

| | | Significant changes | | |
|---|---|---|---|---|
| P. | Description | Time before button press (in sec) | Frequency band | Topographic area (▲ increase, ▼ decrease) |
| 4 | Mindfulness | −52 to 0 | Gamma2 | Right central ▲ |
| 4 | Bodily pain | −47 to 1 | 40 Hz | Right: central, parietal ▼ |
| | | | Gamma2 | Central, right: central, parietal ▼ |
| 8 | Emptiness | – | – | – |
| 8 | Spiritual energy | −12/−11 | Beta2 | Virtually all ▲ |
| 8 | Mental difficulty | −10 to 14 | 40 Hz | Prefrontal, right frontal; ▲ |
| | | −9 to 11 | Gamma2 | Prefrontal, right frontal ▲ |
| | | −10 to 11 | Gamma1 | Prefrontal, left frontal, right frontal ▲ |
| | | −9 | Beta1, Alpha, Theta | Virtually all ▲ |
| 8 | Emotional relaxation | −42 to 30 | 40 Hz | Prefrontal, right frontal, right temporal ▼ |
| | | | Gamma2 | Prefrontal, (right frontal) |
| | | −30 to 20 | Gamma1 | Prefrontal, right frontal ▲ |
| 8 | Bodily relaxation | – | – | Look for time before −40 |
| 10 | Spiritual energy (level of light-beings) | −14 to −12 (−12) (−13/−12) | 40 Hz, Gamma2, Gamma1, Beta2; (Theta; Delta) | Virtually all; ▲ (virtually all; all?, left temporal) |
| 10 | Spiritual energy (level of normally deceased) | Until −15 | 40 Hz, Gamma2, Gamma1 | Left temporal, left central maybe look at sequence before −60 to −15; |
| 10 | Spiritual contact | −24 to −22 | 40 Hz | Virtually all, especially: left+right frontal, temporal, and central, left parietal, occipital ▲; |
| | | | Gamma2, Gamma1 | Left frontal, right frontal, left temporal, right temporal, left central ▲ |
| | | | Beta2 | Left frontal, left temporal, left central ▲ |
| | | −22 | Alpha | Left temporal, right temporal ▲ |
| | | −23/−22 | Delta | Left temporal ▲ |
| | | From 0 on | | Then eyes open and speaking |
| 14 | Imagery (war & peace) | −16 to −2 | 40 Hz | Right temporal ▼ |

(continued)

**Table 3** (continued)

| P. | Description | Significant changes: Time before button press (in sec) | Frequency band | Topographic area (▲ increase, ▼ decrease) |
|---|---|---|---|---|
| 14 | Spiritual contact (higher being) | −15/−14 | 40 Hz | Prefrontal, left: frontal, right: frontal, central, and parietal ▲ |
| | | | Gamma2 | Right temporal, right central ▲ |
| | | | Gamma1 | Right frontal, right temporal, right central ▲ |
| | | −17 to −7 | Beta2 | Right parietal, occipital ▼ |
| 16 | Bodily sensation (warmth) | | | Reported until 19:00, nothing found |
| 18 | Spiritual energy (open head chakras) | −34 to −12 | 40 Hz | Left parietal, and at −27 more ▲ |
| | | −33, −27 | Gamma2 | Left parietal ▲ |
| | | −33 to −12 | Gamma1 | Left parietal ▲ |
| | | −34 to −12 | Beta2 | Left parietal ▲ |
| | | −35 to −33 | Beta1 | Left parietal ▲ |
| | | −29 to −25 | | |
| | | −20 to −15 | | |
| | | −34 to −12 | Alpha | Left parietal ▲ |
| | | | Theta | Left parietal, and at −28 virtually all ▲ |
| | | | Delta | Left parietal ▲ |
| 18 | Spiritual energy (stream from above) | −10 to −2; peak −5 to −2 | All | Left parietal ▲ |
| 18 | Bodily imagery (energy from floor into belly) | −12 | All (especially Theta, Delta) | All ▲ |
| 18 | Bodily imagery (consciousness from head into solar plexus) | – | – | – |
| 20 | Emotional spiritual energy (deep happiness) | – | – | – |
| 30 | Bodily sensation (warmth) | −30 to 0 | Beta2, Beta1, Alpha | All ▲ |
| | | −30 to 20 | Theta | Virtually all ▲, then ▼ |
| | | −28 to 0 | Delta | Virtually all ▼ |
| 34 | Spiritual energy light relaxation | −33/−32 | Beta2, Beta1, Alpha, Theta | Virtually all ▲ |

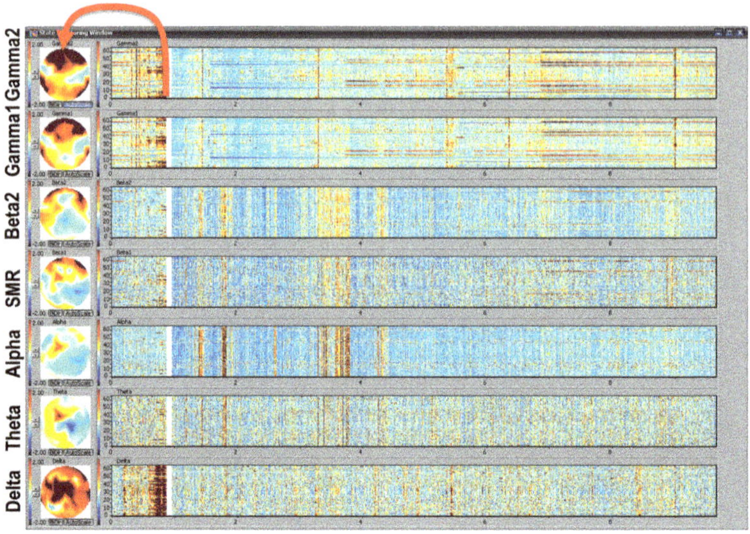


**Fig. 7** This is the screen-shot of the real-time state monitoring window representing the EEG data of one session of one participant which is provided by the State Monitoring Device. Seven frequency bands are displayed namely delta, theta, alpha, beta1, beta2, gamma1, and gamma2 from bottom to top. The topographic views on the left show the current spatial distribution of the corresponding normalized band power which is updated every 0.5 s. Each of the coloured bars contains the 64 channels of EEG in the vertical dimension while in the horizontal dimension the user can overlook a time span of about 10 min in this setting. *Blue/green* areas indicate a decreased activity while *yellow/red* indicates an increase. Here, the colour graph covers a range of +/−3 standard deviations

As one can see in this table, there is a wide diversity both in the reported events and in the corresponding changes in brain activity which we assumed to be accompanied with the event. Unfortunately, many meditators did not make use of the button press option because it would have distracted them from their meditation experience. Therefore, despite measuring 50 meditators, we only have a relatively small number of about 20 events to show. These 20 events were heterogeneous and highly individual. This diversity reflects the behavior of each individuality which makes it impossible to classify a specific response into a subjective category. Despite having only a few reported state events, some of the subjective reports upon a button press seem to be accompanied by significant changes in physiology. This result also suggests the development of an online State Monitoring Device which may resolve many more events during a session because the researcher can interfere immediately after the occurrence of an event. The changes could be recognized and directly classified into a subjective experience if meditators were supervised and asked for their experiences during the session. As a disadvantage, an immediate feedback would mean that a meditator might get distracted in the meditation. However, the collection of several spontaneous reports will provide a data basis for finding physiological commonalities of similar subjective experiences. In the following, our approach of a successful implementation of such a State Monitoring Device is described.

## Monitoring and Visualization of Individual Events

So far we have reported individual data of spontaneous events which were marked during recording or defined by the participant's task. The data were statistically reduced in order to grasp a specific correlative component. In the following we seek a method of displaying a full set of EEG parameters over an entire recording period of meditation. This would provide an overview of all significant signal changes during an entire session.

A display of a whole meditation session in one diagram is illustrated in Fig. 7 showing a time series of 35 min with PSD measures from 7 frequency bands and 64 electrodes with a time resolution of down to 1 s. Each of the 7 vertically stacked bars depicts a frequency band including 64 stacked time-lines corresponding to 64 electrodes displayed over a time range of several minutes. Thus, vertical pixels within a bar reflect a topographic map at each moment in time as exemplified in the left circle-shaped maps. The spectral intensity is color-coded according to the color bars. In order to be highly sensitive in the visualization of small but relevant changes in the spectral power, an autoadaptive standardization algorithm was applied to the PSD data for its display. Therefore, PSD values were standardized to the standard deviation of the mean PSD in a session run after subtraction of the mean PSD of the recording run.

This process could be applied online using the brain-computer interface Thought Translation Device (TTD) as a software framework (Birbaumer et al. 1999; Hinterberger et al. 2003a, b). The resulting tool was termed State Monitoring Device. For visualization of standardized PSD independently of previous recordings an approach was implemented that calculated an on-going reference from past samples of the current recording run which is updated with each calculation step of the PSD. Further, a real-time topographic visualization module was coded allowing for the display of the brain signal topography using the Delaunay triangulation algorithm. Thus, the current topographic signal distribution was displayed and updated every half second. It should be noted that this visualization is a powerful qualitative tool for visual data inspection; however, more complex quantitative information still has to be gained off-line in further statistical analysis.

## Conclusion

Statistical data analysis is a standard way of presenting results of psychophysiological measurements in a condensed and compact fashion, demonstrating commonalities and variances of a data pool. However, depending on the averaging process, statistical methods mask information about individual properties (after averaging over subjects), temporal information (after averaging over time) or spatial information (after averaging over electrodes). As a consequence, the reader misses information about individual patterns or the unique experiences a participant had at specific times. This chapter attempts to deal with these issues from various aspects.

After presenting topographic gamma band activities as a group average, these gamma topographies were also presented separately for each of the 50 individual meditators. Even though the averages across highly proficient meditators show significant decreases of gamma 1 activity in central areas, the individual gamma maps offer a huge variability in the gamma topographic amplitudes and patterns. As the maps were sorted by proficiency, it could be easily demonstrated that gamma is hardly a measure for identification of meditation proficiency. Neither could a clearly visible gamma increase during meditation compared to a resting state in highly proficient meditators, as reported in the literature (Lutz et al. 2008), be detected (see Fig. 5). The graphs also demonstrate that resting state gamma seems to be highly individual and not correlated with meditation experience as such. Due to the heterogeneity of results with respect to gamma, the questions of how we can trust a finding of gamma effects and what their real underlying sources are, still seem unanswered. These graphs demonstrate that even significant overall changes in EEG activity cannot be used for generalization. Therefore, it is important to keep in mind that averaging statistics does not allow for generalizable statements. It is rather a mean to express tendencies of brain activity to change in one or the other direction.

A similar conclusion has to be drawn from the standardized visualization of time series EEG data. Comparing the averaged activity over whole periods of measurement such as resting state or meditation, one can only find tendencies about what is changing in the brain from one state to the other. The state monitoring approach uncovers a huge number of significant changes in EEG activity during a meditation session. Those changes are completely ignored in common averaging statistics. However, when knowing from subjective reports that a person actually observed several meaningful state changes and events of conscious aberrations during a meditation session, the question arises, how are those subjective observations linked to the objectively measured EEG state changes? Unfortunately, the data reported in this chapter can only be regarded as a piloting first step in the direction of an individual state of consciousness analysis. With the monitoring and classification approaches which became available within these studies, however, we can learn to read those individual patterns by comparing those prominent brain state changes with the individual experience.

Real-time state monitoring could allow therapists and teachers in meditation and clinical hypnosis to trace the states of consciousness in order to guide their clients in their process better. Therefore, in a further stage, it is planned to professionalize and use the system to drive a feedback environment that reacts to changes and is conducive to the states of consciousness. This may give meditators, clients or patients the chance to intensify their experiences when obtaining additional feedback about their state changes. Further, the state monitoring approach could be applied to completely paralyzed or comatose patients to show whether and how they are able to react to external stimulation. In case of visible reactions, we could apply brain-computer communication tools such as the TTD.

**Acknowledgments** I thank the Samueli Institute of Information Biology (SIIB, VA, USA) as well as the BIAL foundation and the Beckley foundation for the support of this project. I also acknowledge the support in discussions and data acquisition from Tsutomu Kamei and Harald Walach, and, finally, all the participants worldwide who have contributed to these results.

## References

Birbaumer, N., H. Flor, N. Ghanayim, T. Hinterberger, I. Iverson, E. Taub, et al. 1999. A brain-controlled spelling device for the completely paralyzed. *Nature* 398: 297–298.

Cahn, B.R., and J. Polich. 2006. Meditation states and traits: EEG, ERP, and neuroimaging studies. *Psychological Bulletin* 132(2): 180–211.

Gratton, G., M.G.H. Coles, and E. Donchin. 1983. A new method for off-line removal of ocular artefact. *Electroencephalography and Clinical Neurophysiology* 55: 468–484.

Hinterberger, T., A. Kübler, J. Kaiser, N. Neumann, and N. Birbaumer. 2003a. A brain-computer-interface (BCI) for the locked-in: comparison of different EEG classifications for the Thought Translation Device (TTD). *Clinical Neurophysiology* 114: 416–425.

Hinterberger, T., J. Mellinger, and N. Birbaumer. 2003b. The thought translation device: structure of a multimodal brain-computer communication system. In *Proceedings of the 1st International IEEE EMBS Conference on Neural Engineering*, Capri Island, 20–22 Mar 2003, 603–606.

Hinterberger, T., T. Kamei, and H. Walach. 2011a. Psychophysiological classification and staging of mental states during meditative practice. *Biomedizinische Technik. Biomedical Engineering* 56: 341–350. doi:10.1515/BMT.2011.021.

Hinterberger, T., N. Kohls, T. Kamei, A. Feilding, and H. Walach. 2011b. Neurophysiological correlates to psychological trait variables in experienced meditative practitioners. In *Neuroscience, consciousness and spirituality*, ed. H. Walach, S. Schmidt, and W.B. Jonas, 129–155. Heidelberg/London/New York: Springer.

Kasamatsu, A., and T. Hirai. 1969. An electroencephalographic study of Zen meditation (Zazen). *Psychologia* 12: 205–225.

Kohls, N., and H. Walach. 2006. Exceptional experiences and spiritual practice – a new measurement approach. *Spirituality and Health International* 7(3): 125–150.

Lehmann, D., P.L. Faber, P. Achermann, D. Jeanmonod, L.R. Gianotti, and D. Pizzagalli. 2001. Brain sources of EEG gamma frequency during volitionally meditation-induced, altered states of consciousness, and experience of the self. *Psychiatry Research* 108(2): 111–121.

Lutz, A., H.A. Slagter, J.D. Dunne, and R.J. Davidson. 2008. Attention regulation and monitoring in meditation. *Trends in Cognitive Sciences* 12(4): 163–169.

Walach, H., N. Buchheld, V. Buttenmüller, N. Kleinknecht, and S. Schmidt. 2006. Measuring mindfulness – the Freiburg Mindfulness Inventory (FMI). *Personality and Individual Differences* 40(8): 1543–1555.

West, M.A. 1979. Meditation. *The British Journal of Psychiatry* 135(5): 457–467.

West, M.A. 1980. Meditation and the EEG. *Psychological Medicine* 10(2): 369–375.

# Does Neuroimaging Provide Evidence of Meditation-Mediated Neuroplasticity?

**Shawn S. Clausen, Cindy C. Crawford, and John A. Ives**

**Abstract** Results of recent magnetic resonance imaging studies suggest that meditation may be associated with region-specific structural neuroplasticity. To test the hypothesis that meditation-related brain function predicts site-specific structural changes in meditators, we conducted two meta-analyses: one of studies localizing brain activity during meditation, and a second of studies measuring differences in brain structure between meditators and non-meditators. Activation Likelihood Estimation (ALE) meta-analysis of five studies measuring brain activation during meditation revealed the greatest clusters of activity to be in the left frontal cortex and left precuneus. ALE of four studies measuring the differences in brain structure between meditators and controls revealed that meditators tended to have greater brain volume in the left inferior temporal gyrus. Thus, brain activity during meditation did not predict region-specific structural differences between meditators and non-meditators. This finding may reflect recognized limitations in neuroimaging methodology rather than the refutability of the hypothesis itself. Future efforts aimed at understanding the relationship between brain activity and structural changes in the brain should focus on improving neuroimaging experimental design and incorporating evidence from other branches of neurocognitive science. Progress in these areas promises to elucidate the connection between mind-body practices, and brain structure and function.

S.S. Clausen, M.D., M.PH.
Department of General Preventive Medicine, Uniformed Services University of the Health Sciences, 4301 Jones Bridge Road, Room A1040A, Bethesda, MD 20814-4799, USA
e-mail: shawn.clausen@usuhs.mil

C.C. Crawford
Samueli Institute, Alexandria, VA, USA

J.A. Ives (✉)
Brain, Mind and Healing, Samueli Institute, 1737 King Street Suite 600, Alexandria, VA 22314, USA
e-mail: jives@siib.org

S. Schmidt and H. Walach (eds.), *Meditation – Neuroscientific Approaches and Philosophical Implications*, Studies in Neuroscience, Consciousness and Spirituality 2,
DOI 10.1007/978-3-319-01634-4_7, 

## Introduction

Recent magnetic resonance imaging (MRI) studies demonstrate that training in some skills such as music, language, and spatial navigation is associated with increased gray matter in regions relevant to the learned task (Gaser and Schlaug 2003; Machelli et al. 2004; Maguire et al. 2000). Intensive training in juggling, for example, leads to increased gray matter in regions known to be important in perceptual-motor coordination (Draganski et al. 2004). In addition, increases in gray matter occur in less time than previously recognized. A 2007 study aimed at defining the usage dependent time scale of structural changes found that architectural alterations in the brain were detectable within a week of the intervention (May et al. 2007).

Meditation is a mind-body practice which originated in ancient religious and spiritual traditions. In Western culture, its secular clinical application falls under the rubric of complementary and alternative medicine (CAM). Meditation incorporates a diverse array of practices including various combinations of breathing, mantra recitation, focused attention, and practices related to spiritual beliefs. It has been extensively studied as a method of relaxation and has been associated with numerous physiological benefits (Benson 1984; Davidson et al. 2003; Kabat-Zinn et al. 1992, 1998). It has also been used as a method of fostering attention and emotional self-regulation (Jha et al. 2007; Ospina et al. 2007).

Functional magnetic imaging studies of meditation demonstrate that meditation is associated with brain activation in a number of cortical and sub-cortical regions including the frontal and parietal cortices, known to be important in the mental process of attention, and the insula, known to be important in emotional regulation (Lazar et al. 2000; Holzel et al. 2007). Some authors have found structural differences between the brains of meditators and non-meditator controls and speculate that differences are due to a meditation-associated training effect and structural neuroplasticity (Lazar et al. 2005; Holzel et al. 2008; Luders et al. 2009).

While common areas of activation have been identified among neuroimaging studies exploring the brain during meditation, there is some inconsistency across studies. Moreover, studies exploring differences in the brain structure of meditators versus controls fail to demonstrate consistent differences between the two groups.

Meta-analysis, specifically coordinate-based meta-analysis (CBMA), offers a method of pooling neuroimaging study results, and thus can be used to summarize and clarify these two sub-categories of MRI studies (i.e., studies of brain activation associated with meditation, and studies measuring structural differences between meditators and non-meditators). Results of such meta-analyses are amenable to comparison and would help determine if brain activity during meditation predicts differential brain structure between meditators and non-meditators. To date, no such studies have been published.

To explore the hypothesis that meditation mediates region-specific structural neuroplasticity, we reviewed the literature to identify fMRI studies measuring brain activity during meditation, and performed Activation Likelihood Estimation (ALE) meta-analysis of the pooled coordinates. Studies measuring brain structure in meditators were considered separately and coordinates were pooled in a second ALE meta-analysis (Laird et al. 2005; Turkeltaub et al. 2002). We then compared the results of these meta-analyses to see if brain activation predicted region-specific structural differences between meditators and non-meditators.

## Methods

Two independent reviewers completed a multilevel systematic literature search. Using the key words "meditation AND mri", databases (PubMed, EMBASE, Cinahl, Cochrane Database for Systematic Reviews, and PsychINFO) were searched for MRI studies of meditation published since 2000. (Because "yoga" is synonymous to "meditation" in Cinahl, the key words "yoga AND mri" were also used to search this database). Our gray literature search included searches of metaRegister of Controlled Trials, Cochrane Central Register of Controlled Trials, and Google Scholar. References of identified studies were hand searched.

For a study to be included in our review, it had to meet the following inclusion/exclusion criteria: (1) the study had to be an interventional or observational study; case reports, case series, letters, and editorials were excluded; (2) the study had to use magnetic resonance imaging to measure neural activity during meditation (state), or to measure brain structure of individuals with meditation experience versus controls (trait); (3) studies measuring differential responses to stimuli (pain, sound, etc.) during meditation versus rest, or in meditators versus non-meditators were not included in this review. In instances where studies focused primarily on stimulus–response but also included information on activation during meditation, pertinent information was extracted from the study and included in our review; (4) studies had to be published between 2000 and 2009 (2000 documents the first study of this kind); and (5) while subject description was required, there was no restriction on the study population. Disagreements between reviewers with regard to inclusion/exclusion of individual studies were resolved by consensus.

Activation Likelihood Estimation (ALE) meta-analysis was performed using GingerALE v1.1 software and the following parameters: a full-width half maximum (FWHM) of 10 mm, permutations of 5000, q-value of 0.05, and a minimum cluster size of 100 $mm^3$ (Laird et al. 2005). Once the thresholded ALE map was generated using GingerALE v1.1, an anatomical underlay (in Talairach space) was downloaded and images were generated using Multi-Image Analysis Graphical User Interface (Mango) (Kochunov et al. 2002) (Lancaster and Martinez Accessed 24 May 2009).

## Results

Using the inclusion criteria outlined, a total of 1,116 titles/abstracts were reviewed and 1,065 were excluded. From the remaining 51 abstracts, the full articles were reviewed in more detail and 13 articles were considered for inclusion. Four

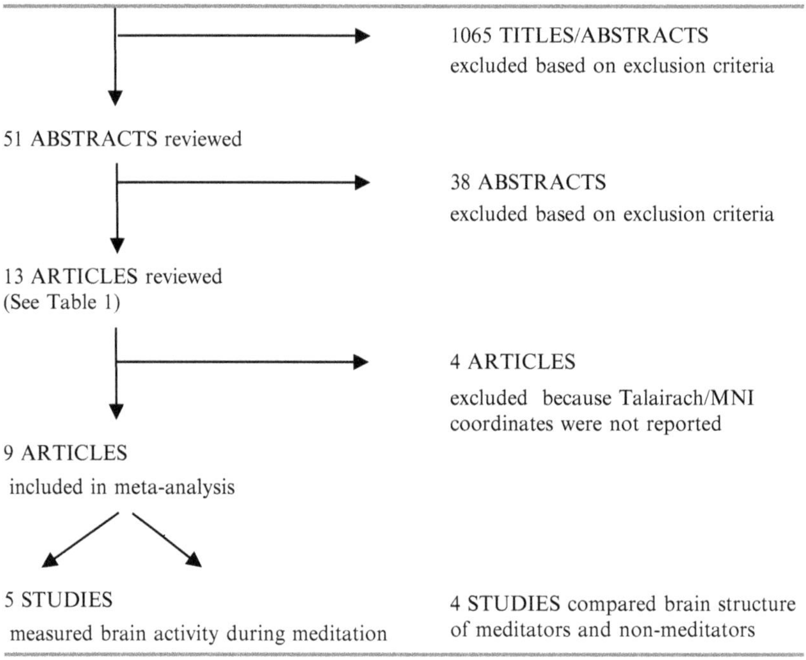


**Fig. 1** Study selection flow chart

studies were excluded from meta-analysis because Montreal Neurologic Institute (MNI) coordinates were not reported. This resulted in nine studies eligible for meta-analysis, five were studies of brain activation during meditation, and four were studies comparing brain structure between meditators and non-meditators (see Fig. 1).

Systematic review of the 13 included studies, which used MRI to study meditation/meditators, revealed that such studies are generally small, ranging in size from 5 subjects to 44 subjects. Most studies included controls. However, two of the studies measuring brain activation during meditation used the non-meditative state as the control. The age of participants ranged from 22 years to 71 years (mean 34–53 years). Overall, there was a preponderance of males studied. In the 11 of 13 experiments where both sexes were represented, investigators took care to ensure that equal numbers of each sex were included in the experimental and control groups. Most participants were right handed. Handedness was matched in studies reviewed. The type of meditation practiced and the duration of practice varied between studies (2–46 years). While attention was given to meditation type and the possible impact that this might have on MRI results, from the perspective of neuroimaging, meaningful subcategories have yet to be delineated and meditation type was not considered here (i.e., all meditation types were included in meta-analysis).

In addition to the potential covariates listed in Table 1, inconsistently reported covariates include ethnicity/race, and level of education. Very few studies reported on the general health of participants and controls. Only 1 out of 13 studies included information about medication use. With regard to image acquisition and statistical analysis, the majority of studies reviewed included information about scanners and software utilized. Only studies which reported statistically significant coordinates were included in meta-analysis (see Table 1).

## Brain Activation During Meditation

Two of the studies reviewed measured brain activation during meditation as compared to rest (Shimomura et al. 2008; Lazar et al. 2000); these studies revealed consistently greater activation in the frontal cortex during meditation. The parietal and cingulate cortices were also activated during meditation, but this finding was less consistent. (Lazar et al.) found additional areas of activation in the midbrain and putamen (Lazar et al. 2000).

Three studies measured the difference in brain activity during meditation between meditators and non-meditators. (Lutz et al. 2008); (Brefczynski-Lewis et al. 2007); and (Hölzel et al. 2007) found consistently greater activation during meditation in meditators versus controls in the frontal, parietal, occipital/temporal cortices and hippocampus and, less consistently, greater activation in the cingulate cortex and other subcortical regions (Lutz et al. 2008; Brefczynski-Lewis et al. 2007; Holzel et al. 2007). These authors infer that activation of specific brain areas is related to attention and emotional regulation and that the differential activation between meditators and non-meditators represents a measurable training effect.

Activation Likelihood meta-analysis of studies measuring brain activation during meditation revealed the greatest clusters of activity to be in the left superior frontal gyrus, left medial frontal gyrus, left middle frontal gyrus, and the left precuneus. Areas reported to demonstrate activation in more than one of the individual studies but which did not result in significant clusters of activity on ALE include the cingulate and insula (see Table 2, Figs. 2 and 3). (See Appendix 1 for details of extracted coordinates.)

## Differences in Brain Structure Between Meditators and Controls

Pagnoni and Cekic (2007) found that the total gray matter volume showed a marginally significant negative correlation with age in the control group but not the meditators. Meditators showed significantly less age-related gray matter volume decline in the left putamen than controls (Pagnoni Gand Cekic 2007).

**Table 1** Magnetic resonance imaging studies meeting inclusion criteria for systematic review

| | Author | Year | Meditators/controls | Age range [mean] | Sex | Type of meditation | Duration of practice (mean) | Meta-analysis included/excluded (reason) |
|---|---|---|---|---|---|---|---|---|
| Structural differences between meditators and non-meditators | Luders et al. | 2009 | 22/22 | 30–71 years [53 years] | 9 male/13 female | Varied (Vipassana, Zazen, Samatha, others) | 5–46 years (24 years) | Included |
| | Lazar et al. | 2005 | 20/15 | [38 years] | 13 male/7 female | Buddhist | 9 ± 7 years | Excluded (MNI coordinates not available) |
| | Hölzel et al. | 2008 | 20/20 | [34 years] | 16 male/4 female | Vipassana | 2–16 years (9 years) | Included |
| | Vestergaard-Poulsen et al. | 2009 | 10/10 | [55 years] | 6 male/4 female | Buddhist | 14–31 years (16 years) | Included |
| | Pagnoni et al. | 2007 | 13/13 | [37 years] | 10 male/3 female | Buddhist | > 3 years | Included |

| | | | | | | | | |
|---|---|---|---|---|---|---|---|---|
| Activation during meditation | Shimomura et al. | 2008 | 8/0 | [41 years] | 8 male/0 female | Buddhist | > 10 years | Included |
| | Brefczynski-Lewis et al. | 2007 | 14 expert meditators/16 novice meditators/11 incentivized novice meditators | [47 years] | Not given | Buddhist | 10,000–54,000 total hours | Included |
| | Hölzel et al. | 2007 | 15/15 | [34 years] | 12 male/3 female | Vipassana | 2–16 years (8 years) | Included |
| | Lazar et al. | 2000 | 5/0 | 22–45 years | 4 male/1 female | Kundalini | > 4 years | Included |
| | Short et al. | 2007 | 13/0 | 18–60 years [43 years] | 5 male/8 female | Varied | > 4 years | Excluded (MNI coordinates not available) |
| | Lutz et al. | 2008 | 16/16 | 29–64 years | 14 male/2 female | Buddhist | (45 years) | Included |
| | Khushu et al. | 2005 | 11/0 | 32–65 | 10 male/1 female | Brahmakumari Rajayoga | > 3 years | Excluded (MNI coordinates not available) |
| | Ritskes et al. | 2003 | 11/0 | 32–62 | 8 male/3 female | Zen | 15–25 years (8 years) | Excluded (MNI coordinates not available) |

**Table 2** ALE meta-analysis: brain activation during meditation

| | Author | Year | Meditators/controls | Age range [mean] | Sex | Type of meditation | Duration of practice (mean) | Meta-analysis included/excluded (reason) |
|---|---|---|---|---|---|---|---|---|
| Structural differences between meditators and non-meditators | Luders et al. | 2009 | 22/22 | 30–71 years [53 years] | 9 male/13 female | Varied (Vipassana, Zazen, Samatha, others) | 5–46 years (24 years) | Included |
| | Lazar et al. | 2005 | 20/15 | [38 years] | 13 male/7 female | Buddhist | 9 ± 7 years | Excluded (MNI coordinates not available) |
| | Hölzel et al. | 2008 | 20/20 | [34 years] | 16 male/4 female | Vipassana | 2–16 years (9 years) | Included |
| | Vestergaard-Poulsen et al. | 2009 | 10/10 | [55 years] | 6 male/4 female | Buddhist | 14–31 years (16 years) | Included |
| | Pagnoni et al. | 2007 | 13/13 | [37 years] | 10 male/3 female | Buddhist | > 3 years | Included |

| | | | | | | | | |
|---|---|---|---|---|---|---|---|---|
| Activation during meditation | Shimomura et al. | 2008 | 8/0 | [41 years] | 8 male/0 female | Buddhist | > 10 years | Included |
| | Brefczynski-Lewis et al. | 2007 | 14 expert meditators/16 novice meditators/11 incentivized novice meditators | [47 years] | Not given | Buddhist | 10,000–54,000 total hours | Included |
| | Hölzel et al. | 2007 | 15/15 | [34 years] | 12 male/3 female | Vipassana | 2–16 years (8 years) | Included |
| | Lazar et al. | 2000 | 5/0 | 22–45 years | 4 male/1 female | Kundalini | > 4 years | Included |
| | Short et al. | 2007 | 13/0 | 18–60 years [43 years] | 5 male/8 female | Varied | > 4 years | Excluded (MNI coordinates not available) |
| | Lutz et al. | 2008 | 16/16 | 29–64 years | 14 male/2 female | Buddhist | (45 years) | Included |
| | Khushu et al. | 2005 | 11/0 | 32–65 | 10 male/1 female | Brahmakumari Rajayoga | > 3 years | Excluded (MNI coordinates not available) |
| | Ritskes et al. | 2003 | 11/0 | 32–62 | 8 male/3 female | Zen | 15–25 years (8 years) | Excluded (MNI coordinates not available) |

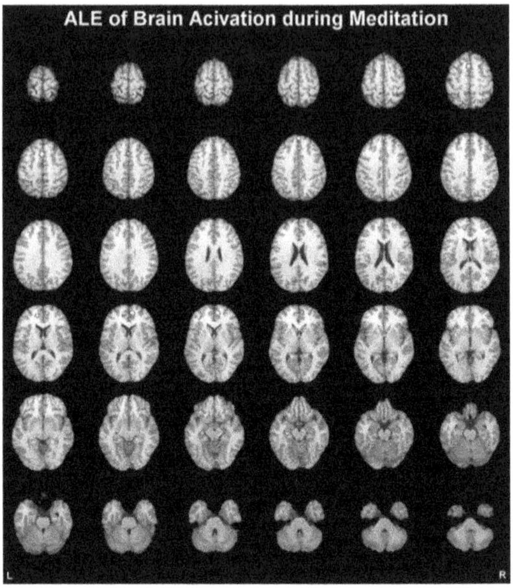


**Fig. 2** Mango images of ALE activation clusters (Table 2): brain activation during meditation – axial view

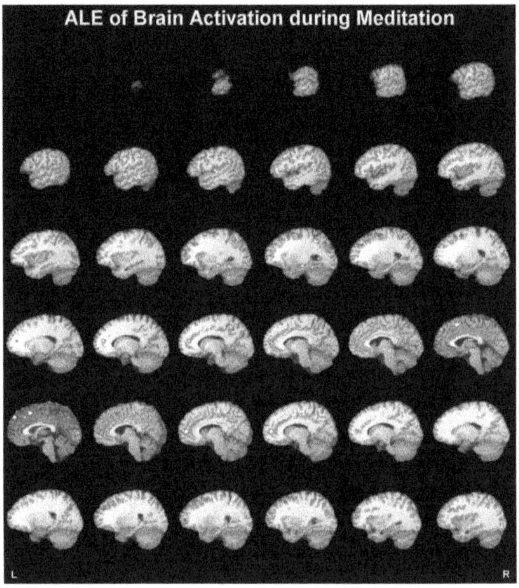


**Fig. 3** Mango images of ALE activation clusters (Table 2): brain activation during meditation – sagittal view

Hölzel et al. (2008) used voxel-based morphometry to measure the gray matter concentration in meditators versus non-meditators. Gray matter concentration (GMC) was found to be greater in meditators than non-meditators in the right hippocampus and the right anterior insula. Concentration in the left inferior temporal gyrus showed a trend toward significance. The GMC of the medial orbitofrontal

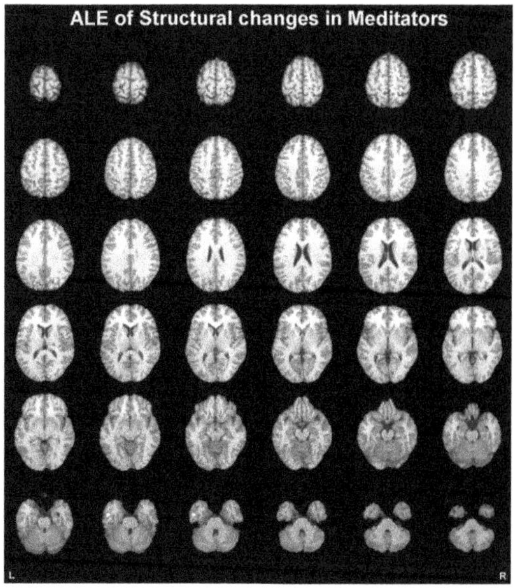


**Fig. 4** Mango images of ALE volume clusters (Table 3): differences in brain structure between meditators and controls – axial view

cortex was positively correlated with hours of practice. There was a positive correlation between total hours of training in meditation and mean GMC within the left inferior temporal gyrus. This suggests a causal relationship (Holzel et al. 2008).

Vestergaard-Poulsen et al. (2009) found greater gray matter density in meditators versus controls in the medulla oblongata, left superior frontal gyrus, left inferior frontal gyrus, anterior cerebellum, and the left fusiform gyrus. Structural differences were also seen in the dorsal medulla gray matter/dorsal motor nucleus of the vagal nerve (regions of autonomic respiratory control and vagal tone). There was no correlation between structure and hours of practice (Vestergaard-Poulsen et al. 2009).

Luders et al. (2009) reported that there were significantly greater gray matter volumes in meditators versus controls in the right orbitofrontal cortex, thalamus, posterior superior parietal lobule, and inferior temporal gyrus as well as the right and left paracentral lobule. Significantly larger hippocampal volumes were also seen. Global cerebral measurements were the same for meditators and controls. Type of meditation practiced had no impact on results. Number of years practiced did not correlate with GM volume (Luders et al. 2009).

In summary, among studies measuring differences in brain structure between meditators and controls, differences are noted most frequently in the frontal and temporal cortices as well as subcortical regions including the hippocampus, thalamus, brainstem, and cerebellum.

ALE meta-analysis of studies measuring the difference between meditators and non-meditators revealed the greatest cluster to be in the left inferior temporal gyrus. The second greatest cluster was found in the brainstem (noted visually) but fell outside of the normalized brain used here. Multiple other areas of differential cluster size were found, but these were much smaller than the two largest clusters (see Figs. 4 and 5, Table 3). (See Appendix 2 for details of extracted coordinates.)

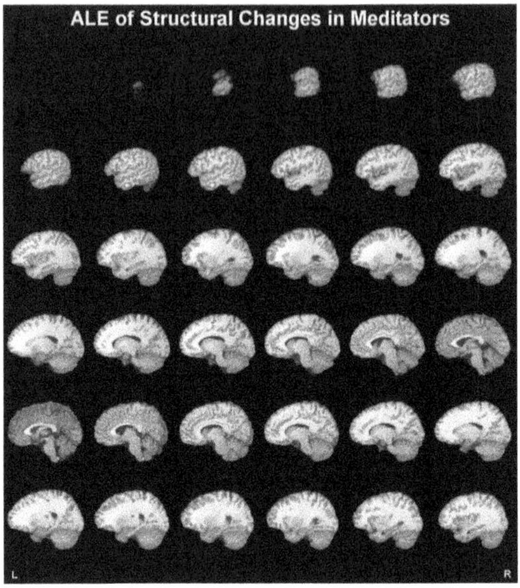


**Fig. 5** Mango images of ALE volume clusters (Table 3): differences in brain structure between meditators and controls – sagittal view

## Discussion

Neuroplasticity refers to the ability of the nervous system to adapt to internal and external stimuli. Magnetic resonance imaging allows assessment of meditation-associated brain activity, differential brain activation in meditators versus non-meditator controls (a training effect), and subtle in vivo structural differences between meditators and controls.

In the studies of brain activation reviewed here, fMRI demonstrates that meditation is associated with activation in a variety of areas, most consistently the frontal and parietal cortices, and the hippocampus. Meta-analysis revealed the most significant clusters of brain activation during meditation to be in the left frontal and parietal cortices. This result is consistent with the individual studies. Authors of the studies reviewed here speculate that meditation is associated with a training effect and that activation in reported areas is associated with the cognitive tasks required during meditation: attention/focus, management of discursive thoughts, and emotional regulation.

Four studies (130 subjects' total) measured structural differences in the brains of meditators and controls (Vestergaard-Poulsen et al. 2009; Holzel et al. 2008; Luders et al. 2009; Pagnoni and Cekic 2007). These studies found greater volumes of cortical, hippocampal, and subcortical gray matter in meditators as compared to controls. Specific structural differences varied between studies. Meta-analysis appeared to adequately summarize individual study findings and demonstrated the largest structural difference to be in the inferior temporal gyrus.

**Table 3** ALE meta-analysis: differences in brain structure between meditators and controls

| | Author | Year | Meditators/controls | Age range [mean] | Sex | Type of meditation | Duration of practice (mean) | Meta-analysis included/excluded (reason) |
|---|---|---|---|---|---|---|---|---|
| Structural differences between meditators and non-meditators | Luders et al. | 2009 | 22/22 | 30–71 years [53 years] | 9 male/13 female | Varied (Vipassana, Zazen, Samatha, others) | 5–46 years (24 years) | Included |
| | Lazar et al. | 2005 | 20/15 | [38 years] | 13 male/7 female | Buddhist | 9 ± 7 years | Excluded (MNI coordinates not available) |
| | Hölzel et al. | 2008 | 20/20 | [34 years] | 16 male/4 female | Vipassana | 2–16 years (9 years) | Included |
| | Vestergaard-Poulsen et al. | 2009 | 10/10 | [55 years] | 6 male/4 female | Buddhist | 14–31 years (16 years) | Included |
| | Pagnoni et al. | 2007 | 13/13 | [37 years] | 10 male/3 female | Buddhist | > 3 years | Included |
| Activation during meditation | Shimomura et al. | 2008 | 8/0 | [41 years] | 8 male/0 female | Buddhist | > 10 years | Included |
| | Brefczynski-Lewis et al. | 2007 | 14 expert meditators/16 novice meditators/11 incentivized novice meditators | [47 years] | Not given | Buddhist | 10,000–54,000 total hours | Included |
| | Hölzel et al. | 2007 | 15/15 | [34 years] | 12 male/3 female | Vipassana | 2–16 years (8 years) | Included |
| | Lazar et al. | 2000 | 5/0 | 22–45 years | 4 male/1 female | Kundalini | > 4 years | Included |
| | Short et al. | 2007 | 13/0 | 18–60 years [43 years] | 5 male/8 female | Varied | > 4 years | Excluded (MNI coordinates not available) |
| | Lutz et al. | 2008 | 16/16 | 29–64 years | 14 male/2 female | Buddhist | (45 years) | Included |
| | Khushu et al. | 2005 | 11/0 | 32–65 | 10 male/1 female | Brahmakumari Rajayoga | > 3 years | Excluded (MNI coordinates not available) |
| | Ritskes et al. | 2003 | 11/0 | 32–62 | 8 male/3 female | Zen | 15–25 years (8 years) | Excluded (MNI coordinates not available) |

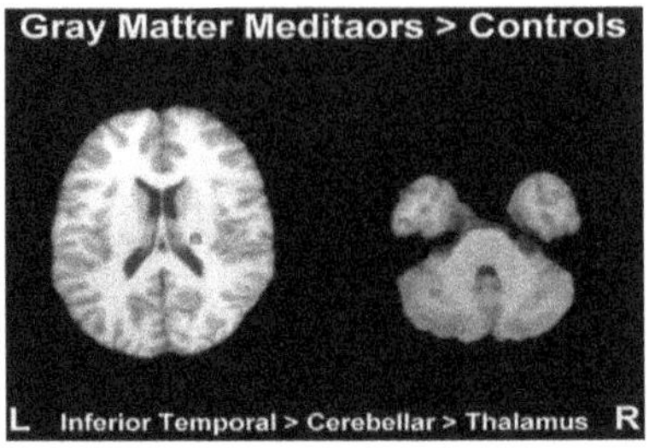


**Fig. 6** Structural differences between meditators and controls: representative volume clusters (Inferior Temporal Lobe, Cerebellum, and Thalamus)

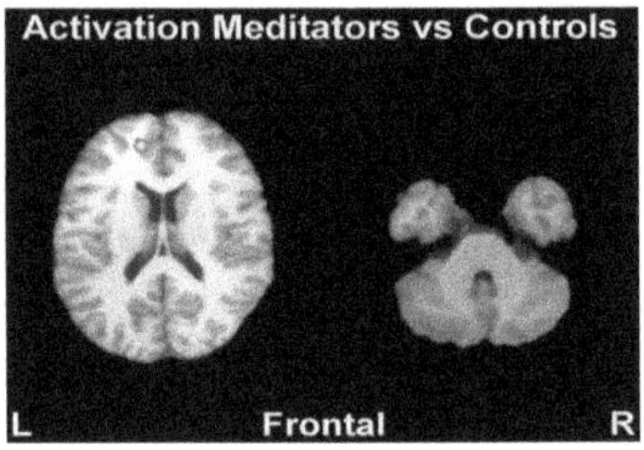


**Fig. 7** Brain activity during meditation: representative activity cluster (*Left Frontal Region*)

In several domains, intensive training has been found to increase gray matter in regions relevant to a learned task – increased gray matter in the posterior intraparietal sulcus (known to be important perceptual-motor coordination and visual attention) in jugglers, for example. In the current analysis, regional activation during meditation (most prominent in the frontal cortex) did not predict structural differences between meditators and controls (most prominent in the inferior temporal gyrus) (see Figs. 6 and 7).

There are several potential reasons for failure of meditation-associated brain activity to predict region-specific structural differences. First, the exact relationship between neural activity and the BOLD signal has yet to be elucidated. Limits in knowledge of precisely what neuronal activity (i.e., excitation or inhibition) is reflected by the hemodynamically-dependent BOLD signal limits the degree to which inferences can be drawn from its measurement. Second, as an extension of this principle, the quantity of information processing by the brain does not always correlate with metabolic activity. Task practice can result in increased performance efficiency and simultaneous decrease in brain activity as measured by fMRI. This results in confusion as to how to interpret fMRI results. For example, in the current review Brefcyznski-Lewis, et al. found that meditators with more hours of practice showed less activation than meditators with fewer hours of practice. They attribute this to a "...quieter mental state, such that tasks...become more effortless" (Brefczynski-Lewis et al. 2007). On the other hand, Short, et al. found that long-term meditators demonstrated more activity in attention-related brain structures than short-term meditators and interpreted this to reflect greater expertise in meditation (Short et al. 2007). Thus, opposite changes in brain activation have been interpreted as reflecting the same thing – greater task proficiency. Until the relationship between brain activity and cognitive processes is better defined, interpreting the cognitive processes represented by fMRI-based brain activity will remain obscure. Third, results from fMRI studies of meditation potentially reflect experimental design more than the cognitive process of meditation itself. Specifically, the switching of attention necessary with the boxcar design used in these studies could be as responsible for activity in attention-related brain structures as meditation itself. Fourth, neuroscience has yet to fully characterize neuroplasticity at the cellular level. Some brain regions and cell types may respond to training with region specific change in volume while others do not. Fifth, it is also possible that neuroplasticity associated with meditation is not related to a region-specific training effect, but rather to a generally healthier physical environment created by another meditation related factor such as down-regulation of the hypothalamic-pituitary-adrenal (HPA) axis. The findings of Pagnoni et al., that total gray matter volume showed a negative correlation with age in controls but not in mediators, supports this notion. Finally, it is important to remember that meditation is a psychological construct. It can be difficult to operationalize and exists potentially without a definable physical locus or network of loci. Whether or not such complicated thought processes can be characterized by localized brain changes has yet to be determined.

This study includes a number of limitations. First, the total number of studies analyzed here is small. Additional data from future experiments has the potential to significantly impact these results. Second, while ALE meta-analysis is appropriate for evaluation of fMRI studies measuring brain activation and MRI based structural studies separately, there is no precedence in the literature for comparing results from different study subtypes. This practice should be considered speculative at this time. Third, meta-analysis depends on the quality and reporting of constituent studies. Studies evaluated here were inconsistent in reporting potentially important covariates such as general health, and medication use.

The notion that meditation could train the brain in such positive mental attributes as attentiveness, self-regulation, and emotional descrimination and that there attributes could be instantiated structurally has profound implications that warrant further exploration. The study of meditation-associated neuroplasticity would be moved forward most expediently by performing a randomized controlled trial of participants before and after introduction of a meditation program. Recent data indicating shorter usage dependent time-scales for structural change makes such studies more feasible than previously recognized. Simultaneous collection of neurophysiological data including measures of attention and emotional regulation would help to corroborate or refute conclusions drawn from neuroimaging studies, and avoid the need (and trap) of reverse inference. While the authors of the studies reviewed here focus on the cognitive process of attention and emotional regulation during meditation, surprisingly little attention is paid to brain activity related to the hypothalamic-pituitary-adrenal (HPA) axis and relaxation during meditation. Building on the neurophysiological model of the HPA axis proposed by (Newberg et al. 2003) would help to identify relevant brain structures and allow correlation of neuroimaging and hormonal test results (Newberg and Iverson 2003). Improved modeling would also help to determine whether possible meditation-mediated neuroplasticity is related to a general improvement in health or to activation-specific stimuli.

In conclusion, training in some domains leads to measurable structural plasticity in the brain. Previous authors have suggested that meditation is a form of mental training which is associated with region specific structural changes. While it was theorized that studies measuring brain activation during meditation would predict region-specific neuroplasticity, findings here refute this hypothesis. This may have more to do with methodology than the refutability of the hypothesis itself. The notion that mental practices can enhance neuroplasticity warrants further investigation. Such exploration promises to suggest new ways of training the brain and promoting its longevity. And, it may resurrect some old methods along the way.

**Acknowledgments** Many thanks to Angela Laird, PhD, University of Texas, San Antonio and her staff for support in the use of GingerALE and Mango; Allison Rollins, Librarian, Uniformed Services University of the Health Sciences (USUHS) for assistance in our literature search; and to Wayne Jonas, MD, and Joan Walter of The Samueli Institute, as well as Roger Gibson DVM, Tomoko Hooper, MD, Cara Olsen, PhD, and Daniel Burnett, MD, of USUHS for their support and direction in this project. This work is supported by Uniformed Services University of the Health Sciences under Contract No MDA 905-03-C-0003. The views, opinions and/or findings contained in this report are those of the author(s) and should not be construed as an official USU position, policy or decision unless so designated by other documentation.

## Appendix 1 Extracted Coordinates: Brain Activation During Meditation

| Author | Year | Hemisphere | Anatomic Region | x | y | z | Z max t | p value |
|---|---|---|---|---|---|---|---|---|
| Shimomura et al. | 2008 | Left | Superior frontal gyrus | 0 | 5 | 62 | 4.71 | <0.05 |
| | | Left | Superior frontal gyrus | −8 | 1 | 63 | 3.89 | <0.05 |
| | | Left | Superior frontal gyrus | −4 | −8 | 67 | 3.41 | <0.05 |
| | | Left | Superior frontal gyrus | −2 | 6 | 51 | 4.14 | <0.05 |
| | | Left | Medial frontal gyrus | −2 | 14 | 49 | 4.56 | <0.05 |
| | | Left | Medial frontal gyrus | 0 | 27 | 41 | 3.82 | <0.05 |
| | | Left | Middle frontal gyrus | −44 | 14 | 42 | 5.2 | <0.05 |
| | | Left | Middle frontal gyrus | −55 | 19 | 29 | 4.05 | <0.05 |
| | | Left | Middle frontal gyrus | −50 | 6 | 44 | 3.49 | <0.05 |
| | | Right | Supramarginal gyrus | 51 | −53 | 36 | 4.95 | <0.05 |
| | | Right | Supramarginal gyrus | 61 | −43 | 33 | 4.12 | <0.05 |
| | | Right | Angular gyrus | 44 | −61 | 33 | 4.14 | <0.05 |
| | | Right | Medial frontal gyrus | 2 | 14 | 47 | 4.59 | <0.05 |
| | | Left | Inferior frontal gyrus | −32 | 35 | 7 | 4.11 | <0.05 |
| | | Left | Inferior frontal gyrus | −44 | 43 | −2 | 3.59 | <0.05 |
| | | Left | Inferior frontal gyrus | −35 | 24 | 6 | 3.58 | <0.05 |
| | | Left | Middle frontal gyrus | −40 | 12 | 51 | 3.82 | <0.05 |
| | | Left | Middle frontal gyrus | −32 | 12 | 55 | 3.72 | <0.05 |
| | | Left | Middle frontal gyrus | −28 | 5 | 55 | 3.61 | <0.05 |

(continued)

(continued)

| Author | Year | Hemisphere | Anatomic Region | x | y | z | Z max | t | p value |
|---|---|---|---|---|---|---|---|---|---|
| Brefczynski-Lewis et al. | 2007 | Left | Middle frontal gyrus/ Inferior frontal gyrus | −49 | 29 | 19 | | 3.2 | <0.005 |
| | | Right | Superior frontal gyrus | 31 | 42 | 31 | | 2.4 | <0.05 |
| | | Left | Middle frontal gyrus/ dorsal lateral prefrontal cortex | −21 | 6 | 50 | | 2.5 | <0.05 |
| | | Left | Rectal gyrus | −0.5 | 43 | −26 | | 3.4 | <0.005 |
| | | Left | Precentral, dorsal lateral prefrontal cortex | −34 | −2 | 36 | | 3 | <0.01 |
| | | Left | Intraparietal sulcus, superior parietal, supramarginal gyrus | −24 | −61 | 46 | | 3.2 | <0.005 |
| | | Right | Superior parietal | 14 | −62 | 54 | | 3.8 | <0.005 |
| | | Right | Cuneus | 22 | −85 | 11 | | 4 | <0.005 |
| | | Left | Middle temporal gyrus, inferior frontal gyrus | −38 | −7 | −26 | | 5.1 | <0.005 |
| | | Right | Middle temporal gyrus | 54 | −12 | −8 | | 3.2 | <0.005 |
| | | | Fusiform | −42 | −55 | −16 | | 3.5 | <0.005 |
| | | Left | Putamen | −30 | −20 | 3 | | 2.8 | <0.01 |
| | | Right | Lentiform, parahippocampus | 29 | −42 | 11 | | 2.9 | <0.01 |
| | | | Cerebellum, declive, culmen | −4 | −56 | −14 | | 3.3 | <0.005 |
| | | Left | Cerebellar tonsil | −22 | −39 | −40 | | 3.3 | <0.005 |
| | | Left | Anterior middle frontal gyrus | −26 | 43 | 7 | | −3.17 | <0.01 |
| Hölzel et al. | 2007 | Left | Anterior cingulate cortex | −12 | 42 | 12 | | 5.11 | 0.002 |
| | | Right | Anterior cingulate cortex | 9 | 48 | 9 | | 3.97 | 0.025 |
| | | Left | Dorsal medial prefrontal cortex | −12 | 45 | 15 | | 4.48 | 0.017 |
| | | Left | Dorsal medial prefrontal cortex | 0 | 48 | 39 | | 4.2 | 0.031 |
| | | Right | Dorsal medial prefrontal cortex | 6 | 51 | 3 | | 4.36 | 0.017 |
| | | Right | Dorsal medial prefrontal cortex | 3 | 48 | 39 | | 4.1 | 0.029 |
| | | Left | Inferior temporal | −51 | −3 | −42 | | 13.43 | 0.000 |
| | | Left | Inferior orbital frontal | −45 | 36 | −21 | | 8.26 | 0.000 |
| | | Right | Cerebellum | 36 | −84 | −42 | | 7.98 | 0.000 |
| | | Right | Rectus | 15 | 21 | −21 | | 7.92 | 0.000 |
| | | Left | Cerebellum | −24 | −93 | −36 | | 7.73 | 0.000 |
| | | Left | Superior medial frontal | −9 | 63 | 6 | | 7.45 | 0.000 |

(continued)

(continued)

| Author | Year | Hemisphere | Anatomic Region | x | y | z | Z max | t | p value |
|---|---|---|---|---|---|---|---|---|---|
| Lazar et al. | 2000 | | Anterior cingulum | 6 | 33 | 0 | | | <0.001 |
| | | | Basal ganglia (putamen) | 28 | −15 | −6 | | | <0.001 |
| | | | Midbrain | −15 | −15 | −15 | | | <0.001 |
| | | | Midbrain | 0 | −12 | −9 | | | <0.001 |
| | | | Parahippocampal gyrus | −25 | −24 | −15 | | | <0.001 |
| | | | Superior frontal gyrus | −6 | 24 | 50 | | | <0.001 |
| | | | Middle frontal gyrus | −40 | 30 | 37 | | | <0.001 |
| | | | Medial frontal gyrus | 12 | 48 | 9 | | | <0.001 |
| | | | Parietal lobule | −21 | −48 | 53 | | | <0.001 |
| | | | Superior parietal lobule | −21 | −63 | 53 | | | <0.001 |
| | | | Superior parietal lobule | −31 | −57 | 53 | | | <0.001 |
| | | | Superior parietal lobule | −28 | −54 | 43 | | | <0.001 |
| | | | Superior/inferior parietal lobule | 40 | −60 | 46 | | | <0.001 |
| | | | Inferior parietal lobule | −34 | −36 | 43 | | | <0.001 |
| | | | Superior temporal gyrus | 59 | −60 | 28 | | | <0.001 |
| | | | Middle temporal gyrus | 59 | −57 | 3 | | | <0.001 |
| | | | Parahippocampal gyrus | −28 | −21 | −12 | | | <0.001 |
| | | | Precentral gyrus | 46 | −12 | 53 | | | <0.001 |
| | | | Postcentral gyrus | −25 | −39 | 62 | | | <0.001 |
| | | | Paracentral lobule | −6 | −33 | 65 | | | <0.001 |
| Lutz et al. | 2008 | L/R | Precuneus | −5 | −55 | 52 | | 5.8 | <0.0005 |
| | | Right | Supramarginal gyrus | 52 | −43 | 37 | | 6.8 | <0.0005 |
| | | L/R | Inferior parietal lobule | 46 | −40 | 47 | | 6.1 | <0.0005 |
| | | L/R | Anterior insula | 37 | 15 | 1 | | 4.8 | <0.0005 |
| | | L/R | Superior temporal sulcus | 54 | −38 | 14 | | 5.7 | <0.0005 |
| | | L/R | Superior temporal gyrus | 54 | 5 | −7 | | 5.2 | <0.0005 |
| | | Right | Superior parietal lobule | 34 | −60 | 46 | | 6.1 | <0.0005 |
| | | L/R | Posterior cingulate gyrus | 6 | −40 | 40 | | 5.5 | <0.0005 |
| | | Right | Middle temporal gyrus | 58 | −47 | −3 | | 6.4 | <0.0005 |
| | | L/R | Parahippocampus | 12 | −39 | 4 | | 5.3 | <0.0005 |
| | | L/R | Fusiform gyrus | 48 | −38 | −17 | | 4.93 | <0.0005 |
| | | L/R | Cerebellum | 16 | −48 | −12 | | 6.1 | <0.0005 |
| | | Right | Anterior cingulate gyrus | 5 | 24 | 37 | | 4.1 | <0.005 |
| | | Right | Medial frontal gyrus | 9 | 6 | 42 | | 4.1 | <0.005 |
| | | Right | Mid frontal gyrus | 22 | 9 | 58 | | 3.4 | <0.005 |
| | | L/R | Brainstem | 3 | −22 | −6 | | | |

## Appendix 2 Extracted Coordinates: Meta-Analysis of Differences in Brain Volume Between Meditators and Controls

| Author | Year | Study measure | Hemisphere | Anatomic region | x | y | z | Z max | t | p value |
|---|---|---|---|---|---|---|---|---|---|---|
| Luders et al. | 2009 | GM volume | Right | Orbito-frontal cortex | 28 | 41 | −3 | | | 0.0001 |
| | | | Left | Inferior temporal lobe | −45 | −8 | −28 | | | 0.0003 |
| | | | Right | Thalamus | 21 | −22 | 14 | | | 0.0003 |
| | | | Left | Paracentral lobe | −12 | −9 | 54 | | | 0.0004 |
| | | | Right | Paracentral lobe | 22 | −23 | 48 | | | 0.0005 |
| Hölzel et al. | 2008 | GM concentration | Right | Hippocampus | 38 | −32 | −12 | | | 0.027 |
| | | | Right | Anterior insula | 36 | 12 | 6 | | | 0.022 |
| | | | Left | Inferior temporal gyrus | −49 | −9 | −28 | | | 0.058 |
| | | | Medial | Orbitofrontal cortex | 1 | 45 | −16 | | | 0.023 |
| Vestergaard-Poulsen et al. | 2009 | GM density | Right | Medulla oblongata | 3 | −37 | −58 | 5.24 | | <0.05 |
| | | | Left | Superior frontal gyrus | −24 | 50 | 4 | 4.29 | | <0.05 |
| | | | Left | Inferior frontal gyrus | −44 | 28 | 6 | 3.96 | | <0.05 |
| | | | Bilateral | Cerebellum | 33 | −59 | −36 | 4.14 | | <0.05 |
| | | | Right | Medulla Oblongata | 5 | −45 | −51 | 4.89 | | <0.05 |
| | | | Left | Medulla Oblongata | 5 | −46 | −51 | 5.49 | | <0.05 |
| | | | Left | Fusiform gyrus | −52 | −19 | −29 | 4.75 | | <0.05 |
| | | | | Cerebellum | −29 | −58 | −37 | 4.14 | | <0.05 |
| Pagnoni et al. | 2007 | GM volume | Left | Putamen | −33 | −5 | 2 | | 5.45 | 0.001 |

## References

Benson, H. 1984. *The relaxation response*. New York: Avon.

Brefczynski-Lewis, J.A., A. Lutz, H.S. Schaefer, D.B. Levinson, and R.J. Davidson. 2007. Neural correlates of attentional expertise in long-term meditation practitioners. *PNAS* 104(27): 11483–11488.

Davidson, R.J., J. Kabat-Zinn, J. Schmacher, M. Rosenkranz, D. Muller, and S.F. Santorelli. 2003. Alterations in brain and immune function produced by mindfulness meditation. *Psychosomatic Medicine* 65: 564–570.

Draganski, B., C. Gaser, V. Busch, G. Schuierer, U. Bogdahn, and A. May. 2004. Changes in grey matter induced by training. *Nature* 427: 311–312.

Gaser, C., and G. Schlaug. 2003. Brain structures differ between musicians and nonmusicians. *Journal of Neuroscience* 23(27): 9240–9245.

Hölzel, B.K., U. Ott, H. Hempel, A. Hackl, K. Wolf, R. Stark, et al. 2007. Differential engagement of anterior cingulate and adjacent medial frontal cortex in adept meditators and non-meditators. *Neuroscience Letters* 421(1): 16–21. doi:S0304-3940(07)00451-X [pii] 10.1016/j.neulet.2007.04.074 [doi].

Hölzel, B.K., U. Ott, T. Gard, H. Hempel, M. Weygandt, K. Morgen, et al. 2008. Investigation of mindfulness meditation practitioners with voxel-based morphometry. *Social Cognitive and Affective Neuroscience* 3(1): 55–61.

Jha, A.P., J. Krompinger, and M.J. Baime. 2007. Mindfulness training modifies subsystems of attention. *Cognitive, Affective, & Behavioral Neuroscience* 7(2): 109–119.

Kabat-Zinn, J., A.O. Massion, J. Kristeller, L.G. Peterson, K.E. Fletcher, L. Pbert, et al. 1992. Effectiveness of a meditation-based stress reduction program in the treatment of anxiety disorders. *The American Journal of Psychiatry* 149(7): 936–943.

Kabat-Zinn, J., E. Wheeler, T. Light, A. Skillings, M.J. Scharf, T.G. Cropley, et al. 1998. Influence of a mindfulness meditation-based stress reduction intervention on rates of skin clearing in patients with moderate to severe psoriasis undergoing phototherapy (UVB) and photochemotherapy (PUVA). *Psychosomatic Medicine* 60(5): 625–632.

Khushu, S., S.S. Kumaran, S. Telles, R.P. Tripathi, K.V. Naveen, and T.L. Matthew. 2005. Functional mapping of human brain during meditation: An fMRI study. *Biomedicine Trivandrum Then Taramani* 25(2): 40–47.

Kochunov, P., J.L. Lancaster, P. Thompson, A.W. Toga, P. Brewer, J. Hardies, et al. 2002. An optimized individual target brain in the Talaraich coordinate system. *NeuroImage* 17(2): 922–927.

Laird, A.R., M. Fox, C.J. Price, D.C. Glahn, A.M. Uecker, J.L. Lancaster, et al. 2005. ALE meta-analysis: Controlling the false discovery rate and performing statistical contrasts. *Human Brain Mapping* 25: 155–164.

Lancaster, J.L., and M.J. Martinez. Multi-image analysis GUI. http://ricuthscsaedu/mango/. Accessed 24 May 2009.

Lazar, S.W., G. Bush, R.L. Gollub, G.L. Fricchione, G. Khalsa, and H. Benson. 2000. Functional brain mapping of the relaxation response and meditation. *NeuroReport* 11(7): 1581–1585.

Lazar, S.W., C.E. Kerr, R.H. Wasserman, J.R. Gray, D.N. Greve, M.T. Treadway, et al. 2005. Meditation experience is associated with increased cortical thickness. *NeuroReport* 16(17): 1893–1897. doi:00001756-200511280-00005 [pii].

Luders, E., A.W. Toga, N. Lepore, and C. Gaser. 2009. The underlying anatomical correlates of long-term meditation: Larger hippocampal and frontal volumes. *NeuroImage* 45(3): 672–678.

Lutz, A., J. Brefcyznyski-Lewis, T. Johnstone, and R.J. Davidson. 2008. Regulation of the neural circuitry of emotion by compassion meditation: Effects of meditative expertise. *PLoS One* 3(3): 1–10.

Machelli, A., J.T. Crinion, J.T. Noppeney, J. O'Doherty, J. Ashburner, R.S. Frakowiak, et al. 2004. Neurolinguistics: Structural plasticity in the bilingual brain. *Nature* 431(7010): 757.

Maguire, E.A., D.G. Gadian, I.S. Johnsrude, C.D. Good, J. Ashburner, R.S. Frackowiak, et al. 2000. Navigation-related structural change in the hippocampi of taxi drivers. *Proceedings of the National Academy of Sciences* 97(8): 4398–4403.

May, A., G. Hajak, S. Gansbauer, T. Steffens, B. Langguth, T. Kleinjung, et al. 2007. Structural brain alterations following 5 days of intervention: Dynamic aspects of neuroplasticity. *Cerebral Cortex* 17: 205–210.

Newberg, A.B., and J. Iverson. 2003. The neural basis of the complex task of meditation:neurotransmitter and neurochemical considerations. *Medical Hypotheses* 61(2): 282–291.

Ospina, M.B., K. Bond, M. Karkhaneh, L. Tjosvold, B. Vandermeer, Y. Liang, et al. 2007. Meditation practices for health: State of the research. *Evidence Report/Technology Assessment* 155: 1–263.

Pagnoni, G., and M. Cekic, 2007. Age effects on gray matter volume and attentional performance in Zen meditation. *Neurobiology Aging* 28(10): 1623–1627. doi:S0197-4580(07)00243-6 [pii]10.1016/j.neurobiolaging.2007.06.008 [doi].

Ritskes, R., M. Ritskes-Hoitinga, H. Stodkilde-Jorgensen, K. Baerentsen, T. Hartman, and S. Denmark. 2003. MRI scanning during Zen meditation: The picture of enlightenment. *Constructivism in the Human Sciences* 8(1): 85–89.

Shimomura, T., M. Fujiki, J. Akiyoshi, T. Yoshida, M. Tabata, H. Kabasawa, et al. 2008. Functional brain mapping during recitation of Buddhist scriptures and repetition of the Namu Amida Butsu: A study in experienced Japanese monks. *Turkish neurosurgery* 18(2): 134–141.

Short, E.B., S. Kose, Q. Mu, J. Borckhart, A. Newberg, M.S. George, et al. 2007 Regional activation during meditation shows time and practice effects: An exploratory fMRI study. http://creativecommonsorg:1–7.

Turkeltaub, P.E., G.F. Eden, K.M. Jonesm, and T.A. Zeffirom. 2002. Meta-analysis of the functional neuroanatomy of single-word reading: Method and validation. *NeuroImage* 16: 765–780.

Vestergaard-Poulsen, P., M. van Beek, J. Skewes, C.R. Bjarkam, M. Stubberup, and J. Bertelsen, et al. 2009. Long-term meditation is associated with increased gray matter density in the brain stem. *NeuroReport* 20(2): 170–174. doi:10.1097/WNR.0b013e328320012a [doi].

# Opening Up Meditation for Science: The Development of a Meditation Classification System

**Stefan Schmidt**

**Abstract** Defining meditation in order to conduct meditation research turns out to be close to impossible. This is, amongst others, related to the fact that the Western term "meditation" entails a wide range of different practices. From this perspective it does not make sense to use the term meditation or to perform research on such a generic concept without any further descriptions. So far contemplative science has taken a pragmatic approach towards this problem, mainly by referencing to certain spiritual traditions. More detailed descriptions of meditation practices are difficult to obtain since meditation is mainly about first person experiences and these experiences cannot be shared adequately. We try to overcome this problem here by developing a four dimensional modular descriptive system of meditation which draws only on behavioral and mental concepts of western psychology with the four dimensions (i) attention regulation, (ii) motivation (iii) attitude and (iv) practical context.

## Introduction: What Is Meditation?

The practice of meditation has overcome the reputation of being a strange spiritual technique related to sitting around and doing nothing mainly conducted by outsiders interested in altered states of consciousness. Nowadays meditation is more and more considered also a secular technique for achieving goals, benefits and virtues needed in our society and in our personal lives. And both, this practice and the related benefits find more and more acceptance in the public. Thus, meditation and mindfulness-based approaches are applied in many different contexts such as in clinical contexts or education.

---

S. Schmidt (✉)
European University Viadrina, Frankfurt (Oder), Germany

University Medical Center Freiburg, Freiburg, Germany
e-mail: stefan.schmidt@uniklinik-freiburg.de

S. Schmidt and H. Walach (eds.), *Meditation – Neuroscientific Approaches and Philosophical Implications*, Studies in Neuroscience, Consciousness and Spirituality 2,
DOI 10.1007/978-3-319-01634-4_8, 

This change in perspective is driven, amongst others, by the results of many scientific studies on meditation and its effects. Science is one of the most dominant practices of our culture to provide meaning and value for our modern society. Meditation came to the West through different paths (see e.g. Nattier 1995 for an account on Buddhist paths to the West) but the ones that are mainly responsible for the start of the scientific inquiry are mainly related to the cultural changes in the 1960s with the Vietnam war and the hippie movement (see Schmidt 2011 for more details). This resulted in a first wave of meditation research in the 1970s which then declined in the 1980s. The second wave followed in the 1990s and was initiated by multiple forces. One was the rise of neuroscience through more developed techniques revealing new insights into the brain and brain processes. The other being that Western sciences finally acknowledged consciousness as a serious topic for scientific inquiry. A third important force was the development of the Mind and Life Dialogues by Adam Engle and Francisco Varela with the Dalai Lama in the late 1980s. Interestingly, meditation practices coming from our western Christian contexts, e.g. out of the Christian mystical tradition or the exercises of Saint Ignatius of Loyola practiced in the Jesuit tradition, had hardly any influence on the scientific inquiry of meditation so far.

The Western scientific approach to any topic of inquiry requires a clear definition of the object to be investigated. Such a definition has the function to enable communication among scientists, i.e. to ensure that everybody is speaking about the same object, construct or concept. But it also has the function to differentiate the object or concept from similar objects and concepts. So the question is, whether there is a clear cut and well-known and accepted definition of what *meditation* is, at least within the context of scientific inquiry? And the answer is no. So far most of the research is based on a pragmatic approach. A person who reports that s/he has had meditation experience for several years is considered a meditator. Often the tradition of that meditation technique is recorded and subsequently reported in the respective publication. Alternatively, novices to meditation are taught certain meditation techniques over a certain time period and then the techniques are named and described in the according scientific report.

On the other hand all researchers active in the field are well aware that there is a multitude of different meditation approaches and techniques, and many of them are subject to scientific inquiry. The questions arising here are: can we consider them to be the same? Can we compare them? Can we even differentiate them? Where is the borderline for considering a certain practice still meditation? What about dancing meditation or body oriented techniques like Hatha Yoga or Tai Chi? The current pragmatic approach to summarize a large variety of techniques under the word meditation would be, according to Lutz et al. (2008), similar to using the generic word sports and thus implying that all kinds of sport are more or less the same.

The problem of this approach is obvious. The more we understand about the differential effects of many different approaches to the mind the more we see the limitation of the pragmatic term meditation (Awasthi 2013). Before I will suggest a new approach to deal with this problem I will give a short overview about other approaches to define and conceptualize meditation. These are differentiated in classical eastern and modern western concepts.

## What Is Meditation in the Ancient Eastern Traditions?

Interestingly there is no single term for 'meditation' in the Eastern traditions but rather a loosely connected array of notions. In the *Śvetāśvatara Upanishads* (after 500 BCE) the three terms of *asana* (sitting), *pratyāhāra* (withdrawl of the senses) and *prānāyāma* (breath restraint), were used, and this already demonstrates that single mental techniques were more in the foreground than an overarching concept as we would expect. Later Upanishads such as the *Maitri* taps the term *dhyāna* (roughly meditation), *dhāranā* (concentration) and *samādhi* (absorption). The later Yoga Sutras uses *dhyāna*, which comes closest to our western notion of *meditation*. Early Buddhist texts use *bhāvanā* (mental development or contemplation) and *jhāna* (from *dhyāna*; which becomes the stages of absorption). Also from the term *dhyāna* comes the Chinese term *ch'an* and then subsequently the Japanese *Zen*. In the Tibetan Buddhist tradition the Tibetan word *sgom* is used in order to describe meditation. A direct translation of this term would be *to familiarize*. This translation nicely demonstrates the view that meditation is a procedure in which one trains the mind in certain attitudes, states and ways of being in order to have these beneficial states also available outside the formal meditation practice.

Alexander Berzin (2009) a scholar in Tibetan Buddhism offers a nice definition of meditation as:

> The repeated practice of generating and focusing on a beneficial state of mind in order to build it up as a habit.

## How Is Meditation Defined in the Western Scientific Context?

Based on the above analysis that there is no unique word in the Ancient Eastern scriptures for what Westerners mean when they talk about meditation, it is no surprise to see that the term meditation itself is also of Western origin. It derives from the Latin *meditari* 'to consider' or 'to think over' and was used in the Middle Ages to describe the continuous mental dwelling on a spiritual or other topic, and later a discourse about a certain subject. Famous examples are Saint Augustines *Soliloquii* or *Talks with Oneself* and Descartes' *Meditationes de prima philosophia*. The relationship to our modern understanding may derive from the fact that such a mental dwelling often was done in solitude.

With respect to modern approaches to define meditation operationally for the use of scientific studies we can find several approaches. Among the many suggestions there are also two proposals for sets of criteria in order to call a technique "meditation". Cardoso et al. (2004, p. 59) call a practice meditation if it

> Utilizes (1) a specific technique (clearly defined), involving (2) muscle relaxation somewhere during the process and (3) "logic relaxation": a necessarily (4) self-induced state, using a (5) "self-focus" skill (coined "anchor").[1]

[1] The use of the term 'logic relaxation' here refers to a practice without analyzing, judging or expecting anything from the process.

One can immediately see the problems resulting in such a list of criteria. For instance what about a meditation induction by a teacher or coming from a CD? And has meditation necessarily to be linked to muscle relaxation? A Canadian group published in 2007 a large health technology assessment report on meditation based interventions (Ospina et al. 2007). In order to have inclusion and exclusion criteria for that report they defined a set of 'demarcation criteria' for meditation in a Delphi-process with experts (Bond et al. 2009). These iterative expert interviews resulted in three essential criteria: (1) a defined technique, (2) logic relaxation, (3) a self-induced state/mode, as well as five important but not essential criteria: (4) involving a state of psychophysical relaxation somewhere in the process, (5) using a self-focus skill or anchor, (6) involving an altered state/mode of consciousness, mystic experience, enlightenment or suspension of logical thought processes, (7) in a religious/spiritual/philosophical context, (8) involving an experience of mental silence. Also with these criteria problems regarding specificity will come up. The three essential criteria would also apply if somebody plays the guitar in a relaxed mood just for her private pleasure or if somebody jogs. Also with all these criteria I see some conceptual mixing up of the intentions of the meditation and potential results of it. E.g. some of the additional mystical experience or mental silence may be the consequence of meditation but they are not necessary conditions in order to call a practice meditation. A third definition by Jaseja (2009, p. 483) I would like to mention tries to conceptualize meditation solely on a neuroscientific account, i.e. exclusively from a third person perspective:

> …a complex neural practice that induces changes in neurophysiology and neurochemistry of brain resulting in altered neurocognition and behavior in the practitioner.

The shortcoming of such a definition is obvious and demonstrates the general problems of a third person approach towards mental techniques relating to inner experience. Here also the intended consequences of the practice are already included and again the definition does not fulfill the criteria of specificity. Getting deliberately drunk would comply with this definition as well.

One of the most cited definitions of meditation was made by Dean Shapiro in 1982 and refers especially to the process of attention regulation within meditation. We will come back to that definition in the following paragraph where it will be used as a minimal description for the development of a more refined meditation classification system.

Overall it can be seen that it is almost impossible to draw a straight 'demarcation' line which separates the practice of meditation from other techniques. And with such a wide scope of different approaches included it may not even make sense to do so. Let's think again about Lutz's et al. (2008) analogy with sports. Does it make sense to define sports in order to study rowing or playing chess? Certainly not. With a growing body of empirical research regarding meditation we see more and more that different meditative approaches lead to different effects. So what is more needed than a definition is a way to differentiate and describe different meditative approaches.

If we look how researchers approach this topic, we see again a pragmatic solution. Often scientists just name the respective traditions in which their participants were trained or which they applied as an intervention in their studies. The description of these respective meditation techniques then refer mostly to the concepts of the context they were taken from. If those are spiritual then references to this spiritual system are made, but often (as e.g. in the program mindfulness-based stress reduction, MBSR) these techniques are secularized and then just the techniques which were taught in the course are shortly described. In the end it all looks similar but in fact we hardly know, whether meditators participating in one study have used similar or rather different techniques to those in other studies. A good example is the narrative review on morphometric studies comparing the brain structures of experienced meditators with non-meditating controls by Ott et al. (2011). At that time the authors found five such publications which studied *Insight Meditation, Zen, Vipassanā, Tibetan Buddhist Meditation,* or *Zazen, Vipassanā, Samatha and others*. Overall differences between meditators and non-meditators were found in seven different brain structures, but in no case more than two studies reported the same structure. Four structures were only reported by one study. The interesting question here is where do these differences come from? Can they be tracked to specific tasks and practices within these different meditation techniques?

In order to approach such questions in more detail it seems advisable not to insist on a definition which separates meditation from other practices but rather to develop a system to describe in detail what a certain practice entails. Important here seems to be that we no longer draw on notions and descriptions stemming from spiritual traditions. If a meditation practice is described as *Vipassana* or *Zazen* we do not know what exactly the meditators have done during the meditation session until we have ourselves a deep understanding and practice of the respective tradition. Furthermore, there are often also differences in practices within a certain tradition. For example the description *Tibetan Buddhist Meditation* encompasses a multitude of different meditative approaches with a wide range including even unusual techniques such as analytical meditation. Often the ways these techniques are taught is also dependent on the respective teacher and we cannot assume that everybody trained in any of these techniques is exactly practicing the same from a (neuro-) scientific point of view.

Moreover, we cannot even assume that all meditators taught by a certain teacher in a certain meditation technique are practicing the same. The practice of these students may vary in relation to their personal abilities and capacities, their understanding of the method taught and will also relate to the individual learning history of each person. In a qualitative study we have interviewed several participants of an MBSR course held by the same teacher. One aspect of that investigation was how each of the participants understood the concept of *mindfulness*. We could find a wide range of different conceptualizations and understandings. Interestingly, many aspects of these individual concepts could be meaningfully interpreted when the individual biography and background (job, family situation, life situation etc.) was taken into account (Fattah 2009). What can be seen from this example is that learning always means to connect something new with something existing (novelty and confirmation) and thus is different for every person.

## A Classification System for Meditation

Based on the above considerations I suggest developing a meditation classification system. The goal of this approach is to be able to describe a given meditation practice in purely descriptive terms. These terms should reference behavioral descriptions (what is the person actually *doing*?) in a wider sense and this will also include mental 'activities'. The reference frame for these descriptions should be the Western scientific context, more specifically the field of psychology. This means in turn that references to traditional philosophical or spiritual systems and the respective terms should be avoided. This orientation towards western psychology does not imply any hierarchy between different contexts or philosophical systems. It just expresses that contemplative science is conducted within a context and reference frame of the western scientific paradigm. Thus, it should be possible to describe the object of inquiry adequately and comprehensively within this reference frame.

If we imagine a person being in meditation and we want to describe the situation as sketched above, it is obvious that there are different classes of description. A good starting point is the already mentioned statement by D. H. Shapiro (1982, p. 268) which describes the activity of meditation as follows:

> Meditation refers to a family of techniques which have in common a conscious attempt to focus attention in a non-analytical way, and an attempt not to dwell on discursive, ruminating thought.

The core aspect of this description (next to the wise use of the word 'attempt') is the conscious approach towards attention. The expression to *focus* attention may be too limiting and the term *attention regulation* might be more appropriate here (see e.g. Bishop et al. 2004), since not all forms of meditation rely on focused attention. So one class of descriptors that needs to be addressed is how the meditator deals with or regulates his or her attention during the meditation.

Another important class is the question why is the person doing this? What is his or her *motivation* to engage in this particular practice? It was shown elsewhere that the motivation to practice meditation is a crucial aspect of the practice itself (Schmidt 2011; Shapiro et al. 2006; Walsh and Shapiro 2006) which in turn determines many other aspects of this process such as experience or outcome (Shapiro 1992). So the second category of description has to describe the *motivational component of the meditation practice*. Slightly related to this aspect is the category of *attitude*. This aspect is also addressed in the approaches to conceptualize mindfulness in terms of western psychology by Bishop et al. (2004), Shapiro et al. (2006), and Shapiro and Schwartz (1999). By attitude we mean a certain mental stance a person is adopting during his or her meditation practice. Examples are openness, curiosity or warm-heartedness. A fourth category refers to the *context of the meditation* situation. Is the person sitting, lying or walking? Are there external instructions? From a teacher or prerecorded? Is the meditation performed in a group or by the person him- or herself? These seem to be trivial aspects regarding the inner experience but should be nevertheless considered since these context variables are known to limit or to enable certain processes or experiences.

Thus, overall we can identify the following four categories of description: *Attention regulation, motivation, attitude* and *practical context*. This list is not necessarily complete and I consider it necessary for a good descriptive system that it is open to adaptations as well. This might be necessary whenever some meditation approach is raising aspects which cannot be covered by the above mentioned categories.

## *Attention Regulation*

Attention regulation with respect to meditation can be assessed from a first person perspective as well as from a third person perspective, since there is a multitude of methods to assess attentional performance and capacities. But for the development of a classification system it will be necessary to address the first person perspective which can only be experienced and reported by the meditator herself.

The most frequently addressed modes of attention regulation refer to the differences in the Buddhist approaches of *Samādhi* and *Vipassanā*. Samādhi stands for a meditation technique with a very narrow and highly selective focus of attention. Here, the meditator should only attend to the object of meditation and try not to get distracted. This procedure is mostly practiced in order to calm the mind (quiescence), and allows, if practiced regularly, to reach deep states of consciousness (Shankman 2008; Solé-Leris 1986; Wallace 2006). Vipassanā on the other stands for 'insight' and is often also translated as 'Insight Meditation'. Here the meditator should observe the nature of all phenomena coming into his or her mental space, with the overarching aim of receiving insights into the 'true nature'[2] of these phenomena. So, in contrast to Samādhi there is no such thing as distraction in Vipassanā. Rather the idea is to be aware of every phenomenon coming into the focus of attention and to observe it with detachment. In order not to get carried away, usually an anchor or object, for instance the breath, is used: The meditator can return to this anchor whenever s/he notices that his or her mind was wandering. This form of attention regulation with a wider focus can be practiced within a certain phenomenal area, e.g. with emotions, thoughts, or bodily sensations (e.g. in the body-scan). But in its most extreme form it can be applied to all phenomenal realms at once (sometimes called 'choiceless awareness'). In a more secular terminology these two different ways of paying attention within meditation are often also described as *concentrative* and as *mindfulness meditation* (Dunn et al. 1999; Kornfield 1993; Shapiro 1982; Valentine and Sweet 1999). An even more descriptive terminology for the use of contemplative science was suggested by Lutz et al. (2008) with the terms *focused attention* and *open monitoring*. The conceptualization of these terms is slightly different and is not discussed here in detail. However, this often reported dichotomy

[2] See the chapter of Edge (this volume) for a refined discussion of the epistemological status of such an assertion.

between two major attentional strategies is somewhat misleading, since it implies that they are completely different and independent. But the opposite is true.

An unfocused attention to all mental or other content, such as in *Vipassanā* meditation, presupposes the capacity to focus, else mindfulness is lost quickly. In that sense mindfulness presupposes focused attention. In the same sense, attention-focusing meditation techniques also foster mindfulness, as the capacity to stay present generalizes ideally to all events and situations.

Furthermore, in a three dimensional descriptive model of consciousness Walach (2011) also uses the dimension concentration vs. expansion. He demonstrates that states of deep concentration and quiescence have also a quality of expansion and argues that such states can be reached via both strategies – those of expanding and those of concentrating the mind. This makes clear that we need a more refined description of attention regulation processes in order to arrive at the above mentioned aim.

Here is a fresh start. The most predominant feature of attention regulation every meditator is well aware of is that attention lapses. Attention is sometimes described as a semi-automatic process. While we are able to direct it willfully to a focus of our choice we don't have full control about our attention. For an untrained person the process of attention regulation cannot be maintained over time. The most basic level of assessment regarding attention regulation is whether the meditator is currently willfully regulating her attention to a focus or whether the attention has lapsed and is driven by more autonomous and unregulated processes. Wendy Hasenkamp (this volume but see also Hasenkamp and Barsalou 2012; Hasenkamp et al. 2012) has conceptualized the cyclic process of attention regulation during a meditation with fixed focus into the four stages (a) *focus* (b) *mind wandering* (lapsed attention), (c) *aware* of the moment the meditator realizes that he or she has lost focus, and (d) *shift,* meaning the process of shifting the attention back to the selected task or focus. Based on this model she can differentiate neural correlates for each of these four stages (Hasenkamp, this volume). The major difficulty of this approach is that we are by definition not aware of stage (b) and also not of the shift from stage (a) to (b). In many studies participants are asked to press a button whenever they became aware of mind wandering or lapsed attention and shift their attention back (stage c and d) (Burg and Michalak 2011; Schmidt 2012; Hinterberger this volume; see also Mrazek et al. this volume for methods of the assessment of mind wandering). Such measures are somewhat conclusive especially if the meditator is prompted from time to time and asked whether her attention in this specific moment was distracted or not as Burg and Michalak (2011) do. However, if a meditator only presses the button once during a 20 min meditation session this could be an indicator of a well-focused mind or the product of 20 min of unwanted mind-wandering with just one short moment of awareness. Furthermore it could be shown that the individual threshold when exactly to press the button can vary in different cultural contexts (Schmidt 2012).

Once a basic attentional stability can be maintained as a prerequisite for a deeper meditation practice more dimensions of attention regulation are at hand. These may be the object or focus of attention, the width of the attentional focus, the stability of attention, the mode to deal with distractions, the mental faculty in which attention regulation takes place and whether the attention focus changes during that respective meditation practice (or stays the same). We have developed a detailed questionnaire for the retrospective assessment of these dimension immediately at the end of a given meditation session. The development of this scale took place in an iterative process. At the beginning we conducted several expert interviews regarding this topic (B. Alan Wallace, Fred v. Allmen, Jonathan Shear) which were recorded and served as a basis to construct a first questionnaire. Next, this questionnaire was filled in by several meditators from different traditions which gave us extensive feedback. After several discussions of this feedback within our team we developed an improved version of the questionnaire and this procedure was continued several times. In the final version we assess the following dimensions with 12 questions.

- *Attitude* (item 1 and 2). Here the meditator is asked whether at the beginning of the session she is intentionally getting aware of an initial attitude (item 1) and whether she is intentionally adopting a certain attitude (item 2). Answer format for both items yes – no.
- *Mental Faculty* (item 3) Here the mediator is asked whether the practiced meditation technique is predominantly within a certain mental faculty (multiple choice: emotional, cognitive, body-oriented, visualization, memories, others (please name)).
- *Object of meditation* (item 4). Here the meditator is asked whether she chooses a central object for her meditation (multiple choice: yes (one central object), yes (several objects), no). Furthermore several objects are offered (multiple choice, multiple answers possible: breath, body or body sensations, object of the external world, visualization (of an object), mantra, koan, sound, other (please name)).
- *Breath* (item 5). If 'breath' in item 4 was chosen the meditator is asked to indicate where exactly the focus on the breath is taken (multiple choice: nostrils, tip of the nose, upper lip, breast, belly, the whole area were the breath can be felt).
- *Breath regulation* (item 6). The meditator is asked whether she is regulating her breath during the session (VAS, anchors no regulation, strong regulation).
- *Maintaining the same focus* (item 7). Here the meditator is asked whether she stays with the same focus of attention during the whole session (yes – no).
- *Width of attention focus* (item 8). Here a short explanatory text on what is meant by width of attention focus is presented (two sentences). Then the meditator is asked on the width of the predominant focus (VAS, anchors narrowest possible focus – widest possible focus). In the second question the meditator is asked to indicate on the same VAS an area in which she varies her focus, if she does so intentionally during her meditation.
- *Stability of attention* (item 9). Here the meditator is asked for the ability to keep the attention stable on the chosen focus (VAS, anchors no all – very good).
- *Effort* to maintain focus (item 10). Here the meditator is asked regarding the effort made in order to maintain the focus of attention (VAS, anchors no effort at all – very strong effort).

- Strategy to deal with mental *distractions* (item 11). Here the meditator is asked how she is dealing when she notices that other sensations, emotions or thoughts are showing up. Five strategies are offered ranging from immediate return to focus of meditation up to choose the new content as a new focus (separate five step Likert scale for all five strategies).
- *Meditation without object* (item 12). Finally the meditator is asked whether she is meditating without any inner or outer object/focus of her meditation (yes – no). If she answers with yes there is space for a short verbal description aiming at lay persons.

With respect to a third person perspective on attention regulation during meditation a large body of recent research results is available. There are two major models for the conceptualization of attention. One is the two factor model by Corbetta et al. (Corbetta and Shulman 2002; Corbetta et al. 2008) based on the two neurophysiological mechanism of a dorsal and ventral attention system (see Austin, this volume for a detailed description connecting with meditation practice). The other one is the more psychological and functional based system by Poser et al. (Fan et al. 2002a, 2005). Here attention is conceptualized in three different networks termed alerting, orienting and executive attention.

In many of these studies (but not in all) a positive relationship between meditation experience and various indicators of attentional performance can be found (Brefczynski-Lewis et al. 2007; Chan and Woollacott 2007; Jha et al. 2007; Lutz et al. 2008; Tang et al. 2007; Valentine and Sweet 1999; van den Hurk et al. 2010). However, please note that most of the paradigms applied here rely on reaction time based assessments. This is a rather indirect assessment of mental capacities developed through meditation since meditation practice mostly does not imply aspects of reacting quickly. The form of continuous attention regulation practiced in almost all meditations comes closer to what is in psychology termed *sustained attention* or *vigilance* (Warm et al. 2008). Thus, future research should point more into this direction and develop the adequate tests. A good example can be seen in the work of MacLean et al. (2010).

## Motivation

Possible motivations to practice meditation have a wide range today. While meditation was initially only performed in a spiritual context we have now also secular practices. These secular contexts entail many new motivations which are sometimes related to the different areas of application in which meditation is now explored, e.g. clinical, education, or work related contexts. I have argued elsewhere that the respective motivation of meditation practice is crucial in order to understand the meditation process in total since it displays the major orientation of the whole practice (Schmidt 2011). Dean Shapiro demonstrated in a study on long term meditators that the initial goals of the meditators are related to the respective outcome of their practice and he coined the phrase "what you get is related to what you want" (Shapiro 1992, p. 25). He describes the various intentions to meditate on a continuum ranging from *self-regulation* (e.g. calming, relaxation, stress reduction) via *self-exploration* (e.g. studying one's psychological patterns) to *self-transformation* which is referring to spiritual motives.

One needs to consider that there will be several layers or levels of motivation regarding time range and specificity of the desired goal. Imagine a person who has learned meditation out of a spiritual motivation. After a busy and socially very demanding meeting or conference day this person deliberately retreats from the next social activity in order to practice some silent meditation with the short term goal of relaxation and regaining an inner balance. This example shows that short-term and long-term goals can be different and both will be important to understand a specific meditation practice.

With respect to long-term goals and primary motivation of meditation we have developed a meditation motivation scale (Netz 2012; Schmidt and Netz 2011) in order to assess empirically this category of the meditation classification system.

Initially a large range of meditators was asked for their motivation to meditate in order to retrieve items on an empirical basis. Based on their answers we constructed a first scale in German consisting of 58 items offering a broad range of motivations. These items had to be rated on a five point Likert scale and were all semantically transformed into statements of the form: "I am meditating because …..". In an online survey 549 meditators filled in the scale. First, we performed a psychometric assessment. During this process 27 items were discarded because of low variance, low discriminatory power, skewed distributions or high/low item difficulty. In the next step we performed an exploratory factor analysis (Varimax) with orthogonal scale rotation. We finally arrived at a scale with 31 items and four factors. These were able to explain a total of 55.3 % of variance with an internal consistency of Cronbach's $\alpha = 0.90$. Motivations could be grouped into four distinct factors (see Table 1).

One can see from this table that most of the variance is explained by secular motives. But this does not necessarily mean that these motives are exclusively secular. As could be seen in the above example it may be possible that a primary spiritual motivation also encompasses secular aspects. However, a predominantly secular motivation, on the other hand, will rarely include spiritual motives as well. Also one needs to consider that these long-term goals assessed here might change over time. In our cross-sectional study we have tried to assess this aspect by correlating the four factors with meditation experience (in months). There was a negative correlation with the first factor wellbeing ($r = -0.22$, $p < 0.01$) and a positive correlation with the fourth factor self-transformation ($r = 0.20$, $p < 0.01$). The other two factors were uncorrelated. This indicates indeed that people with longer meditation experience have stronger spiritual motivation compared to more motivation for relaxation and wellbeing, which is more prevalent in short term meditators. However no causal inference can be drawn from such a cross-sectional study.

This scale can be used for assessing and describing the motivation and long-term goals of meditators participating in a meditation study.

**Table 1** Overview on the factor structure of the motivation meditation scale

| Factor | Title | Marker item | Items | %Variance explained | Cronbach's alpha |
|---|---|---|---|---|---|
| All | Motivation for meditation | I am meditating because… | 31 | 55.3 | 0.897 |
| 1 | Wellbeing | …I want to feel good | 9 | 17.6 | 0.893 |
| 2 | Emotion-regulation | ….to deal with pain | 9 | 14.3 | 0.895 |
| 3 | Self exploration | … to see who I am in this particular moment | 7 | 12.5 | 0.845 |
| 4 | Self-trans-formation | …to reach enlightenment | 6 | 11.0 | 0.813 |

## *Attitude*

Meditation as an intentional activity is often practiced by adopting a certain attitude (e.g. openness, joy, curiosity etc.) at the beginning of the session, and in our questionnaire we explicitly ask about this aspect. However, setting out with a certain attitude does not allow for the conclusion that this attitude was also maintained during the session. On the other hand there is often the discussion whether certain attitudes (e.g. acceptance or equanimity) are a result of a certain meditation practice and develop as a consequence of meditation, or whether they need a more active and intentional cultivation. With respect to the meditation classification system especially the latter aspect is of interest while the former one is perhaps more adequate for assessment systems targeting meditation effects (see e.g. Reavley and Pallant 2009). In the section on meditation definitions I have demonstrated that the Eastern perspective of familiarizing the mind with a certain state of mind is very fruitful. According to this view certain qualities, like for instance compassion, can be cultivated by adopting them during a respective meditation.

In Western approaches mainly Jon Kabat-Zinn (1990) and Shauna Shapiro (Shapiro and Schwartz 1999; Shapiro et al. 2006) have addressed the issue of adopting an attitude during meditation (see also Schmidt 2004). These intentional attitudes can be easily assessed by providing meditators with a list of possible attitudes and some space to add more individual ones. We are currently developing such a list by identifying terms which show some semantic consistencies with regard to meditators coming from different meditation traditions.

## *Practical Context*

Compared to dimensions only accessible by the meditator herself, such as attitude or attention regulation strategies, the practical context of a given meditation situation is easier to assess. The justification for adding these aspects to the classification system is that many of these aspects influence meditation practice, such as the sitting position, or whether eyes are closed, but are also important with respect to science. One can also imagine that this dimension may contain important confounding variables relevant to research, which would go unnoticed otherwise.

We constructed a short assessment tool again in an iterative process with the help of many meditators. This asks for the following contextual factors (below). All questions are presented in multiple choice fashion with an open category for free text. In the beginning we also assess meditation experience in a structured way in order to allow for an estimate of total meditation experience in months and for an average daily time period (in minutes) maintained within the last 8 weeks.

- Body position (sitting/lying/walking/standing/other)
  - If sitting, which position
  - If lying, which position
  - If walking, which walking speed
- Eyes (open/closed)
  - If open, focused on an object (yes/no)
- Setting (group/alone)
- Instructions (none/by teacher/by CD)
  - If CD music as background (yes/no)
- Timing of meditation session (external timer/self-timed)
- Location (at home/meditation center/other)
- Regular meeting a meditation teacher? (yes/no)
- Attending meditation groups? (yes/no)
- Is your practice aligned to a specific meditation tradition (yes/no)
  - If yes, please name tradition as detailed as possible.

## Applications

The described system is modular in its outlay and purely descriptive. Dimensions can be added or dropped with respect to the objectives of the research. Also it is open to changes in the sense that questions asking for important information which are missing can easily be added. However, this is not true for the motivation dimension which is assessing the construct of meditation motivation psychometrically using multiple items to enhance reliability of assessment.

The system can be applied to assess meditation practice in general. For instance in a specific study on the effects of meditation experience on attention regulation one could assess the predominant meditation practice of the participants within the last week or the last 3 months with this tool. Then several aspects could be correlated with (neuro-)physiological or behavioral data. One can imagine that meditators performing routinely body-oriented mindfulness techniques would likely show a different neurophysiological pattern than others who rely predominantly on visualization techniques.

A more specific approach targets only one specific session. In this case only some slight reformulations need to be made in the introductory parts of the single questionnaires. Such an approach could be interesting for example in a neuroscientific state study assessing different brain activities in relation to different attentional strategies (wide vs. narrow focus).

Another approach would be to develop descriptive profiles for certain meditation techniques. One can imagine that the profile of somebody performing a body-scan (changing focus, focus body, wide attention, mental faculty body-oriented, relaxed strategies with distraction etc.) will be different from a metta meditation (i.e. loving kindness meditation) using metta sentences (constant focus, focus mantra, narrow attention, mental faculty emotional, tight strategies with distractions etc.).

Our first application of the description system was in a study which investigated whether the regular practice of a wide or narrow focus could be related to different performances regarding the three attentional networks of alerting, orienting, and executive attention (Markowiak 2011). In this study 24 meditators with an average meditation experience of 12.9 years and intensive practice within the last 8 weeks were compared with a sex and age matched control group regarding their performance in the Attention Network Test (ANT) (Fan et al. 2002b). Additionally we correlated the width of the attention focus assessed by our meditation description system with the performance in the three attention networks. There were no substantial correlations for the orienting and alerting network. In the executive attention network meditators significantly outperformed controls. These results were slightly improved with a narrower focus although the correlation of $r=0.34$ was not significant for the sample.

Overall this newly developed system seems to be feasible but its empirical relevance needs still to be demonstrated.

## Conclusion

The meditation classification system proposed here was constructed with the idea to overcome the problem of defining the term meditation operationally and to stimulate more research looking into the details of meditation practices and techniques rather than taking meditation as a monolithic unit. Whether it is really possible to grasp the key differences in meditation practices by such a descriptive system has still to be explored. The principal problem that we cannot share our experiences directly (see Schmidt and Walach in this volume), while meditation is all about experience remains. It might be that the classification system is not fine grained enough to get the relevant issues or it might even be that this approach via verbal formulations is inadequate. But all this can only be found out by empirical research. Meditation has finally arrived in science, now it is time to build the adequate tools to assess it.

**Acknowledgments** I would like to thank Robert Forman for help in clarifying the different Eastern notions referring to contemplative practices.

# References

Awasthi, B. 2013. Issues and perspectives in meditation research: In search for a definition. *Frontiers in Psychology* 3: 613. doi:10.3389/fpsyg.2012.00613.

Berzin, A. 2009. The Berzin archives. Glossary English terms. http://www.berzinarchives.com/web/en/about/glossary/glossary.html#xM. Accessed 24 June 2013.

Bishop, S.R., M. Lau, S. Shapiro, L.E. Carlson, N.D. Anderson, J. Carmody, et al. 2004. Mindfulness: A proposed operational definition. *Clinical Psychology: Science and Practice* 11: 230–241.

Bond, K., M.B. Ospina, N. Hooton, L. Bialy, D.M. Dryden, N. Buscemi, D. Shannahoff-Khalsa, J. Dusek, and L.E. Carlson. 2009. Defining a complex intervention: The development of demarcation criteria for "meditation". *Psychology of Religion and Spirituality* 1: 129–137.

Brefczynski-Lewis, J.A., A. Lutz, H.S. Schaefer, D.B. Levinson, and R.J. Davidson. 2007. Neural correlates of attentional expertise in long-term meditation practicioners. *Proceedings of the National Academy of Science* 104(27): 11483–11488.

Burg, J.M., and J. Michalak. 2011. The healthy quality of mindful breathing: Associations with rumination and depression. *Cognitive Therapy and Research* 35(2): 179–185.

Cardoso, R., E. De Souza, L. Camano, and J.R. Leite. 2004. Meditation in health: An operational definition. *Brain Research Protocols* 14(1): 58–60.

Chan, D., and M. Woollacott. 2007. Effects of level of meditation experience on attentional focus: Is the efficiency of executive or orientation networks improved? *Journal of Alternative and Complementary Medicine* 13(6): 651–657.

Corbetta, M., and G.L. Shulman. 2002. Control of goal-directed and stimulus-driven attention in the brain. *Neuroscience* 3: 201–215.

Corbetta, Maurizio, G. Patel, and G.L. Shulman. 2008. The reorienting system of the human brain: From environment to theory of mind. *Neuron* 58(3): 306–324. doi:10.1016/j.neuron.2008.04.017.

Dunn, B.R., J.A. Hartigan, and W.L. Mikulas. 1999. Concentration and mindfulness meditations: Unique forms of consciousness? *Applied Psychophysiology and Biofeedback* 24(3): 147–165.

Fan, J., B.D. McCandliss, T. Sommer, A. Raz, and M.I. Posner. 2002a. Testing the efficiency and independence of attentional networks. *Journal of Cognitive Neuroscience* 14(3): 340–347. doi:10.1162/089892902317361886.

Fan, J., B.D. McCandliss, T. Sommer, A. Raz, and M.I. Posner. 2002b. Testing the efficiency and independence of attentional networks. *Journal of Cognitive Neuroscience* 14(3): 340–347.

Fan, J., B.D. McCandliss, J. Fossella, J.I. Flombaum, and M.I. Posner. 2005. The activation of attentional networks. *NeuroImage* 26(2): 471–479.

Fattah, S. 2009. Kommunikative Konstruktion von Achtsamkeit in Erzählinterviews. Qualitative Studie mit Teilnehmern eines MBSR-Kurses. Unpublished thesis, Institute for Psychology, University Freiburg.

Hasenkamp, W., and L.W. Barsalou. 2012. Effects of meditation experience on functional connectivity of distributed brain networks. *Frontiers in Human Neuroscience* 6: 38. doi:10.3389/fnhum.2012.00038.

Hasenkamp, W., C.D. Wilson-Mendenhall, E. Duncan, and L.W. Barsalou. 2012. Mind wandering and attention during focused meditation: A fine-grained temporal analysis of fluctuating cognitive states. *NeuroImage* 59(1): 750–760.

Jaseja, H. 2009. Definition of meditation: Seeking a consensus. *Medical Hypotheses* 72(4): 483. doi:10.1016/j.mehy.2008.11.015.

Jha, A.P., J. Krompinger, and M. Baime. 2007. Mindfulness training modifies subsystems of attention. *Cognitive, Affective and Behavioral Neuroscience* 7(2): 109–119.

Kabat-Zinn, J. 1990. *Full catastrophe living: Using the wisdom of your body and mind to face stress, pain, and illness.* New York: Bantam Dell.

Kornfield, J. 1993. The seven factors of enlightment. In *Paths beyond ego: The transpersonal vision*, ed. R. Walsh and F. Vaughan, S. 56–S. 59. Los Angeles: J.P. Tarcher/Perigee.
Lutz, A., H.A. Slagter, J.D. Dunne, and R.J. Davidson. 2008. Attention regulation and monitoring in meditation. *Trends in Cognitive Science* 12: 163–169.
MacLean, K.A., E. Ferrer, S.R. Aichele, D.A. Bridwell, A.P. Zanesco, T.L. Jacobs, et al. 2010. Intensive meditation training improves perceptual discrimination and sustained attention. *Psychological Science* 21(6): 829–839.
Markowiak, M. 2011. Meditation im Fokus. Wie beeinflussen verschiedene Meditationstechniken das Aufmerksamkeitssystem? Institut für Psychologie. Albert-Ludwigs-Universität Freiburg.
Nattier, J. 1995. Visible and invisible: Jan Nattier on the politics of representation in buddhist America. *Tricycle* 5: 42–49.
Netz, B. 2012. Meditation und Motivation- Konstruktion eines Fragebogens zur Motivation von Meditierenden. Unpublished thesis, Institute for Psychology, University Freiburg.
Ospina, M.B., T.K. Bond, M. Karkhaneh, L. Tjosvold, B. Vandermeer, Y. Liang. et al. 2007. Meditation practices for Health: State of the Research Evidence Report/Technology Assessment No. 155. Rockville: AHRQ Publication No. 07-E010.
Ott, U., B.K. Hölzel, and D. Vaitl. 2011. Brain structure and meditation: How spiritual practice shapes the brain. In *Neuroscience, consciousness and spirituality*, ed. Walach Harald, S. Schmidt, and W.B. Jonas, 119–128. New York: Springer.
Reavley, N., and J.F. Pallant. 2009. Development of a scale to assess the meditation experience. *Personality and Individual Differences* 47(6): 547–552. doi:10.1016/j.paid.2009.05.007.
Schmidt, S. 2004. Mindfulness and healing intention: Concepts, practice, and research evaluation. *The Journal of Alternative and Complementary Medicine* 10(Supplement 1): S7–S14.
Schmidt, S. 2011. Mindfulness in east and west – is it the same? In *Neuroscience, consciousness and spirituality*, ed. H. Walach, S. Schmidt, and W.B. Jonas, 23–38. New York: Springer.
Schmidt, S. 2012. Can we help just by good intentions? A meta-analysis of experiments on distant intention effects. *Journal of Alternative and Complementary Medicine* 18(6): 529–533.
Schmidt, S., and B. Netz. 2011. Why are people meditating? – the meditation motivation scale. Fourth European conference on integrated medicine, Berlin, 7–8 Oct 2011.
Shankman, R. 2008. *The experience of samadhi. An in-depth exploration of Buddhist meditation*. Boston: Shambala.
Shapiro, D. 1982. Overview: Clinical and physiological comparison of meditation with other self-control strategies. *American Journal of Psychiatry* 139(3): 267–274.
Shapiro, D.H. 1992. A preliminary study of long-term meditators: Goals, effects, religious orientation, cognitions. *The Journal of Transpersonal Psychology* 24(1): 23–39.
Shapiro, S.L., and G.E. Schwartz. 1999. Intentional systemic mindfulness: An integrative model for self-regulation and health. *Advances in Mind-Body Medicine* 15: 128–134.
Shapiro, Shauna L., L.E. Carlson, J.A. Astin, and B. Freedman. 2006. Mechanisms of mindfulness. *Journal of Clinical Psychology* 62(3): 373–386.
Solé-Leris, A. 1986. *Tranquility and insight. An introduction to the oldest form of Buddhist meditation*. Boston: Shambala.
Tang, Y.Y., Y. Ma, J. Wang, Y. Fan, S. Feng, Q. Lu, et al. 2007. Short-term meditation training improves attention and self-regulation. *PNAS Proceedings of the National Academy of Sciences of the United States of America* 104(43): 17152–17156.
Valentine, E.R., and P.L.G. Sweet. 1999. Meditation and attention: A comparison of the effects of concentrative and mindfulness meditation on sustained attention. *Mental Health, Religion and Culture* 2(1): 59–70.
Van den Hurk, P.A.M., F. Giommi, S.C. Gielen, A.E.M. Speckens, and H. Barendregt. 2010. Greater efficiency in attentional processing related to mindfulness meditation. *The Quarterly Journal of Experimental Psychology* 63(6): 1168–1180.
Walach, H. 2011. *Spiritualität*. Klein Jasedow: Drachen Verlag.
Wallace, B.A. 2006. *The attention revolution*. Boston: Wisdom Publications.
Walsh, Roger, and S.L. Shapiro. 2006. The meeting of meditative disciplines and western psychology: A mutually enriching dialogue. *American Psychologist* 61(3): 227–239.
Warm, J.S., R. Parasuraman, and G. Matthews. 2008. Vigilance requires hard mental work and is stressful. *Human Factors* 50(3): 433–441.

# The Neurobiology of Meditation and Mindfulness

**Tobias Esch**

**Abstract** Neurobiological effects of meditation and mindfulness can be detected in the brain as functional and also structural alterations in grey and white matter, particularly in areas related to attention and memory, interoception and sensory processing, or self- and auto-regulation (including control of stress and emotions). On the molecular level, dopamine and melatonin are found to increase, serotonin activity is modulated, and cortisol as well as norepinephrine have been proven to decrease. These findings are reflected in functional and structural changes documented by imaging techniques such as fMRI or EEG. They may be relevant for medicine and health care, especially with reference to therapeutic strategies for behavior change and life-style modification, or in association with stress regulation and the treatment of addiction. Neuronal mechanisms of mindfulness can be divided into four areas: attention regulation, body awareness, emotion regulation and self-perception.

## Abbreviations

| | |
|---|---|
| ACC | Anterior cingulated cortex |
| BDNF | Brain-derived neurotrophic factor |
| E | Epinephrine |
| EEG | Electroencephalogram |
| fMRI | Functional magnetic resonance imaging |

T. Esch, M.D. (✉)
Division of Integrative Health Promotion, Healthy University, Coburg University of Applied Sciences, Friedrich-Streib-Str. 2, Coburg, D-96450, Germany

Neuroscience Research Institute, State University of New York, New York, USA
e-mail: tesch@bidmc.harvard.edu

S. Schmidt and H. Walach (eds.), *Meditation – Neuroscientific Approaches and Philosophical Implications*, Studies in Neuroscience, Consciousness and Spirituality 2,
DOI 10.1007/978-3-319-01634-4_9, 

NE Norepinephrine
PCC Posterior cingulated cortex
PFC Prefrontal cortex

## Introduction

Meditation and mindfulness techniques are gaining importance in medicine and health care, because they have been shown to be effective in various clinical conditions (see Ernst et al. 2009; Fjorback et al. 2011; Grossman et al. 2004; Vøllestad et al. 2012). As a consequence, research has been focused on potential mechanisms of action and first reports on models for the underlying molecular principles and their neurobiological foundations are evolving. The new paradigm of a brain that regulates itself and adapts to stimuli from the environment but also from the internal milieu of the individual consciousness it supports, is useful for understanding these processes. At the center of the interest lies a potential for physiological, psychological, and neurobiological stress reduction (e.g., Esch et al. 2003, 2013; Esch and Stefano 2010; Jung et al. 2010, 2012; Manoch et al. 2009; Marchand 2012; Mohan et al. 2011; Stefano et al. 2005, 2006). Stress is an important factor for morbidity and mortality (e.g., see Rosengren et al. 2004; Russ et al. 2012), and thus methods for stress reduction are gaining importance as well. Knowledge about the facts and connections we already know, particularly with regard to current neuroscience, could be helpful for physicians, therapists and health professionals. Therefore, a summary of the most important neurobiological findings regarding meditation is presented here.

## Biological Principles of Meditation and Mindfulness

Meditation can be defined as willfully and purposefully regulating one's own attention, either for the purpose of relaxation, exploring oneself or personal growth and transcendence (see the chapter by Schmidt). Meditation can operationally be divided into two categories: Either one is focusing attention on a *changing* object such as physical sensations in the body scan, progressive muscle relaxation, autogenic training, or movements in yoga, tai chi, qigong, or mental content as in guided imagery, or mindfulness meditation proper. Or an *unchanging* or repetitive object is constantly held in focus, such as in mantra meditations like Transcendental meditation (TM) or Benson meditation, breathing meditation like Zazen, ostinato drumming, rhythmic dancing, jogging/flow, etc. (see Benson and Klipper 2000; Ott 2010; Petermann and Vaitl 2009). However, in practice both categories frequently overlap, and normally the ability to keep one's attention focused on a steady object is the precondition for the capacity to constantly attend to moving objects. As a common ground, all techniques are usually conducted with an attitude of intentionally directed or focused concentration and attention, which is called 'mindful awareness' or 'mindfulness'. As a consequence, they have the potential to elicit the so-called 'relaxation response' (Benson and Klipper 2000; Esch et al. 2003). This physiological reaction is the biological – that is, natural – antagonist of stress, thus antagonizing the 'stress response'. This property is, supposedly, responsible for some of the observed clinical, medical or therapeutic effects of meditation, particularly in stress-associated diseases like cardiovascular, immune, proinflammatory or neurodegenerative diseases, including anxiety and depression (for an overview see, e.g., Esch et al. 2003 or Stefano et al. 2005).

Mindfulness refers to the specific or formal practice of mindfulness training and meditation (e.g., Ernst et al. 2009), as well as to the outcome of such a training, the general capacity to be mindful, or being fully present. Jon Kabat-Zinn, who brought the principles of mindfulness to Western medicine, describes it as a specific kind of paying attention characterized by a non-judgmental, purposeful and continuing awareness of all mental and physical or bodily states and processes, from one moment to the next. It is the awareness that arises from paying attention on purpose, in the present moment, and without evaluating or judging what comes to mind (Kabat-Zinn 1990). Under stress frequently automatic mental patterns are formed and are part of the cognitive stress response. Such patterns can be observed as automatic negative thoughts or constant rumination, especially about the past or assumptions about the future, and as the tendency of the mind to wander away in boring or challenging as well as stressful situations. If mindfulness is practiced systematically, such automatic chains can be broken. A person skilled in meditation can intentionally suppress automatic thoughts by actively keeping the mind in the present, for instance, through staying with the actual experience as it happens here and now, or through focusing on the breath. This alone can be enough to reduce psychological and physiological stress (see, for example, Ernst et al. 2008; Esch and Esch 2013; Esch and Stefano 2010; Esch et al. 2013; Stefano et al. 2005). While being mindful, we usually experience that our lives are not instantaneously threatened, that is, in the real present, right here and now, there is no life-threatening danger, and consequently there is actually no specific 'reason' to be stressed or alarmed (Esch 2002, 2003, 2008; Thees et al. 2012). This active and intentional turn towards the non-stressful present moment is an innate capacity or attitude, like an inner 'mindset' that humans possess, biologically, and that can be facilitated and trained by systematic mental practice like mindfulness meditation (cf., Rossano 2007).

For a uniform and more exact definition of mindfulness for scientific purposes, Bishop et al. developed a working model consisting of two aspects (Bishop et al. 2004): the first aspect is represented by self-regulation of attention, i.e. actively regulating attention towards an awareness of the present and the experiences therein, and the second by the sustained and active maintenance of an attitude of abiding curiosity, openness, and acceptance in the face of these very experiences. Thereby, qualities such as acceptance are developed, that is, non-judging and 'letting go', as well as presence and 'connectivity': contact with the momentary inner sensual, mental and bodily experiences on the one hand, and an empathic and attentive 'outreach' to the surrounding world and environment on the other hand. Together with the capacity to ‘grow from the inside’ and to better deal and cope with challenges and stress, these aspects resemble very much what is also known from Positive Psychology and happiness research as requirements or ingredients for a successful or happy and healthy life, i.e. high level of satisfaction with one’s life and self-contentment, including resiliency (e.g., see Esch 2011). Thus, it is not surprising that the underlying concepts and different models widely overlap. This is particularly due to a shared neurobiology between the diverse concepts, and most importantly the likely relevance of brain mechanisms that involve endogenous reward and motivation processes which are imbedded in the brain for nearly all these models and processes. They also share aspects of attention regulation – from self- and auto-regulation to an activation of regions in the brain that correlate with empathic and compassionate behaviors, also embracing regions with mirror neuron activity (Esch and Stefano 2011; see below).

Kabat-Zinn (1990) differentiates two ways of practice necessary to make mindfulness become a steady and integral part of daily life: Formal mindfulness training, such as sitting or walking meditation, consists of specific exercises and techniques to stabilize the state of attentive and mindful awareness in the present moment. Informal practice, on the other hand, includes maintenance and persistence of mindfulness as a ‘state of being’ during daily routines and activities such as dish washing, shopping, speaking, brushing the teeth, etc. These informal practices serve to integrate a mindful attitude into a way of living. In principle, both forms can be trained and both are, rather confusingly, sometimes called ‘mindfulness meditation’. However, usually the term ‘meditation’ refers to the more formal aspects of mindfulness training, e.g., body scan, observing and bringing attention to the breath, etc. Yet this complex training, typically delivered within a structured process, e.g. a mind-body program or a behavioral group intervention, usually contains at the same time formal and informal practices, as critical and important ingredients. Here, the informal aspects particularly relate to the experience and activities of daily life. Thus, mindfulness training such as Mindfulness Based Stress Reduction (MBSR) training, or mind-body medical training including mindfulness as a core element are now successfully offered and practiced within medicine and the health care system, or within other settings, including kindergarten, schools, occupational health, or even nursing homes (e.g., see Ernst et al. 2008; Esch and Esch 2013; Mendelson et al. 2010).

Such multifaceted programs train participants in critical elements of mindfulness meditation, aiming at reducing physical or mental ailments and distress, or increasing self-efficacy (Ernst et al. 2009). In this context, Kabat-Zinn and colleagues see their work as a complementary or mind-body medical approach within behavioral medicine, where patients and participants, in contrast to conventional medicine, are usually not separated along indications or different diseases, indicating different treatments, except where such programs have been adapted for specific groups. Hence, in mindfulness training, the element of mindfulness is key, more or less regardless of the actual indication. And yet assumed health effects of being mindful and thereby reducing stress lie at the center of all measures as a common ‘denominator’ (Salzberg and Kabat-Zinn 2000). In fact, for this generic method across diseases, mobilizing inner salutogenic resources and potentials, and modifying illness-prone behavior patterns appears to be critical (Ernst et al. 2009; Esch 2010). In this way, mindfulness meditation relates to the inner capability to self-help, self-care or heal. Thus, the biological principle of self- and auto-regulation is the center piece and target that is, originally, rather independent of external circumstances and conditions. In that sense, mindfulness-based approaches help individuals to regain their autoregulative capacities that have been lost, e.g., through repeated challenges of stress.

## Neuroscience of Mindfulness and Mindfulness Meditation

At the center of mindfulness-based approaches is the deliberate focus of attention on feelings and sensations perceptible in the present moment. The feelings are observed, but not evaluated or actively changed. For example, through mindful observation of breathing, posture, tension, pain etc., self-perception is trained and a connection can be made to 'alienated' parts of the body (Kabat-Zinn 1990). In addition, to primarily psychological approaches at explaining it, physiological and, in particular, neuroscience and brain-related effects and mechanisms of action of meditation and mindfulness are increasingly being examined today.

### *Scientific Evidence*

According to the model of frontal alpha-symmetry of cortical activation, a relative increase in left-anterior brain activity is associated with positive feelings and improved immune function (Davidson 1998). Based on this assumption, Davidson et al. (2003) sought and found evidence of significantly stronger activation of left frontal areas of the brain, associated with a greater increase in antibody titer after a flu shot in participants at the end of a mindfulness course in comparison with those on the waiting list. The extent of the increase in cortical activity correlated with the increase in the antibody titer. This again indicates in this context the inseparability of mind and body and the cross-linking of the systems involved, as well as the continuity between central and peripheral processes. In other studies, a regional increase in the activity of numerous specific areas of the brain, such as motivation and reward areas and regions that control attention and the 'emotionality' of physiological (physical) responses, like the orbitofrontal cortex, interoception, like the insular cortex, and (central) autonomic functions, as well as controlling and evaluating emotions and linking them to memory formation, like the hippocampus, was found in participants during or after meditation (Critchley et al. 2001; Davidson 1998; Esch et al. 2004a; Hölzel et al. 2007, 2011a; Kang et al. 2013; Lazar et al. 2005; Linden 2000; Newberg et al. 2001; Newberg and Iversen 2003; Vestergaard-Poulsen et al. 2009). However, it appears that global brain activity tends to be reduced in experienced meditators (Lazar et al. 2000). It could be said that this reflects increased 'efficiency' (Esch and Stefano 2010). Newberg and Iversen (2003) identified a neurophysiological pattern of changes that occurred during meditative states that included certain key cerebral structures and hormonal and autonomic reactions, indicating changes in cognition, sensory perception, as well as affect (affective state), and an overall (i.e., brain-rooted) influence on endocrine-hormonal and autonomic activity. There are also increased functional – and structural – asymmetries, in particular lateralization and a tendency to shift activity 'forward', a so called 'anteriorization' seem to be relevant (see Davidson et al. 2003; Kang et al. 2013; Newberg et al. 2010; Yu et al. 2011). Creswell et al. (2007) found that subjects with a higher level of self-attributed mindfulness display greater prefrontal cortical activation and deactivation of the amygdala when naming emotions. Recent studies confirm this finding and correlate it with structural plastic adaptation effects in the area of the amygdala, which was 'shrinking' during a typical 8-week mindfulness training event (Desbordes et al. 2012; Hölzel et al. 2010). At the same time, there is a positive correlation between the structural and thus objective changes in the brain on the one hand, and the subjective perception of stress on the other hand: stress is perceived to a lesser extent or is coped with better (Hölzel et al. 2010). These findings confirm the assumption that mindfulness potentially has a positive effect on neuronal pathways that regulate affect and emotions (see below).

To summarize the *structural* effects of meditation on the brain, it can be stated that in those who meditate regularly (in comparison with control groups), the thickness of various areas of the cortex potentially increases. This can be specifically observed in those areas associated with attention and memory, interoception, and sensory processing as well as with self-regulation and auto-regulation (see Hölzel et al. 2008, 2011a; Kang et al. 2013; Lazar et al. 2005; Newberg et al. 2010; Pagnoni and Cekic 2007; Wang et al. 2011).

With respect to *functional* changes in brain activity, Lutz et al. (2004) show that many years of meditation lead to an increase in high-frequency, synchronized gamma waves in the EEG during a certain kind of meditation that involved the cultivation of compassion. Such a pattern also occurs during increased attention and functional learning processes and is associated with higher order conscious and cognitive processes, among other things. In this particular study, however, the coherence and synchronization of the high-frequency 40 Hz oscillation covered an extraordinarily large part of the brain. Cahn et al. (2010) confirmed these observations. Apparently, gamma activation in particular is an indication of the 'quality of meditation', i.e., the quality and experience of the person practicing it. But relevant changes in the EEG can also be observed after a relatively short period of practice. In addition to the alpha and gamma waves described, changes in the beta band and especially in the theta band can occur – here again in the relevant regions of the brain (Aftanas and Golocheikine 2001, 2002; Baijal and Srinivasan 2010; Cahn et al. 2012; Hinterberger et al. 2011; Jacobs et al. 1996; Kerr et al. 2011; Kjaer et al. 2002; Kubota et al. 2001; Lagopoulos et al. 2009; Yu et al. 2011). However, it can generally be assumed that there are significant differences in the EEG and physiology of the brain between long-term and short-term meditators or novices. Therefore, experience, individual length of practice, and also personal characteristics, as well as differences in the meditation techniques used must be considered (see, e.g., Brefczynski-Lewis et al. 2007; Farb et al. 2007). From an EEG perspective and for the experimental differentiation and estimation of the quality of the individual meditation experience, we can use the theta band (pronounced frontal, near midline) for depth or relaxation, connotation of ‘inner reflection’ and ‘self-contemplation’ on the one hand. The gamma band in the parietal or parieto-occipital and temporal region, i.e. more lateral, seems to refer to the experience of ‘dissolution of boundaries’ or transcendence, and also the ‘quality of compassion’, including the perception of unity and coherence, on the other hand. Finally, the occurrence of synchronicity or synchronization in the regions mentioned, including a spatial expansion of those synchronous waves and frequencies beyond the places of origin seems to be a ‘good’ signature. Characteristic increases in those bands are found regularly in experienced meditators; the alpha changes described above are usually observed only in beginners, as they are associated nonspecifically with relaxation and the closing of the eyes.

It is clear that, depending on individual ‘expertise’ and also on the momentary state of mood or general trait, and the meditator’s initial state of happiness, meditation can be associated with short peaks or exceptional moments or even ‘rapture’ but also with sustained feelings of deep joy, satisfaction, and ‘inner peace’ (Esch 2011; Esch and Stefano 2004, 2005). Today some of the neuromolecular correlates of these feelings are known, such as the involvement of dopamine and others (Esch and Stefano 2010; see below). Thus, these subjective feelings undoubtedly have objectifiable neurobiological correlates. For example, in an extreme case, the gamma oscillations can even extend to the ‘whole’ brain and generate very high amplitudes (i.e., involving many neurons) thus displaying strong synchronicity, which probably gives the subjective impression of fusion of subject and object or conveys a ‘mystical experience’, and is sometimes also termed *global binding* (Lutz et al. 2004; Ott 2010). In such moments, spatial representation of the ‘ego’ and proprioceptive input from the body (i.e., the subjective body image) are modified (see the chapter of Farb in this volume). Although these states are actually observed only in long-term meditators, analogous states can possibly occur in the area of pharmacotherapy as well as ‘side effects’, in drug use, or in psychopathologies as so called productive symptoms (Esch 2011). The consequences of these states – adverse effects, contraindications or desired aspects in therapeutic utilization – of a mindfulness- or meditation-based process are currently the subject of considerable controversy (see Lustyk et al. 2009).

## *Classification System for Mechanisms of Action*

Britta Hölzel and Ulrich Ott propose the following system for the assessment and classification of the neuronal mechanisms of action of mindfulness (see Hölzel et al. 2011b).

## Attention Regulation

Mindfulness techniques can help to focus or broaden attention. On the one hand, regular practice makes it easier to focus and keep attention on an object, to notice more quickly when thoughts drift away, and to deal with disruptions more effectively, by either blocking them out or accepting them. On the other hand, it is easier to disperse attention or broaden the 'light beam of consciousness' (Esch 2011), expanding the window of the present as it were and thus noticing more if desired. This can be confirmed in several ways in an experimental setting (e.g., van Leeuwen et al. 2009), but can also be seen in everyday life in heightened perception and awareness of breathing and a reduced tendency to be distracted. In the brain, as mentioned, the prefrontal cortex (PFC) and the anterior cingulated cortex (ACC) are especially involved. True to the slogan *use it or lose it*, over time, attention training, as it seems, also results in a slower aging process of the brain (compared with control persons, non-meditators; e.g., Pagnoni and Cekic 2007). This is probably caused by the functional and structural differentiation and morphological strengthening of networks in the areas of the brain described, as well as by the overall improvement in connectivity, involving not only increased density or size of the grey matter, but of the white matter as well (see Esch 2011; Esch and Stefano 2010; Kang et al. 2013; Luders et al. 2011; Ott 2010). Cultivating a 'beginner's mind' that remains in the present could also slow down the aging of the brain. This seems to be associated with a reduction in baseline activation of the so called default mode network (e.g., see Pagnoni et al. 2008). This statement is surely speculative due to the complexity of the findings. However, we know that with improved attention, memory is generally 'strengthened', and the functions, as well as the activity in the prefrontal working memory and in the declarative memory associated with the hippocampus, including the general ability to learn and remember, are improved (see Hölzel et al. 2010, 2011a; Mohan et al. 2011; Zeidan et al. 2010). Hence, meditation training may induce learning that is not stimulus- or task-specific, but process-specific, e.g. pattern recognition, and thereby may result in enduring changes in mental function (Desbordes et al. 2012).

## Body Awareness

It is generally true that mindfulness meditation increases functional and structural activity in the somatosensory and insular cortex. It can therefore be assumed that, over time, meditation improves the ability for interoception as well as exteroception (Esch 2011), refining body awareness, in particular, through greater differentiation of 'inner maps' (Ott 2010). The cingulum is also involved in this process, where relevant aspects of the current experience and detected 'errors' are filtered and thus recognized more rapidly. This means that the person who meditates on a regular basis feels 'good', because he or she is in close contact with his or her own feelings. This in turn leads to improved 'intuition' and a 'gut feeling' that is in fact associated with an enhanced representation of feelings from the inner body. Motion and skills memory and the corresponding areas in the dorsal striatum are also strengthened (see Newberg et al. 2010; Pagnoni and Cekic 2007). But it is likely that this improved ability to achieve 'attunement' is useful not only in order to recognize more quickly what is happening in one's *own* body, e.g., whether stress must be modulated, or what would constitute useful or appropriate reactions and actions (Esch and Stefano 2010; Hölzel et al. 2010). These also include effectively recognizing warning signals in time. In addition, mindfulness may also be useful for coming into closer contact with *others*. This is so, because the regions for 'body awareness' activated by mindfulness include areas and modalities that are needed for resonance with others as well (see also Siegel 2007). In addition to the classic mirror neuron areas, e.g., in the PFC, the associated areas in the temporal lobe and in the region of the temporoparietal junction should be mentioned as well (Hölzel et al. 2011a; Kang et al. 2013). Empathy and compassion, that is the emotional ability to empathize, but also the cognitive ability to perceive the perspective of the other (see *Theory of Mind*), are presumably reinforced (e.g., Desbordes et al. 2012). Hence, compassion and altruism, which can be trained through meditation and involve, among others, the prefrontal or orbitofrontal brain, also activate relevant dopaminergic midbrain structures and enhance the overall PFC-limbic (dopaminergic) connectivity (see below).

## Emotion Regulation

In neurobiology today we distinguish three levels of the limbic system or, functionally speaking, of the limbic auto-regulation, which is responsible for endogenous control of emotion and motivation (see also Esch 2011). The lower and middle levels that are involved in generating, and usually unconsciously 'evaluating', affects and emotions, through structures such as the amygdala, are generally difficult to control consciously. However, the upper limbic level has the ability to influence the regulation of emotions (and, maybe with stronger limitations, affects). It is sometimes also termed 'paralimbic' and comprises part to the prefrontal cortex, namely, in addition to the ACC, the orbitofrontal cortex and, in my view, parts of the insula as well (Esch 2011). This area of motivation and function of auto-regulation, which can be seen as a 'bridge' between the cortex and the limbic system or, figuratively speaking, between reason and cognition on the one hand, and affects and emotions on the other hand, can be strengthened by practicing mindfulness (Creswell et al. 2007; Esch and Stefano 2010; Hölzel et al. 2010, 2011a; Wang et al. 2011). Arguably, it may be easier to recognize and actively break the 'downwards spiral' or the vicious cycle between negative cognitions and the emotions linked with them, or vice versa, if those bridging functions are strong (see Ott 2010). In other words: a technique that activates these regions may also have a central influence on integrative processes in the brain, because it connects mind and reason, with affect and soma, or physiology. Thus, the *mind-body connection* becomes real, can be experienced and measured. In addition, self-regulation, the modulation of emotions and perceived stress is improved. These processes become neuronally objectifiable, controllable, and locatable. Interestingly, in the center of this assumedly enhanced connectivity between the PFC and limbic areas are those areas that 'subjectively' contribute to the experience of connectedness, affiliation, and acceptance. In other words, mindfulness training possibly leads to a systematic desensitization of negative affects and emotions, and at the same time to an increase in compassion, openness, and equanimity. Thereby, in addition to the inhibitory effect on the amygdala, the hippocampus is strengthened and the temporoparietal junction (see *Embodiment, Perspective Taking*) and the posterior cingulated cortex (PCC) are activated. This can be interpreted as relativizing the significance of contents and promoting recognition of what is important (see also Desbordes et al. 2012; Hölzel et al. 2011a; Khalsa et al. 2009; Ott 2010). However, findings concerning the PCC are still inconsistent (Kang et al. 2013). What can be seen in any event is that, in principle, affective circuits and 'autonomic' functions linked with them can be shaped by training, similar to attention (Newberg et al. 2010). Thus, it is possible to actively integrate higher neuronal structures, thus inhibiting inappropriate or 'undesired' automatic patterns and reactions to unpleasant stimuli like stress. In this way, new behavior patterns can arise, including those that lead to a reduction in or improved regulation of stress. It is also possible that other functions can be improved such as better coping with pain (see also Gard et al. 2012; Schmidt et al. 2011).

## Self-Perception

Studies have shown that practicing mindfulness can potentially lead to a more 'refined' self-perception. This means that ideas of the self and the self image on the one hand, and actual self-perception on the other hand can be better distinguished (see, e.g., Farb et al. 2007 and the chapter by Farb in this volume). Similar processes have already been observed in the context of pain modulation. Here, the practitioner learns *not* to identify with thoughts, perceptions, and feelings ('you are not the pain', 'pain and suffering are not one'). This attitude is also termed *de-centering* or *dis-identification* and can possibly be trained by mindfulness, which leads to a more mindful and distanced handling of stressful thoughts, emotions, psychosocial or mental stress, reduces 'over-identifying with the ego' (negative ego beliefs), and so prevents the occurrence of negative consequences (Ernst et al. 2009; Linden 2000; Plews-Ogan et al. 2005; Schmidt et al. 2011; Teasdale et al. 1995, 2002; Walach et al. 2007; Williams et al. 2000). This would then be, in the truest sense of the words, a 'self-efficacy experience' or experience of authentic internal control (see, e.g., Sonntag et al. 2010). Neurobiologically, it appears to be less a process of primarily cognitive control ('I do not want to have any more pain' or 'I can cope with it'), but in fact a change in pain processing (Gard et al. 2012) ('I feel less pain', 'The pain I feel is not bothering me'). Mindfulness again appears to be a means of modulating self and pain perception and pain anticipation (physical, but also regarding 'experiencing unfairness' and mental pain) by involving the sensory and interoceptive areas, such as the ACC and the insula. Such positive control experiences can generally help the individual in assuming responsibility for his or her own health (Esch 2002, 2003; Sonntag et al. 2010). Some authors also state that the constructs of the ego on the one hand and the self on the other are experienced more clearly and separately from one another when, with the help of mindfulness training, for example, inner evaluations like judgments or appraisal are reduced and, instead, autonomy, self-awareness, authenticity, and integrity are experienced, and simultaneously, feelings of coherence, connectedness and consistency arise (Esch 2011; Ott 2010). Here, it is important to state that in the relevant literature, the self includes aspects of self-attribution, such as self-reference, ego image, and corresponding 'appraisal structures', that is, an egocentric perspective, neuronally attributed more to midline structures in the brain including the medial PFC, as well as the non-judgmental focus on the present experience *of* the self, for which more lateral structures, including in the insula and the somatosensory cortex, are activated (Farb et al. 2007; Lazar 2011). Sometimes the 'allocentric perspective' is also mentioned in the context of this lateralization or activation of the lateral networks (see Hanson 2009 and the chapter by Austin in this volume).

Although mindfulness, obviously, has a tendency to strengthen the 'self' in contrast with the ego, and to reduce the tendency to judge and evaluate experiences, it appears that, in the context of the various aspects of the self, particularly the *flexible* and nuanced *handling* and recognition of the different modes, and the possibility of 'switching back and forth' are trained (see Farb et al. 2007; Malinowski 2012; Ott 2010). The fact that the *default mode*, detectable near the midline structures, or *resting state network*, which is associated with 'leisure' and inner reflection , as well as with self-related 'daydreaming', is modulated by meditation and mindfulness also fits with this auto-regulatory approach (Lazar 2011; McAvoy et al. 2008; Ott 2010; Pizoli et al. 2011; Schnabel 2010). On the one hand it is said that mindfulness is that which would arise, or be 'left over', if we would not do or think anything, the 'natural condition of our mind' (Walach 2010) as a *resting state*, on the other hand, mindfulness also appears to be able to inhibit the activity in the *default mode network*, i.e. to prevent daydreaming (Ott 2010). We need to await the results of further studies, but we can probably assume that a common denominator may be that mindfulness improves self-perception ability as well as increases psychological, inner flexibility. Very likely this is achieved in the context of an improved ability to regulate one's own brain activity. Mindfulness thus ultimately results in more degrees of freedom in behavior control (see, e.g., Malinowski 2012). An associated feeling of increased internal control ('I can') potentially strengthens self-efficacy and the feeling of 'inner strength' (see also Esch 2003, 2010; Sonntag et al. 2010). This could be important for the healthcare system and in particular for medical lifestyle modification or addiction treatment. In addition, mindfulness has a positive effect on resonance circuits in the brain regarding attunement and empathy through mirror neurons, which then in turn functionally integrate near-midline and lateral networks (see Esch 2011; Siegel 2007). So it is certainly correct to emphasize the aspect of integration as the central 'pivotal element' of the observed effects in the context of brain-related adaptation reactions in practitioners of mindfulness.

The classification system of the mechanisms of action of mindfulness can also be understood as an interwoven pyramid of various states, which – in the sense of a continuum of practice or experience, but also in the course of a therapeutic process – are potentially followed through during meditation. A certain hierarchy, e.g. depending on the individual meditation quality or practitioner's experience, can be construed (Esch 2011; Ott 2010). Thus, regulating attention is a prerequisite for other exercises and further 'steps'. Body awareness and the acceptance associated with it, as a result of which the body or the mind is no longer, for example, perceived as an 'enemy' or merely as sick, is in turn an important prerequisite for an integral healing process. The regulation of emotion, which is also about alternative behaviors and coping with pain, sorrow, stress, anxiety, depression, addiction, craving, etc., in other words about breaking negative vicious circles and emotional and cognitive self-regulation, is undoubtedly a part of this, especially in the context of an in-depth psychotherapeutic approach. Finally, in this process it is important to scrutinize one's own ego beliefs, not necessarily in order to question them, but to acknowledge them, and to identify and activate potentials and resources that may be blocked by unfavorable attributions. So one may succeed in taking the subject out of a mode of turmoil and disharmony, i.e., out of the *reactive mode* and finally to transfer it to a *resting* or *responsive mode* (c.f., Hanson 2009). This can then give rise to happiness (Esch 2011; Hanson 2009). In the end, a transcendental or spiritual experience may arise, sometimes called 'resolution of the ego', or an experience of unity. In this experience a primarily cognitive and clearly defined self-perception, which would be spatially and sensually contained, does no longer exist, but rather a perception of an utterly unlimited connectedness accompanied by deep compassion. However, these states also appear to depend greatly on the individual's prior experience or predispositions – as well as on the meditation method practiced. This is reflected in the fact that the areas for spatial-sensory location or the spatial memory tend to be down-regulated in some methods, at least in long-term meditators, while up-regulated, but then relatively consistently, in others.

## Neuromolecular Aspects of Meditation

For several years attempts have been made to discover, aside from the macroscopic and morphological changes in the function and structure of the brain and central nervous system, changes and mechanisms at the molecular level that correspond with the observations made in the context of practicing mindfulness and meditation. In many areas, however, little progress has been made due to the difficulty of examining the human brain in real time during meditation while making valid and reliable measurements at the same time. Moreover, many methods are still in the experimental stage, so relatively well substantiated methods and conclusive models are used to merge the findings from various areas to an overall picture (see, e.g., Esch and Stefano 2010).

In any event, there is presumably involvement of the central limbic and mesolimbic or mesostriatal mechanisms. Thus, the brain's motivation and reward systems – with dopamine as the leading messenger substance – appear to be involved (overview see Esch et al. 2004b; Esch and Stefano 2004). So it is not surprising that various authors were able to directly prove the presence of dopamine both in the brain and in plasma in connection with meditation (e.g., Jung et al. 2010; Kjaer et al. 2002). There have also been repeated reports of the involvement of enzymes for the production and release of norepinephrine (NE) and epinephrine (E), therefore establishing a direct connection with stress physiology and with stress modulation at the molecular level (e.g., Jung et al. 2012). Other factors sensitive to stress are apparently also involved, such as the brain-derived neurotrophic factor (BDNF), although there are, as so often in this context, genetic predispositions (e.g. polymorphisms) that make a response more likely (see also Esch and Stefano 2010; Jung et al. 2012).

Lower values are found for plasma NE in meditators in comparison with control persons (e.g., Infante et al. 2001). The enzyme systems responsible for this (see above) are also involved in dopamine and morphine metabolism, although the real-time involvement of endogenous morphine during meditation has not been reliably proven yet, probably for the reasons mentioned above (see, e.g., Mantione et al. 2008, 2010a, b). The situation is also somewhat unclear with respect to serotonin. It appears that meditation has a positive effect on the modulation of peripheral and central serotonin levels, with a tendency to raise them, but this effect is not consistently verifiable (Bujatti and Riederer 1976; Liou et al. 2010; Solberg et al. 2004; Walton et al. 1995; Yu et al. 2011). This may be due to circadian rhythms and interaction with other hormone systems, as there is, for instance, a 'competition' between melatonin and tryptophan.

By contrast, meditation increases peripheral melatonin levels (Liou et al. 2010; Solberg et al. 2004) and reduces cortisol levels (Brand et al. 2012; Esch et al. 2007; Walton et al. 1995) quite reliably. At the same time it can be seen that sympathetic *responsivity* is reduced (e.g., Hoffman et al. 1982) in favor of increased parasympathetic tone (Bujatti and Riederer 1976). Interestingly, in this context it has long been known and quite thoroughly proven that stress – especially chronic stress – lowers acetylcholine levels by increasing acetylcholinesterase activity and activating the respective genes (Evron et al. 2005), which again illustrates the stress/anti-stress regulation mechanism at the molecular level. Also, relaxation/meditation training activates discriminative gene patterns (Dusek et al. 2008). Both, acetylcholine and morphine increase the activity of the constitutive nitric oxide-producing enzymes (overview, e.g., Stefano and Esch 2005), which is presumably one reason why elevated nitric oxide levels have been found in the breath of meditators (e.g., Dusek et al. 2006; Mantione et al. 2007). Interestingly, constitutive nitric oxide has an anti-inflammatory effect, e.g., by inhibiting the pro-inflammatory nuclear transcription factor NFkappaB, and chronic stress does the opposite and induces inflammation (Esch et al. 2002). This could represent a neurobiological core element of the stress-modulating capacity of meditation and associated health benefits (Esch and Stefano 2010; Stefano and Esch 2005). However, this last point is still speculative.

In summary, it can be established that meditation can counteract stress at the mental, physical, physiological, and molecular level. Some of these effects are subjective by definition, but can definitely be objectified using suitable methods. Regarding the improvement in mood and affect and subjective feelings of relaxation, happiness or 'simply' pleasure that have sometimes been reported by meditators, it can be assumed that the endogenous motivation and reward systems are involved and a direct relationship with self-regulation can thus again be assumed.

## Conclusions and Outlook

Meditative experiences and mindfulness are rooted not only in psychology, but in neuroscience and neurobiology as well. They can be detected at the level of the brain in the area of functional, but also structural changes in grey and white matter, especially in those areas and networks associated with attention and memory, interoception, and sensory processing as well as with self- and auto-regulation. This includes emotion and stress regulation to which, in addition to the integration of central autonomic regions, the limbic system and endogenous motivation and reward centers belong. Anxiety and 'sensitivity to stress' can be reduced and the ability to learn and remember is presumably improved.

At the level of neurobiological auto-regulation and molecular control, it has been proven that dopamine and melatonin are increased and cortisol and norepinephrine are reduced during meditative states, i.e., classic stress hormones are inhibited. However, increases in serotonin, nitric oxide, acetylcholine, and endogenous morphine are still speculative, which is due in part to the complexity and difficulties with measuring methods. Thus, there is a need for further research. But in any case, the findings are already interesting for medicine and the healthcare system, especially for therapeutic behavior change, lifestyle modification, and addiction treatment. They can also be used to address the issue of and appraise the risks and 'side effects' of meditation.

Some authors relate meditation/mindfulness in this context to elements of motivation physiology and biology. The *approach system* and the *avoidance system* must be differentiated here (e.g., Esch 2011; Hanson 2009; McColl and Singer 2012; Sauer et al. 2011). Avoidance and aversion also includes the fight-or-flight and stress response, and approach or appetite also includes binding and reward perception and the anticipation of it (e.g., Bisaz and Sullivan 2012; McCall and Singer 2012). The binding or affiliation system is also described as a social relief and welfare system, and at the neurobiological level presumably includes a central role of oxytocin signaling pathways, a connection that awaits further study (Esch and Stefano 2005; Groppe et al. 2013; McCall and Singer 2012). The desire and reward system, on the other hand, is also described as the *wanting* or *pleasure system* and it incorporates dopamine and opioidergic signal pathways as central neurotransmitters (Esch and Stefano 2004; Singer 2010). Thereby, there is a bridge to the aversion system, e.g., when there are pharmacological, but also mental or psychosocial withdrawal symptoms. In such a situation the stress hormones assume a dominant role. Thus all of these systems are connected neurobiologically and neuromolecularly, and mindfulness is believed to be able to return a 'system in turmoil', i.e., when regulation is in the reactive mode, and stress, allostasis, and depletion prevail, to the *responsive mode*, with the primary function of down-regulation or even homeostasis and the goal of regeneration, restoration and 'recharge', associated with feelings of inner peace, non-wanting or 'quiescence' (McCall and Singer 2012).

There is apparently a correlation between motivation regulation and self-regulation on the one hand, and controlling affects and emotions and stress regulation on the other – and all of these aspects appear to respond quite well to mindfulness therapy. Moreover, in a broader sense, mindfulness leads functionally and systemically to an increase in the degrees of 'freedom', i.e., internal flexibility, since adaptation to reality, perceived control, self-efficacy, and self-management skills are strengthened. This is expressed neurobiologically, in the brain, e.g., by an increase in asymmetry in EEG and fMRI findings: neurological imaging shows asymmetric activity shifting and lateralization, and in addition frequently 'anteriorization'. Especially, the relevant networks are activated via mindfulness practice, and this ultimately means that there is a network potential for growth, but also for acceptance, affiliation, and happiness or deep inner satisfaction. Of course, outcomes are speculative.

Chronic stress over the lifetime has been proven to lead to degenerations in parts of the PFC, leading, among others, to a decrease in the ability to regulate emotions, but also to deteriorated executive functions and a reduced performance of the working memory in the dorsolateral PFC. Furthermore, there may also be degeneration in the hippocampus, which is very stress-sensitive, while the amygdala may gain in size, with presumed effects such as elevated anxiety, or continuously feeling stressed. These are developments that the regular practice of meditation and mindfulness, including compassion training, appear to counteract, potentially, although there are still many questions with respect to therapeutic approaches.

In conclusion, we state again that the neuronal mechanisms of action of mindfulness can be systematically classified in four areas: attention regulation, body awareness, emotion regulation, and self-perception. Britta Hölzel and Ulrich Ott propose this classification and it appears to be useful and practicable.

## References

Aftanas, L.I., and S.A. Golocheikine. 2001. Human anterior and frontal midline theta and lower alpha reflect emotionally positive state and internalized attention: High-resolution EEG investigation of meditation. *Neuroscience Letters* 310: 57–60.

Aftanas, L.I., and S.A. Golocheikine. 2002. Non-linear dynamic complexity of the human EEG during meditation. *Neuroscience Letters* 330: 143–146.

Baijal, S., and N. Srinivasan. 2010. Theta activity and meditative states: Spectral changes during concentrative meditation. *Cognitive Processing* 11: 31–38.

Benson, H., and M.Z. Klipper. 2000. *The relaxation response*. New York: Harper Collins.

Bisaz, R., and R.M. Sullivan. 2012. Developmental neurobiology of the rat attachment systems and its modulation by stress. *Behavioral Sciences* 2: 79–102.

Bishop, S.R., M. Lau, S. Shapiro, L. Carlson, N.D. Anderson, J. Carmody, et al. 2004. Mindfulness. A proposed operational definition. *Clinical Psychology: Science and Practice* 11(3): 230–241.

Brand, S., E. Holsboer-Trachsler, J.R. Naranjo, and S. Schmidt. 2012. Influence of mindfulness practice on cortisol and sleep in long-term and short-term meditators. *Neuropsychobiology* 65: 109–118.

Brefczynski-Lewis, J.A., A. Lutz, H.S. Schaefer, D.B. Levinson, and R.J. Davidson. 2007. Neural correlates of attentional expertise in long-term meditation practitioners. *Proceedings of the National Academy of Sciences of the United States of America* 104: 11483–11488.
Bujatti, M., and P. Riederer. 1976. Serotonin, noradrenaline, dopamine metabolites in transcendental meditation-technique. *Journal of Neural Transmission* 39: 257–267.
Cahn, B.R., A. Delorme, and J. Polich. 2010. Occipital gamma activation during Vipassana meditation. *Cognitive Processing* 11: 39–56.
Cahn, B.R., A. Delorme, and J. Polich. 2012. Event-related delta, theta, alpha, and gamma correlates to auditory oddball processing during Vipassana meditation. *Social Cognitive and Affective Neuroscience* 8: 100–111.
Creswell, J.D., B.M. Way, N.I. Eisenberger, and M.D. Lieberman. 2007. Neural correlates of dispositional mindfulness during affect labeling. *Psychosomatic Medicine* 69: 560–565.
Critchley, H.D., R.N. Melmed, E. Featherstone, C.J. Mathias, and R.J. Dolan. 2001. Brain activity during biofeedback relaxation: A functional neuroimaging investigation. *Brain* 124: 1003–1012.
Davidson, R.J. 1998. Anterior electrophysiological asymmetries, emotion, and depression: Conceptual and methodological conundrums. *Psychophysiology* 35: 607–614.
Davidson, R.J., J. Kabat-Zinn, J. Schumacher, M. Rosenkranz, D. Muller, S.F. Santorelli, et al. 2003. Alterations in brain and immune function produced by mindfulness meditation. *Psychosomatic Medicine* 65: 564–570.
Desbordes, G., L.T. Negi, T.W.W. Pace, B.A. Wallace, C.L. Raison, and E.L. Schwartz. 2012. Effects of mindful-attention and compassion meditation training on amygdala response to emotional stimuli in an ordinary, non-meditative state. *Frontiers in Human Neuroscience* 6: 292.
Dusek, J.A., B.H. Chang, J. Zaki, S. Lazar, A. Deykin, G.B. Stefano, et al. 2006. Association between oxygen consumption and nitric oxide production during the relaxation response. *Medical Science Monitor: International Medical Journal of Experimental and Clinical Research* 12: CR1–CR10.
Dusek, J.A., H.H. Otu, A.L. Wohlhueter, M. Bhasin, L.F. Zerbini, M.G. Joseph, et al. 2008. Genomic counter-stress changes induced by the relaxation response. *PLoS One* 3: e2576.
Ernst, S., J. Welke, C. Heintze, R. Gabriel, A. Zöllner, S. Kiehne, et al. 2008. Effects of mindfulness-based stress reduction on quality of life in nursing home residents: A feasibility study. *Forschende Komplementärmedizin* 15: 74–81.
Ernst, S., S.M. Esch, and T. Esch. 2009. Die Bedeutung achtsamkeitbasierter Interventionen in der medizinischen und psychotherapeutischen Versorgung [The impact of mindfulness-based interventions on health care]. *Forschende Komplementärmedizin* 16: 296–303.
Esch, T. 2002. Gesund im Stress: Der Wandel des Stresskonzeptes und seine Bedeutung für Prävention, Gesundheit und Lebensstil [Health in stress: Change in the stress concept and its significance for prevention, health and life style]. *Gesundheitswesen* 64: 73–81.
Esch, T. 2003. Stress, Anpassung und Selbstorganisation: Gleichgewichtsprozesse sichern Gesundheit und Überleben [Stress, adaptation, and self-organization: Balancing processes facilitate health and survival]. *Forschende Komplementärmedizin* 10: 330–341.
Esch, T. 2008. Mind-body-Medizin: Stress, Stressmanagement und Gesundheitsförderung [Mind-body medicine: Stress, stress management and health promotion]. *Komplementäre und Integrative Medizin* 49: 35–39.
Esch, T. 2010. Von der Kuration zur Gesundheitsförderung: Paradigmenwechsel in der ärztlichen Beratung. In *Adipositas und Public Health. Rahmenbedingungen, interdisziplinäre Zugänge und Perspektiven für erfolgreiche Präventionsstrategien*, ed. C. Heintze. München: Juventa.
Esch, T. 2011. *Die Neurobiologie des Glücks. Wie die Positive Psychologie die Medizin verändert.* Stuttgart: Thieme.
Esch, T., and S.M. Esch. 2013. *Stressbewältigung mithilfe der Mind-Body-Medizin: Trainingsmanual zur Integrativen Gesundheitsförderung*. Berlin: MWV.
Esch, T., and G.B. Stefano. 2004. The neurobiology of pleasure, reward processes, addiction and their health implications. *Neuroendocrinology Letters* 25: 235–251.
Esch, T., and G.B. Stefano. 2005. The neurobiology of love. *Neuroendocrinology Letters* 26: 175–192.

Esch, T., and G.B. Stefano. 2010. The neurobiology of stress management. *Neuroendocrinology Letters* 31: 19–39.
Esch, T., and G.B. Stefano. 2011. The neurobiological link between compassion and love. *Medical Science Monitor: International Medical Journal of Experimental and Clinical Research* 17: RA65–RA75.
Esch, T., G.B. Stefano, G.L. Fricchione, and H. Benson. 2002. An overview of stress and its impact in immunological diseases. *Modern Aspects of Immunobiology* 2: 187–192.
Esch, T., G.B. Stefano, and G.L. Fricchione. 2003. The therapeutic use of the relaxation response in stress-related diseases. *Medical Science Monitor: International Medical Journal of Experimental and Clinical Research* 9: RA23–RA34.
Esch, T., M. Guarna, E. Bianchi, and G.B. Stefano. 2004a. Meditation and limbic processes. *Biofeedback* 32: 22–32.
Esch, T., M. Guarna, E. Bianchi, W. Zhu, and G.B. Stefano. 2004b. Commonalities in the central nervous system's involvement with complementary medical therapies: Limbic morphinergic processes. *Medical Science Monitor: International Medical Journal of Experimental and Clinical Research* 10: MS6–MS17.
Esch, T., J. Duckstein, J. Welke, and V. Braun. 2007. Mind/body techniques for physiological and psychological stress reduction: Stress management via Tai Chi training – a pilot study. *Medical Science Monitor: International Medical Journal of Experimental and Clinical Research* 13: CR488–CR497.
Esch, T., U. Sonntag, S.M. Esch, and S. Thees. 2013. Stress management and mind-body medicine: A randomized controlled longitudinal evaluation of students' health and effects of a behavioral group intervention at a middle-size German University (SM-MESH). *Forschende Komplementärmedizin* 20: 129–137. doi:10.1016/j.ctcp.2009.01.003.
Evron, T., L.B. Moyal-Segal, N. Lamm, A. Geffen, and H. Soreq. 2005. RNA-targeted suppression of stress-induced allostasis in primate spinal cord neurons. *Neurodegenerative Diseases* 2: 16–27.
Farb, N.A.S., Z.V. Segal, H. Mayberg, J. Bean, D. McKeon, Z. Fatima, et al. 2007. Attending to the present: Mindfulness meditation reveals distinct neural modes of self-reference. *Social Cognitive and Affective Neuroscience* 2: 313–322.
Fjorback, L.O., M. Arendt, E. Ornbøl, P. Fink, and H. Walach. 2011. Mindfulness-based stress reduction and mindfulness-based cognitive therapy: A systematic review of randomized controlled trials. *Acta Psychiatrica Scandinavica* 124: 102–119.
Gard, T., B.K. Hölzel, A.T. Sack, H. Hempel, S.W. Lazar, D. Vaitl, et al. 2012. Pain attenuation through mindfulness is associated with decreased cognitive control and increased sensory processing in the brain. *Cerebral Cortex* 22: 2692–2702.
Groppe, S.E., A. Gossen, L. Rademacher, A. Hahn, L. Westphal, G. Gründer, et al. 2013. Oxytocin influences processing of socially relevant cues in the ventral tegmental area of the human brain. *Biological Psychiatry* 74(3): 172–179. pii: S0006-3223(13)00039-5.
Grossman, P., L. Niemann, S. Schmidt, and H. Walach. 2004. Mindfulness-based stress reduction and health benefits. A meta-analysis. *Journal of Psychosomatic Research* 57: 35–43.
Hanson, R. 2009. *Buddha's brain. The practical neuroscience of happiness, love and wisdom.* Oakland: New Harbinger.
Hinterberger, T., T. Kamei, and H. Walach. 2011. Psychophysiological classification and staging of mental states during meditative practice. *Biomedizinische Technik/Biomedical engineering* 56: 341–350.
Hoffman, J.W., H. Benson, P.A. Arns, G.L. Stainbrook, G.L. Landsberg, J.B. Young, et al. 1982. Reduced sympathetic nervous system responsivity associated with the relaxation response. *Science* 215: 190–192.
Hölzel, B.K., U. Ott, H. Hempel, A. Hackl, K. Wolf, R. Stark, et al. 2007. Differential engagement of anterior cingulate and adjacent medial frontal cortex in adept meditators and non-meditators. *Neuroscience Letters* 421: 16–21.
Hölzel, B.K., U. Ott, T. Gard, H. Hempel, M. Weygandt, K. Morgen, et al. 2008. Investigation of mindfulness meditation practitioners with voxel-based morphometry. *Social Cognitive and Affective Neuroscience* 3: 55–61.

Hölzel, B.K., J. Carmody, K.C. Evans, E.A. Hoge, J.A. Dusek, L. Morgan, et al. 2010. Stress reduction correlates with structural changes in the amygdala. *Social Cognitive and Affective Neuroscience* 5: 11–17.

Hölzel, B.K., J. Carmody, M. Vangel, C. Congleton, S.M. Yerramsetti, T. Gard, et al. 2011a. Mindfulness practice leads to increases in regional brain gray matter density. *Psychiatry Research* 191: 36–43.

Hölzel, B.K., S.W. Lazar, T. Gard, Z. Schuman-Olivier, D.R. Vago, and U. Ott. 2011b. How does mindfulness meditation work? Proposing mechanisms of action from a conceptual and neural perspective. *Perspectives on Psychological Science* 6: 537–559.

Infante, J.R., M. Torres-Avisbal, P. Pinel, J.A. Vallejo, F. Peran, F. Gonzalez, et al. 2001. Catecholamine levels in practitioners of the transcendental meditation technique. *Physiology and Behavior* 72: 141–146.

Jacobs, G.D., H. Benson, and R. Friedman. 1996. Topographic EEG mapping of the relaxation response. *Biofeedback and Self Regulation* 21: 121–129.

Jung, Y.H., D.H. Kang, J.H. Jang, H.Y. Park, M.S. Byun, S.J. Kwon, et al. 2010. The effects of mind-body training on stress reduction, positive affect, and plasma catecholamines. *Neuroscience Letters* 479: 138–142.

Jung, Y.H., D.H. Kang, M.S. Byun, G. Shim, S.J. Kwon, G.E. Jang, et al. 2012. Influence of brain-derived neurotrophic factor and catechol O-methyl transferase polymorphisms on effects of meditation on plasma catecholamines and stress. *Stress* 15: 97–104.

Kabat-Zinn, J. 1990. *Full catastrophe living: Using the wisdom of your body and mind to face stress, pain, and illness*. New York: Delta.

Kang, D.H., H.J. Jo, W.H. Jung, S.H. Kim, Y.H. Jung, C.H. Choi, et al. 2013. The effect of meditation on brain structure: cortical thickness mapping and diffusion tensor imaging. *Social Cognitive and Affective Neuroscience* 8: 27–33.

Kerr, C.E., S.R. Jones, Q. Wan, D.L. Pritchett, R.H. Wasserman, A. Wexler, et al. 2011. Effects of mindfulness meditation training on anticipatory alpha modulation in primary somatosensory cortex. *Brain Research Bulletin* 85: 96–103.

Khalsa, D.S., D. Amen, C. Hanks, N. Money, and A. Newberg. 2009. Cerebral blood flow changes during chanting meditation. *Nuclear Medicine Communications* 30: 956–961.

Kjaer, T.W., C. Bertelsen, P. Piccini, D. Brooks, J. Alving, and H.C. Lou. 2002. Increased dopamine tone during meditation-induced change of consciousness. *Brain Research. Cognitive Brain Research* 13: 255–259.

Kubota, Y., W. Sato, M. Toichi, T. Murai, T. Okada, A. Hayashi, et al. 2001. Frontal midline theta rhythm is correlated with cardiac autonomic activities during the performance of an attention demanding meditation procedure. *Research. Cognitive Brain Research* 11: 281–287.

Lagopoulos, J., J. Xu, I. Rasmussen, A. Vik, G.S. Malhi, C.F. Eliassen, et al. 2009. Increased theta and alpha EEG activity during nondirective meditation. *Journal of Alternative and Complementary Medicine* 15: 1187–1192.

Lazar, S.W. 2011. The meditating brain. Berlin: 4. Europäischer Kongress für Integrative Medizin.

Lazar, S.W., G. Bush, R.L. Gollub, G.L. Fricchione, G. Khalsa, and H. Benson. 2000. Functional brain mapping of the relaxation response and meditation. *Neuroreport* 11: 1581–1585.

Lazar, S.W., C.E. Kerr, R.H. Wasserman, J.R. Gray, D.N. Greve, M.T. Treadway, et al. 2005. Meditation experience is associated with increased cortical thickness. *Neuroreport* 16: 1893–1897.

Linden, W. 2000. Meditation. In *Handbuch der Entspannungsverfahren, Band 1: Grundlagen und Methoden*, ed. D. Vaitl and F. Petermann, 256–268. Weinheim: Psychologie Verlags Union.

Liou, C.H., C.W. Hsieh, C.H. Hsieh, D.Y. Chen, C.H. Wang, J.H. Chen, et al. 2010. Detection of nighttime melatonin level in Chinese Original Quiet Sitting. *Journal of the Formosan Medical Association* 109: 694–701.

Luders, E., K. Clark, K.L. Narr, and A.W. Toga. 2011. Enhanced brain connectivity in long-term meditation practitioners. *Neuroimage* 57: 1308–1316.

Lustyk, M.K., N. Chawla, R.S. Nolan, and G.A. Marlatt. 2009. Mindfulness meditation research: Issues of participant screening, safety procedures, and researcher training. *Advances in Mind Body Medicine* 24: 20–30.

Lutz, A., L.L. Greischar, N.B. Rawlings, M. Ricard, and R.J. Davidson. 2004. Long-term meditators self-induce high-amplitude gamma synchrony during mental practice. *Proceedings of the National Academy of Sciences of the United States of America* 101: 16369–16373.

Malinowski, P. 2012. Wirkmechanismen der Achtsamkeitspraxis. Wie fördert Meditation positive psychologische Veränderungen? In *Achtsamkeit: Ein buddhistisches Konzept erobert die Wissenschaft*, ed. M. Zimmermann, C. Spitz, and S. Schmidt, 91–100. Bern: Hans Huber.

Manoch, R., A. Gordon, D. Black, G. Malhi, and R. Seidler. 2009. Using meditation for less stress and better wellbeing: A seminar for GPs. *Australian Family Physician* 38: 454–458.

Mantione, K.J., T. Esch, and G.B. Stefano. 2007. Detection of nitric oxide in exhaled human breath: Exercise and resting determinations. *Medical Science Monitor: International Medical Journal of Experimental and Clinical Research* 13: MT1–MT5.

Mantione, K.J., P. Cadet, W. Zhu, R.M. Kream, M. Sheehan, G.L. Fricchione, et al. 2008. Endogenous morphine signaling via nitric oxide regulates the expression of CYP2D6 and COMT: Autocrine/paracrine feedback inhibition. *Addiction Biology* 13: 118–123.

Mantione, K.J., W. Zhu, R.M. Kream, T. Esch, and G. Stefano. 2010a. Regulation of the transcription of the catechol-O-methyltransferase gene by morphine and epinephrine. *Activitas Nervosa Superior Rediviva* 52: 51–56.

Mantione, K.J., R.M. Kream, and G.B. Stefano. 2010b. Variations in critical morphine biosynthesis genes and their potential to influence human health. *Neuroendocrinology Letters* 31: 11–18.

Marchand, W.R. 2012. Mindfulness-based stress reduction, mindfulness-based cognitive therapy, and zen meditation for depression, anxiety, pain, and psychological distress. *Journal of Psychiatric Practice* 18: 233–252.

McAvoy, M., L. Larson-Prior, T.S. Nolan, S.N. Vaishnavi, M.E. Raichle, and G. d'Avossa. 2008. Resting states affect spontaneous BOLD oscillations in sensory and paralimbic cortex. *Journal of Neurophysiology* 100: 922–931.

McColl, C., and T. Singer. 2012. The animal and human neuroendocrinology of social cognition, motivation and behavior. *Nature Neuroscience* 15: 681–688.

Mendelson, T., M.T. Greenberg, J.K. Dariotis, L.F. Gould, B.L. Rhoades, and P.J. Leaf. 2010. Feasibility and preliminary outcomes of a school-based mindfulness intervention for urban youth. *Journal of Abnormal Child Psychology* 38: 985–994.

Mohan, A., R. Sharma, and R.L. Bijlani. 2011. Effect of meditation on stress-induced changes in cognitive functions. *Journal of Alternative and Complementary Medicine* 17: 207–212.

Newberg, A.B., and J. Iversen. 2003. The neural basis of the complex mental task of meditation: neurotransmitter and neurochemical considerations. *Medical Hypotheses* 61: 282–291.

Newberg, A., A. Alavi, M. Baime, M. Pourdehnad, J. Santanna, and E. d'Aquili. 2001. The measurement of regional cerebral blood flow during the complex cognitive task of meditation: A preliminary SPECT study. *Psychiatry Research* 106: 113–122.

Newberg, A.B., N. Wintering, M.R. Waldman, D. Amen, D.S. Khalsa, and A. Alavi. 2010. Cerebral blood flow differences between long-term meditators and non-meditators. *Consciousness and Cognition* 19: 899–905.

Ott, U. 2010. *Meditation für Skeptiker*. München: O.W. Barth.

Pagnoni, G., and M. Cekic. 2007. Age effects on gray matter volume and attentional performance in Zen meditation. *Neurobiology of Aging* 28: 1623–1627.

Pagnoni, G., M. Cekic, and Y. Guo. 2008. "Thinking about not-thinking": Neural correlates of conceptual processing during Zen meditation. *PLOS One* 3(9): e3083.

Petermann, F., and D. Vaitl. 2009. *Entspannungsverfahren: Das Praxishandbuch*. Beltz: Weinheim.

Pizoli, C.E., M.N. Shah, A.Z. Snyder, J.S. Shimony, D.D. Limbrick, M.E. Raichle, et al. 2011. Resting-state activity in development and maintenance of normal brain function. *Proceedings of the National Academy of Sciences of the United States of America* 108: 11638–11643.

Plews-Ogan, M., J.E. Owens, M. Goodman, P. Wolfe, and J. Schorling. 2005. Brief report: A pilot study evaluating mindfulness-based stress reduction and massage for the management of chronic pain. *Journal of General Internal Medicine* 20: 1136–1138.

Rosengren, A., S. Hawken, S. Ounpuu, K. Sliwa, M. Zubaid, W.A. Almahmeed, et al. 2004. INTERHEART investigators. Association of psychosocial risk factors with risk of acute myocardial infarction in 11119 cases and 13648 controls from 52 countries (the INTERHEART study): Case–control study. *Lancet* 364: 953–962.

Rossano, M. 2007. Did meditating make us human? *Cambridge Archaeological Journal* 17(1): 47–58.

Russ, T.C., E. Stamatakis, M. Hamer, J.M. Starr, M. Kivimäki, and G.D. Batty. 2012. Association between psychological distress and mortality: Individual participant pooled analysis of 10 prospective cohort studies. *British Medical Journal* 345: E4933. doi:10.1136/bmj.e4933.

Salzberg, S., and J. Kabat-Zinn. 2000. Achtsamkeit als Medikament. In *Die heilende Kraft der Gefühle*, ed. D. Goleman, 134–181. München: Dt. Taschenbuch-Verl.

Sauer, S., H. Walach, and N. Kohls. 2011. Gray's behavioural inhibition system as a mediator of mindfulness towards well-being. *Personality and Individual Differences* 50: 506–511.

Schmidt, S., P. Grossman, B. Schwarzer, S. Jena, J. Naumann, and H. Walach. 2011. Treating fibromyalgia with mindfulness-based stress reduction: Results from a 3-armed randomized controlled trial. *Pain* 152: 361–369.

Schnabel, U. 2010. *Muße – Vom Glück des Nichtstuns*. München: Blessing.

Siegel, D.J. 2007. *The mindful brain*. New York: Norton.

Singer, T. 2010. Motivational systems in the brain. Mind & Life XX conference: altruism and compassion in economic systems. Session: Altruism: Evolutionary origins and modern expressions. Host: Mind & Life International, Co-Sponsored by The Mind & Life Institute and The University of Zürich, Kongresshaus, Zürich, 9 April.

Solberg, E.E., A. Holen, Ø. Ekeberg, B. Østerud, R. Halvorsen, and L. Sandvik. 2004. The effects of long meditation on plasma melatonin and blood serotonin. *Medical Science Monitor: International Medical Journal of Experimental and Clinical Research* 10: CR96–CR101.

Sonntag, U., T. Esch, L. von Hagen, B. Renneberg, V. Braun, and C. Heintze. 2010. Locus of control, self-efficacy and attribution tendencies in obese patients – implications for primary care consultations. *Medical Science Monitor: International Medical Journal of Experimental and Clinical Research* 16: CR330–CR335.

Stefano, G.B., and T. Esch. 2005. Integrative medical therapy: Examination of meditation's therapeutic and global medicinal outcomes via nitric oxide. *International Journal of Molecular Medicine* 16: 621–630.

Stefano, G.B., H. Benson, G.L. Fricchione, and T. Esch. 2005. *The stress response: Always good and when it is bad*. New York: Medical Science International.

Stefano, G.B., G.L. Fricchione, and T. Esch. 2006. Relaxation: Molecular and physiological significance. *Medical Science Monitor: International Medical Journal of Experimental and Clinical Research* 12: HY21–HY31.

Teasdale, J.D., Z.V. Segal, and J.M. Williams. 1995. How does cognitive therapy prevent depressive relapse and why should attentional control (mindfulness training) help? *Behaviour Research and Therapy* 33: 25–39.

Teasdale, J.D., R.G. Moore, H. Hayhurst, M. Pope, S. Williams, and Z.V. Segal. 2002. Metacognitive awareness and prevention of relapse in depression: Empirical evidence. *Journal of Consulting and Clinical Psychology* 70: 275–287.

Thees, S., J. Gobel, G. Jose, R. Bohrhardt, and T. Esch. 2012. Die Gesundheit von Studierenden im Bologna-Prozess: Untersuchungen zu Gesundheitsverhalten, Stress und Wohlbefinden zeigen Handlungsbedarf. *Prävention und Gesundheitsförderung* 3: 196–202.

van Leeuwen, S., N.G. Müller, and L. Melloni. 2009. Age effects on attentional blink performance in meditation. *Consciousness and Cognition* 18: 593–599.

Vestergaard-Poulsen, P., M. van Beek, J. Skewes, C.R. Bjarkam, M. Stubberup, J. Bertelsen, et al. 2009. Long-term meditation is associated with increased gray matter density in the brain stem. *Neuroreport* 20: 170–174.

Vøllestad, J., M.B. Nielsen, and G.H. Nielsen. 2012. Mindfulness- and acceptance-based interventions for anxiety disorders: A systematic review and meta-analysis. *British Journal of Clinical Psychology* 51: 239–260.

Walach, H. 2010. *Wie Achtsamkeit und Meditation unsere Arbeitskultur verändern können*. Berlin: Meditation und Wissenschaft.

Walach, H., E. Nord, C. Zier, B. Dietz-Waschkowski, S. Kersig, and H. Schüpbach. 2007. Mindfulness-based stress reduction as a method for personnel development: A pilot evaluation. *International Journal of Stress Management* 14: 188–198.

Walton, K.G., N.D. Pugh, P. Gelderloos, and P. Macrae. 1995. Stress reduction and preventing hypertension: preliminary support for a psychoneuroendocrine mechanism. *Journal of Alternative and Complementary Medicine* 1: 263–283.

Wang, D.J., H. Rao, M. Korczykowski, N. Wintering, J. Pluta, D.S. Khalsa, et al. 2011. Cerebral blood flow changes associated with different meditation practices and perceived depth of meditation. *Psychiatry Research* 191: 60–67.

Williams, J.M., J.D. Teasdale, Z.V. Segal, and J. Soulsby. 2000. Mindfulness- based cognitive therapy reduces overgeneral autobiographical memory in formerly depressed patients. *Journal of Abnormal Psychology* 109: 150–155.

Yu, X., M. Fumoto, Y. Nakatani, T. Sekiyama, H. Kikuchi, Y. Seki, et al. 2011. Activation of the anterior prefrontal cortex and serotonergic system is associated with improvements in mood and EEG changes induced by Zen meditation practices in novices. *International Journal of Psychophysiology* 80: 103–111.

Zeidan, F., S.K. Johnson, and B.J. Diamond. 2010. Mindfulness meditation improves cognition: evidence of brief mental training. *Consciousness and Cognition* 19: 597–605.

# Meditation Effects in the Social Domain: Self-Other Connectedness as a General Mechanism?

**Fynn-Mathis Trautwein, José Raúl Naranjo, and Stefan Schmidt**

**Abstract** Recent theories and findings in psychology and neuroscience suggest that *self* and *other* are interconnected, both on a conceptual and on a more basic bodily-affective representational level. Such self-other connectedness is supposed to be fundamental to empathy, social bonding and compassion. Meditation techniques – in particular mindfulness and loving-kindness meditation – have been found to foster these social capacities. Therefore, this contribution brings together both fields of research. In a first step, we examine self and other from the perspective of psychology and neuroscience, integrating findings from these fields into a dimension of mental functioning anchored to self-centeredness and self-other-connectedness, respectively. In a second step, we explore how mindfulness and loving-kindness meditation may act differentially upon this dimension. Finally, by referring to a recent experiment from our lab, it is illustrated how research hypotheses can be derived from this framework. Such investigations could help to comprehend meditation effects in the social domain, and more generally, further the scientific understanding of self and other.

## Introduction

Meditation can be characterized as a kind of attention regulation which involves focusing on one's inner experience while refraining from social interactions. During the last decade, the practice of meditation has attracted considerable research interest,

F.-M. Trautwein (✉)
Department of Social Neuroscience, Max Planck
Institute of Human Cognitive and Brain Sciences, Leipzig, Germany
e-mail: trautwein@cbs.mpg.de

J.R. Naranjo • S. Schmidt
Department of Psychosomatic Medicine and Psychotherapy,
University Medical Center, Albert-Ludwigs-University Freiburg, Freiburg, Germany
e-mail: joseraul.naranjo@uniklinik-freiburg.de; stefan.schmidt@uniklinik-freiburg.de

S. Schmidt and H. Walach (eds.), *Meditation – Neuroscientific Approaches and Philosophical Implications*, Studies in Neuroscience, Consciousness and Spirituality 2,
DOI 10.1007/978-3-319-01634-4_10, 

primarily investigating effects on attentional, emotional and cognitive processes (for a recent review, see Sedlmeier et al. 2012). However, there is increasing evidence that meditation also shapes intersubjective experience and behavior, intensifying empathy, compassion, and altruism (e.g. Birnie et al. 2010; Klimecki et al. 2012; Leiberg et al. 2011). Thus, it seems an important question, how such a solitary practice as meditation can foster these social capacities (cf. Kristeller and Johnson 2005).

Especially in the Buddhist tradition, a core aim of meditation is to gain experiential insight into the purportedly illusory nature of the self – often referred to as wisdom. Interestingly, according to this contemplative tradition, the cultivation of prosocial mental qualities such as loving-kindness and compassion is closely related to the development of wisdom: both are said to naturally support and complement each other.

This chapter will draw on psychological and neuroscientific research as well as contemplative accounts of the *self* and it's relation to the *other*. An important notion will be their interconnected nature, both in subjective experience and on neuronal and functional levels. Thus, our aim is to incorporate both a first person perspective and a third person perspective on these phenomena. This will provide a theoretical background in order to understand the intertwined nature of a change in the experience of the self and the capacity to connect more closely with others. This will also give an answer to the puzzling question how the solitary retreat into meditation may bring about interpersonal benefits.

## Self and Other

The nature of the *self* and it's relation to the *other* have been of long-standing interest in philosophy and, more recently, in psychology and neuroscience. While the focus in Western philosophy has often been on being individualistic, seperated, and autonomous, some authors have considered self and other as intimately connected, for example in humanistic and phenomenological approaches (cf. Buber 1923/2009; Husserl 1950/1991; Zahavi 2006). The individualistic view is reflected in a wealth of Western psychological research, which has drawn a picture of the self as a central, or even "totalitarian" (Greenwald 1980) cognitive construct (for a review, see Mischel and Morf 2003). However, the second view – self and other as intimately connected – has recently been endorsed by developmental, social, and intercultural psychology and neuroscience (Han et al. 2011; Decety and Sommerville 2003). Accordingly, on a bodily and affective level of representation, sensations, actions, and emotions of self and other are represented within common representational networks, which could provide a basis for social capacities such as empathizing with and taking the perspective of others (Decety and Chaminade 2003; Gallese 2003). Furthermore, based on ample evidence, some authors have claimed that self and other also overlap on a conceptual level of representation (Aron et al. 2004; Cross et al. 2002). In this view, individuals tend to integrate their social context, and especially close others, into their self concept.

In the following, we will review evidence for self-other connectedness on both levels, starting with the conceptual level of the self. Furthermore, we will point out that considerable interindividual variability exists in self-other connectedness and also that conceptual and bodily-affective levels of self-other representation are closely interwoven. As a result, we integrate these findings into a framework which assumes a single dimension of mental functioning, spanned between self-centeredness on the one end, and self-other connectedness on the other.

## *The Conceptual Level: Self as a Cognitive Construct*

The term *self* is used in manifold ways, that include diverse but interrelated aspects such as self-awareness, identity, and regulation of behavior (for a recent review see Leary and Tangney 2011). A widely accepted view is that one aspect of the self, named the "me self" by William James (1890), can be conceived as a constructed cognitive entity – an organized knowledge, or theory of self-related information – which is reflected in the perception of having a particular identity (e.g. Epstein 1973; Higgins 1996).

In the last decade, neuroimaging studies pointed to certain areas within the cortical midline of the brain to be activated when reflecting upon the self, for example when rating whether certain trait adjectives are descriptive of oneself or not (Qin and Northoff 2011; van der Meer et al. 2010). Furthermore, imaging studies consistently found a set of brain areas which are active during the resting state (i.e. when no task engagement is required). These have been labeled the "default mode of brain function" (Fox et al. 2005; Raichle et al. 2001). These areas include the medial prefrontal cortex (MPFC) and anterior and posterior cingulate cortex,, whose activity is correlated to mind-wandering (Christoff et al. 2009). Interestingly, they strongly overlap with those areas recruited during explicit self-reflection (Gusnard et al. 2001; Qin and Northoff 2011; Whitfield-Gabrieli et al. 2011). Thus, the default mode of brain function seems to be closely tied to the perception of "being a self".

As evidenced by a long history of behavioral experiments, mental functioning appears to be centered on the cognitive construct of the self in several ways. Regarding attention, it is well known that self-relevant information is preferentially processed in a highly automatic mode (Bargh 1982; Geller and Shaver 1976; Moray 1959; Wood and Cowan 1995). In memory, the superiority of self-related material is well supported (for a review see Symons and Johnson 1997). Finally, much evidence has been reported on the motivation to protect and aggrandize the self in order to feel valuable (reviewed in Crocker and Park 2011). In the pursuit of self-esteem, individuals distort reality by means of self-serving biases, which led Greenwald (1980) to speak of the "totalitarian ego". Well known examples are biased causal attributions (Campbell and Sedikides 1999) and unrealistically optimistic self-judgments (e.g. Alicke et al. 1995; Weinstein and Lachendro 1982). All this literature supports the notion of the self as a "special" cognitive construct, that reinforces and aggrandizes itself pervasively, while differentiating itself from others. Some have taken this as a fundamental characteristic of healthy and effective human functioning (Greenwald 1980; S. E. Taylor 1988).

However, some authors have stressed the social origin of the self (e.g. Higgins 1996; Mead 1934; Neisser 1988; Tomasello 1993). Concordantly, depending on the cultural context, people relate to the self in different ways (Heine 2001; Oyserman et al. 2002). Markus and Kitayama (1991) stated that in some cultures – most prominently east Asian cultures – people tend to perceive themselves as interdependent, connected to others and part of a collective, striving to fit-in and relate harmoniously to the social context. This view of the self has been termed the *interdependent self construal*. In contrast, American and other Western cultures are characterized by an *independent self construal*, which emphasizes autonomy, consistency, and distinctiveness of the self.[1] Markus and Kitayama (1991, p. 224) posited that "these construals can influence, and in many cases determine, the very nature of the individual experience, including cognition, emotion and motivation". Indeed, there is increasing evidence for an influence on various self-related processes, for example, the need to have a positive view of the self (reviewed in Heine et al. 1999), the use of self-serving attributional biases (reviewed in Mezulis et al. 2004), and the effect of self-relevance on memory (e.g. Zhu and Zhang 2002). Strengthening the assumption of a causal role of self-construal and demonstrating its dynamic nature, studies that prime a specific kind of construal show effects on perceived closeness, self-concept, cognition, and emotion (reviewed in Oyserman and Lee 2008).

Recently, fMRI studies demonstrated neural correlates of stronger interconnectedness in self-definition characteristic of interdependent cultures (reviewed in Han et al. 2011): In Chinese participants, higher activity in the medial prefrontal cortex – a region typically linked to self-referential processing – was consistently associated with trait judgments regarding oneself and a close other compared to judgments regarding a familiar other. In Westerners, this was only the case for self judgments but not for judgments regarding close others (Zhu et al. 2007). Furthermore, a recent meta-analysis integrated results from 25 brain imaging studies, which investigated trait evaluation related to self and others. The meta-analysis systematically controlled for the degree of self-relatedness of the other, that is, whether a personally close other or a familiar, but not personally close other was used in a study. The results yielded significant differences in MPFC activation between self and familiar others, but not between self and close-others (Murray et al. 2012).

This indicates that also Westerners do not only define themselves as isolated individuals but also, even if to a smaller extent, in terms of their relationships and group memberships. This has been pointed out in detail by social psychologists (Brewer and Gardner 1996; Chen et al. 2006; Turner 1986). Concerning close relationships, Aron and colleagues have argued and demonstrated convincingly that the other becomes, to some extent, integrated into the self (Aron et al. 1991, 1992, 2004). Furthermore, considerable individual variability exists in the extent to which individuals define themselves in terms of their relationships by including others into the self (Cross et al. 2000, 2002, 2003). Cross et al. (2002, p. 414) concluded that for some individuals "the relational self seems to be the default level of

[1] For a comparison of Eastern and Western concepts of the 'self' see also the chapter by Edge in this volume.

self-representation". This has persistent effects on cognitive processes. For example, Cross et al. (2002) compared participants high in interdependent-relational self-definition to highly independent individuals and found that the former group had a better memory for relational information about others and described themselves and a close friend more similarly.

Depending on the situation, state shifts occur at the conceptual level of self-other connectedness: For example, intentionally taking the perspective of another person has been found to involve a conceptual self-other overlap (Davis et al. 1996, 2004; Laurent and Myers 2011). In general, the idea that people include others and groups of others into the self is widely supported (Coats et al. 2000; Goldstein and Cialdini 2007; Smith and Henry 1996; Smith et al. 1999) and discussed as a mechanism for reducing social biases (Galinsky and Moskowitz 2000), fostering social bonds (Galinsky et al. 2005), and acting prosocially (Cialdini et al. 1997).

## *The Bodily-Affective Level: Self as Embodied Agent*

Various authors have suggested that a "minimal self" (Gallagher 2000) or "core self" (Damasio 2010) underlies and precedes the development of the conceptual self. This notion of the self refers to the "consciousness of oneself as an immediate subject of experience, unextended in time" (Gallagher 2000, p. 15).[2] According to Damasio (2010), the core self involves four aspects: a specific spatial standpoint anchored to one's body, a sense of agency, a sense of ownership, and primordial bodily feelings. Thus, this self is closely linked to the body and, by that, to the experience of affective states.

As will be outlined below, research in diverse fields, including phenomenology, developmental psychology, and neuroscience, indicates that this level of selfhood is characterized by a coupling, or connectedness with others. This link seems to be a hard-wired basis of various human social abilities. According to phenomenological accounts, the perception of the other as a bodily being is inherently linked to the experience of one's own self on this minimal, or bodily level. (Thompson 2001; Zahavi 2006). That is, the perception of the other's lived body differs from the perception of mere physical objects, because we directly experience the other "as an embodied subject of experience like oneself" (Thompson 2001, p. 17). Put differently, the perception of the other's body always conveys the sense that this body is another center of orientation in space, that it is a source of voluntary action, and that it is infused with sensations and feelings. Thus, the perception of the other is always, in a very basic sense, empathic, because we experience the other as a sentient, expressive, and intending being.

[2] "Unextended in time" is meant here in opposition to accounts of the self on the conceptual level, which also involve the notion of being an autobiographically extended self. However, as phenomenologists have pointed out, also the moment-to-moment subjective experience does involve temporality, in the sense of being aware of the immediate past and future (Zahavi 2006).

These phenomenological observations are supported by findings from developmental psychology. A primordial coupling of the bodily perception of self and other is strikingly demonstrated by the ability and proneness of newborns less than 72 h old to imitate facial expressions (Meltzoff and Moore 1989). Because in doing so infants use body parts not visible to themselves, their inborn body schema must be structured in a way that allows to map the other's bodily appearance to their own bodily self. According to Meltzoff (2007), this self-other equivalence in perception and action soon develops into the capability to understand others as having mental states such as intentions and emotions. Therefore, Meltzoff (2007, p. 126) argued that "the like me' nature of others is the starting point for social cognition". This embodied link between self and others and its contribution to creating social connections has also been demonstrated in adults: Imitation of gestures and expressions occurs automatically and has therefore been termed the"chameleon effect". It fosters affiliation, empathic responses such as emotional contagion, and prosocial behavior (Chartrand and Bargh 1999; Lakin and Chartrand 2003; Stel et al. 2008; van Baaren et al. 2004).

Pointing to an underlying neural mechanism, brain imaging studies on perspective taking, empathy, and imitation have demonstrated an overlap of brain areas – so-called "shared networks" – involved in representing one's own intentions, emotions, and actions and those of others (reviewed in Decety and Sommerville 2003; Hein and Singer 2008). Echoing Meltzoff's "like me" notion, it has been argued that psychological identification – the (innate) notion that others are like the self – is fundamental to shared neuronal networks and thus lays the ground for human social capacities (Decety and Chaminade 2003; Gallese 2003). Most importantly, a range of empathy related phenomena are supposed to be based on self-other overlap, both on the neuronal and the psychological level (Preston and Hofelich 2012; Preston and de Waal 2002; de Waal 2008). These phenomena include emotional contagion, which refers to the automatic sharing of emotions, empathy in the sense of understanding emotions of others, and compassion as the resulting concern for the wellbeing of others. The role of neuronal self-other overlap in empathic concern for suffering others has been demonstrated by a recent fMRI study: When seeing others in pain, activity in the anterior insula, a part of the network that is also active during first-hand experience of pain, was correlated with self-reported empathic concern and predicted costly helping (Hein et al. 2010).

Evidently, complex social phenomena such as the empathic understanding of others and compassionate concern for others must also, at some level, involve differentiations of self and other in order to avoid a complete conflation. Otherwise, being confronted with suffering would result only in distress, impeding effective communication, prosocial motivation, and action. Some evidence points to the right temporoparietal junction (TPJ) as a crucial node for self-other differentiations (Decety and Lamm 2006). However, little is known about the involved mechanisms. For example, which point of reference is used to create a self/no-self distinction? Nor have the phenomenological dimensions of these differentiations been thoroughly investigated.

Nevertheless, our aim here is to show that at a very basic level an individual must establish a connection to the other as being "like the self" (Decety and Chaminade 2003). But under which conditions does this occur? On the one hand, the situational context seems to have an influence, for example when it allows for cognitive reappraisal (Lamm et al. 2007). On the other hand, individual characteristics play a role: In a study by (Chartrand and Bargh 1999, experiment 3), participants high in trait perspective taking (Davis 1983), compared to those low on that measure, tended to display more automatic imitation. Similarly, neuroscientific studies of empathy yielded higher resonance in shared networks in individuals with high self-reported trait empathy (Avenanti et al. 2009; Jabbi et al. 2007; Lamm et al. 2007). Thus, self-other connectedness at this bodily-affective level seems to vary between persons and situations.

## *Interactions Between Embodied and Conceptual Levels*

Self-other coupling on the bodily level seems to interact with conceptual forms of self-other connectedness. For example, several experiments (reported in Ashton-James et al. 2007) indicated that, when being unobtrusively mimicked by another person, self-construals shifted towards interdependence (see above), and perceived closeness with others increased (as measured by the Inclusion of Other in the Self Scale; Aron et al. 1992). Conversely, in another study, interdependent self-construal was associated with more non-conscious mimicry compared to independent self-construal (van Baaren et al. 2003). Hence, bidirectional interactions seem to occur between the bodily-affective and the conceptual level. Interestingly, the above mentioned study by Hein et al. (2010) revealed different neuronal patterns in response to another person in pain depending on whether a shared social identity existed. Participants of this study were recruited from a local soccer team, while the person in pain was a confederate who posed as an ingroup member, i.e. as a fan of the same group, or as an outgroup member, i.e. as a fan of a rival team. When the person belonged to the outgroup compared to the ingroup, reduced activity in the anterior insula, a part of the shared pain network, and enhanced activity in the nucleus accumbens, a reward related are was observed. Thus, when participants conceptually differentiated between themselves and the person in pain based on group membership, they seemed to reduce empathic responding to the pain while deriving pleasure from the situation. Similarly, a recent study found activity in the shared pain network when seeing a close other experiencing social exclusion, but not in response to a stranger's social suffering (Meyer et al. 2012). Moreover, the shared pain activity in the close other condition was correlated with self-reported self-other overlap. From our point of view, these studies indicate that self-other connectedness at the conceptual level modulates connectedness of self and other at the bodily-affective level, and vice versa.

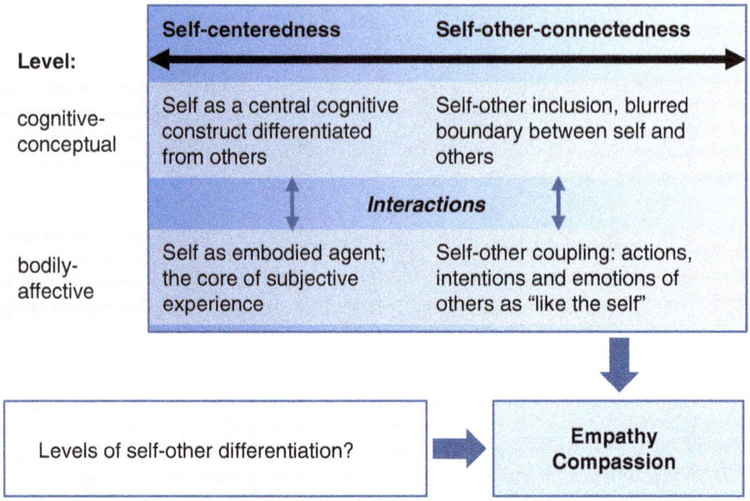


**Fig. 1** A framework of mental functioning based upon a common distinction of a cognitive-conceptual and bodily-affective level of self representation. Because of their tight interactions, both levels can be conceptualized as a dimension which is anchored to self-centeredness on one side and to self-other-connectedness on the other side. Self-other-connectedness is necessary for the experience of empathy and compassion; however, self-other differentiation at a different level might also contribute to the experience of these states

## *Self-Centeredness and Self-Other-Connectedness*

Taken together, the presented literature indicates that the conceptual self as a cognitive unit sharply delimited from others may be, at least in its extreme form, a particular characteristic of Western cultures. However, considerable situational and individual variability does also exist within Western culture. Also at a more basic, embodied level, the tendency to identify with others, to connect, and to empathize with them seems to involve individual variability. Hence, we integrate both levels into a framework of mental functioning (see Fig. 1).

Similar suggestions have been made by others. As an umbrella concept, (Wayment and Bauer 2008) introduced the term *quiet ego* and contrasted it to a loud ego. They see in the quiet ego "a self-identity that is not excessively self-focused but also not excessively other focused – an identity that incorporates others without losing the self" (p. 8). According to the authors, characteristics of a quiet ego include non-defensive awareness in terms of a detachment from egoistic appraisals, interdependence, i.e. conceptual integration of self and others, and compassion for self and others. A related concept is *self compassion* (Neff 2003b). According to Neff, de-emphasizing the individual self in favor of interdependent and shared aspects of identity reduces the importance of self-esteem and opens the possibility for self-compassion – taking a kind and understanding attitude towards the self. Thus, in self-compassionate individuals "the boundaries between self and other are softened" (Neff 2008, p. 95).

Dambrun and Ricard (2011) have proposed a theory, which assumes that psychological functioning varies between self-centeredness and selflessness. In self-centered functioning, the self is perceived as existing continuously and independently from others, and serves as a central point of reference for many psychological activities. High importance is given to the self relative to others, resulting in cognitive biases and the need to aggrandize and protect the self. Thus, this conception of self-centeredness is highly similar to the one we have provided above. Dambrun and Ricard (2011) state that the more psychological functioning shifts towards selflessness, the more the self is perceived as a changing, impermanent, and interdependent element of its social and natural environment. Therefore, motivation and cognition is directed more towards achieving harmony within the elements of this context, resulting in altruism and benevolent emotion and motivation (e.g. kindness, empathy, compassion). Hence, the conception of selflessness in this model involves interdependence or connectedness of self and other, as we have described it. Based on the literature discussed above we complement this notion of selfless functioning by highlighting the non-conceptual levels of self-other connectedness, which have been linked to empathy and compassion in recent research (reviewed in Decety and Sommerville 2003; Hein and Singer 2008).

By drawing on Buddhist philosophy, Dambrun and Ricard (2011) describe another feature of selflessness, i.e. impermanence of the self in the sense that the self is not regarded as a "real", permanently existing entity. This Buddhist view of the self will be discussed in the next section. We will point out that loving-kindness meditation (LKM) and mindfulness meditation (MM) can be differentiated by their aim to either foster self-other connectedness (LKM) or to dis-identify from a reified, permanent self (MM). In order to comprehend the effects of both practices and their interplay from a scientific point of view, we will now apply the framework of self-centeredness and self-other connectedness developed above.

## Buddhist Meditation and the Self

### *Mindfulness Meditation*

At the core of Buddhist teachings lies the idea that there is no such thing like a permanent, truly existing entity called "a self" (Olendzki 2006). Instead, the ordinary experience of having a self that "owns" one's body, emotions, and perceptions and which is the origin of one's thoughts and actions is the outcome of certain mental processes.[3] These processes involve grasping, i.e. the craving for pleasant and the rejection of unpleasant feelings, both of which seem to imply a permanent self which could actually be affected by these feelings. Through repeated grasping, the

[3] In this respect the Buddhist approach to the self resembles the view of cognitive science, however, it takes another step and tries to transform the everyday experience of a "real" self through meditative practice, an endeavor not approached by Western science (cf. Varela et al. 1991).

perception of having a personal identity is instantiated, which results into a style of mental functioning similar to what we have described above as self-centered functioning: "Grasping merely consists of regarding any aspect of experience with the stance This is mine; this is me; this is my self" (Olendzki 2006, p. 257). According to Buddhism, grasping unavoidably leads to psychological distress.

To achieve enduring happiness, one needs to become aware of these processes and to suspend them by disidentifying from the ordinary sense of self. To this aim, MM can be employed. Typically, this practice involves focusing on current somato-sensory and mental events in a non-conceptual, non-judgmental manner, thereby suspending ongoing self-referential thoughts (Farb et al. 2007).

While most of the research devoted to this practice focused on processes of attention and emotion regulation, some support is available that this practice does, as suggested by Buddhist theory, involve a change in the self (see also the chapter by Edge in this volume). One example involves the concept of *decentering* (or *reperceiving*), which refers to a dis-identification from mental contents and has been suggested as a core mechanism of MM (Shapiro et al. 2006). Specifically, the concept designates a shift from "being immersed in the drama of our personal narrative or life story" towards being able "to stand back from (witness) our story' about who and what *we* ultimately are" (Shapiro et al. 2006, p. 379). Decentering, as assessed with self-report scales (Fresco et al. 2007; Lau et al. 2006), has been found to be an outcome of MM, both in terms of trait and state changes (Feldman et al. 2010; Orzech et al. 2009).

More direct evidence comes from brain imaging research. In an fMRI study (Farb et al. 2007), participants who had attended an 8-week mindfulness training were compared with novices. Brain activity was recorded while reading trait words in a mindful self-focus (being aware of body sensations, i.e. the embodied self) or in a narrative self-focus (reflecting upon what the trait words meant for oneself, i.e. the conceptual self). During experiential self-focus, reduced activity in the medial prefrontal cortex, an area associated with self-referential processing, and enhanced activity in viscerosomatic areas was observed in comparison to the narrative self-focus in participants trained in MM. In contrast, this dissociation of narrative and experiential self-focus was weaker in untrained participants. In another study, trait mindfulness scores were negatively correlated with resting state activity in the medial prefrontal cortex, indicating that mindfulness does habitually attenuate self-referential mental activity (Way et al. 2010). In a similar vein, changes in default mode network connectivity have been associated to intensive MM (V. A. Taylor et al. 2013; J. A. Brewer et al. 2011).

In summary, this evidence suggests that MM does in fact lead to a decrease in self-centered functioning. However, these results hardly reflect the radical transformation of one's sense of self, which is the ultimate goal of Buddhist meditation practices.[4] Rather, they seem to reflect dis-identification from some mental contents on the conceptual level, which is likely to occur already after less meditation practice.

[4] Several authors have approached these states of advanced or even complete spiritual transformation from a scientific stance. However, they differ in their scientific approach and in the phenomenologies assumed to be present in these states (for examples, see Albahari 2011; and chapters by Austin and Edge in this volume).

## *Loving-Kindness Meditation*

In Buddhism, the insight into the illusory nature of the self is regarded as wisdom, while compassion and loving-kindness are regarded as necessary complements to it (Wallace 2001b). This is often expressed by a metaphor: "Wisdom and compassion are like the two wings of a bird: Both are necessary for the bird to soar" (C. Feldman 2005, p. 15). While MM is aimed at wisdom, with loving-kindness meditation (LKM) one intends to cultivate unconditional and impartial kindness towards the self and others.[5] This quality, originally named *mettā* in Pali and *maitrī* in Sanskrit, and often translated as *loving-kindness*, is described by feelings of connectedness to others and a heartfelt wish for their well-being. The experience of this state is said to result in prosocial motivation and behavior (Wallace 2001b).

Usually, loving-kindness is cultivated together with three other qualities, namely compassion, empathetic joy and equanimity (Salzberg 1995). These qualities constitute the *four immeasurables*. They can be practiced separately, but because of their interrelated nature, development in one will also advance the others. While loving-kindness yearns for the well-being of others, compassion is understood as the strong wish to end the suffering of others (Wallace 2001b). Evidently, both are strongly related and thus "really two sides of the same coin" (Wallace 2001a, p. 219). Empathetic joy means taking part in other people's delight. Finally, equanimity means dismantling separations between self, friends, or enemies so that loving-kindness and compassion can be experienced without bias. Typically, in LKM one directs positive wishes at specific persons, including oneself, groups, human beings in general, or even at all sentient beings. The practitioner can also imagine the warmth or light of the "radiated" love or visualize the person whom it is "sent" to. Typical wishes are: "May you be free from danger", "may you have mental and physical happiness". These wishes are thought to help establish feelings of love and kindness. Thus, the core of the practice is not the recitation of these phrases but the mindful awareness of the feelings connected to them. Practitioners often begin by focusing on themselves. With further progress in the meditative development and also during the course of one meditation session, the practitioner expands the feelings, usually in the following order: (1) self, (2) close other, (3) neutral person (4) difficult person or "foe" (5) groups of others, (6) all sentient beings. The aim is to experience the same degree of loving-kindness in all of these instances.

Based on the phenomenology of the practice, we suggest that a core element of it is to increase self-other connectedness, both at the conceptual and at the bodily-affective level. Impacting the conceptual level, the practice involves the realization

---

[5] Across Buddhist traditions, several techniques exist which are aimed at cultivating compassion and loving-kindness (e.g. Rinpoche and Mullen 2005). We focus on an account from Theravada Buddhism, which is based on the Mettā Sutta of the Pali Canon and taught in the modern Vipassana movement (Buddharakkhita 1995; Salzberg 1995). Beginning research on these practices is reviewed in (Hofmann et al. 2011).

that all humans, and eventually, all sentient beings, strive to gain happiness and reduce suffering. However, a strengthening of the embodied link between self and other might even be more important. This assumption is based on two observations: (i) Developing loving-kindness for oneself involves looking at the self from the perspective of another person. Thus, during this practice a kind of exchange of self and other occurs, or as Wallace (Wallace 2001b, p. 10) puts it: "one has entered into an I-you relationship with oneself". (ii) When developing loving-kindness for a widening circle of others until extending it to all sentient beings, one empathizes with their striving for happiness and their wish to be free of suffering. By that, we assume, meditators extend and generalize the "like me" link of empathy discussed above. Even though this might involve conceptual "top-down" influences, for example by thinking of a shared humanity, the core is a bodily-affective resonance – often described as a "*heartfelt* wish" – with imagined others. Thus, this practice seems to involve an empathic coupling of self and an extending range of others, increasingly blurring distinctions between them.

A recent study (Colzato et al. 2012) yielded evidence for the embodied nature of this process by employing the social Simon task, a spatial compatibility task performed jointly by two persons (Sebanz et al. 2003). Response costs induced by joint action in this task have been taken to evidence shared representation of others and one's own actions. In Colzato et al. (2012), practicing Buddhists demonstrated stronger interferences in the joint task compared to matched controls. The authors suggested that this effect represents a larger self-other integration and relate it to the concepts of compassion and connectedness endorsed by Buddhists. However, it is not possible to relate this finding to specific meditative practices, as these are not reported by the authors. More conclusive in this respect is a laboratory experiment with meditation novices (Hutcherson et al. 2008). A short, guided LKM was compared to a neutral imagery task. LKM lead to larger increases in explicit ratings of connectedness and positivity and in implicit positive evaluations measured with an affective priming task. A recent experiment addressed the question whether LKM would actually increase prosocial behavior (Leiberg et al. 2011). The latter was assessed during an interactive computer game in an implicit and ecologically valid manner. Relative to baseline, helping behavior increased after a 1-day training in LKM, but not after 1-day training in a memorization technique. Finally, an fMRI study (Lutz et al. 2008) measured neuronal responses to neutral and emotional sounds in expert meditators (10,000–50,000 h of meditation practice including compassion meditation) and in age-matched controls. During a loving-kindness-compassion state versus a resting state, responses in networks associated with the experience and sharing of affective states were increased. This effect was stronger in meditators than in controls. In conclusion, when assuming that LKM increases bodily-affective resonance between self and others, these results are in line with recent accounts of empathy, compassion and altruism, which regard self-other coupling as the basis of these prosocial phenomena (Preston and Hofelich 2012; Preston and de Waal; de Waal 2008).

### *The Interplay Between Mindfulness and Loving-Kindness*

Reflecting the close relationship of mindfulness and loving-kindness, or compassion, in Buddhist thought, there seems to be a bidirectional interplay between them. On the one hand, mindfulness is said to naturally result in connectedness with others (Salzberg 2011). On the other hand, the practice of LKM supposedly helps to refrain from engaging in self-referential thought and thus supports mindfulness (Hofmann et al. 2011).

Our theoretical framework allows conceiving both directions of the interplay. Firstly, conceptual boundaries between self and other seem to prevent individuals from experiencing empathy (as indicated by the studies from Hein et al. 2010; Meyer et al. 2012). Simply reducing self-referential mental activity may therefore increase empathic receptivity (for detailed discussion of a similar point, see Schuster 1979). Secondly, the focus on somatosensory sensations often involved in MM presumably leads to an increase of connectedness between self and others through awareness of the bodily coupling. For example, when perceiving others' suffering, mindfulness may increase awareness of co-perceived bodily and affective empathic reactions. By being mindfully open the meditator's own potential for suffering can resonate with the other person's suffering (see also Schmidt 2004). Lastly, mindfulness does already entail an attitude which exchanges self and other: "By *observing* one's own body, rather than simply *identifying* with it, one cultivates a kind of self-alterity, by experiencing one's own body simply as a matrix of phenomena, rather than as a self" (Wallace 2001b, p. 6). Taken together, these mechanisms might explain why mindfulness contributes to changes in the social domain: Mindfulness inhibits self-centeredness, primarily on the conceptual level, thereby enabling to experience connectedness to others at the bodily-affective level. In fact, some correlational and longitudinal studies suggest that both, trait mindfulness and MM, are associated with empathy, concern for others, and interpersonal closeness (summarized in Block-Lerner et al. 2007; see also Dekeyser et al. 2008). Moreover, the concept of self-compassion comprises the separate but intercorrelated subcomponents self-kindness, mindfulness, and common humanity or a sense of connectedness with others (Neff 2003a, b). In line with our suggestions, self-compassion was correlated with greater self-reported compassion for humanity, empathic concern, perspective taking, and altruism in meditators and community adults; however, some of these correlations were not observed among college students (Neff and Pommier 2013).

Our theoretical suggestions agree with a 2-stage model of the effects of meditation on empathy, compassion and altruism proposed by Kristeller and Johnson (2005). The first step involves heightening awareness of and disengaging from dysfunctional reaction patterns through basic meditation practice. This includes disengagement from self-centered and self-protective patterns, facilitating the ability to experience needs of others. Thus, this step entails the effects of MM described above. According to the model, a second step is necessary to increase empathy and compassion: an explicit focus on love and compassion towards others as in LKM. As explained in detail above, the framework of self-other connectedness and its application to LKM allows to propose an underlying mechanism.

Additionally, the framework also allows for conceiving the other direction of the interplay between LKM and MM: According to interactions between conceptual and embodied levels, an increase in self-other connectedness in the affective, embodied domain through LKM should reduce self-centeredness on the conceptual level. Indirect evidence comes from a randomized, longitudinal field experiment on LKM (Fredrickson et al. 2008). Over the course of a LKM training, a range of daily assessed positive emotions increased in the meditation group but not in the control group. These gains in positive emotions were associated with increases in trait mindfulness.

More direct evidence for a close relationship between the development of loving-kindness and a change in the self comes from a recent study conducted at our lab (Trautwein et al., submitted). In the next section we will shortly discuss this study in order to illustrate how the conception of a dimension of self-other connectedness can be used to derive research hypotheses on meditation effects in the social domain.

## Investigating the Impact of Loving-Kindness Meditation on Self-Other Connectedness

Is the cultivation of loving-kindness through LKM accompanied by a decrease in self-centeredness and an increase in self-other connectedness? We addressed this question, which is a consequence of the framework developed above, in a recent experiment (Trautwein et al., submitted). As an indicator of self-other connectedness, event-related potentials (ERP) of the EEG elicited by one's own and a close other's face were assessed in long-term practitioners of LKM and in a closely matched control group. Of specific interest for our study was the P300 component in the ERP, which is usually assessed in oddball paradigms, i.e. in response to stimuli, which appear infrequently against a background of standard stimuli. The P300 is a positive potential in the EEG occurring approximately 300–500 ms after presentation of stimuli which engage an individual's attentional resources; and it has maximal amplitudes at frontal to parietal midline sites (Comerchero and Polich 1999; Polich 2007). Previous studies consistently found larger P300 amplitudes for self-related stimuli including one's own name, face, autobiographical information and self-referent pronouns compared to not self-related stimuli (e.g. Gray et al. 2004; Tacikowski and Nowicka 2010; Zhao et al. 2009, 2010, 2011). Therefore, the self-relevance effect, defined as the difference between self- and other-related P300 amplitudes, might reflect the high priority given to the individual self, which is characteristic of self-centered functioning, at an implicit level. Thus, if LKM leads to a stronger integration, or connectedness, of self-other representations, a smaller self-relevance effect in terms of reduced differences between self- and other-related P300 should be associated with this practice. This hypothesis was tested in our study.

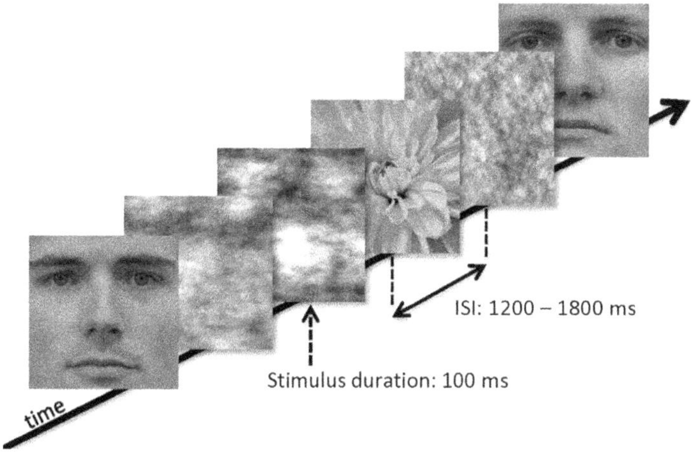


**Fig. 2** The oddball task in Trautwein et al. (submitted) consisted of a fast stimulus sequence, containing a flower (20 % of the trials) which required a button press (the target stimulus), self and close other's face (distracter stimuli, 20 % of the trials), and scrambled versions of these stimuli (standard stimuli, 60 % of the trials). Two blocks of a total of 375 trials were presented. ISI = inter-stimulus-interval

Specifically, the study investigated trait and state effects of LKM on self-other connectedness. State effects refer to transient shifts in self-other connectedness which may occur during and directly after the meditative practice, whereas trait effects correspond to lasting changes due to long-term practice of LKM. As the study's main variable of interest, P300 amplitudes in response to self and a close other's face were measured during an oddball task (see Fig. 2 for a description). To measure trait effects, 11 long-term practitioners of LKM were compared to 11 control participants matched for age, sex, handedness, and education. Meditators came from different Buddhist traditions, but all had a regular LKM practice of, on average, 9 years (SD = 8 years) and they had practiced meditation on average for 12 years (SD = 9 years). In order to investigate state effects of LKM, meditators were measured additionally after a short LKM state and after a closely matched control state. The former involved 10 min of LKM directed at the close other, whose picture was used to assess P300 amplitudes. In the control task, participants were asked to think about the close other in a neutral manner. The order of these two priming tasks was counterbalanced and they were always followed by P300 assessments.

As additional outcomes of LKM, the study assessed self-compassion (Bartel 2009; Neff 2003a) and compassionate love (Sprecher and Fehr 2005) with questionnaires. Compassionate love is a concept resembling the Buddhist notion of compassion and loving-kindness. It is defined as "an attitude toward other(s), either close others or strangers or all of humanity; containing feelings, cognitions, and behaviors that are focused on caring, concern, tenderness, and an orientation toward supporting, helping, and understanding the other(s)" (Sprecher and Fehr 2005, p. 630).

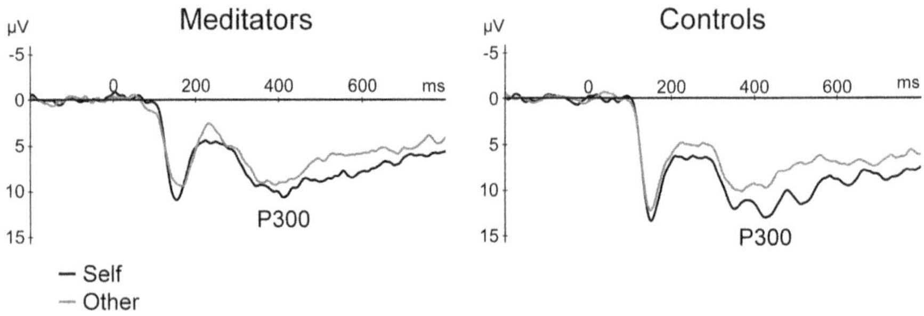


**Fig. 3** Exemplary illustration of results: event-related potentials elicited by self-face and other-face at Pz electrode site for meditators and controls

The results of the study yielded significant differences in questionnaire scores: compared to the control group, meditators reported to experience more compassionate love towards strangers and all of humanity. Similarly, meditators described themselves as being more compassionate towards themselves. In agreement with these results, other studies found associations between meditation practice and compassionate love (Leiberg et al. 2011; Sprecher and Fehr 2005) and self-compassion (reviewed in Barnard and Curry 2011).

As an indicator of self-centeredness vs. self-other connectedness, the study assessed P300 amplitudes. A three factorial ANOVA of P300 amplitudes [group (meditators, controls) × stimulus (self, other) × electrode (Fz, FCz, Cz, CPz, Pz)] yielded a significant main effect for stimulus. Thus, concurring with previous studies (e.g. Gray et al. 2004), larger P300 amplitudes were elicited by self faces compared to close other faces (see Fig. 3 for an exemplary illustration). Most importantly, a significant group x stimulus x electrode interaction supported the hypothesis that a change in self- and other-related processing is associated with meditation practice. Post-hoc analysis indicated that at frontal and central sites, differences between self and other tended to be similar in both groups (or slightly higher in controls). At a posterior site (Pz), however, the self-relevance effect was more pronounced in controls than in meditators.

At Pz electrode location, P300 differences (self minus other) were correlated with the individual duration of meditation experience, suggesting that this effect was in fact an outcome of meditation practice. Furthermore, in agreement with the framework developed above, a close relationship between decentering of the self and a stronger affective connectedness with others was indicated by a negative correlation between the self-relevance effect and self-reported compassion for others. This correlation was driven by P300 amplitudes elicited by the self face, that is, they tended to be smaller in those participants who reported experiencing more compassion for strangers and humanity.

A crucial point of our theoretical framework was that individuals may differ depending on whether they represent themselves more in a way that is inherently connected to others (self-other-connectedness), or rather as an entity separate from others (self-centeredness). If the self-relevance effect is regarded as a particular characteristic of self-centeredness, these findings give support to the idea that LKM shifts an individual's mode of representation towards self-other connectedness.

However, an alternative explanation of this finding must be considered: As the P300 component elicited by distracter stimuli (P3a) presumably reflects attentional resource allocation (Polich 2007), and meditation training has been associated with changes in attentional processes (Brefczynski-Lewis et al. 2007), one might argue that this finding reflects a change in attentional domains, but not in the self. In fact, effects of a meditative state on auditory P3a have been reported and interpreted as a decreased attentional engagement elicited by distracting stimuli (Cahn and Polich 2009). The counter-argument to this alternative explanation is that the change in P300 amplitudes in our study was specific to the type of stimulus (self vs. other) as indicated by the group × stimulus × electrode interaction. A general reduction in attention allocation to distracters would have been reflected in a main effect of group and not mediated by the stimulus type. Therefore, it is likely that the findings of the study do represent a reduced self-centeredness in meditators.

Furthermore, as predicted by the framework, the results suggest that reduced self-centeredness is related to increased empathic and compassionate connectedness with others. This assumption is supported by the finding that the reduction in self-centeredness, as assessed by the ERP data, was accompanied by increased compassion for strangers and humanity, as measured by the compassionate love scale (Sprecher and Fehr 2005). More specifically, one implication of our framework was that sharp conceptual boundaries between self and other (self-centeredness) might inhibit the hard-wired bodily-affective coupling between them. The inverse relationship between P300 amplitudes elicited by one's own face and unspecific compassionate love provide support for this assumption.

However, due to their cross-sectional nature these findings do not provide direct evidence for a causal link between LKM and increased self-other connectedness. In order to test for such a causal effect of LKM, the study also assessed P300 components directly after two priming conditions (LKM and a control task). However, the results did not yield evidence for a state effect of LKM: After meditators had completed a short LKM, P300 amplitudes did not differ from measurements taken after a structurally similar control task involving other-related thinking. Several explanations may account for this lack of state effects: For example, the P300 measure might not be sensitive enough to the bodily-affective connectedness induced by LKM; or the state effect might be too transient in order to be captured by the paradigm, which was applied only after the meditation itself. Future studies could clarify this by employing measures which can be taken during the meditation or which tap more into the embodied level of the self.

Taken together, the results in the group comparison of our study must be interpreted carefully with regard to their causal nature. Although both groups were matched closely regarding potentially confounding factors, the relationship between LKM practice, reduced self-centeredness and increased engagement with others might not be as direct as assumed, but moderated by other variables. Similarly, the direction of causality cannot be derived from these findings. Finally, another reason for cautious interpretation of the results is the small sample size of this study.

Nevertheless, it seems a promising endeavor to further investigate the relationship between neurophysiological and behavioral measures that tap into the structure of the self – it's centrality or openness to others – and outcomes in the interpersonal domain, including empathy, compassion and prosocial behavior.

## Conclusion

In this chapter we integrated findings on the structure of the self and it's connectedness with others and pointed out how this theoretical background can improve the understanding of meditation effects in the social domain. Converseley, we hope that the consideration of contemplative accounts may also complement the scientific understanding of self and other.

According to recent work in social psychology and neuroscience, connectedness, or the overlap of self and other representations, is vital for a variety of interpersonal capabilities such as empathy, compassion, and the creation and maintenance of social bonds. Buddhist contemplative accounts highlight how the concept of a permanent, individual self causes distress, while disengagement from it through MM will increase happiness and a loving and caring connectedness with others. The proposed framework allows for conceiving aspects of both of these approaches at once: While, on a conceptual level, mental functioning tends to be centered on an individual self, people do also include others – to some extent – into their concept of self. Furthermore, a hard-wired bodily and affective link provides a basis for intersubjective sharing of mental states. Because both of these levels seem to be closely interconnected, the centeredness on the conceptual self does also limit the ability to affectively connect with others. Thus, a decrease in self-centeredness regarding the conceptual self leads to increased affective self-other-connectedness. Inversely, reinforcing the affective link to others does also foster self-other connectedness on the conceptual level.

Based on this framework, we suggest that meditation practices such as MM and LKM lead to a change in the self via two intertwined roads:

- MM decreases self-centeredness primarily on the conceptual level; or in other words, helps disengaging from conceptual self-related contents.
- LKM increases self-other connectedness on the bodily-affective level; that is, LKM increases identification with others building upon a hard-wired bodily-affective link between self and other.

A recent study from our lab (Trautwein et al., submitted) was discussed in order to illustrate how empirical research questions can be derived from this framework. The results of this study support our theoretical considerations, however they must be regarded as preliminary due to their cross-sectional nature and the small sample size of that study.

In order to comprehend phenomena such as empathy and compassion, which essentially contribute to individual and societal well-being, the investigation of the self-other relation seems to be a worthwhile endeavor. The meeting of psychological and neuroscientific approaches with contemplative techniques such as MM and LKM might contribute here, especially to understanding processes involved in constructing and dissolving boundaries between self and other. From a theoretical perspective, the scrutiny of these dynamics is a crucial challenge towards a more unified view of selfhood and intersubjectivity. From a practical stance, this might contribute to the development of scientifically informed strategies for reducing social conflict and fostering human flourishing in relation to others.

## References


Albahari, M. 2011. Nirvana and ownerless consciousness. In *Self, no self?: Perspectives from analytical, phenomenological, and Indian traditions*, ed. M. Sidertis, E. Thompson, and D. Zahavi. Oxford, UK: Oxford University Press.

Alicke, M.D., D.L. Breitenbecher, T.J. Yurak, and D.S. Vredenburg. 1995. Personal contact, individuation, and the better-than-average effect. *Journal of Personality* 68(5): 804–825.

Aron, A., E.N. Aron, M. Tudor, and G. Nelson. 1991. Close relationships as including other in the self. *Journal of Personality and Social Psychology* 60(2): 241–253.

Aron, A., E.N. Aron, and D. Smollan. 1992. Inclusion of other in the self scale and the structure of interpersonal closeness. *Journal of Personality and Social Psychology* 63(4): 596–612.

Aron, A., T. McLaughlin-Volpe, D.J. Mashek, G. Lewandowski, S. Wright, and E.N. Aron. 2004. Including others in the self. *European Review of Social Psychology* 15(1): 101–132.

Ashton-James, C., R.B. van Baaren, T.L. Chartrand, J. Decety, and J. Karremans. 2007. Mimicry and me: The impact of mimicry on self-construal. *Social Cognition* 25(4): 518–535.

Avenanti, A., I. Minio-Paluello, I. Bufalari, and S.M. Aglioti. 2009. The pain of a model in the personality of an onlooker: Influence of state-reactivity and personality traits on embodied empathy for pain. *NeuroImage* 44(1): 275–283.

Bargh, J.A. 1982. Attention and automaticity in the processing of self-relevant information. *Journal of Personality and Social Psychology* 43(3): 425–436.

Barnard, L.K., and J.F. Curry. 2011. Self-compassion: Conceptualizations, correlates, & interventions. *Review of General Psychology* 15(4): 289–303.

Bartel, C. 2009. Self-compassion: Ein Neues Konzept Positiver Selbstzuwendung. Eine Deutsche Adaptierung Der "Self Compassion Scale". Diplomarbeit, Albert-Ludwigs-Universität Freiburg.

Birnie, K., M. Speca, and L.E. Carlson. 2010. Exploring self-compassion and empathy in the context of mindfulness-based stress reduction (MBSR). *Stress and Health* 26(5): 359–371.

Block-Lerner, J., C. Adair, J.C. Plumb, D.L. Rhatigan, and S.M. Orsillo. 2007. The case for mindfulness-based approaches in the cultivation of empathy: Does nonjudgmental, present-moment awareness increase capacity for perspective-taking and empathic concern? *Journal of Marital and Family Therapy* 33(4): 501–516.

Brefczynski-Lewis, J., A. Lutz, H.S. Schaefer, D.B. Levinson, and R.J. Davidson. 2007. Neural correlates of attentional expertise in long-term meditation practitioners. *Proceedings of the National Academy of Sciences of the United States of America* 104(27): 11483–11488.

Brewer, M.B., and W. Gardner. 1996. Who is this "we"? Levels of collective identity and self representations. *Journal of Personality and Social Psychology* 71(1): 83–93.

Brewer, J.A., P.D. Worhunsky, J.R. Gray, Y.-Y. Tang, J. Weber, and H. Kober. 2011. Meditation experience is associated with differences in default mode network activity and connectivity. *Proceedings of the National Academy of Sciences of the United States of America* 108(50): 20254–20259.

Buber, M. 1923/2009. *Ich und Du (Text)*. Ditzingen: Reclam.

Buddharakkhita, A. 1995. *Metta: The philosophy and practice of universal love*. Kandy: Buddhist Publication Society.

Cahn, B.R., and J. Polich. 2009. Meditation (Vipassana) and the P3a event-related brain potential. *International Journal of Psychophysiology* 72(1): 51–60.

Campbell, W.K., and C. Sedikides. 1999. Self-threat magnifies the self-serving bias: A meta-analytic integration. *Review of General Psychology* 3(1): 23–43.

Chartrand, T.L., and J.A. Bargh. 1999. The chameleon effect: The perception-behavior link and social interaction. *Journal of Personality and Social Psychology* 76(6): 893–910.

Chen, S., H.C. Boucher, and M.P. Tapias. 2006. The relational self revealed: Integrative conceptualization and implications for interpersonal life. *Psychological Bulletin* 132(2): 151–179.

Christoff, K., A.M. Gordon, J. Smallwood, R. Smith, and J.W. Schooler. 2009. Experience sampling during fMRI reveals default network and executive system contributions to mind wandering. *Proceedings of the National Academy of Sciences of the United States of America* 106(21): 8719–8724.

Cialdini, R.B., S.L. Brown, B.P. Lewis, C. Luce, and S.L. Neuberg. 1997. Reinterpreting the empathy-altruism relationship: When one into one equals oneness. *Journal of Personality and Social Psychology* 73(3): 481–494.

Coats, S., E.R. Smith, H.M. Claypool, and M.J. Banner. 2000. Overlapping mental representations of self and in-group: Reaction time evidence and its relationship with explicit measures of group identification. *Journal of Experimental Social Psychology* 36(3): 304–315.

Colzato, L.S., H. Zech, B. Hommel, R. Verdonschot, W.P.M. van den Wildenberg, and S. Hsieh. 2012. Loving-kindness brings loving-kindness: The impact of Buddhism on cognitive self-other integration. *Psychonomic Bulletin and Review* 19(3): 541–545.

Comerchero, M.D., and J. Polich. 1999. P3a and P3b from typical auditory and visual stimuli. *Clinical Neurophysiology* 110(1): 24–30.

Crocker, J., and L.E. Park. 2011. Seeking Self-Esteem. In *Handbook of self and identity*, 2nd ed, ed. M.R. Leary and J.P. Tangney. New York: Guilford Press.

Cross, S.E., P.L. Bacon, and M.L. Morris. 2000. The relational-interdependent self-construal and relationships. *Journal of Personality and Social Psychology* 78(4): 791–808.

Cross, S.E., M.L. Morris, J.S. Gore, R. Kruger, N.D. Katter, J. Neal, et al. 2002. Thinking about oneself and others: The relational-interdependent self-construal and social cognition. *Journal of Personality and Social Psychology* 82(3): 399–418.

Cross, S.E., J.S. Gore, and M.L. Morris. 2003. The relational-interdependent self-construal, self-concept consistency, and well-being. *Journal of Personality and Social Psychology* 85(5): 933–944.

Damasio, A.R. 2010. *Self comes to mind: Constructing the conscious brain*. New York: Random House.

Dambrun, M., and M. Ricard. 2011. Self-centeredness and selflessness: A theory of self-based psychological functioning and its consequences for happiness. *Review of General Psychology* 15(2): 138–157.

Davis, M.H. 1983. Measuring individual differences in empathy: Evidence for a multidimensional approach. *Journal of Personality and Social Psychology* 44(1): 113–126.

Davis, M.H., L. Conklin, A. Smith, and C. Luce. 1996. Effect of perspective taking on the cognitive representation of persons: A merging of self and other. *Journal of Personality and Social Psychology* 70(4): 713–726.

Davis, M.H., T. Soderlund, J. Cole, E. Gadol, M. Kute, M. Myers, et al. 2004. Cognitions associated with attempts to empathize: How do we imagine the perspective of another? *Personality and Social Psychology Bulletin* 30(12): 1625–1635.

de Waal, F.B.M. 2008. Putting the altruism back into altruism: The evolution of empathy. *Annual Review of Psychology* 59: 279–300.

Decety, J., and T. Chaminade. 2003. When the self represents the other: A new cognitive neuroscience view on psychological identification. *Consciousness and Cognition* 12(4): 577–596.

Decety, J., and C. Lamm. 2006. Human empathy through the lens of social neuroscience. *The Scientific World Journal* 6: 1146–1163.

Decety, J., and J.A. Sommerville. 2003. Shared representations between self and other: A social cognitive neuroscience view. *Trends in Cognitive Sciences* 7(12): 527–533.

Dekeyser, M., F. Raes, M. Leijssen, S. Leysen, and D. Dewulf. 2008. Mindfulness skills and interpersonal behaviour. *Personality and Individual Differences* 44(5): 1235–1245.

Epstein, S. 1973. The self-concept revisited. Or a theory of a theory. *The American Psychologist* 28(5): 404–416.

Farb, N.A.S., Z.V. Segal, H. Mayberg, J. Bean, D. McKeon, Z. Fatima, et al. 2007. Attending to the present: Mindfulness meditation reveals distinct neural modes of self-reference. *Social Cognitive and Affective Neuroscience* 2(4): 313–322.

Feldman, C. 2005. *Compassion: Listening to the cries of the world*. Berkeley: Rodmell Press.

Feldman, G., J. Greeson, and J. Senville. 2010. Differential effects of mindful breathing, progressive muscle relaxation, and loving-kindness meditation on decentering and negative reactions to repetitive thoughts. *Behaviour Research and Therapy* 48(10): 1002–1011.

Fox, M.D., A.Z. Snyder, J.L. Vincent, M. Corbetta, D.C. Van Essen, and M.E. Raichle. 2005. The human brain is intrinsically organized into dynamic, anticorrelated functional networks. *Proceedings of the National Academy of Sciences of the United States of America* 102(27): 9673–9678.

Fredrickson, B.L., M.A. Cohn, K.A. Coffey, J. Pek, and S.M. Finkel. 2008. Open hearts build lives: Positive emotions, induced through loving-kindness meditation, build consequential personal resources. *Journal of Personality and Social Psychology* 95(5): 1045–1062.

Fresco, D.M., M.T. Moore, M.H.M. van Dulmen, Z.V. Segal, S.H. Ma, J.D. Teasdale, et al. 2007. Initial psychometric properties of the experiences questionnaire: Validation of a self-report measure of decentering. *Behavior Therapy* 38(3): 234–246.

Galinsky, A.D., and G.B. Moskowitz. 2000. Perspective-taking: decreasing stereotype expression, stereotype accessibility, and in-group favoritism. *Journal of Personality and Social Psychology* 78(4): 708–724.

Galinsky, A.D., G. Ku, and C.S. Wang. 2005. Perspective-taking and self-other overlap: Fostering social bonds and facilitating social coordination. *Group Processes and Intergroup Relations* 8(2): 109–124.

Gallagher, S. 2000. Philosophical conceptions of the self: Implications for cognitive science. *Trends in Cognitive Sciences* 4(1): 14–21.

Gallese, V. 2003. The manifold nature of interpersonal relations: The quest for a common mechanism. *Philosophical Transactions of the Royal Society of London B* 358: 517–528.

Geller, V., and P. Shaver. 1976. Cognitive consequences of self-awareness. *Journal of Experimental Social Psychology* 12(1): 99–108.

Goldstein, N.J., and R.B. Cialdini. 2007. The spyglass self: A model of vicarious self-perception. *Journal of Personality and Social Psychology* 92(3): 402–417.

Gray, H., N. Ambady, W.T. Lowenthal, and P. Deldin. 2004. P300 as an index of attention to self-relevant stimuli. *Journal of Experimental Social Psychology* 40(2): 216–224.

Greenwald, A.G. 1980. The totalitarian ego: Fabrication and revision of personal history. *American Psychologist* 35(7): 603–618.

Gusnard, D.A., E. Akbudak, G.L. Shulman, and M.E. Raichle. 2001. Medial prefrontal cortex and self-referential mental activity: Relation to a default mode of brain function. *Proceedings of the National Academy of Sciences of the United States of America* 98(7): 4259–4264.

Han, S., L. Zhang, and Z. Ying. 2011. The relation between the self and others: A transcultural neuroimaging approach. In *Culture and neural frames of cognition and communication*, ed. S. Han and E. Pöppel, 77–91. Heidelberg: Springer.

Hein, G., and T. Singer. 2008. I feel how you feel but not always: The empathic brain and its modulation. *Current Opinion in Neurobiology* 18(2): 153–158.

Hein, G., G. Silani, K. Preuschoff, C.D. Batson, and T. Singer. 2010. Neural responses to ingroup and outgroup members' suffering predict individual differences in costly helping. *Neuron* 68(1): 149–160.

Heine, S.J. 2001. Self as cultural product: An examination of East Asian and North American selves. *Journal of Personality* 69(6): 881–906.

Heine, S.J., D.R. Lehman, H.R. Markus, and S. Kitayama. 1999. Is there a universal need for positive self-regard? *Psychological Review* 106(4): 766–794.

Higgins, E.T. 1996. The "self digest": Self-knowledge serving self-regulatory functions. *Journal of Personality and Social Psychology* 71(6): 1062–1083.

Hofmann, S.G., P. Grossman, and D.E. Hinton. 2011. Loving-kindness and compassion meditation: Potential for psychological interventions. *Clinical Psychology Review* 31(7): 1126–1132.

Husserl, E. 1950/1991. *Husserliana (1): Cartesianische Meditationen und Pariser Vorträge*. Dordrecht: Nijhoff.

Hutcherson, C.A., E.M. Seppala, and J.J. Gross. 2008. Loving-kindness meditation increases social connectedness. *Emotion* 8(5): 720–724.

Jabbi, M., M. Swart, and C. Keysers. 2007. Empathy for positive and negative emotions in the gustatory cortex. *NeuroImage* 34(4): 1744–1753.

James, W. 1890. *The principles of psychology*. New York: Henry Holt and Company.

Klimecki, O.M., S. Leiberg, C. Lamm, and T. Singer. 2013. Functional neural plasticity and associated changes in positive affect after compassion training. *Cerebral Cortex* 23(7): 1552–1561.

Kristeller, J.L., and T. Johnson. 2005. Cultivating loving kindness: A two-stage model of the effects of meditation on empathy, compassion, and altruism. *Zygon* 40(2): 391–408.

Lakin, J.L., and T.L. Chartrand. 2003. Using nonconscious behavioral mimicry to create affiliation and rapport. *Psychological Science* 14(4): 334–339.

Lamm, C., C.D. Batson, and J. Decety. 2007. The neural substrate of human empathy: Effects of perspective-taking and cognitive appraisal. *Journal of Cognitive Neuroscience* 19(1): 42–58.

Lau, M.A., S.R. Bishop, Z.V. Segal, T. Buis, N.D. Anderson, L.E. Carlson, et al. 2006. The toronto mindfulness scale: Development and validation. *Journal of Clinical Psychology* 62(12): 1445–1467.

Laurent, S.M., and M.W. Myers. 2011. I know you're me, but who am I? Perspective taking and seeing the other in the self. *Journal of Experimental Social Psychology* 47(6): 1316–1319.

Leary, M.R., and J.P. Tangney. 2011. The self as an organizing construct in the behavioral and social sciences. In *Handbook of self and identity*, 2nd ed, ed. M.R. Leary and J.P. Tangney, 3–14. New York: The Guilford Press.

Leiberg, S., O. Klimecki, and T. Singer. 2011. Short-term compassion training increases prosocial behavior in a newly developed prosocial game. *PLoS One* 6(3): e17798.

Lutz, A., J. Brefczynski-Lewis, T. Johnstone, and R.J. Davidson. 2008. Regulation of the neural circuitry of emotion by compassion meditation: Effects of meditative expertise. *PLoS One* 3(3): e1897.

Markus, H.R., and S. Kitayama. 1991. Culture and the self: Implications for cognition, emotion, and motivation. *Psychological Review* 98(2): 224.

Mead, G.H. 1934. *Mind self and society from the standpoint of a social behaviorist*. Chicago: University of Chicago.

Meltzoff, A.N. 2007. 'Like me': A foundation for social cognition. *Developmental Science* 10(1): 126–134.

Meltzoff, A.N., and M.K. Moore. 1989. Imitation in newborn infants: Exploring the range of gestures imitated and the underlying mechanisms. *Developmental Psychology* 25(6): 954–962.

Meyer, M.L., C.L. Masten, Y. Ma, C. Wang, Z. Shi, N.I. Eisenberger, et al. 2012. Empathy for the social suffering of friends and strangers recruits distinct patterns of brain activation. *Social Cognitive and Affective Neuroscience*. doi:10.1093/scan/nss019.

Mezulis, A.H., L.Y. Abramson, J.S. Hyde, and B.L. Hankin. 2004. Is there a universal positivity bias in attributions? A meta-analytic review of individual, developmental, and cultural differences in the self-serving attributional bias. *Psychological Bulletin* 130(5): 711–747.

Mischel, W., and C.C. Morf. 2003. The self as a psycho-social dynamic processing system: A meta-perspective on a century of the self in psychology. In *Handbook of self and identity*, 2nd ed, ed. M.R. Leary and J.P. Tangney, 15–43. New York: The Guilford Press.

Moray, N. 1959. Attention in dichotic listening: Affective cues and the influence of instructions. *Quarterly Journal of Experimental Psychology* 11(1): 56–60.

Murray, R.J., M. Schaer, and M. Debbané. 2012. Degrees of separation: A quantitative neuroimaging meta-analysis investigating self-specificity and shared neural activation between self- and other-reflection. *Neuroscience and Biobehavioral Reviews* 36(3): 1043–1059.

Neff, K.D. 2003a. The development and validation of a scale to measure self-compassion. *Self and Identity* 2: 223–250.

Neff, K.D. 2003b. Self-compassion: An alternative conceptualization of a healthy attitude toward oneself. *Self and Identity* 2: 85–101.

Neff, K.D. 2008. Self-compassion: Moving beyond the pitfalls of a separate self-concept. In *Transcending self-interest: Psychological explorations of the quiet ego*, ed. J. Bauer and H.A. Wayment, 95–106. Washington, DC: American Psychological Association.

Neff, K.D., and E. Pommier. 2013. The relationship between self-compassion and other-focused concern among college undergraduates, community adults, and practicing meditators. *Self and Identity* 12: 160–176.

Neisser, U. 1988. Five kinds of self-knowledge. *Philosophical Psychology* 1(1): 35–59.

Olendzki. 2006. The transformative impact of non-self. In *Buddhist thought and applied psychological research: Transcending the boundaries*, ed. D.K. Nauriyal, M.S. Drummond, and Y.B. Lal, 250–261. New York: Routledge.

Orzech, K.M., S.L. Shapiro, K.W. Brown, and M. McKay. 2009. Intensive mindfulness training-related changes in cognitive and emotional experience. *The Journal of Positive Psychology* 4: 212–222.

Oyserman, D., and S.W.S. Lee. 2008. Does culture influence what and how we think? Effects of priming individualism and collectivism. *Psychological Bulletin* 134(2): 311–342.

Oyserman, D., H.M. Coon, and M. Kemmelmeier. 2002. Rethinking individualism and collectivism: Evaluation of theoretical assumptions and meta-analyses. *Psychological Bulletin* 128(1): 3–72.

Polich, J. 2007. Updating P300: An integrative theory of P3a and P3b. *Clinical Neurophysiology* 118(10): 2128–2148.

Preston, S.D., and F.B.M. de Waal. 2002. Empathy: Its ultimate and proximate bases. *The Behavioral and Brain Sciences* 25(1): 1–72.

Preston, S.D., and A.J. Hofelich. 2012. The many faces of empathy: Parsing empathic phenomena through a proximate, dynamic-systems view of representing the other in the self. *Emotion Review* 4(1): 24–33.

Qin, P., and G. Northoff. 2011. How is our self related to midline regions and the default-mode network? *NeuroImage* 57(3): 1221–1233.

Raichle, M.E., A.M. MacLeod, A.Z. Snyder, W.J. Powers, D.A. Gusnard, and G.L. Shulman. 2001. A default mode of brain function. *Proceedings of the National Academy of Sciences of the United States of America* 98(2): 676–682.

Rinpoche, R.T., and K. Mullen. 2005. The Buddhist use of compassionate imagery in mind healing. In *Compassion: Conceptualisations, research and use in psychotherapy*, ed. P. Gilbert, 403. New York: Routledge.

Salzberg, S. 1995. *Lovingkindness. The revolutionary art of happiness*. Bosten: Shambala.

Salzberg, S. 2011. Mindfulness and loving-kindness. *Contemporary Buddhism* 12(1): 177–182.

Schmidt, S. 2004. Mindfulness and healing intention: Concepts, practice, and research evaluation. *Journal of Alternative and Complementary Medicine* 10(1): 7–14.

Schuster, R. 1979. Empathy and mindfulness. *Journal of Humanistic Psychology* 19(1): 71–77.

Sebanz, N., G. Knoblich, and W. Prinz. 2003. Representing others' actions: Just like one's own? *Cognition* 88(3): 11–21.

Sedlmeier, P., J. Eberth, M. Schwarz, D. Zimmermann, F. Haarig, S. Jaeger, et al. 2012. The psychological effects of meditation: A meta-analysis. *Psychological Bulletin*. doi:10.1037/a0028168.

Shapiro, S.L., L.E. Carlson, J.A. Astin, and B. Freedman. 2006. Mechanisms of mindfulness. *Journal of Clinical Psychology* 62(3): 373–386.

Smith, E.R., and S. Henry. 1996. An in-group becomes part of the self: Response time evidence. *Personality and Social Psychology Bulletin* 22(6): 635–642.

Smith, E.R., S. Coats, and D. Walling. 1999. Overlapping mental representations of self, in-group, and partner: Further response time evidence and a connectionist model. *Personality and Social Psychology Bulletin* 25(7): 873–882.

Sprecher, S., and B. Fehr. 2005. Compassionate love for close others and humanity. *Journal of Social and Personal Relationships* 22(5): 629–651.

Stel, M., R.B. van Baaren, and R. Vonk. 2008. Effects of mimicking: Acting prosocially by being emotionally moved. *European Journal of Social Psychology* 38: 965–976.

Symons, C.S., and B.T. Johnson. 1997. The self-reference effect in memory: A meta-analysis. *Psychological Bulletin* 121(3): 371–394.

Tacikowski, P., and A. Nowicka. 2010. Allocation of attention to self-name and self-face: An ERP study. *Biological Psychology* 84(2): 318–324.

Taylor, S.E. 1988. Illusion and well-being: A social psychological perspective on mental health. *Psychological Bulletin* 103(2): 193–210.
Taylor, V.A., V. Daneault, J. Grant, G. Scavone, E. Breton, S. Roffe-Vidal, et al. 2013. Impact of meditation training on the default mode network during a restful state. *Social Cognitive and Affective Neuroscience* 8(1): 4–14.
Thompson, E. 2001. Empathy and consciousness. *Journal of Consciousness Studies* 8: 1–32.
Tomasello, M. 1993. The interpersonal origins of self concept. In *The perceived self: Ecological and interpersonal sources of self knowledge*, ed. U. Neisser, 174–184. Cambridge, UK: Cambridge University Press.
Turner, J. 1986. The significance of the social identity concept for social psychology with reference to individualism, interactionism and social influence. *British Journal of Social Psychology* 25: 237–252.
van Baaren, R.B., W.W. Maddux, T.L. Chartrand, C. de Bouter, and A. van Knippenberg. 2003. It takes two to mimic: Behavioral consequences of self-construals. *Journal of Personality and Social Psychology* 84(5): 1093–1102.
van Baaren, R.B., R.W. Holland, K. Kawakami, and A. van Knippenberg. 2004. Mimicry and prosocial behavior. *Psychological Science* 15(1): 71–74.
van der Meer, L., S. Costafreda, A. Aleman, and A.S. David. 2010. Self-reflection and the brain: A theoretical review and meta-analysis of neuroimaging studies with implications for schizophrenia. *Neuroscience and Biobehavioral Reviews* 34(6): 935–946.
Varela, F.J., E. Thompson, and E. Rosch. 1991. *The embodied mind*. Cambridge, MA: MIT Press.
Wallace, B.A. 2001a. *Buddhism with an attitude: The Tibetan seven-point mind-training*. Ithaca: Snow Lion Publications.
Wallace, B.A. 2001b. Intersubjectivity in Indo-Tibetan Buddhism. *Journal of Consciousness Studies* 8(5–7): 209–230.
Way, B.M., J.D. Creswell, N.I. Eisenberger, and M.D. Lieberman. 2010. Dispositional mindfulness and depressive symptomatology: Correlations with limbic and self-referential neural activity during rest. *Emotion* 10(1): 12–24.
Wayment, H.A., and J.J. Bauer. 2008. *Transcending self-interest: Psychological explorations of the quiet ego*. Washington, DC: American Psychological Association.
Weinstein, N.D., and E. Lachendro. 1982. Egocentrism as a source of unrealistic optimism. *Personality and Social Psychology Bulletin* 8(2): 195–200.
Whitfield-Gabrieli, S., J.M. Moran, A. Nieto-Castañón, C. Triantafyllou, R. Saxe, and J.D.E. Gabrieli. 2011. Associations and dissociations between default and self-reference networks in the human brain. *NeuroImage* 55(1): 225–232.
Wood, N., and N. Cowan. 1995. The cocktail party phenomenon revisited: How frequent are attention shifts to one's name in an irrelevant auditory channel? *Journal of Experimental Psychology: Learning, Memory, and Cognition* 21(1): 255–260.
Zahavi, D. 2006. *Subjectiviy and selhood: Investigating the first-person perspective*. Cambridge, MA: The MIT Press.
Zhao, K., J. Yuan, Y. Zhong, Y. Peng, J. Chen, L. Zhou, et al. 2009. Event-related potential correlates of the collective self-relevant effect. *Neuroscience Letters* 464(1): 57–61.
Zhao, K., Q. Wu, H.D. Zimmer, and X. Fu. 2011. Electrophysiological correlates of visually processing subject's own name. *Neuroscience Letters* 491(2): 143–147.
Zhou, A., Z. Shi, P. Zhang, P. Liu, W. Han, H. Wu, et al. 2010. An ERP study on the effect of self-relevant possessive pronoun. *Neuroscience Letters* 480(2): 162–166.
Zhu, Y., and L. Zhang. 2002. An experimental study on the self-reference effect. *Science in China Series C: Life Series* 45(2): 120–128.
Zhu, Y., L. Zhang, J. Fan, and S. Han. 2007. Neural basis of cultural influence on self-representation. *NeuroImage* 34(3): 1310–1316.

# Mindfulness Meditation and the Experience of Time

**Marc Wittmann and Stefan Schmidt**

*Walking is very calming. One step after another, one foot moving into the future and one in the past. Did you ever think about this? It's like our bodies are caught in the middle. The hard part is staying in the present. Really being here.*

Janet Cardiff (2005, p. 75)

**Abstract** Many personal reports from experienced meditators exist on how subjective time slows down in meditation practice as well as in everyday life. However, hardly any empirical work exists regarding this exceptional experience. In this theoretical chapter we discuss cognitive and neural models of time perception. We aim at showing how the subjective passage of time and duration are modified by functional states of mindfulness, i.e. by attention regulation, body awareness and emotion regulation. The ability of expert mindfulness meditators to focus more strongly on sensory experiences and to be more strongly aware of feelings and of body states leads to a slowing down of time in the present moment. Moreover, as a consequence of more efficient attention regulation capacities, memory formation is enhanced which in

M. Wittmann (✉)
Institute for Frontier Areas of Psychology and Mental Health, Wilhelmstr. 3a, Freiburg 79098, Germany
e-mail: wittmann@igpp.de

S. Schmidt
Department of Psychosomatic Medicine and Psychotherapy, University Medical Center Freiburg, Hauptstr. 8, Freiburg 79104, Germany

Institute for Transcultural Health Studies, European University Viadrina, Frankfurt (Oder), Germany
e-mail: stefan.schmidt@uniklinik-freiburg.de

S. Schmidt and H. Walach (eds.), *Meditation – Neuroscientific Approaches and Philosophical Implications*, Studies in Neuroscience, Consciousness and Spirituality 2, DOI 10.1007/978-3-319-01634-4_11, 

retrospect leads to a subjective lengthening of past duration. Empirical studies concerning time perception in meditation practitioners would help to understand meditative states and at the same time would foster knowledge on cognitive-emotional as well as neural processes underlying the experience of time.

## Introduction

In experienced meditators conscious awareness is altered during meditation while attention is being focused on the experience of the present moment. Essentially, mindfulness meditation is understood as bringing awareness to each moment in time. Being mindful entails attending to sensory experiences in the outside world, thoughts, emotions and body sensations by curiously observing them with a warm-hearted attitude of openness and acceptance (Kabat-Zinn 2005). One aspect that is noticed by individuals during mindfulness meditation as well as by experienced meditators in everyday experience is that subjective time slows down considerably – in the present moment as well as experienced for longer periods of time (see the chapter "The 'sense' of time passing" in Kabat-Zinn 2005).

In phenomenological analyses of subjective time by Edmund Husserl, Maurice Merleau-Ponty or William James the dual experience of temporality is prominently discussed as encompassing (1) the sense of a present, the feeling of *nowness* and (2) the feeling of a flow of time leading to the experience of duration (Wittmann 2009a). On the one hand, we perceive the unity of the present, a fundamental temporal window of consciousness that defines our sense of an extended *now*. On the other hand, we experience the passage of time, best described by a stream or flow. The flow constitutes itself as an event which is anticipated, then experienced, and later remembered, and by this way generating a sense of duration. The sense of the present moment and of the flow of time constitutes our experience of temporality as complementary phenomena. Whereas there are indications that the subjective present might be expanded in experienced meditators (Carter et al. 2005; Sauer et al. 2012a), systematic studies on the passage of time and of subjective duration have hardly been undertaken so far (but see Chihara 1989). In the following passages we will concentrate mainly on this second aspect of temporality, on the perception of the passage of time leading to the feeling of duration, although the two aspects of temporality – the sense of the present moment and the feeling of the flow of time – are intertwined phenomena.

However essential the dimension of time is for human experience, by viewing the research literature on time perception it becomes clear that no consensus exists on how humans experience the passage of time; many different psychological and neurophysiological models exist (Wittmann and van Wassenhove 2009). Therefore, we address the issue of time perception in meditation practice for two reasons: For one, a better understanding of the phenomenon that time slows down for meditators – both during meditation and in everyday experience – could lead to a better understanding of the meditation practice itself. Second, knowledge about this phenomenon could foster the understanding of how we as humans perceive time. The two aspects underlie the motivation for this theoretical account of bringing together research on time perception and on the effects of meditation practice. In the following paragraphs, essential features of mindfulness meditation will be highlighted. It will be shown how these characteristics are decisive for changes in the subjective experience of time in meditators.

## Functional Characteristics of Mindfulness Meditation as Modulators of Subjective Time

At least four functional aspects encompass the state of mindfulness, as summarized by Hölzel et al. (2011), namely attention regulation, body awareness, emotion regulation, and changes in the perspective of the self. As will be shown, these aspects are also dominantly involved in modulating and generating our experience of time. Therefore, a discussion on the relationship between meditative states and time perception will be grouped according to these domains.

1. Attention regulation

The state of meditation can be described as deliberately and continuously focusing attention on the present moment and some authors even take attention regulation as one of the defining features of meditation and mindfulness (Bishop et al. 2004; Shapiro 1982). One way of learning attention regulation is to keep the focus of attention on a single object, e.g. the breath, by observing the breathing in and breathing out. The ability to focus attention on an object continuously encompasses the disregard of distractors such as upcoming thoughts and sensations for an extended period of time. In experimental studies experienced meditators indeed perform with higher accuracy and speed in many components of attention such as focused attention, sustained attention and alertness as assessed with computerized attention tests. That is, regular meditation practice enhances the ability of attention regulation in everyday non-meditative states (e.g., Chan and Woollacott 2007; Jha et al. 2007; Lutz et al. 2009; MacLean et al. 2010; Valentine and Sweet 1999; van den Hurk et al. 2010; van Leeuwen et al. 2009; Zeidan et al. 2010). This ability of meditators to maintain focused attention over time is assumed to lead to an increased awareness of external and internal processes and events. Thus, mindfulness practice leads to a stronger awareness of sensory experience in the *here* and *now* as well as to an increased ability of sustained attention (Brown and Ryan 2003; Sauer et al. 2012b).

In phenomenological analyses, consciousness is understood as an island of presence in the continuous flow of time (Husserl 1928). In more mechanistic terms, mental presence is based on working memory function integrating perceived events over time to form a unified experience (Wittmann 2011). Since meditation practice leads to an increased awareness of sensory and mental experience through heightened attentional capacities, this ability will also influence memory formation because there are fewer slips of attention and distractions from the focus of attention.

Interestingly, the ancient Pali word for mindfulness *sati* has also the meaning of 'to remember' (Analayo 2004), and this is interpreted by the fact that full attention to the present moment will facilitate memory function. As has indeed been shown, working memory capacity is enhanced in individuals who regularly meditate; meditators recall more learned items after a certain time span during which distractor items are interfering (Jha et al. 2010; Zeidan et al. 2010). These attention and memory effects of meditative practice have consequences for subjective time.

Two fundamental perspectives in time perception can be discerned: prospective and retrospective estimates of duration (Zakay and Block 1997). In prospective time experiments participants have to judge the duration of an interval they are presently experiencing. They direct more or less attention to the passage of time while a particular duration has to be judged. If more attention is directed to time during an explicit duration estimation task, duration is experienced as being comparatively longer than when attention is distracted from time. The time spent waiting for a bus (when one strongly attends to the passage of time) feels much longer than the same time interval chatting with a friend (when being distracted from time). Studies show how an increased working memory capacity and greater attention regulation abilities are associated with greater accuracy in many different temporal judgment tasks (Brown 1997; Pütz et al. 2012; Ulbrich et al. 2007, 2009). That is, the ability of attention regulation in mindfulness meditators is the key factor for explaining performance accuracy in explicit time estimation tasks in the seconds-to-minutes range – similar to meditation effects on general perceptual discrimination accuracy (MacLean et al. 2010).

In retrospective time, an observer estimates a time span that has already elapsed. In this case, subjective duration is reconstructed from memory. That is, retrospective duration depends on the amount of experienced contextual changes stored in memory. The more experienced changes have been stored, the longer an interval is judged to have lasted (Bailey and Areni 2006; Zakay and Block 2004). This relation between memory storage and the estimation of duration holds for retrospectively judged intervals in the seconds-to-minutes range related to working memory. But it is also assumed to hold for long-term episodic memory where the amount of changes during a life span is linked to the judgment of duration, i.e. how fast the last 10 years passed (Wittmann and Lehnhoff 2005). Since individuals trained in mindfulness meditation techniques are more strongly aware of sensory events and they process and store more items in working memory, their retrospective judgement of past duration will likely lead to the impression of longer duration and a slowing down of subjective time. Being more strongly aware of what is happening now causes more experienced changes to be stored in memory which in turn leads to a relative expansion of retrospective duration.

2. Body awareness

The body is the vehicle or frame of reference for sensory and mental experience. Many meditation techniques therefore incorporate instructions to focus on body sensations. Attention is for example directed to the breathing process or to specific body parts (such as in the "body scan"). Although not many studies exist in which body awareness skills in meditators were objectively tested, experienced practitioners report having a greater awareness of their body states (for an overview of empirical findings, see Hölzel et al. 2011).

Neuroimaging studies suggest that, among other regions, the insular cortex is modulated through mindfulness meditation practice. The insula is the primary receptive area for visceral input and for monitoring physiological states of the body (Craig 2002). In two studies, experienced meditators, as compared to controls, showed greater gray matter concentration (Hölzel et al. 2008) and greater cortical thickness (Lazar et al. 2005) in the right anterior insula. In another study, the instruction to concentrate on respiratory sensations lead to a greater increase in neural activation in the posterior insula in trained meditators as compared to control subjects (Farb et al. 2012). Moreover, experienced meditators, who generated a loving-kindness-compassion meditation state, showed neural activation in the right anterior insula that was related to the degree of successfully entering the meditative state (Lutz et al. 2008). Mindfulness practice can also be seen as training of interoceptive awareness, and is thus related to changes in brain regions representing and integrating body states.

The insular cortex, which integrates body signals over time, has in recent conceptualizations been suggested to form the anatomical and functional basis for the creation of the sense of time and of awareness in general (Craig 2009a; Wittmann 2009b). In his model, Craig (2002, 2009b) assumes that body signals received in the dorsal posterior insula are processed and integrated in a posterior-to-anterior progression, this progressive integration culminating in the anterior insula, when brain activation becomes associated with the conscious awareness of bodily feelings. Subjective feelings depend upon bodily signals, as visceral and somatosensory feedback from the peripheral nervous system is integrated with contextual, i.e. perceptual, motivational, social and cognitive information leading to the awareness of complex feeling states (Damasio 1999; Craig 2009b). A unified meta-representation of homeostatic feelings is built in the anterior insula which generates an experience of the self at one moment. A succession of these meta-representations across time would provide the continuity of subjective awareness, a series of elementary emotional moments. Important for our context here, the experience of time would be created by successive moments of self-realization (Craig 2009a). The sense of time thereafter would be related to the temporal integration of signals from the interoceptive system.

In fact, the insular cortex has repeatedly been shown to be activated in neuroimaging studies of time perception (for a meta-analysis, see Wiener et al. 2010). The insula might specifically be implicated in the feeling of the passage of time of several seconds duration: in an event-related functional magnetic resonance imaging (fMRI) study using temporal intervals of 9 and 18 s, activation in the dorsal posterior insular cortex was linked to the temporal encoding of these intervals: neural activation in this region increased with increasing interval length and peaked at the end of the interval (Wittmann et al. 2010, 2011). When these intervals had to be reproduced as indicated by a button press, a similar linear increase of activation was seen in the anterior insula as well as regions of the frontal cortex which peaked shortly before the actual button press. In line with the model by Craig (2009a), it was suggested that this accumulator-like activity represented the integration of body signals over time that was used to represent duration.

The conjecture that interoception might be at the base of time perception is supported by a psychophysiological study showing a positive relationship between time estimation accuracy and the slope of cardiac slowing during the perception of temporal intervals with durations between 8 and 20 s (Meissner and Wittmann 2011). This study is probably the first observation of a direct association between changes in heart rate, a measure of the autonomic body function, and the accuracy in interval timing. That is, changes in cardiac activity, as registered in the brain, might be used as a measure for subjective duration. Moreover, also conscious awareness of the own heart beat as assessed with the heartbeat perception test (Pollatos et al. 2005) was related to time perception accuracy. That is, individuals who are more sensitive to their heart beat, i.e. who more accurately count the number of heart beats for a given interval, were also more accurate in time estimation (Meissner and Wittmann 2011).

This summary of findings implies a clear hypothesis related to meditators who are trained to be aware of their body processes. First of all, if the sense of time is so strongly related to the awareness of body processes, then the focus on subjective time will inevitably lead to a slowing down of the passage of time in meditators who are more aware of their body processes. Two factors of meditation procedures are dispositional for becoming aware of the flow of time, possibly through body awareness. The instruction to focus on the present moment is inherently time related. The meditator deliberately guides his or her awareness to the temporality of experience. One could even say, during meditation the sense of time passing is heightened, extremely so especially for beginners who often experience time as passing much too slowly, sometimes provoking inner restlessness and boredom. Moreover, the rhythmical nature of breathing and the heart beat constitutes a bodily meter that refers to the passing of time. In fact, the breathing cycle of a relaxed state with a period of approximately 3 s corresponds to the duration of an experienced moment in the flow of time (Wittmann 2011, 2012). Moreover, there are some tasks in meditation which explicitly direct present moment attention to bodily signals, i.e. performing the so-called *body scan* in mindfulness meditation, or full bodily attention in a wide-focused mindfulness state. In this sense meditation is a state of body awareness in the present moment. Understanding the state of being mindful will lead to an understanding of the sense of time as embodied experience.

3. Emotion regulation

Mindfulness meditation enhances emotional well-being. Healthy individuals who meditate regularly report fewer negative mood states, fewer ruminative thoughts as well as more positive states of mind (Jain et al. 2007; Jha et al. 2010). In clinical settings, structured group programs of mindfulness-based stress reduction reduce negative affect in patients with physical, psychosomatic and psychiatric disorders (Fjorback and Walach 2012; Grossman et al. 2004; Piet and Hougaard 2011) and the same can be seen in non-clinical populations (Sedlmeier et al. 2012). Experiential openness and acceptance of thoughts and feelings in mindfulness meditation, in essence, means that upcoming mood states are in the focus of attention, even when negative emotions are experienced (Hölzel et al. 2011). Mindfulness practitioners expose themselves to their affective experiences in a non-reactive manner and thus learn to cope with their mood states. Sustained attention to these states leads to a situation of exposure, which in turn and over time may lead to the decrease or even extinction of an emotional reaction. However, emotion regulation in this sense does not necessarily imply a down-regulation of emotion but a greater awareness of mood states and an increased ability of emotional control while experiencing these states. Indeed, one of the mechanisms of mindfulness seems to be the reduction of avoidance behavior (Sauer et al. 2011). An individual learns to accept his or her automatic emotional reactions and thus not to enhance them by aversion or attraction. As a consequence, higher levels of positive affect can be experienced (Ekman et al. 2005; Jha et al. 2010).

The sense of time is intimately linked to affective states. In periods of severe mental distress such as in depression or anxiety the passage of time slows down considerably and time intervals are overestimated (Bschor et al. 2004; Wittmann et al. 2006). In everyday life, boredom is the prototypical mood state in which time passes too slowly. The feeling of an existential vacuum directs attention to time since no meaningful distraction is experienced which in turn leads to a (painful) feeling that subjective time slows down. Moreover, in experimental settings, subjects overestimate the duration of emotional stimuli which last several hundred milliseconds to a few seconds, bodily arousal, attention, and sentience being discussed for explaining these effects of time experience (Droit-Volet and Gil 2009; Schirmer 2011; Wittmann 2009b). Specifically, it is assumed that increased physiological arousal leads to a faster speed of an internal pacemaker underlying time perception (Gil and Droit-Volet 2012). A faster speed of such an internal clock would lead to the accumulation of more pulses emitted by a hypothetical pacemaker operating in the seconds range (Zakay and Block 1997). There is ample evidence of an overestimation of duration in the sub- and supra-second range of mostly negative emotional stimuli which is related to higher physiological arousal. However, also the presentation of positive emotional stimuli, associated with higher physiological arousal, can lead to a relative overestimation of duration as compared to neutral stimuli (Lambrechts et al. 2011).

By trying to integrate the aforementioned findings on the relationship between emotion and time perception, a tentative hypothesis would be that experts in meditation practice who experience a positive emotional tone and who are more sensitive to their upcoming mood states, while attending to the temporal aspects of perception, will feel a general slowing of the pace of time. A stronger awareness of emotional states is related to an increase in accumulated pulses of an assumed pacemaker-accumulator and thus would lead to a relative prolongation of subjective time. In addition, also a retrospective account of time perception can explain why time slows down for experienced meditators. A stronger attentional focus on present experience coupled with a positive emotional tone will lead to an enhanced encoding of events. This in turn results in more items collected in memory. When looking back in time more memories can be retrieved which amounts to a relative prolongation of a perceived time span as opposed to a situation where less memorable events are retrieved.

## Summary

Contemporary philosophical conceptions of the self which take into account findings from the neurosciences and psychology have a striking resemblance to Buddhist teachings concerning the perception of a self (Hölzel et al. 2011). The programmatic title of the book *Being no one. The self-model theory of subjectivity* (Metzinger 2004) summarizes the notion that there is no static self as a permanent entity or substance, but that ongoing mental processes are engaged in producing a transient model of the self. On a basic level, the bodily self, as created by the continuous visceral and proprioceptive input from the body, is the functional anchor of phenomenal experience (Damasio 2003). According to the model proposed by Craig (2009a, b) integration processes of body signals culminating in the anterior insula as meta-representations of successive emotional moments are associated with the experience of an embodied and extended self in time. Importantly, phenomenological conceptualizations have put forward the idea that subjective time emerges through the existence of the self across time as embodied entity (Kiverstein 2009; Zahavi 2005). Mindfulness meditation is the enhanced awareness of the present moment and this psychological state is associated with brain areas involved in processing of the bodily self (Farb et al. 2007).

In mindfulness meditation a practitioner focuses on the experience of the embodied self at the present moment. Being mindful in everyday life is equivalent to being conscious of one's body states and feelings, of oneself at this particular moment in time. Because the feeling of time is created through attending to the embodied self at the present moment, being exceptionally mindful slows down the passage of time. Moreover, subjective time slows down in retrospect because greater awareness of one's experiences leads to enriched memory contents, which in turn expands subjective duration. An increased focus on an *experienced* self at the present moment slows down the subjective passage of time – now and in retrospect.

## References

Analayo, V. 2004. *Satipatthana: The direct path to realization*. Cambridge: Windhorse.

Bailey, N., and C.S. Areni. 2006. Background music as a quasi clock in retrospective duration judgments. *Perception and Motor Skills* 102: 435–444.

Bishop, S.R., M. Lau, S. Shapiro, L.E. Carlson, N.D. Anderson, J. Carmody, Z.V. Segal, S. Abbey, M. Speca, D. Velting, and G. Devins. 2004. Mindfulness: A proposed operational definition. *Clinical Psychology: Science and Practice* 11: 230–241.

Brown, S.W. 1997. Attentional resources in timing: Interference effects in concurrent temporal and nontemporal working memory tasks. *Perception and Psychophysics* 59: 1118–1140.

Brown, K.W., and R.M. Ryan. 2003. The benefits of being present: Mindfulness and its role in psychological well-being. *Journal of Personality and Social Psychology* 84: 822–848.

Bschor, T., M. Ising, M. Bauer, U. Lewitzka, M. Skerstupeit, B. Müller-Oerlinghausen, and C. Baethge. 2004. Time experience and time judgment in major depression, mania and healthy subjects. A controlled study of 93 subjects. *Acta Psychiatrica Scandinavica* 109: 222–229.

Cardiff, J. 2005. The walk book. By Mirjam Schaub, edited by Thyssen-Bornemisza Art Contemporary Vienna in collaboration with Public Art Fund, New York.
Carter, O.L., D.E. Presti, C. Callistemon, Y. Ungerer, G.B. Liu, and J.D. Pettigrew. 2005. Meditation alters perceptual rivalry in Tibetan Buddhist monks. *Current Biology* 15: R412–R413.
Chan, D., and M. Woollacott. 2007. Effects of level of meditation experience on attentional focus: Is the efficiency of executive or orientation networks improved? *Journal of Alternative and Complementary Medicine* 13: 651–657.
Chihara, T. 1989. Zen meditation and time-experience. *Psychologia* 32: 211–220.
Craig, A.D. 2002. How do you feel? Interoception: The sense of the physiological condition of the body. *Nature Reviews Neuroscience* 3: 655–666.
Craig, A.D. 2009a. Emotional moments across time: A possible neural basis for time perception in the anterior insula. *Philosophical Transactions of the Royal Society B* 364: 1933–1942.
Craig, A.D. 2009b. How do you feel – now? The anterior insula and human awareness. *Nature Reviews Neuroscience* 10: 59–70.
Damasio, A. 1999. *The feeling of what happens: body and emotion in the making of consciousness*. San Diego: Harcourt.
Damasio, A. 2003. Feelings of emotion and the self. *Annals of the New York Academy of Science* 1001: 253–261.
Droit-Volet, S., and S. Gil. 2009. The time-emotion paradox. *Philosophical Transactions of the Royal Society B* 364: 1943–1954.
Ekman, P., J.R. Davidson, M. Ricard, and B.A. Wallace. 2005. Buddhist and psychological perspectives on emotions and well-being. *Current Directions in Psychological Science* 14: 59–63.
Farb, N.A., Z.V. Segal, H. Mayberg, J. Bean, D. McKeon, Z. Fatima, and A.K. Anderson. 2007. Attending to the present: Mindfulness meditation reveals distinct neural modes of self-reference. *Social Cognitive and Affective Neuroscience* 2: 313–322.
Farb, N.A., Z.V. Segal, and A.K. Anderson. 2012. Mindfulness meditation training alters cortical representations of interoceptive attention. *Social Cognitive and Affective Neuroscience* 8: 15–26.
Fjorback, L.O., and H. Walach. 2012. Meditation based therapies – a systematic review and some critical observations. *Religions* 3: 1–18.
Gil, S., and S. Droit-Volet. 2012. Emotional time distortions: The fundamental role of arousal. *Cognition and Emotion* 26: 847–862.
Grossman, P., L. Niemann, S. Schmidt, and H. Walach. 2004. Mindfulness-based stress reduction and health benefits. A meta-analysis. *Journal of Psychosomatic Research* 57: 35–43.
Hölzel, B.K., U. Ott, H. Hempel, A. Hackl, K. Wolf, R. Stark, and D. Vaitl. 2008. Investigation of mindfulness meditation practicioners with voxel-based morphometry. *Social Cognitive and Affective Neuroscience* 3: 55–61.
Hölzel, B.K., S.W. Lazar, T. Gard, Z. Schuman-Olivier, D.R. Vago, and U. Ott. 2011. How does minfulness work? Proposing mechanisms of action from a conceptual and neural perspective. *Perspectives on Psychological Science* 6: 537–559.
Husserl, E. 1928. *Vorlesungen zur Phänomenologie des inneren Zeitbewußtseins*. Halle: Max Niemeyer Verlag.
Jain, S., S.L. Shapiro, S. Swanick, S.C. Roesch, P.J. Mills, I. Bell, and G.E. Schwartz. 2007. A randomized controlled trial of mindfulness meditation versus relaxation training: Effects of distress, positive states of mind, rumination, and distraction. *Annals of Behavioral Medicine* 33: 11–21.
Jha, A.P., J. Krompinger, and M.J. Baime. 2007. Mindfulness training modifies subsystems of attention. *Cognitive, Affective, and Behavioral Neuroscience* 7: 109–119.
Jha, A.P., E.A. Stanley, A. Kiyonaga, L. Wong, and L. Gelfand. 2010. Examining the protective effects of mindfulness training on working memory capacity and affective experience. *Emotion* 10: 54–64.
Kabat-Zinn, J. 2005. *Coming to our senses*. New York: Hyperion.
Kiverstein, J. 2009. The minimal sense of self, temporality and the brain. *Psyche* 15: 59–74.

Lambrechts, A., N. Mella, V. Pouthas, and M. Noulhiane. 2011. Subjectivity of time perception: A visual emotional orchestration. *Frontiers in Integrative Neuroscience* 5: 73.
Lazar, S.W., C.E. Kerr, R.H. Wasserman, J.R. Gray, D.N. Greve, M.T. Treadway, M. McGarvey, B.T. Quinn, J.A. Dusek, H. Benson, S.L. Rauch, C.I. Moore, and B. Fischl. 2005. Meditation experience is associated with increased cortical thickness. *NeuroReport* 16: 1893–1897.
Lutz, A., J. Brefczynski-Lewis, T. Johnstone, and R.J. Davidson. 2008. Regulation of the neural circuitry of emotion by compassion meditation: Effects of meditative expertise. *PLoS ONE* 3(3): e1897.
Lutz, A., H.A. Slagter, N.B. Rawlings, A.D. Francis, L.L. Greischar, and J.R. Davidson. 2009. Mental training enhances attentional stability: Neural and behavioral evidence. *The Journal of Neuroscience* 29: 13418–13427.
MacLean, K.A., E. Ferrer, S.R. Aichele, D.A. Bridwell, A.P. Zanesco, T.L. Jacobs, B.G. King, E.L. Rosenberg, B.K. Sahdra, P.R. Shaver, B.A. Wallace, G.R. Mangun, and C.D. Saron. 2010. Intensive meditation training improves perceptual discrimination and sustained attention. *Psychological Science* 21: 829–839.
Meissner, K., and M. Wittmann. 2011. Body signals, cardiac awareness, and the perception of time. *Biological Psychology* 86: 289–297.
Metzinger, T. 2004. *Being no one. The self-model theory of subjectivity*. Cambridge, MA: MIT Press.
Piet, J., and E. Hougaard. 2011. The effect of mindfulness-based cognitive therapy for prevention of relapse in recurrent major depressive disorder: A systematic review and meta-analysis. *Clinical Psychology Review* 31: 1032–1040.
Pollatos, O., W. Kirsch, and R. Schandry. 2005. On the relationship between interoceptive awareness, emotional experience, and brain processes. *Cognitive Brain Research* 25: 948–962.
Pütz, P., P. Ulbrich, J. Churan, M. Fink, and M. Wittmann. 2012. Duration discrimination in the context of age, sex, and cognition. *Journal of Cognitive Psychology* 24: 893–900.
Sauer, S., H. Walach, and N. Kohls. 2011. Gray's behavioural inhibition system as a mediator of mindfulness towards well-being. *Personality and Individual Differences* 50: 506–511.
Sauer, S., J. Lemke, M. Wittmann, N. Kohls, U. Mochty, and H. Walach. 2012a. How long is now for mindfulness meditators? *Personality and Individual Differences* 52: 750–754.
Sauer, S., H. Walach, S. Schmidt, T. Hinterberger, S. Lynch, A. Büssing, and N. Kohls. 2012b. Assessment of mindfulness: Review on state of the art. *Mindfulness*. doi:10.1007/s12671-012-0122-5.
Schirmer, A. 2011. How emotions change time. *Frontiers in Integrative Neuroscience* 5: 58.
Sedlmeier, P., J. Eberth, M. Schwarz, D. Zimmermann, F. Haarig, S. Jaeger, and S. Kunze. 2012. The psychological effects of meditation: A meta-analysis. *Psychological Bulletin* 138: 1139–1171.
Shapiro, D. 1982. Overview: Clinical and physiological comparison of meditation with other self-control strategies. *American Journal of Psychiatry* 139: 267–274.
Ulbrich, P., J. Churan, M. Fink, and M. Wittmann. 2007. Temporal reproduction: Further evidence for two processes. *Acta Psychologica* 125: 51–65.
Ulbrich, P., J. Churan, M. Fink, and M. Wittmann. 2009. Perception of temporal order: The effects of age, sex, and cognitive factors. *Aging, Neuropsychology and Cognition* 16: 183–202.
Valentine, E.R., and P.L.G. Sweet. 1999. Meditation and attention: A comparison of the effects of concentrative and mindfulness meditation on sustained attention. *Mental Health, Religion and Culture* 2: 59–70.
van den Hurk, P.A.M., F. Giommi, S.C. Gielen, A.E.M. Speckens, and H. Barendregt. 2010. Greater efficiency in attentional processing related to mindfulness meditation. *The Quarterly Journal of Experimental Psychology* 63: 1168–1180.
van Leeuwen, S., N.G. Müller, and L. Melloni. 2009. Age effects on attentional blink performance in meditation. *Consciousness and Cognition* 18: 593–599.
Wiener, M., P. Turkeltaub, and H.B. Coslett. 2010. The image of time: A voxel-wise meta-analysis. *NeuroImage* 49: 1728–1740.
Wittmann, M. 2009a. The subjective flow of time. In *The encyclopedia of time*, ed. H.J. Birks, 1322–1324. Thousand Oaks: SAGE Publications.

Wittmann, M. 2009b. The inner sense of time. *Philosophical Transactions of the Royal Society B* 364: 1955–1967.

Wittmann, M. 2011. Moments in time. *Frontiers in Integrative Neuroscience* 5: 66.

Wittmann, M. 2012. *Gefühlte Zeit: Kleine Psychologie des Zeitempfindens*. München: C.H. Beck.

Wittmann, M., and S. Lehnhoff. 2005. Age effects in perception of time. *Psychological Reports* 97: 921–935.

Wittmann, M., and V. van Wassenhove. 2009. The experience of time: Neural mechanisms and the interplay of emotion, cognition and embodiment. *Philosophical Transactions of the Royal Society B* 364: 1809–1813.

Wittmann, M., T. Vollmer, C. Schweiger, and W. Hiddemann. 2006. The relation between the experience of time and psychological distress in patients with hematological malignancies. *Palliative and Supportive Care* 4: 357–363.

Wittmann, M., A.N. Simmons, J. Aron, and M.P. Paulus. 2010. Accumulation of neural activity in the posterior insula encodes the passage of time. *Neuropsychologia* 48: 3110–3120.

Wittmann, M., A.N. Simmons, T. Flagan, S.D. Lane, J. Wackermann, and M.P. Paulus. 2011. Neural substrates of time perception and impulsivity. *Brain Research* 1406: 43–58.

Zahavi, D. 2005. *Subjectivity and selfhood: Investigating the first-person perspective*. Cambridge, MA: MIT Press.

Zakay, D., and R.A. Block. 1997. Temporal cognition. *Current Directions in Psychological Science* 6: 12–16.

Zakay, D., and R.A. Block. 2004. Prospective and retrospective duration judgments: An executive-control perspective. *Acta Neurobiologiae Experimentalis* 64: 319–328.

Zeidan, F., S.K. Johnson, B.J. Diamond, Z. David, and P. Goolkasian. 2010. Mindfulness training improves cognition: Evidence of brief mental training. *Consciousness and Cognition* 19: 597–605.

# Meditation and Hypnosis at the Intersection Between Phenomenology and Cognitive Science

**Michael Lifshitz, Emma P. Cusumano, and Amir Raz**

**Abstract** Cognitive scientists increasingly turn to contemplative practices such as hypnosis and meditation to shed light on consciousness and cognition. By their very nature, such practices call scientists to address the qualitative, lived experience of the subject. Yet, while the rise of contemplative techniques in neuroscience research has highlighted the importance of incorporating subjective experience within the empirical sciences of mind, the practical reality of marrying first- and third-person methods remains largely unactualised. Given that hypnosis and meditation exert powerful influence on subjective experience, we propose that they can serve as potent instruments for elucidating the structures and mechanisms of conscious experience in cognitive science settings. Here we discuss the motivation for a so-called 'neurophenomonological' approach and outline recent findings from the domains of hypnosis and meditation. Concrete examples illustrate how such contemplative practices can go beyond their place as objects of investigation to emerge as complementary experimental tools, thereby advancing the synthesis of scientific and phenomenological studies of mind (This article draws on ideas and expositions that ML and AR authored in the introduction of a 2012 special issue on hypnosis and meditation in *The Journal of Mind-Body Regulation* (see volume 2, issue 1)).

## Neurophenomenology and the Gesture of Awareness

Following a few early studies that highlighted the inaccuracy of introspection (e.g., Nisbett and Wilson 1977), throughout much of the twentieth century cognitive scientists largely eschewed subjective reports. Efforts shifted instead toward

M. Lifshitz • E.P. Cusumano • A. Raz (✉)
Lady Davis Institute for Medical Research, Institute of Community and Family Psychiatry, McGill University and the Jewish General Hospital, 4333 Cote Ste. Catherine Road, Montréal QC H3T 1E4, Canada
e-mail: amir.raz@mcgill.ca

S. Schmidt and H. Walach (eds.), *Meditation – Neuroscientific Approaches and Philosophical Implications*, Studies in Neuroscience, Consciousness and Spirituality 2,
DOI 10.1007/978-3-319-01634-4_12, 

understanding the workings of the brain, leading to major progress in regard to neuroimaging technology and brain models (Bandettini 2012). Alongside the emergence of the neuroimaging era, however, some cognitive researchers began to reflect on the value of refining scientific accounts of conscious experience. They realised that if they were to elucidate how and why experiences emerged from biological processes in the brain, they needed careful descriptions of those very experiences they sought to explain (Lutz 2004). However, while we are now equipped with a plethora of advanced methods for imaging and modelling the brain, first-person methods for describing and theorizing about phenomenal experience continue to lag behind.

In response to this practical discrepancy, in the 1990s Francisco Varela notably proposed "a quest to marry modern cognitive science and a disciplined approach to human experience" (Varela 1996). He coined this approach 'neurophenomenology,' a mission to establish a method within the cognitive sciences for acquiring descriptions of conscious experience such that they could be integrated with third-person cognitive and biological accounts. Neurophenomenology is based on the notion that with appropriate training, participants can reliably generate and sustain specific experiential states and provide accurate reports on those states (Lutz and Thompson 2003). Researchers can then analyse such reports alongside behavioural and neurobiological data, eventually establishing "strong reciprocal constraints between phenomenological accounts of experience and cognitive-scientific accounts of mental processes" (Lutz and Thompson 2003, p. 48).

Importantly, neurophenomenology does not propose a return to the introspective techniques of early twentieth-century psychology. Such "naïve" introspection assumes that we have the ability to observe our experience as we might uncritically observe an "inner visual field" (Varela 1998, p. 32). Instead, neurophenomenology situates itself as an extension of various contemplative lineages including Husserlian Continental Phenomenology and Eastern meditative traditions such as Buddhism. In contrast to the inwardly reflective, meta-cognitive turn of psychological introspection, these contemplative traditions aim to study subjective life by putting aside the presuppositions that suffuse our ordinary encounters with experience and engaging in a more reflexive and embodied meta-awareness of the experiential field. Although the terms designating such awareness and the subtleties of its application differ between traditions, for the purposes of this paper we refer to the overarching concept of meta-awareness (for a detailed comparison between Buddhist notions of mindfulness and Continental descriptions of phenomenological reduction, please see (Varela et al. 1992)). In a recent account, phenomenologist Daniel Schmicking sketches this gesture of meta-awareness as a practice of setting aside commonsensical and scientific beliefs about experience, metaphysical realities, and an extended narrative self (2010). Although Schmicking says that the function of this practice is "to step back from the situation" (p. 42), clearly we can never step wholly outside of our experience to some impartial position. Instead, the movement of awareness must involve stepping back from our usual *absorption* in the mental contents under investigation. The difficult task of phenomenology, therefore, is to observe and describe experiences impartially, without cognitive bias, whilst living through them subjectively. For the scientist advancing a neurophenomenological approach, moreover, the crucial challenge is to enable the research participant to foster this gesture of awareness.

While traditional Continental Phenomenology hardly addressed the critical issue of how to actually practice and achieve meta-awareness, recent accounts have begun to outline an embodied method of phenomenological awareness, often drawing on contemplative accounts of Buddhist mindfulness practices (Depraz et al. 2000, 2003; Steinbock 2004; Thompson 2006). Inspired by these recent integrative efforts, here we characterise the practice of meta-awareness as the reflexive skill of bringing experiential *processes* into focus without being swept up by absorption in particular experiential *contents*. Contents include perceptual phenomena such as sights and sounds, as well as conceptual items such as discursive thoughts and internal imagery, and perhaps also more vague and pervasive embodied experiences such as moods and affective states. Processes, on the other hand, refer to the structural and temporal relationships that determine the mode of appearance of these different contents – i.e., more fundamental experiential structures such as selfhood, intersubjectivity, temporality, and spatiality. The gesture of meta-awareness involves adjusting the aperture of attention to reframe experiential contents in the wider focus of the structural and temporal relationships that hold between them.

The crucial question, then, is how do researchers actually go about incorporating meta-awareness practices into their experimental paradigms? Meta-awareness involves an experiential shift more radical than simply trying to think impartial thoughts, because this latter strategy would merely lead to absorption in new conceptual contents rather than a holistic observation of the process of experience. For example, when we try to think about how a thought appears, we become engrossed in a thought about our thinking and therefore miss *how* this new thought gives itself to us. Rather than rearranging our conceptual content, therefore, meta-awareness likely requires a profound shift in our mode of attention: withholding the habitual propagation of discursive elaboration and opening a bare attentional space in which the relationships between phenomena can reveal themselves as they are. Opening such an attentional space presents a formidable challenge, however, because our patterns of discursive overlay and absorption in experiential contents are overlearned and highly automatic. Genuine meta-awareness, therefore, would require powerful first-person tools for overriding habitual preoccupation with phenomenal contents and holding attention open to the process of experience.

Here, we will argue that hypnosis and meditation are strong candidates for such phenomenological tools. Historical accounts as well as recent scientific investigations have documented the power of these practices to radically influence how people experience themselves and their environments. We propose that cognitive scientists can harness hypnosis and meditation to enable experimental subjects to suspend presuppositions, loosen deep-seated patterns of absorption, and sustain unencumbered meta-awareness in an experimental setting. These practices may allow researchers to generate specific and reliable alterations in consciousness, and perhaps achieve more fine-grained and unbiased phenomenological reports. Thus, meditation and hypnosis offer practical avenues for actualizing the neurophenomonological project.

## Meditation and Hypnosis as Empirical and Experiential Tools

While there are many ways to envision the gesture of meta-awareness in practice, hypnosis and meditation may have much to offer, not least of all because they come along with a broad and growing body of psychological and neuroscientific literature exploring their mechanisms and effects. As research tools, hypnosis and meditation have already made important contributions to cognitive neuroscience (for reviews see Oakley and Halligan 2009; Slagter et al. 2011). Although hypnosis and meditation represent distinct domains of practice, they appear to overlap in phenomenology, cognitive mechanisms, neural substrates, and potential therapeutic merits. We recently published a special issue addressing the merits of juxtaposing hypnosis and meditation to advance our understanding of their underlying mechanisms and help elucidate salient topics in cognitive neuroscience (Lifshitz and Raz 2012). Here we expound on how these powerful practices can incorporate and illuminate a neurophenomenological approach.

Cognitive scientists typically distinguish between mental processes that are controlled and those that are automatic. Whereas controlled processes are voluntary, slow, and effortful, automatic processes are involuntary, fast, and effortless (Schneider and Shiffrin 1977). Achieving literacy, for example, is a controlled and deliberate process requiring attention. Once learned and sufficiently practiced, however, reading becomes an automatic process, proceeding quickly and without effort (MacLeod 1991). A common view posits that extensive practice can render effortful processes more automatic. Once automatised, these processes become resistant to control and largely imperturbable (MacLeod and Dunbar 1988). Modifying such ballistic processes is central to the neurophenomenological project for at least two reasons: (i) These automatic processes often reflect the profound experiential structures that phenomenological investigation seeks to elucidate, and (ii) Overriding specific overlearned patterns of discursive conceptual thinking may allow participants to achieve more open and refined states of meta-awareness and provide more accurate and sensitive phenomenological reports. With this framework in mind, we will now explore how hypnosis and meditation can modulate deep-seated cognitive structures and occupy an important place in the broader landscape of neurophenomenology.

## Meditation

Meditative practices may provide a potent means of overriding habitual assumptions and adopting a radical meta-awareness toward experience. Largely originating in Buddhist traditions, mindfulness meditation refers to a broad range of mental practices geared at training attention to disengage from undesirable patterns of mental absorption. Although the aims of these practices range from mundane relaxation to spiritual enlightenment, most forms of meditation emphasize non-discursive meta-awareness of moment-to-moment experience (Lutz et al. 2006). Whereas some techniques involve focused attention on a particular experiential object such as the breath or a mantra, other practices involve non-discriminatory widening of attention to include the whole field of present-moment experience (Lutz et al. 2008). Both of these overarching meditative styles train the flexible skills of (i) noticing the tendency to become lost in thoughts and feelings, (ii) disengaging such habitual patterns of absorption, and (iii) shifting attention to lived experience unmediated by conceptualization. Although the specific instructions for achieving such receptive states of attention differ between traditions, formal meditation generally involves sitting silently with an upright, alert, and yet relaxed posture. The eyes may be open or closed, and traditions place varying degrees of emphasis on bodily stillness. The point of adopting a quiet and motionless posture is not to escape the usual tossing and turning of the mind, but rather to create an experiential space free from distraction wherein one can observe the full activity of the mind – tossing, turning, and all.

Newcomers to meditation are often surprised to find that sitting quietly in meditation is hardly relaxing. At first, practitioners can observe only coarse mental processes such as fully formed thoughts and overwhelming affective states. With practice, however, attention becomes refined such that seasoned meditators report observing subtle experiential processes including the arising, dwelling and decaying of mental phenomena and the constitution of the apparent duality of self and other. Although such accounts are difficult to corroborate scientifically, numerous empirical reports have demonstrated the impact of meditation on a wide range of attention processes and associated brain functions, including sustained attention (MacLean et al. 2010; Brefczynski-Lewis et al. 2007) and executive control (Tang et al. 2007; Moore et al. 2012). Furthermore, meditation seems to render attention more flexible. When presented with two visual stimuli in rapid succession, people commonly demonstrate an "attentional blink" and fail to attend to the second stimulus. Following a three month meditation retreat, however, practitioners readily detected the second stimulus (Slagter et al. 2007). Moreover, electroencephalography (EEG) recordings showed that meditation participants allocated less attention to processing the first target and thus had more neural resources left over to perceive the second stimulus. These findings suggest that meditation promotes heightened temporal sensitivity to experience by improving the flexibility and efficiency of attention. A recent study extended this notion using a local–global competition task (van Leeuwen et al. 2012). When viewing large font digits (global level) visually composed of smaller font digits (local level), participants typically give precedence to the global level and show delayed reaction times when reporting the number at the local level. Compared to controls, long-term meditators and newly trained novices responded more easily and quickly to the local digits. Among long-term practitioners, moreover, EEG data revealed deeper information processing of both local and global target information. Thus, by cultivating the ability to disengage, reorient, and sustain attention, meditative practices appear to improve the flexibility, sensitivity, and stability of awareness.

Traditional contemplative accounts suggest that the refined attention developed through meditative training provides a powerful means of gaining control over automatic cognitive processes. Moreover, in contrast to the rapid and transient alterations induced by hypnotic suggestion, the cognitive changes brought about through meditative training typically manifest more gradually over several sessions, and in some cases reflect enduring transformations (see Slagter et al. 2011). Crucial to the goal of developing meta-awareness, meditation appears to alter habits of spontaneous mind-wandering (Mrazek et al. 2012; Brewer et al. 2011) and involuntary reactivity in response to strong emotions (Allen et al. 2012; Taylor et al. 2011) and pain (Grant et al. 2011, 2012; Zeidan et al. 2011). One study showed that long-term Zen practitioners demonstrate reduced activation in neural structures related to spontaneous thought (Pagnoni et al. 2008). While lying in a functional magnetic resonance imaging (fMRI) scanner, subjects were instructed to pay attention to their breathing and return to it whenever they noticed distracting thoughts, memories, or sensations. Throughout this meditative condition, they performed a lexical decision task, wherein they pushed a button to indicate whether letters on a screen constituted real English words (e.g., "apple") or strings of letters with plausible readings but no semantic content (e.g., "nabol"). When presented with real words, long-term meditators, compared to controls, demonstrated a "reduced duration of neural response linked to conceptual processing … suggesting that meditative training may foster the ability to control the automatic cascade of semantic associations and, by extension, to voluntarily regulate the flow of spontaneous mentation" (Pagnoni et al. 2008). Certain meditative practices, therefore, appear to allow individuals to spend less time dwelling on conceptual content, perhaps opening a pre-reflective space for observing processes of experience in a more embodied, reflexive mode of meta-awareness.

While empirical research on meditation is still relatively young, cognitive scientists have already amassed a broad evidence base highlighting the power of meditative training to refine attention and override automatic conceptual and affective processes. These findings indicate that meditative practices may lead to more receptive and sensitive observation of experiential structures and to more accurate introspective reports. While such claims are hardly new, few empirical accounts have directly addressed the question of whether meditation actually improves phenomenological awareness. One study showed that meditators performed comparably to non-meditators when asked to estimate the rate of their heartbeat (Khalsa et al. 2008). Ironically, despite their equivalent performance, meditators reported greater confidence in their judgments than controls, intimating that meditation may bias participants to inflate their introspective abilities. While these findings warrant careful consideration, the authors of the study note that the heart rate counting task may not properly reflect the embodied meta-awareness cultivated in most forms of meditation. On the flip side of the coin, another recent study found that meditators compared to non-meditating controls showed greater introspective accuracy following a body-scan meditation and that the amount of meditation experience predicted introspective aptitude (Fox et al. 2012). Thus, whereas direct evidence that meditative training improves meta-awareness remains scarce and inconsistent, traditional claims and recent findings from cognitive neuroscience indicate that meditation will likely make a strong ally for neurophenomenology.

## Hypnosis

Whereas meditation has long occupied a central space in discussions surrounding neurophenomenology, hypnosis has received relatively little attention in this domain. Yet, hypnosis and other forms of suggestion hold great promise for advancing the synthesis of phenomenology and cognitive neuroscience. In recent decades, a mounting body of scientific evidence has demonstrated that, among responsive individuals, hypnotic suggestions can produce remarkable alterations in subjective experience as well as cognitive and brain function. One fruitful empirical approach employs specific hypnotic suggestions to produce "virtual patients" with transient syndromes nearly identical to genuine clinical psychopathologies in terms of experiential substrates and in some cases also neurobiological correlates (Woody and Szechtman 2011; Oakley and Halligan 2009). For example, a recent study used hypnotic suggestion to produce compelling experiences of mirrored-self misidentification – a clinical condition wherein patients no longer recognize their own reflection in a mirror. Following a suggestion that "the person you see in the mirror will not be you, it will be a stranger", highly hypnotizable subjects failed to recognize their own reflection and retained their delusional beliefs in the face of verbal challenges (e.g., "How is it possible that the person in the mirror looks just like you?") as well as behavioural demands (e.g., being asked to touch their nose whilst staring in the mirror) (Barnier et al. 2008). We speculate that closer phenomenological investigation of such hypnotically induced distortions of self-perception could reveal nuances concerning the relation between embodied subjectivity and the self viewed as external object (cf Rochat and Zahavi 2011). Beyond delusions of mirrored-self recognition, cognitive scientists have employed hypnotic suggestion to generate a wide range of virtual syndromes including obsessive-compulsive disorder (Woody and Szechtman 2011), synaesthesia (Cohen Kadosh et al. 2009), alien-hand syndrome (Blakemore et al. 2003), and visuospatial neglect (Priftis et al. 2011). In addition to the obvious practical advantages of studying virtual rather than genuine clinical patients, hypnotic analogues have the added benefit of allowing researchers to generate subtle nuances in symptomology, as well as design and implement novel delusions and psychopathologies that suit their specific research questions. Hypnotic clinical analogues, therefore, constitute an important untapped resource for neurophenomenology. More broadly, this approach points to the great flexibility afforded by hypnosis to modulate deep-rooted structures of experience with just a few brief words of suggestion.

Suggestion can derail processes previously considered ballistic and impervious to wilful intervention (Lifshitz et al. 2013). For example, a suggestion to view coloured images in black and white produced the experience of greyscale vision among highly responsive individuals, with concomitant dampening of low-level brain regions associated with colour processing (Kosslyn et al. 2000). Another example involves the classic Stroop paradigm, wherein participants typically demonstrate a lag when asked to report the ink colour of incongruent colour words (e.g., the word "blue" printed in red) (Stroop 1935). Based on the robustness of this

Stroop interference effect, most cognitive scientists consider processing printed linguistic stimuli inevitable for skilled readers (MacLeod 1991); however, a string of reports from multiple independent laboratories demonstrate that a suggestion to view the stimulus words as meaningless symbols of a foreign language allows participants to override the automaticity of reading and substantially reduce, or in some cases even eliminate, the Stroop interference effect (Raz and Campbell 2011; Raz et al. 2002, 2003, 2006, 2007; Augustinova and Ferrand 2012; Parris et al. 2012). Neuroimaging assays have begun to unravel the mechanisms of de-automatization as a function of suggestion (Casiglia et al. 2010; Raz et al. 2005; Terhune et al. 2010), while behavioural accounts have extended these effects to related cognitive paradigms probing automatic visual attention (Iani et al. 2006, 2009) as well as ballistic multimodal perceptual integration (Lifshitz et al. 2013).

At first glance, such striking cognitive-perceptual changes may seem more like enthralling curiosities than demonstrations of useful phenomenological tools. Yet, in addition to advancing our understanding of controlled and automatic processes as well as their interactions, such subjective alterations can be adapted to derail habitual patterns of conceptual judgment and support non-judgmental meta-awareness of the present moment. Investigators have already started experimenting with using suggestion-based approaches to foster mindful states of awareness for therapeutic purposes (Lynn et al. 2006, 2010). To be sure, however, efforts to induce mindfulness via targeted suggestion are still new and the precise wording of such suggestions would require fine-tuning based on the specific goals of the investigator. Classical descriptions of phenomenological awareness as well as traditional meditation instructions may provide a rich point of departure. Suggestions could be as simple as, for example, noting without judgment the arising and passing of thoughts, emotions, and sensations on a moment-to-moment basis (Lynn et al. 2012).

A subjective sense of effortlessness commonly accompanies hypnotic response, rendering suggestion particularly relevant for promoting meta-awareness. Phenomenologists have often remarked that while one can consciously cultivate a ground ripe for phenomenological insight, the moment of awareness itself involves releasing effortful strategies that would otherwise obscure the phenomena under investigation (Depraz et al. 2000). Acting in accordance with hypnotic suggestions, subjects generally report experiencing their actions and cognitions as effortless and involuntary, as though "the cognitive module that executes the suggestion does so outside of phenomenal awareness" (Kihlstrom 2008). Thus, working below the level of conscious effort, a suggestion for improved phenomenological awareness may allow practitioners to notice their experience while minimizing interference from the wilful act of observation.

One may object, however, that because hypnotic procedures generally involve deep relaxation and mental absorption (Rainville and Price 2003), phenomenological reports following suggestion would conflate the abnormal processes inherent in hypnotic states with the structures of usual waking consciousness. Atypical conscious planes, however, need not accompany response to suggestion; relaxation and fixed attention, although common, are unnecessary for instigating responses typically associated with hypnotic suggestion (Oakley and Halligan 2010). Hypnotic phenomena usually follow even in the absence of an induction ritual or explicit mention of the context of hypnosis (McGeown et al. 2012; Mazzoni et al. 2009; Raz et al. 2006) and responses to suggestions during hypnosis correlate strongly with responses to the same suggestions outside of hypnosis (Kirsch and Braffman 2001). In addition, potential confounding factors associated with the hypnotic ritual can be avoided by means of posthypnotic suggestion – a condition following termination of the hypnotic experience, wherein a subject remains compliant to a suggestion made during hypnosis (Raz and Buhle 2006). Because posthypnotic suggestions function during common wakefulness, they may allow participants to view and describe their experience untarnished by abnormalities surrounding the hypnotic procedure. Hypnotic and posthypnotic suggestion, therefore, constitute potentially fruitful methods of achieving states of receptive observation of experiential processes. Whether hypnosis can really improve meta-awareness and the accuracy of subjective reports, however, remains largely untested. In the following section, we will discuss how meditation and hypnosis can concretely advance a neurophenomenological approach to the mind.

## Meditation and Hypnosis: Neurophenomenology in Action

Some cognitive researchers have already begun to use hypnosis and meditation as phenomenological tools to inform their studies, placing phenomenology and neuroscience in direct contact with one another. Over the past decade, approaches to studying brain function have shifted dramatically from purely task-based paradigms toward methods of investigating the subject at rest, in the absence of external stimulation or goal-directed behaviour (Kelly et al. 2012; Raichle 2010; Callard et al. 2012). In line with this resting-state approach, new paradigms have begun to exploit the benefits of manipulating the attention of participants rather than external task parameters (Raz and Buhle 2006). By allowing researchers to generate profound and highly specific alterations in attention and consciousness without altering the external stimuli, hypnosis and meditation are emerging as valuable tools for investigating spontaneous cognitive activity and the default-mode – a network of brain regions that show increased activity at rest. Default-mode network (DMN) activity correlates with a wide range of internally directed cognitive processes, including mind-wandering, self-oriented thinking, moral reasoning, and episodic memory (Buckner et al. 2008); yet, it is difficult to experimentally manipulate the DMN alongside these processes because the defining feature of the DMN is that it activates spontaneously, in the absence of external task demands. Accordingly, in the past few years, researchers have begun employing contemplative practices in concert with intrinsic connectivity imaging methods to elucidate the psychological correlates of resting-state brain networks such as the DMN (e.g., Brewer et al. 2011; Deeley et al. 2012; Hasenkamp et al. 2012; McGeown et al. 2009; Pagnoni 2012; Pyka et al. 2011; Tang et al. 2012; Taylor et al. 2012).

Harnessing hypnosis together with experiential reports and brain imaging, a recent study showed that hypnotic induction increased subjective ratings of attentional absorption and decreased ratings of mind-wandering, and that these changes were associated with decreased DMN activity and increased activity in prefrontal attention networks (Deeley et al. 2012). Another recent account leveraged a similar neurophenomenological approach to show that subjective ratings of hypnotic depth following an induction were associated with changes in global functional connectivity in the electroencephalography signal. Furthermore, differences in subjective experiential dimensions such as "imagery", "everyday concerns", and "vestibular and other bodily experiences" were associated with distinct patterns of connectivity (Cardeña et al. 2012). These hypnosis studies illustrate how manipulating the experiential state of the subject while collecting subjective reports can enrich and even guide the investigation of intrinsic brain networks and their psychological correlates.

Meditators have extensive practice monitoring and regulating fluctuations of complex cognitive states such as focused attention, distraction, and meta-awareness – states that are tightly linked with intrinsic connectivity networks in the brain. A recent account capitalised on this contemplative ability to better understand the fine-grained temporal vacillations of key attention systems including the DMN, salience network, and executive control network (Hasenkamp et al. 2012). Long-term meditation practitioners performed a simple breath-awareness meditation in an fMRI scanner and pressed a button each time they noticed their mind had wandered from the meditation. By parsing the data surrounding the button press, the researchers were able to distinguish between periods of distraction (in the few seconds prior to the button press), awareness (in the moments immediately surrounding the button press), reorienting of attention (immediately after the press), and finally sustained attention (several seconds after the press). Consistent with their naturalistic model of mind wandering and attention, the researchers found that the cognitive-experiential states inferred via the buttons press reports clearly delineated temporal periods associated with distinct attentional sub-networks. Again, this study underscores the value of yoking phenomenological methods and cutting-edge neuroimaging techniques to arrive at a more complete and, in this case, temporally nuanced view of cognitive and experiential processes.

Let us consider one more example to see how contemplative practice can bring a neurophenomenological approach to bear on one of the most complex and long-standing issues in the sciences of mind: the nature of the self. Contemporary cognitive scientists and phenomenologists often draw a distinction between embodied, pre-reflective subjectivity (the "I" or "minimal self") and higher-order, representational senses of self (the "me" or "narrative self") (Gallagher 2000; Christoff et al. 2011). At first glance, these categories seem to accord with classical Buddhist descriptions of subjectivity, including the tenet of *anatman* or 'no-self,' which refers to the false perception of a permanent self unchanging over time (cf. Siderits et al. 2011). Buddhist sources often describe the goal of meditation as uprooting this illusory sense of an extended self – perhaps similar to the phenomenological notion of narrative self – and uncovering a profoundly embodied, intimate relationship with experience – perhaps akin to the contemporary concept of minimal self. Such phenomenological descriptions can be "front-loaded" into experimental design and explored neuroscientifically (Gallagher and Brøsted Sørensen 2006). One landmark study used functional magnetic resonance imaging to probe the neural correlates of these distinct yet interdependent modes of self-experience in trained meditators (Farb et al. 2007). Participants with 8 weeks of mindfulness training were asked to read personal trait adjectives while engaging in either a "narrative" self-focus, i.e., thinking about how the adjectives related to them as a person, or an "experiential" self-focus, i.e., openly and non-reactively monitoring their moment-to-moment sensory, cognitive, and affective experience.

Consistent with contemplative descriptions, during narrative-focus all participants demonstrated increased activation in neural regions commonly associated with higher-order self-reference and linguistic-semantic processing (medial prefrontal and left lateralized cortices, respectively). During experiential focus, however, only participants trained in meditation demonstrated a pronounced reduction in activity among these higher-order self-reference networks, and a corresponding increase in activity among brain regions associated with internal and external sensory perception as well as body schema (somatosensory cortex, insula, and inferior parietal lobule, respectively). The authors conclude that "these results support distinct, but habitually integrated, aspects of self-reference: (i) higher order self-reference characterised by neural processes supporting awareness of a self that extends across time and (ii) more basic momentary self-reference characterised by neural changes supporting awareness of the psychological present" (Farb et al. 2007). Beyond physiologically corroborating the widespread phenomenological distinction between minimal and narrative selves, these empirical findings expand this understanding by revealing that training can allow people to de-couple these distinct modes of self-reference. Furthermore, the meditative ability to uncouple these usually tightly conjoined self-referential modes was crucial for gaining a better scientific grasp on the nature of the self. The study by Farb et al., therefore, elegantly demonstrates the central thesis of the present paper: that braiding together the strands of contemplative practice, phenomenological description and cognitive neuroscience can lead to a more complete understanding of behaviour, cognition, and the spectrum of human experience.

## Conclusion

Here we explore how contemplative practices such as hypnosis and meditation can advance the emerging interdisciplinary dialogue between cognitive science and phenomenology. We propose that hypnosis and meditation offer powerful tools for instantiating meta-awareness – allowing participants to step back from habitual absorption in experiential contents and attend more openly to fundamental experiential processes. Furthermore, we show how these practices can allow researchers to generate atypical experiential states that may illuminate cognitive structures underlying usual mental life. Although a small number of reports from both the phenomenological and empirical domains have begun to foster the kind of collaborative exchange envisioned here, many basic yet critical questions remain unanswered.

For example, do hypnotic suggestion and mindfulness meditation really improve the accuracy of phenomenological reports, and, if so, to what extent? In addition, how can we meaningfully distil first-person data for analysis alongside behavioural and physiological measurements, and what subjective dimensions do we flatten in the process? We hope that this largely theoretical paper will inspire concrete investigations to answer such crucial questions and help diffuse the boundaries between meditation, hypnosis, cognitive science, and phenomenology.

## References

Allen, M., M. Dietz, K.S. Blair, M. van Beek, G. Rees, P. Vestergaard-Poulsen, et al. 2012. Cognitive-affective neural plasticity following active-controlled mindfulness intervention. *The Journal of Neuroscience* 32(44): 15601–15610.

Augustinova, M., and L. Ferrand. 2012. Suggestion does not de-automatize word reading: Evidence from the semantically based Stroop task. *Psychonomic Bulletin and Review* 19(3): 521–527. doi:10.3758/s13423-012-0217-y.

Bandettini, P.A. 2012. Twenty years of functional MRI: The science and the stories. *NeuroImage* 62(2): 575–588. http://dx.doi.org/10.1016/j.neuroimage.2012.04.026.

Barnier, A.J., R.E. Cox, A. O'Connor, M. Coltheart, R. Langdon, N. Breen, et al. 2008. Developing hypnotic analogues of clinical delusions: mirrored-self misidentification. [Research Support, Non-U.S.Gov't]. *Cognitive Neuropsychiatry* 13(5): 406–430. doi:10.1080/13546800802355666.

Blakemore, S.J., D.A. Oakley, and C.D. Frith. 2003. Delusions of alien control in the normal brain. *Neuropsychologia* 41(8): 1058–1067. doi:10.1016/s0028-3932(02)00313-5.

Brefczynski-Lewis, J.A., A. Lutz, H.S. Schaefer, D.B. Levinson, and R.J. Davidson. 2007. Neural correlates of attentional expertise in long-term meditation practitioners. *Proceedings of the National Academy of Sciences of the United States of America* 104(27): 11483–11488. doi:10.1073/pnas.0606552104.

Brewer, J.A., P.D. Worhunsky, J.R. Gray, Y.-Y. Tang, J. Weber, and H. Kober. 2011. Meditation experience is associated with differences in default mode network activity and connectivity. *Proceedings of the National Academy of Sciences*. doi:10.1073/pnas.1112029108.

Buckner, R.L., J.R. Andrews-Hanna, and D.L. Schacter. 2008. The brain's default network. *Annals of the New York Academy of Sciences* 1124(1): 1–38.

Callard, F., J. Smallwood, and D.S. Margulies. 2012. Default positions: How neuroscience's historical legacy has hampered investigation of the resting mind. *Frontiers in Psychology* 3: 321.

Cardeña, E., P. Jönsson, D.B. Terhune, and D. Marcusson-Clavertz. 2012. The neurophenomenology of neutral hypnosis. *Cortex* 49(2): 375–385. doi:10.1016/j.cortex.2012.04.001.

Casiglia, E., P.S. Schiff, E. Facco, A. Gabbana, V. Tikhonoff, L. Schiavon, et al. 2010. Neurophysiological correlates of post-hypnotic alexia: A controlled study with Stroop test. *American Journal of Clinical Hypnosis* 52(3): 219–233.

Christoff, K., D. Cosmelli, D. Legrand, and E. Thompson. 2011. Specifying the self for cognitive neuroscience. *Trends in Cognitive Sciences* 15(3): 104–112. doi:10.1016/j.tics.2011.01.001.

Cohen Kadosh, R., A. Henik, A. Catena, V. Walsh, and L.J. Fuentes. 2009. Induced cross-modal synaesthetic experience without abnormal neuronal connections. *Psychological Science* 20(2): 258–265. doi:10.1111/j.1467-9280.2009.02286.x.

Deeley, Q., D.A. Oakley, B. Toone, V. Giampietro, M.J. Brammer, S.C.R. Williams, et al. 2012. Modulating the default mode network using hypnosis. *International Journal of Clinical and Experimental Hypnosis* 60(2): 206–228.

Depraz, N., F.J. Varela, and P. Vermersch. 2000. The gesture of awareness: An account of its structural dynamics. In *Investigating phenomenal consciousness: New methodologies and maps*, Advances in consciousness research, vol. 13, ed. M. Velmans, 121–136. Amsterdam: John Benjamins Publishing Company.

Depraz, N., F.J. Varela, and P. Vermersch. 2003. *On becoming aware: A pragmatics of experiencing*. Amsterdam: John Benjamins.

Farb, N.A.S., Z.V. Segal, H. Mayberg, J. Bean, D. McKeon, Z. Fatima, et al. 2007. Attending to the present: Mindfulness meditation reveals distinct neural modes of self-reference. *Social Cognitive and Affective Neuroscience* 2(4): 313–322. doi:10.1093/scan/nsm030.

Fox, K.C.R., P. Zakarauskas, M. Dixon, M. Ellamil, E. Thompson, and K. Christoff. 2012. Meditation experience predicts introspective accuracy. *PLoS One* 7(9): e45370. doi:10.1371/journal.pone.0045370.

Gallagher, S. 2000. Philosophical conceptions of the self: Implications for cognitive science. *Trends in Cognitive Sciences* 4(1): 14–21. doi:10.1016/s1364-6613(99)01417-5.

Gallagher, S., and J. Brøsted Sørensen. 2006. Experimenting with phenomenology. *Consciousness and Cognition* 15(1): 119–134. doi:10.1016/j.concog.2005.03.002.

Gard, T., B.K. Hölzel, A.T. Sack, H. Hempel, S.W. Lazar, D. Vaitl, et al. 2012. Pain attenuation through mindfulness is associated with decreased cognitive control and increased sensory processing in the brain. *Cerebral Cortex* 22(11): 2692–2702. doi:10.1093/cercor/bhr352.

Grant, J.A., J. Courtemanche, and P. Rainville. 2011. A non-elaborative mental stance and decoupling of executive and pain-related cortices predicts low pain sensitivity in Zen meditators. *Pain* 152(1): 150–156. http://dx.doi.org/10.1016/j.pain.2010.10.006.

Hasenkamp, W., C.D. Wilson-Mendenhall, E. Duncan, and L.W. Barsalou. 2012. Mind wandering and attention during focused meditation: A fine-grained temporal analysis of fluctuating cognitive states. *NeuroImage* 59(1): 750–760.

Iani, C., F. Ricci, E. Gherri, and S. Rubichi. 2006. Hypnotic suggestion modulates cognitive conflict: The case of the flanker compatibility effect. *Psychological Science* 17(8): 721–727.

Iani, C., F. Ricci, G. Baroni, and S. Rubichi. 2009. Attention control and susceptibility to hypnosis. *Consciousness and Cognition* 18(4): 856–863.

Khalsa, S.S., D. Rudrauf, A.R. Damasio, R.J. Davidson, A. Lutz, and D. Tranel. 2008. Interoceptive awareness in experienced meditators. *Psychophysiology* 45(4): 671–677. doi:10.1111/j.1469-8986.2008.00666.x.

Kihlstrom, J.F. 2008. The domain of hypnosis, revisited. In *Oxford handbook of hypnosis*, ed. M. Nash and A. Barnier, 21–52. Oxford: Oxford University.

Kirsch, I., and W. Braffman. 2001. Imaginative suggestibility and hypnotizability. *Current Directions in Psychological Science* 10(2): 57–61.

Kosslyn, S.M., W.L. Thompson, M.F. Costantini-Ferrando, N.M. Alpert, and D. Spiegel. 2000. Hypnotic visual illusion alters color processing in the brain. *The American Journal of Psychiatry* 157(8): 1279–1284. doi:10.1176/appi.ajp.157.8.1279.

Lifshitz, M., and A. Raz. 2012. Hypnosis and meditation: Vehicles of attention and suggestion. *The Journal of Mind–Body Regulation* 2(1): 3–11.

Lifshitz, M., N. Aubert-Bonn, A. Fischer, I.F. Kashem, and A. Raz. 2013. Using suggestion to modulate automatic processes: From Stroop to McGurk and beyond. *Cortex* 49(2): 463–473. doi:10.1016/j.cortex.2012.08.007.

Lutz, A. 2004. Introduction—the explanatory gap: To close or to bridge? *Phenomenology and the Cognitive Sciences* 3(4): 325–330. doi:10.1023/b:phen.0000048938.77185.44.

Lutz, A., and E. Thompson. 2003. Neurophenomenology – integrating subjective experience and brain dynamics in the neuroscience of consciousness. *Journal of Consciousness Studies* 10(9–10): 31–52.

Lutz, A., J.D. Dunne, and R.J. Davidson. 2006. Meditation and the neuroscience of consciousness: An introduction. In *Cambridge handbook of consciousness*, ed. P.D. Zelazo, M. Moscovitch, and E. Thompson. New York, NY: Cambridge University Press.

Lutz, A., H.A. Slagter, J.D. Dunne, and R.J. Davidson. 2008. Attention regulation and monitoring in meditation. *Trends in Cognitive Sciences* 12(4): 163–169.
Lynn, S.J., L.S. Das, N.H. Michael, and J.C. Williams. 2006. Mindfulness, acceptance, and hypnosis: Cognitive and clinical perspectives. *International Journal of Clinical and Experimental Hypnosis* 54(2): 143–166.
Lynn, S.J., S. Barnes, A. Deming, and M. Accardi. 2010. Hypnosis, rumination, and depression: Catalyzing attention and mindfulness-based treatments. *International Journal of Clinical and Experimental Hypnosis* 58(2): 202–221.
Lynn, S.J., A. Malaktaris, R. Maxwell, D.I. Mellinger, and D. van der Kloet. 2012. Do hypnosis and mindfulness practices inhabit a common domain? Implications for research, clinical practice, and forensic science. *The Journal of Mind–Body Regulation* 2(1): 12–26.
MacLean, K.A., E. Ferrer, S.R. Aichele, D.A. Bridwell, A.P. Zanesco, T.L. Jacobs, et al. 2010. Intensive meditation training improves perceptual discrimination and sustained attention. *Psychological Science* 21(6): 829.
MacLeod, C.M. 1991. Half a century of research on the Stroop effect: An integrative review. *Psychological Bulletin* 109(2): 163.
MacLeod, C.M., and K. Dunbar. 1988. Training and Stroop-like interference: Evidence for a continuum of automaticity. *Journal of Experimental Psychology. Learning, Memory, and Cognition* 14(1): 126.
Mazzoni, G., E. Rotriquenz, C. Carvalho, M. Vannucci, K. Roberts, and I. Kirsch. 2009. Suggested visual hallucinations in and out of hypnosis. *Consciousness and Cognition* 18(2): 494–499.
McGeown, W.J., G. Mazzoni, A. Venneri, and I. Kirsch. 2009. Hypnotic induction decreases anterior default mode activity. *Consciousness and Cognition* 18(4): 848–855.
McGeown, W.J., A. Venneri, I. Kirsch, L. Nocetti, K. Roberts, L. Foan, et al. 2012. Suggested visual hallucination without hypnosis enhances activity in visual areas of the brain. *Consciousness and Cognition* 21(1): 100–116. doi:10.1016/j.concog.2011.10.015.
Moore, A., T. Gruber, J. Derose, and P. Malinowski. 2012. Regular, brief mindfulness meditation practice improves electrophysiological markers of attentional control. *Frontiers in Human Neuroscience* 6: 18.
Mrazek, M.D., J. Smallwood, and J.W. Schooler. 2012. Mindfulness and mind-wandering: Finding convergence through opposing constructs. *Emotion* 12(3): 442.
Nisbett, R.E., and T.D. Wilson. 1977. Telling more than we can know: Verbal reports on mental processes. *Psychological Review* 84(3): 231–259. doi:10.1037/0033-295x.84.3.231.
Oakley, D.A., and P.W. Halligan. 2009. Hypnotic suggestion and cognitive neuroscience. *Trends in Cognitive Sciences* 13(6): 264–270.
Oakley, D.A., and P.W. Halligan. 2010. Psychophysiological foundations of hypnosis and suggestion. In *Handbook of clinical hypnosis*, 2nd ed, ed. Jay Lynn Steven, Judith W. Rhue, and I. Kirsch. Washington, DC: American Psychological Association.
Pagnoni, G. 2012. Dynamical properties of BOLD activity from the ventral posteromedial cortex associated with meditation and attentional skills. *The Journal of Neuroscience* 32(15): 5242–5249.
Pagnoni, G., M. Cekic, and Y. Guo. 2008. "Thinking about not-thinking": Neural correlates of conceptual processing during Zen meditation. *PLoS One* 3(9): e3083.
Parris, B.A., Z. Dienes, and T.L. Hodgson. 2012. Temporal constraints of the word blindness posthypnotic suggestion on Stroop task performance. *Journal of Experimental Psychology. Human Perception and Performance* 38(4): 833.
Priftis, K., S. Schiff, V. Tikhonoff, N. Giordano, P. Amodio, C. Umiltà, et al. 2011. Hypnosis meets neuropsychology: Simulating visuospatial neglect in healthy participants. *Neuropsychologia* 49(12): 3346–3350. http://dx.doi.org/10.1016/j.neuropsychologia.2011.08.008.
Pyka, M., M. Burgmer, T. Lenzen, R. Pioch, U. Dannlowski, B. Pfleiderer, et al. 2011. Brain correlates of hypnotic paralysis—a resting-state fMRI study. *NeuroImage* 56(4): 2173–2182.
Raichle, M.E. 2010. Two views of brain function. *Trends in Cognitive Sciences* 14(4): 180–190.

Rainville, P., and D.D. Price. 2003. Hypnosis phenomenology and the neurobiology of consciousness. *International Journal of Clinical and Experimental Hypnosis* 51(2): 105–129. doi:10.1076/iceh.51.2.105.14613.

Raz, A., and J. Buhle. 2006. Typologies of attentional networks. *Nature Reviews Neuroscience* 7(5): 367–379.

Raz, A., and N.K.J. Campbell. 2011. Can suggestion obviate reading? Supplementing primary Stroop evidence with exploratory negative priming analyses. *Consciousness and Cognition* 20(2): 312–320. doi:10.1016/j.concog.2009.09.013.

Raz, A., T. Shapiro, J. Fan, and M.I. Posner. 2002. Hypnotic suggestion and the modulation of Stroop interference. *Archives of General Psychiatry* 59(12): 1155.

Raz, A., K.S. Landzberg, H.R. Schweizer, Z.R. Zephrani, T. Shapiro, J. Fan, et al. 2003. Posthypnotic suggestion and the modulation of Stroop interference under cycloplegia. *Consciousness and Cognition* 12(3): 332–346.

Raz, A., J. Fan, and M.I. Posner. 2005. Hypnotic suggestion reduces conflict in the human brain. *Proceedings of the National Academy of Sciences of the United States of America* 102(28): 9978–9983.

Raz, A., I. Kirsch, J. Pollard, and Y. Nitkin-Kaner. 2006. Suggestion reduces the stroop effect. *Psychological Science* 17(2): 91–95. doi:10.1111/j.1467-9280.2006.01669.x.

Raz, A., M. Moreno-Íniguez, L. Martin, and H. Zhu. 2007. Suggestion overrides the Stroop effect in highly hypnotizable individuals. *Consciousness and Cognition* 16(2): 331–338.

Rochat, P., and D. Zahavi. 2011. The uncanny mirror: A re-framing of mirror self-experience. *Consciousness and Cognition* 20(2): 204–213. doi:10.1016/j.concog.2010.06.007.

Schmicking, D. 2010. A toolbox of phenomenological methods. In *Handbook of phenomenology and cognitive science*, ed. S. Gallagher and D. Schmicking, 35–56. New York: Springer.

Schneider, W., and R.M. Shiffrin. 1977. Controlled and automatic human information processing: I. Detection, search, and attention. *Psychological Review* 84(1): 1.

Siderits, M., E. Thompson, and D. Zahavi (eds.). 2011. *Self, no-self? Perspectives from analytical, phenomenological, and Indian traditions*. New York: Oxford University Press.

Slagter, H.A., A. Lutz, L.L. Greischar, A.D. Francis, S. Nieuwenhuis, J.M. Davis, et al. 2007. Mental training affects distribution of limited brain resources. *PLoS Biology* 5(6): e138.

Slagter, H.A., R.J. Davidson, and A. Lutz. 2011. Mental training as a tool in the neuroscientific study of brain and cognitive plasticity. *Frontiers in Human Neuroscience* 5: 17. doi:10.3389/fnhum.2011.00017.

Steinbock, A.J. 2004. Affection and attention: On the phenomenology of becoming aware. *Continental Philosophy Review* 37(1): 21–43. doi:10.1023/B:MAWO.0000049298.44397.be.

Stroop, J.R. 1935. Studies of interference in serial verbal reactions. *Journal of Experimental Psychology* 18(6): 643–662. doi:10.1037/h0054651.

Tang, Y.Y., Y.H. Ma, J. Wang, Y.X. Fan, S.G. Feng, Q.L. Lu, et al. 2007. Short-term meditation training improves attention and self-regulation. *Proceedings of the National Academy of Sciences of the United States of America* 104: 17152–17156.

Taylor, V.A., J. Grant, V. Daneault, G. Scavone, E. Breton, S. Roffe-Vidal, et al. 2011. Impact of mindfulness on the neural responses to emotional pictures in experienced and beginner meditators. *NeuroImage* 57(4): 1524–1533.

Taylor, V.A., V. Daneault, J. Grant, G. Scavone, E. Breton, S. Roffe-Vidal, et al. 2012. Impact of meditation training on the default mode network during a restful state. *Social Cognitive and Affective Neuroscience* 8(1): 4–14. doi:10.1093/scan/nsr087.

Terhune, D.B., E. Cardeña, and M. Lindgren. 2010. Disruption of synaesthesia by posthypnotic suggestion: An ERP study. *Neuropsychologia* 48(11): 3360–3364.

van Leeuwen, S., W. Singer, and L. Melloni. 2012. Meditation increases the depth of information processing and improves the allocation of attention in space. *Frontiers in Human Neuroscience* 6: 133. doi:10.3389/fnhum.2012.00133.

Varela, F.J. 1996. Neurophenomenology: A methodological remedy for the hard problem. *Journal of Consciousness Studies* 3: 330–349.

Varela, F.J. 1998. A science of consciousness as if experience mattered. In *Toward a science of consciousness II: The second Tucson discussions and debates*, vol. 2, ed. S.R. Hameroff, A.W. Kaszniak, and A.C. Scott, 31–44. Cambridge: MIT Press.

Varela, F.J., E.T. Thompson, and E. Rosch. 1992. *The embodied mind: Cognitive science and human experience*. Cambridge, MA: MIT Press.

Woody, E., and H. Szechtman. 2011. Using hypnosis to develop and test models of psychopathology. *Mind-Body Regulation* 1(1): 4–16.

Zeidan, F., K.T. Martucci, R.A. Kraft, N.S. Gordon, J.G. McHaffie, and R.C. Coghill. 2011. Brain mechanisms supporting the modulation of pain by mindfulness meditation. *The Journal of Neuroscience* 31(14): 5540–5548. doi:10.1523/jneurosci.5791-10.2011.

# Insights from Quiet Minds: The Converging Fields of Mindfulness and Mind-Wandering

**Michael D. Mrazek, Benjamin W. Mooneyham, and Jonathan W. Schooler**

**Abstract** Our lives are filled with an endless array of perceptions, thoughts, and feelings, and our attention usually darts back and forth between them. Yet meditative traditions have long valued the capacity to remain undistracted from our immediate experience, and countless individuals make a practice of stabilizing their awareness in the here and now. What are the implications of anchoring our usually restless minds? Could stabilizing our attention provide an informative lens into the dynamics of the human brain? Here we review recent research that situates mindfulness as an opposing construct to mind-wandering and a remedy for wandering minds. We then review empirical intersections between mindfulness and mind-wandering from recent neuroimaging studies.

## Mindfulness and Non-Distraction

The word mindfulness is used with a growing sense of familiarity, but there is ongoing disagreement as to the most privileged and useful definition of this construct (Grossman and Van Dam 2011) or even whether the act of defining mindfulness is appropriate (Schmidt, this volume). Some meditative traditions have defined mindfulness as sustained non-distraction (Brown and Ryan 2003; Wallace and Shapiro 2006; Dreyfus 2011), whereas multifactor conceptualizations of mindfulness emphasize additional qualities as well, such as an orientation toward one's experiences characterized by curiosity, openness, and acceptance (Bishop et al. 2004; Baer et al. 2006).

M.D. Mrazek (✉) • B.W. Mooneyham • J.W. Schooler
Department of Psychological and Brain Sciences, University of California, Santa Barbara, CA 93106-9660, USA
e-mail: mrazek@psych.ucsb.edu

S. Schmidt and H. Walach (eds.), *Meditation – Neuroscientific Approaches and Philosophical Implications*, Studies in Neuroscience, Consciousness and Spirituality 2, DOI 10.1007/978-3-319-01634-4_13, 

These definitions are by no means exhaustive and there are many traditions of mindfulness practice that have evolved over millennia and offer further delineation.

Amid this disagreement, there is nevertheless consensus from meditative traditions that sustained attentiveness represents a fundamental element of mindfulness. Accordingly, we have largely focused our investigations into mindfulness by examining the capacity for non-distraction.[1] Our focus on non-distraction as a central element of mindfulness is not intended to devalue other qualities sometimes ascribed to mindfulness, such as intentionality, non-judgment, awareness, openness, and curiosity. There is continued disagreement as to whether each of these various capacities are sufficient or necessary constituents of mindfulness, or even whether they might be understood as precursors, concomitants, or consequences of mindfulness, rather than aspects of mindfulness per se. Fortunately, empirical investigation into mindfulness can continue despite these disagreements so long as researchers are explicit about their conceptual and operational definitions.

## Mind-Wandering as Task-Unrelated Thought

In direct contrast to mindfulness, which entails a capacity to avoid distraction, mind-wandering is characteristically described as the interruption of task-focus by task-unrelated thought (TUT; Smallwood and Schooler 2006). Unlike the struggle to identify a validated and widely accepted measure of mindfulness, there has been somewhat greater consensus with respect to operational definitions of mind-wandering. The most widely used measure is straightforward: periodically interrupting individuals during a task and asking them to report the extent to which their attention was on the task or on task-unrelated concerns, a procedure known as "thought-sampling", which measures "probe-caught" mind-wandering. There is a broad literature validating the self-report measures of mind-wandering obtained through thought-sampling by using behavioral (Smallwood et al. 2004), event-related potential (ERP; Smallwood et al. 2008c), and fMRI methodologies (Christoff et al. 2009). Such studies suggest that individuals are able to accurately report whether they have been mind-wandering – and even whether they have been aware of it – as revealed by distinct patterns of task performance and neural activation in association with self-reported mind-wandering (for a recent review see Schooler et al. 2011). Additionally, studies using retrospective reports of mind-wandering after a task has been finished typically find results that are similar to those obtained with thought-sampling during the task (Mrazek et al. 2011). This not only provides convergent validity for thought

[1] It is worth noting that this definition of non-distraction always exists with reference to a particular activity. For example, if your goal is to engage in a task, but instead you become deeply focused on off-task concerns, this would not be an example of mindfulness even though your off-task focus may be undistracted.

sampling, but also suggests that asking participants to intermittently report their mind-wandering does not substantially alter their behavior or performance in at least some task contexts (Mrazek et al. 2012; Barron et al. 2011).

Another common measure of mind-wandering involves asking participants to indicate every time they notice that they have been mind-wandering. This measures "self-caught" mind-wandering, providing a straightforward assessment of mind-wandering episodes that have reached meta-awareness as an explicit re-representation of the contents of one's own consciousness (Schooler 2002). By contrast, thought-sampling queries participants at unpredictable intervals and does not require participants to attend to their thoughts independently of an external prompt. However, because thought-sampling probes occur at varying and unpredictable times during a primary task, this method can be used in conjunction with the self-catching measure to catch people mind-wandering before they notice it themselves (Schooler and Schreiber 2004).

Several indirect markers of mind-wandering are also available, including those derived from performance markers of inattention in the Sustained Attention to Response Task (SART; Smallwood et al. 2004, 2007a, b, 2008b; McVay and Kane 2009; Cheyne et al. 2009). The SART is a GO/NOGO task in which participants are asked to respond with a key press as quickly as possible to frequent non-targets and to refrain from responding to rare targets. Different performance markers in this task, such as response times (RTs) or different kinds of errors, have been associated with varying degrees of task disengagement (Cheyne et al. 2009). For example, failures to respond to rare targets (errors of omission) generally indicate a more pronounced state of disengagement than a large coefficient of variability (CV) for RTs (the CV is the standard deviation of RTs divided by the mean). RT CV has been associated with a state of mind-wandering that emerges from a minimally disruptive disengagement of attention characterized by a periodic speeding and slowing of RTs as attention fluctuates slightly (Cheyne et al. 2009; Smallwood et al. 2008b).

## Mindfulness and Mind-Wandering as Opposing Constructs

Many behavioral markers of mind-wandering have a distinctly mindless quality, such as rapid and automatic responding during SART (Smallwood et al. 2004), absent-minded forgetting (Smallwood et al. 2003), and eye-movements during reading that are less sensitive to lexical or linguistic properties of what is being read (Reichle et al. 2010). Furthermore, ERP studies have demonstrated that instances of mind-wandering are characterized by a reduced awareness and/or sensory processing of task stimuli and other objects in the external environment (Barron et al. 2011; Smallwood et al. 2008c; Kam et al. 2010). The ability to remain mindfully focused on a task therefore appears to be in direct opposition to the tendency for attention to wander to task-unrelated thoughts. Starting from this observation, we began our ongoing series of investigations into the relationship between mindfulness and mind-wandering by first examining whether we could find empirical support for this intuitive notion that mind-wandering and mindfulness are opposing constructs.[2]

Existing work that links mindfulness and mind-wandering has relied heavily on the Mindful Awareness Attention Scale (MAAS; Brown and Ryan 2003), the most widely used dispositional measure of mindfulness. This scale addresses the extent to which an individual attends to present experience without distraction (e.g., I find myself listening to someone with one ear, doing something else at the same time; reverse scored). Low self-reported mindfulness as measured by the MAAS is associated with fast and error-prone responding in the SART (Cheyne et al. 2006, 2009). These results show that the measurement of trait-mindfulness by the MAAS can predict behavioral concomitants of real-time mind-wandering observed during performance of a task in the lab.

We recently conducted a more comprehensive investigation into the relationship between the MAAS and several convergent measures of mind-wandering (Mrazek et al. 2011). All participants completed the MAAS, a 10-min mindful breathing task with thought-sampling probes, a 10-min mindful breathing task requiring self-catching of mind-wandering, a 10-min SART, and a self-report measure of trait daydreaming that has been widely used to study mind-wandering (Mason et al. 2007). We found that individuals who reported high levels of mindfulness during daily life also reported less daydreaming. Furthermore, high levels of trait-mindfulness were also associated with less mind-wandering as measured by self-reported TUT during mindful breathing, fewer errors of commission during the SART, and lower RT variability. These results provide converging evidence suggesting that – at least based on their most common operational definitions – mindfulness and mind-wandering are indeed opposing constructs.

## Mindfulness as a Tool for Reducing Mind-Wandering

If mindfulness and mind-wandering are inversely related, it follows that mind-wandering and its disruptive effects on task performance (e.g. Smallwood et al. 2003, 2004, 2007a, b, 2008a; Smallwood 2011a; Reichle et al. 2010) should be reduced by interventions that increase mindfulness. While mindfulness training has been demonstrated to improve executive attention, perceptual sensitivity, and sustained attention (Tang et al. 2007; MacLean et al. 2010), the direct impact of mindfulness training on mind-wandering has until recently been less carefully examined. In fact, to date there has been little progress in developing empirically proven strategies for reducing mind-wandering.

[2]Our selection of non-distraction as a central feature of mindfulness strongly influences our interpretation of the opposing nature of mindfulness and mind-wandering. More inclusive conceptualizations of mindfulness might lead to different perspectives. For instance, although there may be wide agreement that episodes of mind-wandering are not instances of mindfulness, the absence of mind-wandering may not guarantee the presence of mindfulness. One might be undistracted, but lack other qualities sometimes espoused to be essential to mindfulness (non-judgment, openness, curiosity, etc.).

We recently examined whether a brief mindfulness exercise can reduce mind-wandering, thereby both introducing a potential antidote to mind-wandering and establishing a causal relationship between the presence of mindfulness and the absence of mind-wandering. This expectation is consistent with the many well-documented benefits of mindfulness training (for a review see Brown et al. 2007). However, many prior studies have utilized intensive meditation training lasting months or years, limiting the applicability of observed improvements for most societal and educational contexts (Brefczynski-Lewis et al. 2007; MacLean et al. 2010). Furthermore, from a methodological perspective, mindfulness intervention studies typically include so many different aspects of intervention that it is difficult to discern which specific element is responsible for any observed changes. What is useful in discerning the causal role of mindfulness in mitigating mind-wandering is a simple manipulation that directly and specifically targets individuals' ability to remain mindful. Accordingly, we used an 8-min mindful breathing intervention that provides a simple and widely accessible intervention that also affords a high degree of experimental control.

In this investigation, participants were randomly assigned to conditions in which they completed either 8 min of mindful breathing or one of two control conditions: passive relaxation or reading. Expectation effects and demand characteristics were minimized by informing all participants that they were participating in a study designed to examine effects of relaxation on attention. In the mindful breathing condition, participants were instructed to sit in an upright position while focusing their attention on the sensations of their breath without trying to control the rate of respiration. Participants were asked to return their attention to the breath anytime they became distracted. Participants in the reading condition were asked to browse a popular local newspaper, while those in the passive rest condition were asked to relax without falling asleep. Subsequently, all participants completed a 10-min version of the SART. Relative to the two control conditions, those who first completed 8-min of mindful breathing exhibited enhanced performance as measured by behavioral markers of inattention commonly associated with mind-wandering (fewer errors of commission and lower RT variability). The effectiveness of this intervention establishes a causal relationship between the cultivation of mindfulness and subsequent reduction in mind-wandering.

Building on the finding that a brief mindfulness exercise could reduce mind-wandering, we next examined whether a more thorough introduction to mindfulness could reduce mind-wandering in a manner that would enhance working memory capacity (WMC) and reading comprehension. In a recent randomized controlled investigation, we examined whether a 2 week mindfulness training course would be more effective than a comparably demanding nutrition program in decreasing mind-wandering and improving cognitive performance among undergraduates (Mrazek et al. 2013). We found that mindfulness training improved performance on both the measure of WMC and the test of reading comprehension (an adapted version of the Graduate Record Examination; GRE). Mindfulness training also reduced mind-wandering during these tasks as assessed by thought sampling, self-catching, and a validated scale measuring retrospective task-unrelated thought. Notably, improvements in WMC and GRE performance following mindfulness training were mediated by reduced mind-wandering specifically for those who were most prone to distraction at pre-testing. This suggests that mindfulness-based interventions do not only benefit individuals who are already proficient at attentional control, and that training to enhance attentional focus may be a key to unlocking latent cognitive skills that were until recently viewed as immutable.

## Brain Dynamics Through the Lens of Mind-Wandering and Mindfulness

The task-positive and task-negative networks of the human brain – together comprising a substantial portion of the human cerebral cortex – are engaged in an endless back-and-forth. As we engage with a task, task-positive brain regions dedicated to attention and control are activated. This task positive network is also referred to as the dorsal attention network, and it includes brain regions involved in orienting attention and executive control. When our minds wander, a different set of task-negative brain regions activate (these regions are also known as the default mode network). These two networks usually operate in opposition, yet under some circumstances they activate simultaneously (Smallwood et al. 2011b). Understanding the functions and dynamics of these respective networks remains an area of focused investigation. Here we illustrate how the intrinsically related constructs of mindfulness and mind-wandering may provide an informative lens when thinking about the potentially nuanced relationship between the task-positive and task-negative networks.

### *Neural Correlates of Mind-Wandering*

Over the last decade, accumulating evidence has suggested that activation of the task-negative or default-mode network (DMN) may serve as an fMRI marker of mind-wandering (for a review see Gruberger et al. 2011). The DMN is a collection of brain regions that typically show greater activation at rest than during task performance. Direct evidence that the DMN is associated with mind-wandering comes from studies that link this network to reports of task-unrelated thoughts. One approach involves linking retrospective measures of mind-wandering to brain activity (e.g. Andrews-Hanna et al. 2010). Other studies have documented that situations that are associated with greater mind-wandering reports (as assessed outside of the scanner) also lead to greater activity in many of the key elements of the DMN (Mason et al. 2007; McKiernan et al. 2006). Furthermore, Christoff and colleagues (2009) combined experience sampling with fMRI while participants engaged in the Sustained Attention to Response Task. During periods of off-task thought, DMN activity was higher than when participants were focused on the task, an observation that has been replicated by Stawarczyk and colleagues (2011). Importantly, Christoff and colleagues also demonstrated that DMN activity increased prior to performance errors that have themselves been linked to greater mind-wandering. In summary, although DMN activity may underlie more than just task-unrelated thoughts, a growing body of evidence clearly indicates that DMN regions are more active during mind-wandering than during focused task-engagement.

## *Neural Correlates of Mindfulness*

What happens in the brain when someone meditates, and does this depend on expertise? If mind-wandering and mindfulness are opposing constructs, neural markers of mind-wandering should decrease during the practice of mindfulness. This topic has been approached largely through investigations with experienced meditators. The accumulated practice of experienced practitioners might be expected to produce stronger contrasts between rest and meditation, though it is also possible that these individuals' resting state is largely characterized by an ongoing mindfulness (see discussion below of Froeliger et al. 2012). It can also be informative to investigate the underlying neural correlates of mindfulness among individuals without prior training. This second approach is helpful because attempts to identify the neural processes of distinct aspects of mindfulness (like non-distraction) can be obscured by the divergence in how mindfulness is defined and practiced. In fact, researchers often study experienced meditators who have undertaken considerable training in a number of related but potentially dissociable practices: non-distraction, non-judgment, non-reactivity, non-attachment, etc.[3] From this vantage point, investigating the neural mechanisms supporting mindfulness among non-meditators helps control for prior history with mindfulness that may not align with the experimenter's operational definition. Here we review selected research that addresses what happens in the brain during periods of focused-attention meditation in both inexperienced and experienced meditation practitioners, with an emphasis on how this research illuminates the intersection between mind-wandering and mindfulness.[4]

A recent investigation focused specifically on neural correlates of mindfulness among non-meditators (Dickenson et al. 2012). This study used an fMRI block

[3] One could argue that a genuine instance of mindfulness must be characterized by the presence of non-distraction, non-judgment, curiosity, and openness, and therefore, that investigations pertaining to a single quality of mindfulness will be insufficient in their ability to draw conclusions about the construct in its totality. Although more focused research into specific features of mindfulness risks not fully representing the more inclusive characterizations of the construct, they do allow for more tractable empirical investigations. It is considerably more feasible to operationally define and measure non-distraction than it would be to integrate measurements of curiosity, openness, non-judgment, awareness, and non-distraction to ensure that all these elements are present simultaneously.

[4] Focused-attention meditation may not represent an instance of mindfulness within multi-factor frameworks of mindfulness that would require the presence of some constellation of additional qualities besides non-distraction.

design in which participants alternated between 50 s of focused-attention meditation and a control task in which they were provided with the instructions: "let your mind take you wherever it goes as you normally would throughout the day." Relative to the control task, there was increased activation of a variety of regions associated with the task-positive network during meditation. Specifically, significant increases were found in superior parietal lobule (SPL), temporal-parietal junction (TPJ), pre-supplementary motor area, dorsal anterior cingulate gyrus, and the insula. Meditation also led to a decrease in activation of a coherent subset of the default mode network, including medial PFC, dorsomedial PFC, angular gyrus, and precuneus. This study clearly demonstrates that focused-attention meditation can activate brain regions associated with control of attention, even among non-meditators. However, it remains unclear whether the activation of task-positive regions represent focused attention, the *attempt* to focus attention, or both. Regardless, the combination of increased activation of the task-positive network with reduced activation of default-mode network regions is consistent with the notion that individuals are at least attempting to engage their attention on their breath while suppressing the distraction of mind-wandering. This is also broadly consistent with evidence that functional connectivity between task-positive and default mode networks should change during meditation – a topic we return to shortly.

Although Dickenson and colleagues (2012) found clear activation changes during meditation that could be interpreted as reduced mind-wandering, another study employing a similar methodology found only partially consistent findings. Brewer and colleagues (2011) examined differences in neural activation during meditation among both non-meditators and experienced meditators who had an average of over 10,000 h of practice. As part of a larger investigation, participants completed two 4.5-min focused-attention meditations in which they were asked to pay attention to the sensations of breathing. During baseline scans that preceded the meditations, participants were instructed: "please close your eyes and don't think of anything in particular". Unlike Dickenson and colleagues (2012), this investigation did not find meditation-induced reductions in activation of key default mode regions (PCC and mPFC) or any changes in task-positive network regions among non-meditators. By contrast, experienced meditators showed reduced activation in both PCC and mPFC. A between-groups contrast revealed that during focused-attention meditation, meditators showed significantly less activation in PCC and left angular gyrus compared to non-meditators. In sum, advanced meditators showed the expected pattern of reduced default mode network activation, whereas non-meditators showed no significant changes. This complete absence of activation changes that Brewer and colleagues (2011) found in non-meditators is striking, and might be explained by the combination of a small sample size and a long-block design that the authors suggest may have de-optimized their analyses. Additionally, novice meditators may rapidly alternate between attempts to focus, focused attention, and lapses of attention. This could lead to a less consistent activation of brain regions during the focused-attention meditation, although participants in Dickenson et al. (2012) would have been subject to similar fluctuations of attention and yet that investigation revealed several predicted effects.

Although there is some inconsistency with respect to non-meditators, the results across these studies are generally consistent with the notion that brain regions associated with mind-wandering are relatively deactivated during focused-attention meditation. The inconsistency across studies could be a consequence of the variations in block design, meditation instructions, and control/baseline task instructions. It is also worth mentioning that in each study, participants repeatedly alternated between meditation and control tasks. Given that even brief mindfulness inductions can result in subsequently altered performance (Mrazek et al. 2011), there is some risk that meditation-related neural changes could bleed into control tasks. One alternative strategy would be to have all participants first complete a control task and then complete a meditation task, though this design is also not without limitations (i.e. order effects).

## Neural Activation Versus Functional Connectivity

Experienced meditators tend to show greater activation of task-positive regions and decreased activation of DMN regions during meditation, indicating that mindfulness practice influences the magnitude of activity within various brain regions. Given that the brain operates as a highly complex system, it is plausible that these changes in the magnitude of neuronal activity are coupled to changes in the interactions between brain areas (and their associated networks). Functional connectivity analysis of fMRI data has recently gained traction as an informative approach toward understanding the dynamics of the brain and the extent to which different brain areas are functionally connected to one another. Functional connectedness, in the context of fMRI data, is measured by examining the strength of correlations between the time-courses of neuronal activity across different brain regions. Brain regions are functionally connected to the extent that their respective activation patterns are correlated during a particular task or context, regardless of whether those regions are adjacent or structurally connected. Importantly, functional connectivity is present between brain regions even if they exhibit highly anti-correlated (i.e., negatively correlated) patterns of activation; such anti-correlations suggest inhibitory or control processes that are likely just as important to brain function as co-activations of brain areas. Here we review the impact of mindfulness training on functional connectivity both within and between the default-mode and task-positive networks.

### *Mindfulness and Functional Dynamics Within the Default Mode Network*

Jang and colleagues (2011) examined the differences in functional connectivity within the DMN during the resting-state between 35 experienced meditators and 33 controls without prior meditation experience. Separate analyses were conducted using either the medial prefrontal cortex (MPFC) or the posterior cingulate cortex (PCC) as seed regions (both of these regions are considered to be primary hubs of the DMN; see Buckner et al. 2008 for evidence supporting this claim). While the patterns of connectivity were similar for either seed region – including MPFC, PCC, inferior parietal cortices, and lateral temporal cortices for both seed maps – the experienced meditators exhibited greater functional connectivity within the medial prefrontal region than controls. Activation of the medial prefrontal cortex has been associated with concentrating on internal focus and sensations (so-called "internalized attention", Hölzel et al. 2011); these results therefore suggest that meditation practice may produce changes in functional connectivity within anterior DMN regions that may afford enhanced concentration or self-awareness.

Whereas Jang and colleagues (2011) found increased DMN functional connectivity within the MPFC (an anterior DMN region) among experienced meditators, self-reported mindfulness during daily life among non-meditators has also been associated with increased functional connectivity in more posterior portions of the DMN (i.e., the precuneus (Pcu) and the posterior cingulate cortex (PCC)) (Prakash et al. 2012). Although there is a discrepancy regarding the specific region(s) involved, the existing literature therefore indicates that mindfulness is associated with increased functional connectivity within key DMN regions. On appearance, this is difficult to reconcile with the previously reviewed evidence that mindfulness practice leads to less DMN activation. If mindfulness is associated with lower levels of DMN activity, then why should training in mindfulness act to strengthen the functional connections within this network? One potential answer may lie in the relationship between the anterior and posterior regions of the DMN. Jang and colleagues (2011) found greater functional connectivity within an anterior portion of the DMN (i.e., MPFC) while Prakash and colleagues (2012) found greater connectivity within posterior regions (i.e., Pcu and PCC), but recent research suggests that the functional connectivity between these anterior and posterior regions may in fact decrease with mindfulness/meditation training. Hasenkamp and Barsalou (2012), in a study comparing trained vs. novice meditators, found that the trained meditators exhibited decreased functional connectivity between a regional cluster containing portions of the MPFC and anterior cingulate cortex (ACC) and a cluster containing the PCC. Thus, these mindfulness-trained individuals exhibited less anterior-posterior functional coherence within the DMN (a similar connectivity relationship between the ACC and PCC was also observed by Kilpatrick et al. 2011). As such, it is plausible that mindfulness training may improve the intrinsic functioning of this task-negative system overall (as indicated by the observed increase in connectivity within specific regions of the DMN). However, by means of reducing the functional connectivity between anterior and posterior regions, mindfulness may reduce the likelihood of intrusive mind-wandering (e.g., rumination, prospection, etc.), the content of which is provided by the more posterior regions of the DMN (in their DMN review, Buckner et al. (2008) suggest that the DMN, particularly the PCC, is chiefly involved in internal mentations such as episodic remembering, prospection, and theory of mind).

### *Mindfulness and Between-Network Functional Dynamics*

Hypothetically, mindfulness training may be expected to produce several key changes in the way the brain operates, both at rest and also under task constraints. For instance, individuals who have been trained in mindfulness techniques may be able to exert more attentional control over their thoughts, and as such they may be expected to display heightened coherence in brain networks associated with attentional control, especially during task settings in which such control is most critical. In a recent study, Froeliger and colleagues (2012) examined whether mindfulness-trained individuals indeed display these characteristics. They compared meditation-trained individuals with a control group and found that during a resting-state fMRI scan the meditation-trained group displayed higher functional connectivity within the task-positive network, suggesting a heightened control over the direction of attention for these individuals, even outside of the context of meditation. Additionally, those trained in meditation exhibited higher functional connectivity between the task-positive network and the DMN when meditating than when at rest, perhaps indicating a greater level of executive control in these individuals.

Several additional studies have provided converging evidence that mindfulness training enhances connectivity between DMN and task-positive regions. Brewer et al. (2011) reported that meditators (compared to controls) exhibited greater functional connectivity between DMN regions (including the PCC) and task-positive regions (specifically, the DLPFC). Hasenkamp and Barsalou (2012) also found a similar trend in functional connectivity patterns, namely that meditation-trained participants exhibited greater functional connectivity between a key anterior DMN region (e.g., the ventro-medial prefrontal cortex; VMPFC) and a key task-positive region, the inferior parietal lobule (IPL). As the IPL has been functionally associated with attentional disengagement processes (e.g., Posner et al. 1984), the authors suggest that this increased connectivity in meditators better allows for these individuals to disengage from mind-wandering states and re-engage in their meditative practices. This increased coherence between DMN areas and areas involved in attentional control has also been documented by Taylor and colleagues (2012), who compared experienced vs. novice meditators and found, similarly, that experienced meditators displayed higher functional connectivity between the IPL and the VMPFC during the resting-state.

In summary, the existing literature suggests that mindfulness training is associated with increased functional connectivity (i) within key default-mode regions (e.g., MPFC), (ii) within the task-positive network (e.g., DLPFC), and (iii) between default-mode and task-positive regions. How does mind-wandering fit into this picture? In considering this, it is important to keep in mind that functional connectivity does not distinguish between neural recruitment and inhibition. The fact that DMN regions are temporally coupled to task-positive regions in mindful individuals may therefore reflect the fact that these control regions are now, as a result of training, better able to exert inhibitory or reactive influence on these DMN regions, thus preventing mind-wandering from disrupting periods of mindfulness. This interpretation is bolstered by the finding that, overall, DMN activity is lower during meditation for experienced mindfulness practitioners (Brewer et al. 2011).

## Future Directions

This review suggests that one important direction for future research is to integrate these various observations about mindfulness training, mind-wandering, and brain dynamics into a single cohesive training study. We now know that mindfulness training can reduce mind-wandering and that individuals with extensive meditation experience show functional changes within and between task-positive and default-mode regions that are consistent with less mind-wandering. A clear next step would be to determine whether these changes in underlying neural activity mediate the improvements in task-focus.

Future research must also keep potential benefits of mind-wandering in view. After all, there are circumstances in which diverting attention away from the "here and now" is beneficial. Thinking about the past or planning the future can of course be done deliberately, but research has indicated that even spontaneous mind-wandering that occurs when we are occupied with another task can be useful in some circumstances. For instance, recent findings suggest that mind-wandering can promote future planning (Baird et al. 2011) and enhance creative incubation (Baird et al. 2012). Yet the accumulating evidence for the positive outcomes of mindfulness might be interpreted to suggest that mind-wandering is of no benefit, especially within a framework that places these constructs in direct opposition. In contrast, the potential benefits of mind-wandering could be interpreted to suggest a downside to mindfulness. For instance, a practice of mindfulness that eliminated mind-wandering might lead to neglect of distal goals like retirement planning (even though it would not eliminate the opportunity for more deliberate goal-oriented planning). It may therefore be that mindfulness is most helpful when it affords a degree of control over mind-wandering that allows for its benefits while minimizing its costs. With mindfulness individuals might become better able to mind-wander at the right times (e.g. when primary task demands are relatively modest) and on the right topics (e.g. on productive issues that can foster future planning or creativity).

**Acknowledgments** MDM and JWS are supported through United States Department of Education grant R305A110277 awarded to JWS. BWM is supported by a National Science Foundation Graduate Research Fellowship under grant No. DGE-1144085. The content of this article does not necessarily reflect the position or policy of the U. S. Government, and no official endorsement should be inferred.

## References

Andrews-Hanna, J.R., J.S. Reidler, C. Huang, and R.L. Buckner. 2010. Evidence for the default network's role in spontaneous cognition. *Journal of Neurophysiology* 104(1): 322–335.

Baer, R.A., G.T. Smith, J. Hopkins, J. Krietemeyer, and L. Toney. 2006. Using self-report assessment methods to explore facets of mindfulness. *Assessment* 13(1):27–45.doi:10.1177/1073191105283504.

Baird, B., J. Smallwood, and J.W. Schooler. 2011. Back to the future: Autobiographical planning and the functionality of mind-wandering. *Consciousness and Cognition* 20(4): 1604–1611.

Baird, B., J. Smallwood, M.D. Mrazek, J.W. Kam, M.S. Franklin, and J.W. Schooler. 2012. Inspired by distraction mind wandering facilitates creative incubation. *Psychological Science* 23(10): 1117–1122.

Barron, E., L.M. Riby, J. Greer, and J. Smallwood. 2011. Absorbed in thought: The effect of mind-wandering on the processing of relevant and irrelevant events. *Psychological Science* 22(5): 596–601. doi:10.1177/0956797611404083.

Bishop, S.R., M. Lau, S. Shapiro, L. Carlson, N.D. Anderson, J. Carmody, et al. 2004. Mindfulness: A proposed operational definition. *Clinical Psychology: Science and Practice* 11(3): 230–241. doi:10.1093/clipsy.bph077.

Brefczynski-Lewis, J.A., A. Lutz, H.S. Schaefer, D.B. Levinson, and R.J. Davidson. 2007. Neural correlates of attentional expertise in long-term meditation practitioners. *Proceedings of the National Academy of Sciences* 104(27): 11483–11488. doi:10.1073/pnas.0606552104.

Brewer, J.A., P.D. Worhunsky, J.R. Gray, Y.-Y. Tang, J. Weber, and H. Kober. 2011. Meditation experience is associated with differences in default mode network activity and connectivity. *Proceedings of the National Academy of Sciences* 108(50): 20254–20259. doi:10.1073/pnas.1112029108.

Brown, K.W., and R.M. Ryan. 2003. The benefits of being present: Mindfulness and its role in psychological well-being. *Journal of Personality and Social Psychology* 84(4): 822–848. doi:10.1037/0022-3514.84.4.822.

Brown, K.W., R.M. Ryan, and J.D. Creswell. 2007. Mindfulness: Theoretical foundations and evidence for its salutary effects. *Psychological Inquiry* 18(4): 211–237. doi:10.1080/10478400701598298.

Buckner, R.L., J.R. Andrews-Hanna, and D.L. Schachter. 2008. The brain's default network: Anatomy, function, and relevance to disease. *Annals of the New York Academy of Sciences* 1124: 1–38.

Cheyne, J.A., J.S.A. Carriere, and D. Smilek. 2006. Absent-mindedness: Lapses of conscious awareness and everyday cognitive failures. *Consciousness and Cognition* 15(3): 578–592. doi:10.1016/j.concog.2005.11.009.

Cheyne, J., G.J.F. Solman, J.S.A. Carriere, and D. Smilek. 2009. Anatomy of an error: A bidirectional state model of task engagement/disengagement and attention-related errors. *Cognition* 111(1): 98–113. doi:10.1016/j.cognition.2008.12.009.

Christoff, K., A.M. Gordon, J. Smallwood, R. Smith, and J.W. Schooler. 2009. Experience sampling during fMRI reveals default network and executive system contributions to mind wandering. *Proceedings of the National Academy of Sciences of the United States of America* 106(21): 8719–8724.

Dickenson, J., E.T. Berkman, J. Arch, and M.D. Lieberman. 2012. Neural correlates of focused attention during a brief mindfulness induction. *Social Cognitive and Affective Neuroscience* 8(1): 40–47. doi:10.1093/scan/nss030.

Dreyfus, G. 2011. Is mindfulness present-centered and non-judgmental? A discussion of the cognitive dimensions of mindfulness. *Contemporary Buddhism* 12(1): 41–54.

Froeliger, B., E.L. Garland, R.V. Kozink, L.A. Modlin, N.-K. Chen, F.J. McClernon, et al. 2012. Meditation-state functional connectivity (msFC): Strengthening of the dorsal attention network and beyond. *Evidence-Based Complementary and Alternative Medicine* 2012: 1–9. doi:10.1155/2012/680407.

Grossman, P., and N.T. Van Dam. 2011. Mindfulness, by any other name…: Trials and tribulations of sati in western psychology and science. *Contemporary Buddhism* 12(1): 219–239. doi:10.1080/14639947.2011.564841.

Gruberger, M., E. Ben-Simon, Y. Levkovitz, A. Zangen, and T. Hendler. 2011. Towards a neuroscience of mind-wandering. *Frontiers in Human Neuroscience* 5: 56.

Hasenkamp, W., and L.W. Barsalou. 2012. Effects of meditation experience on functional connectivity of distributed brain networks. *Frontiers in Human Neuroscience* 6: 38. doi:10.3389/fnhum.2012.00038.

Hölzel, B.K., S.W. Lazar, T. Gard, Z. Schuman-Olivier, D.R. Vago, and U. Ott. 2011. How does mindfulness meditation work? Proposing mechanisms of action from a conceptual and neural perspective. *Perspectives on Psychological Science* 6(6): 537–559. doi:10.1177/1745691611419671.

Jang, J.H., W.H. Jung, D.-H. Kang, M.S. Byun, S.J. Kwon, C.-H. Choi, et al. 2011. Increased default mode network connectivity associated with meditation. *Neuroscience Letters* 487(3): 358–362. doi:10.1016/j.neulet.2010.10.056.

Kam, J.W.Y., E. Dao, J. Farley, K. Fitzpatrick, J. Smallwood, J.W. Schooler, et al. 2010. Slow fluctuations in attentional control of sensory cortex. *Journal of Cognitive Neuroscience* 23(2): 460–470. doi:10.1162/jocn.2010.21443.

Kilpatrick, L.A., B.Y. Suyenobu, S.R. Smith, J.A. Bueller, T. Goodman, J.D. Creswell, et al. 2011. Impact of mindfulness-based stress reduction training on intrinsic brain connectivity. *NeuroImage* 56(1): 290–298. doi:10.1016/j.neuroimage.2011.02.034.

MacLean, K.A., E. Ferrer, S.R. Aichele, D.A. Bridwell, A.P. Zanesco, T.L. Jacobs, et al. 2010. Intensive meditation training improves perceptual discrimination and sustained attention. *Psychological Science* 21(6): 829–839.

Mason, M.F., M.I. Norton, J.D. Van Horn, D.M. Wegner, S.T. Grafton, and C.N. Macrae. 2007. Wandering minds: The default network and stimulus-independent thought. *Science* 315(5810): 393–395. doi:10.1126/science.1131295.

McKiernan, K.A., B.R. D'Angelo, J.N. Kaufman, and J.R. Binder. 2006. Interrupting the "stream of consciousness": An fMRI investigation. *NeuroImage* 29(4): 1185–1191.

McVay, J.C., and M.J. Kane. 2009. Conducting the train of thought: Working memory capacity, goal neglect, and mind wandering in an executive-control task. *Journal of Experimental Psychology. Learning, Memory, and Cognition* 35(1): 196–204. doi:10.1037/a0014104.

Mrazek, M.D., J.M. Chin, T. Schmader, K.A. Hartson, J. Smallwood, and J.W. Schooler. 2011. Threatened to distraction: Mind-wandering as a consequence of stereotype threat. *Journal of Experimental Social Psychology* 47: 1243–1248.

Mrazek, M.D., J. Smallwood, M.S. Franklin, B. Baird, J.M. Chin, and J.W. Schooler. 2012. The role of mind-wandering in measurements of general aptitude. *Journal of Experimental Psychology: General* 141(4): 788–798. doi:10.1037/a0027968.

Mrazek, M.D., M.S. Franklin, D.T. Phillips, B. Baird, and J.W. Schooler. 2013. Mindfulness training improves working memory capacity and GRE performance while reducing mind-wandering. *Psychological Science* 24(5): 776–781.

Posner, M.I., J.A. Walker, F.J. Friedrich, and R.D. Rafal. 1984. Effects of parietal injury on covert orienting of attention. *The Journal of Neuroscience: The Official Journal of the Society for Neuroscience* 4(7): 1863–1874.

Prakash, R.S., A.A. De Leon, M. Klatt, W. Malarkey, and B. Patterson. 2012. Mindfulness disposition and default-mode network connectivity in older adults. *Social Cognitive and Affective Neuroscience* 8(1): 112–117. doi:10.1093/scan/nss115.

Reichle, E.D., A.E. Reineberg, and J.W. Schooler. 2010. Eye movements during mindless reading. *Psychological Science* 21(9): 1300–1310. doi:10.1177/0956797610378686.

Schooler, J.W. 2002. Re-representing consciousness: Dissociations between experience and meta-consciousness. *Trends in Cognitive Sciences* 6(8): 339–344. doi:10.1016/S1364-6613(02)01949-6.

Schooler, J., and C.A. Schreiber. 2004. Experience, meta-consciousness, and the paradox of introspection. *Journal of Consciousness Studies* 11(7–8): 7–8.

Schooler, J.W., J. Smallwood, K. Christoff, T.C. Handy, E.D. Reichle, and M.A. Sayette. 2011. Meta-awareness, perceptual decoupling and the wandering mind. *Trends in Cognitive Sciences* 15(7): 319–326. doi:10.1016/j.tics.2011.05.006.

Smallwood, J., and J.W. Schooler. 2006. The restless mind. *Psychological Bulletin* 132(6): 946–958. doi:10.1037/0033-2909.132.6.946.

Smallwood, J.M., S.F. Baracaia, M. Lowe, and M. Obonsawin. 2003. Task unrelated thought whilst encoding information. *Consciousness and Cognition* 12(3): 452–484. doi:10.1016/S1053-8100(03)00018-7.

Smallwood, J., J.B. Davies, D. Heim, F. Finnigan, M. Sudberry, R. O'Connor, et al. 2004. Subjective experience and the attentional lapse: Task engagement and disengagement during sustained attention. *Consciousness and Cognition* 13(4): 657–690. doi:10.1016/j.concog.2004.06.003.

Smallwood, J., D.J. Fishman, and J.W. Schooler. 2007a. Counting the cost of an absent mind: Mind wandering as an underrecognized influence on educational performance. *Psychonomic Bulletin and Review* 14(2): 230–236.

Smallwood, J., M. McSpadden, and J.W. Schooler. 2007b. The lights are on but no one's home: Meta-awareness and the decoupling of attention when the mind wanders. *Psychonomic Bulletin and Review* 14(3): 527–533.

Smallwood, J., M. McSpadden, and J.W. Schooler. 2008a. When attention matters: The curious incident of the wandering mind. *Memory and Cognition* 36(6): 1144–1150. doi:10.3758/MC.36.6.1144.

Smallwood, J., M. McSpadden, B. Luus, and J. Schooler. 2008b. Segmenting the stream of consciousness: The psychological correlates of temporal structures in the time series data of a continuous performance task. *Brain and Cognition* 66: 50–56. doi:10.1016/j.bandc.2007.05.004.

Smallwood, J., E. Beach, J.W. Schooler, and T.C. Handy. 2008c. Going AWOL in the brain: Mind wandering reduces cortical analysis of external events. *Journal of Cognitive Neuroscience* 20(3): 458–469. doi:10.1162/jocn.2008.20037.

Smallwood, J., M.D. Mrazek, and J.W. Schooler. 2011a. Medicine for the wandering mind: Mind wandering in medical practice. *Medical Education* 45(11): 1072–1080.

Smallwood, J.M., K. Brown, B. Baird, and J.W. Schooler. 2011b. Cooperation between the default mode network and the fronto-parietal network in the production of an internal train of thought. *Brain Research* 1428: 60–70. doi:10.1016/j.brainres.2011.03.072.

Stawarczyk, D., S. Majerus, P. Maquet, and A. D'Argembeau. 2011. Neural correlates of ongoing conscious experience: both task-unrelatedness and stimulus-independence are related to default network activity. *PloS One* 6(2): e16997.

Tang, Y.Y., Y. Ma, J. Wang, Y. Fan, S. Feng, Q. Lu, et al. 2007. Short-term meditation training improves attention and self-regulation. *Proceedings of the National Academy of Sciences* 104(43): 17152.

Taylor, V.A., V. Daneault, J. Grant, G. Scavone, E. Breton, S. Roffe-Vidal, et al. 2012. Impact of meditation training on the default mode network during a restful state. *Social Cognitive and Affective Neuroscience* 8(1): 4–14. doi:10.1093/scan/nsr087.

Wallace, B.A., and S.L. Shapiro. 2006. Mental balance and well-being: Building bridges between Buddhism and Western psychology. *American Psychologist* 61(7): 690–701.

# Can Contemplative Science Bring Meditation to (Western) Life?

**Norman A.S. Farb**

**Abstract** Secular meditation training (MT) practices such as mindfulness training have sparked recent public and scientific interest, yet it is unknown whether such practices can contribute to lasting cultural change. This chapter argues that widespread cultural integration of meditation depends upon whether the scientific community can provide clear explanations for how meditation promotes well-being. In this spirit, a functional approach for studying MT is proposed: while research can and should be informed by practitioners' subjective reports, meditation research gains scientific ground only by proposing and testing for change in measurable perceptual or regulatory capacities. This empirical commitment need not stand at odds with the beliefs of spiritual practitioners, but can instead reinvigorate discourse and a commitment to deeper understanding of the principles underlying effective meditative practice. A discussion of the author's work illustrates how the functional approach has been applied to investigate MT's purported regulatory benefits, while acknowledging many unresolved mechanistic questions. This chapter is intended to provide the reader with a greater understanding of how a fledgling scientific model of meditation has developed, and suggests some upcoming challenges on the road to achieving a broad social impact.

## Introduction

Contemporary meditation training (MT) refers to a group of contemplative practices aimed at fostering well-being and improving stress tolerance. Representing a secular adaptation of ancient Eastern traditions (Kabat-Zinn 2003), contemplative practices have grown in popularity throughout Western culture over the past three

N.A.S. Farb (✉)
Rotman Research Institute, Baycrest, 3560 Bathurst Street, Toronto, ON M6A 2E1, Canada
e-mail: nfarb@research.baycrest.org

S. Schmidt and H. Walach (eds.), *Meditation – Neuroscientific Approaches and Philosophical Implications*, Studies in Neuroscience, Consciousness and Spirituality 2,
DOI 10.1007/978-3-319-01634-4_14, 

decades. There are many historical reasons for this recent rise in popularity, beginning perhaps with the emergence of meditation practices in the West during the 1960s. At that time, the hippie movement introduced Western youth to the idea of meditation as a form of consciousness transformation (Miles 2004), a practice that was perhaps most famously endorsed by the Beatles following their encounters with the meditation teacher Maharishi (Kroll 1974). Concurrently, the Vietnam war exposed many young Westerners to the Eastern Buddhist practices, some of whom returned to the West to found meditation training programs (Schmidt 2011). While both the hippie movement and Vietnam war have long subsided, their effects in terms of introducing meditation to the West have become increasingly apparent in popular culture: venerable masters in Kung Fu movies have lent a social cachet to meditative practices (Hunt 2003); meditation courses have become widely available through standardized training programs such as mindfulness-based stress reduction (MBSR) (Kabat-Zinn 1982); the Dalai Lama, admired worldwide for his message of peace, has strongly endorsed meditative practice as a path to well-being (Goleman 2004); the scientific community has begun to empirically corroborate reported benefits associated with MT (for recent reviews, see (Hölzel et al. 2011; Shapiro et al. 2006); and mass media endorsement of contemplative science has stoked public imagination, such as a 2003 Time magazine cover on the science of meditation.

Despite this unique confluence of public and scientific interest in techniques such as mindfulness or loving-kindness meditation, MT's future in Western culture is at a crossroads. Public interest is fleeting and shifts from trend to trend; without a more substantive scientific account of *why* MT works, its popularity may wane as the sense of novelty fades. Skeptics of meditation may balk at the mysticism associated with contemplative training, spurning the cultural mystique that first made such practices a source of curiosity in the West. To be fully integrated into Western culture, MT requires description within a Western mode of discourse such as a scientific theory. Subjective reports of training benefits can and will continue to popularize the practice without scientific backing, but deep-rooted institutional change generally requires endorsement from the institution responsible for knowledge discovery, i.e., scientific authority (Bikhchandani et al. 1992). Support from the scientific community gives policy makers the freedom to introduce meditation programs into hospitals and schools, where, if efficacious, their institutionalization more deeply entrenches MT practices in Western culture. With support in clinical and educational settings, similar community programs may follow. For example, this medical integration model has been instrumental in the popularization of MT through 8-week MBSR courses, first introduced to support chronic pain patients through the University of Massachusetts medical school. Following initial demonstrations of clinical efficacy, MBSR course delivery was standardized, allowing forms of the course to be taught today through both health-care facilities and community programs around the world. Yet MBSR requires a lengthy 8-week commitment, including an hour a day of home practice. Why should increasingly busy individuals dedicate their time to such an endeavor?

Despite the recent success of programs such as MBSR, a major challenge on the road to MT's cultural acceptance is that current scientific models do not adequately explain how a solitary meditative practice increases coping and resilience against social stressors. In the author's experience running mindfulness programs, this explanatory gap often leads participants to question the point of meditation practices. These participants seek some relief from worry and suffering, but do not see a connection between hours of silent contemplation and improved relationship quality. For training candidates cautious in committing their time to practice, and for policy makers similarly cautious in committing public resources to funding meditation programs, evidence of the practice's efficacy may not be sufficient, despite robust findings from the clinical sciences (e.g., see Chiesa et al. 2010 for a review). For these more cautious adopters, knowing *why* MT will help may be a critical step in their enrolment. Thus, clinical findings must be supported by basic science evidence for mindfulness' mechanisms of change, evidence which may also have an impact on training practices themselves, refining MT intervention protocols.

This chapter explores the emergence of a Western scientific account of mindfulness, introducing some candidate mechanisms by which MT may promote well-being. The discussion begins by reviewing the different approaches researchers have brought to contemplative science, and how these approaches might contribute to a broad mechanistic theory of mindfulness meditation. The chapter continues with a proposal that MT may best be studied through a functional approach to research: it is argued that while research can and should be informed by practitioners' subjective reports, mindfulness research best contributes to our scientific understanding by proposing and testing for change in specific perceptual or regulatory capacities. Finally, discussion of the author's own work illustrates how the functional approach has been applied to investigate how the meditative practice translates to general benefits, including a discussion of some major outstanding mechanistic questions. This chapter is intended to provide the reader with a greater understanding of how the fledgling scientific model of meditation is developing, and where it still has to go to effect social change.

## Approaches to Contemplative Science

A primary goal of the emerging field of contemplative science is to provide the scientific community with a mechanistic model of how meditation works. In doing so, researchers are constrained by the need to provide empirically-robust research paradigms while faithfully representing the contemplative traditions from which such paradigms are derived. Such derivations still afford researchers' great latitude for exploration: even when studying a particular contemplative practice, research may focus on a wide array of potential mechanisms of change, training interventions, target populations, and outcome measures. This section will review several recent findings in this emerging contemplative science and describe how they fit into the development of a broader theory of MT's mechanisms of action.

One of the broadest distinctions in contemplative research approaches lies in *who* is studied, i.e., whether research focuses on novices or expert practitioners. Thanks to ordinances from the Dalai Lama and leaders of other contemplative traditions, research laboratories have been able to collaborate directly with advanced contemplative practitioners, such as Buddhist monks with more than 10,000 h of meditation experience (e.g., Brefczynski-Lewis et al. 2007). Such research provides insight into the long term effects of meditation practice. At the other end of the spectrum, researchers have begun to study the initial impact of meditation training on novice practitioners (e.g., Tang et al. 2007), hoping to describe how the average person might be affected by engaging in MT. These approaches each have their own advantages and shortcomings: in short-term interventions, findings are more generalizable because the participants are more similar to the population of non-meditators. A weakness of such short-term designs is that it is difficult to know how much of a meditation 'dose' is required to observe a true response as opposed to some sort of expectation effect (Farb 2012). For example, a person first attempting meditation may benefit from attempting a self-improvement practice, even if his or her initial ability to effectively meditate is actually minimal. For this reason, it is important for researchers to follow-up initial findings of MT effects with active-control experiments, so meditation can be measured against practices with known efficacy such as improved nutrition or exercise habits.

Given the issues involved with studying new meditators, an obvious alternative is the study of long term practitioners. The study of experienced meditators avoids the issue of minimal dose, but suffers instead from a lack of experimental control, as such designs are necessarily cross-sectional in nature, and because it is not feasible to randomly assign 10,000 h of meditation to a participant in an experiment! Cross-sectional designs can richly track the relationship between experience and changes in capacity and participant experience, but such designs are not valid models for causal inference, due to the problem of self-selection- perhaps those who achieve such heights of meditation experience were unlike the general population when they began, in disposition or background, and so the accessibility of long term effects to a general population is called into question. Thus the most striking issue in the study of long term practitioners is the question of generalisability: how similar are the people willing to spend decades of intense training to the general population? Are their intentions in pursuing liberation and enlightenment really comparable to that of a secular practitioner engaging in MT as a way to reduce stress? Furthermore, long-term practitioners may become so adept at existing within a meditative state that it becomes difficult to create a baseline or control condition. For these meditators, meditation may be the default state, which may account for the lack of neural activity observed in very long term meditators in comparisons of meditation vs. rest (Brefczynski-Lewis et al. 2007).

As in most dichotomies, both perspectives are valuable in developing a complete longitudinal map of MT effects. In influencing potential participants and policy makers, advanced practitioners can serve as a 'gold standard' for mindfulness outcomes, while the success of new practitioners helps to demonstrate the accessibility and efficacy of these practices in a secular context. Unfortunately, the question of 'who' is engaging in MT is not completely resolved by looking at practitioners with varying levels of experience. Motivations for practice may vary greatly between practitioners, affecting both the quality of the practices and their downstream effects on emotion regulation and well-being. This variation may be most apparent in new practitioners, who come to an MT course for different reasons. For example, some practitioners may be primarily motivated by desire for relaxation and freedom from distressing thoughts, finding relief in the calm of prolonged stillness and attention to body sensation. Other participants may be motivated by a desire for self-insight, and seek to use MT as a vehicle for exploring their relationships with themselves and others. Attentional focus during meditative practice can then vary according to motivation, with the relaxation-oriented group prioritizing attention to feelings of peace and tranquility as an escape from distressing thoughts, while the insight-oriented group views these same thoughts as a learning opportunity, attending to their conceptual progression rather than feelings of relaxation. Critically, the long term effects of engaging in such distinct attentional foci may result in very different outcomes. For example, the relaxation-focused participants might demonstrate improved physiological tone in the face of stress but show little changes in patterns of behavior that create such stressors, whereas the insight-focused participants might show relatively limited benefits to physiological resilience but relatively greater reductions in social stress as they learn to cultivate more skillful responses to perceived social stressors. Indeed, one rare study of the association between expectations and outcomes in long-term practitioners found tentative support for the suggestion that "what you get is related to what you want" in meditation, distinguishing between self-regulatory, self-exploratory, and self-liberation goals and outcomes (Shapiro 1992). The power of expectations to determine perceived benefits in meditation is an important issue in developing a general scientific model of the process. For example, if groups of people with different types of expectations all report benefits from practice, how do investigators make sense of these two different 'flavors' of meditation? Can we say that one is actually better than another?

One approach to studying how different motivations lead to different MT practices is to attempt to validate all motivation, creating an account that permits distinct ways of meditating. On the other hand, some meditative strategies may be superior to others despite variability in individual preferences, and a normative theory of MT would therefore be helpful in distinguishing between what feels comfortable and what is efficacious. In the example above, clinging to positive feelings like relaxation is generally not espoused as the focus of an MBSR course; instead, calm and relaxation are viewed as providing a safe space within which insight into the patterns of emotional and cognitive reactivity can be viewed, leading to personal insight. Meditation may not always be relaxing, especially during stressful periods, and so clinging to a relaxation rather than insight goal may lead to further distress rather than the alleviation of suffering. From this view, it would be a disservice to participants to encourage ill-fated goals of clinging to relaxation. Despite this theoretical concern, it has not been scientifically verified that promotion of relaxation through meditation is unsustainable. If the benefits are real and lasting from such a practice, is it fair to mandate insight as a goal for all MT participants? In negotiating a scientific model of meditation, researchers must show that non-canonical goals such as a relaxation focus do not lead to sustained benefits relative to an insight focus, in accordance with what Buddhist psychological theory suggests. The first step in addressing such questions however involves acknowledging individual differences in participant motivation and including such differences in research models, which to date is rarely done. Furthermore, it will be important to track the change in expectations as participants shift between particular meditation goals. In the Shapiro (1992) study mentioned above, the research on one group of long-term meditators suggests that there may be a systematic progression from self-regulation to self-exploration to self-liberation over long-term practice; as interest in meditation grows, it will be interesting to examine whether increasingly heterogeneous motivations follow a predictable trajectory that supports the idea of discrete, hierarchical stages of meditative insight.

This variability of participant goals points to a broader issue in performing contemplative research: how does an investigator choose the MT mechanisms hypothesized to produce benefits? Does one rely upon well-established stages of meditation gleaned from Buddhist psychological theory and canonized in classic texts, or does one appeal to the idiosyncratic preferences and experiences of contemporary practitioners, whose heterogeneous experiences may better reflect what is actually happening in the training? Furthermore, the Buddhist literature itself is heterogenous – the meditative traditions espoused therein are rich and varied, deriving from thousands of years of consideration and debate on the nature of suffering and a path that leads to liberation from such suffering (Trungpa and Lief 2009). Yet classical Buddhist psychology does offer some homogeneity in describing meditation mechanisms that could be operationalized to form the basis of a research program. For example, a major landmark in the path to liberation from suffering is the recognition of the inherent emptiness to all things, that because all objects are dependent upon our perception of them, they have no independent reality (Tashi and McDougall 2009). Suffering is perpetuated by clinging to these appearances and failing to recognize their illusory nature. Based upon this theory, it is possible then to measure whether MT participants have similar insights, and whether such insights predict increased well being and resilience to stress. A practical concern in this endeavor is what to do if Western participants' reports of experiences do not present such insights. Does this mean that the theory is incorrect, the training is incorrect, the measurement is too early in the training process, or that this theoretical insight does not apply when divorced from a Buddhist cultural context? These thorny issues may stem from an attempt to impose a foreign cultural understanding onto a Westerner's way of relating to his or her personal experience. On the other hand, if these classical theories are accurate, perhaps such insights are universally realized with enough training, but are prematurely sought after following only a few months of training. There is therefore an art to designing research studies seeking out particular benchmarks of progression through meditative practice. What benchmarks are likely to be the most universal? How can they best be measured? In attempting to address these concerns, collaboration with contemplative teachers may be particularly useful, as they will have a subjective sense of these issues from their own teaching experience. At the same time, researchers must be careful not to invalidate practitioner experiences that diverge from canonical texts, as there is no guarantee that the most beneficial effects of meditation in a contemporary Western context will align with ancient teachings.

From a scientific perspective, using classical Buddhist texts as the primary source of hypothesis generation can be useful for supporting and explaining positive findings, but fraught with ambiguity if such findings do not emerge. Furthermore, as many Buddhist constructs are abstract, such as perceiving the inherent emptiness of reality, they can be difficult to operationalize, as we cannot easily create an experiment to manipulate metaphysical perceptions. While a complete theory of meditation must ultimately include such higher order constructs, fledgling efforts to establish a contemplative science may benefit from addressing more tractable steps along the path. An alternative to this top-down investigative approach is to work instead from the bottom-up, attending to capacities specifically recruited by MT interventions that cut across different participant motivations. Such capacities may be general benefits of meditation that occur at the level of attention or sensation. For example, much of MT amounts to a specialized form of attention training, seeking to alter the basic context through which a person perceives the world (Kabat-Zinn 1990). The idea of altered perceptual capacity is one that resonates strongly with Western psychological science: it has long been held that human perceptual acuity is a flexible capacity that can be improved through training (Gibson and Eleanor 1953). Training in a visual task for example results in improved task-performance and altered activity in the brain's visual cortices in the brain (Sasaki et al. 2010). It serves to reason that repeated attention practices associated with MT may yield similar observable capacity changes. This focus on practical capacities rather than more abstract conceptual changes is known as the functionalist approach to scientific research. The functionalist approach may be helpful in characterizing the most universal effects of mindfulness training in that it seeks to establish the effects of training at the most basic perceptual and attentional levels of change, making incremental progress towards changes in higher order insight and personal meaning. By postulating practice effects based on the specific content of the practices, we may begin to forge a chain of evidence between basic perceptual or attentional changes and broader effects on identity and well-being.

It should be emphasized that the proposal of focusing research on functional changes closely related to meditation training is not intended to disparage the importance of contributions from advanced practitioners or through reference to classic Buddhist psychological texts. Indeed, the stages of meditation practice described by contemplative scholars provide an important road map for scientific inquiry, and also serve as a 'quality control' check for secular meditation paradigms. For example, the meditation master Mahasi Sayadaw writes of progressive stages of meditation (Sayadaw 2006) based upon the *Visuddhimagga*, known as 'the great treatise' on Therevada Buddhist meditation practice. After describing meditative practices which focus on awareness of breath sensation and mental labeling of all sensory events, the first stage of insight he describes is a form of mindfulness or meta-awareness of the arising and passing of all sensory and mental events (p. 16). The idea that noticing experience leads to the cultivation of a reliable and consistent form of meta-awareness suggests a testable benchmark for effective training. Further, the discovery of how to cultivate such meta-awareness would constitute a tangible finding in the scientific literature and an important explanatory step in accounting for MT benefits. Thus the classical text provides a direction for research, guiding investigation to try to link practice-specific capacity changes to general changes in meta-awareness.

Contemplative science can therefore progress by testing whether capacity changes related to MT align with canonical benchmarks of meditative progress. Buddhist psychological theory already contains allusions to these benchmarks and what constitutes correct and incorrect practice. To be valid, experimental paradigms or interventions should attempt to promote correct practice, although a study may compare these practices to putatively incorrect practices to test canonical claims. Practices most likely to help participants achieve benchmarks of successful MT can then be included in normative theory. While we are perhaps still far from describing achievement of the first benchmark, i.e., how meta-awareness is developed through attentional practices, several steps towards such a theory have begun to emerge.

## The Functionalist Approach in Contemplative Research

The functionalist approach examines what an object of investigation does, rather than debating its deeper meaning or origins. This perspective has for example been useful in advancing the field of emotion research in its focus on emotion effects rather than seeking to establish an emotion's evolutionary or socially constructed origin (Farb et al. 2013c). In meditation research, the functional approach involves hypothesizing specific observable capacities engendered by meditation training and tracking their cultivation through training, such as the emergence of compassion, equanimity, or body awareness.

From the functionalist perspective, the identification of specific capacities stems from two major sources. First, identification can be based upon practitioner reports, an approach which is most valid from meditation's own philosophy of relying on direct experience, but is also limited by a person's level of insight and the accessibility of meditation mechanisms to consciousness. Identification may also progress by analysis of the meditation practices themselves. For example, MT exercises almost invariably involve directed attention to breath sensation, so one might infer that the ability to accurately and stably monitor respiratory rhythms is a capacity for measuring meditation related change. This focus on observable capacities that are directly related to practice content provides a rich ground for investigation, and unlike more abstract theories of meditation effects, such theories are also falsifiable. For example, it was thought that Vipassana meditators ought to show superior performance in heartbeat detection due to their extensive training in observing subtle body sensations. Empirical research however belies this notion, demonstrating that meditators fared no better in a heartbeat detection task than age-matched controls (Khalsa et al. 2008). Such findings suggest that if there are benefits to sensory detection from meditation practice, these benefits may be localized to the sensory targets and not generalize to other sensory domains. The heartbeat, which is not generally an attentional target in Vipassana practice, therefore was not more readily detected, despite subjective reports of greater ease and accuracy in such detection in advanced meditators. Falsifiability of hypotheses is a cornerstone of good scientific theory (Popper 1962), and an important building block of a convincing mechanistic account of MT. As Vipassana practices are the predominant source of meditative exercises found in mindfulness MT (McCown et al. 2010), the evidence suggests that heartbeat detection may not be an optimal measurement of perceptual plasticity in mindfulness MT, although these findings do not more broadly refute the possibility of enhanced interoceptive accuracy through training.

Despite evidence contradicting increased heartbeat detection through MT, the functionalist perspective has had greater success in identifying meditation-related perceptual changes in surface body awareness. For example, it is scientifically well-established that the brain's representation of the body is biased towards the hands and face, with proportionately greater sensorimotor cortex devoted to these regions than the back or legs (Penfield and Boldrey 1937). The use of attention-based body scans in Vipassana meditation could plausibly give practitioners greater insight into this inequality of sensory resolution for different body parts, and indeed in a recent study, investigators observed that meditators showed greater awareness of this biased mapping than a control group (Fox et al. 2012). Thus, while heartbeat detection may be too subtle a perceptual measure for studying mindfulness training effects, other tests of body awareness may be beneficial in tracing the path from specific capacity training to enhanced well-being.

## Interoception as a Functional Mechanism of Mindfulness MT

Interoception, the capacity to sense visceral and somatic events within the body, is a major attentional focus of early mindfulness MT practices. While more general changes to the attentional system are evident following training interventions (Jha et al. 2007), interoception is a capacity which is specifically related to the training exercises; participants completing mindfulness interventions often report an altered relationship to their breath and body (Hölzel et al. 2011), and early mindfulness instruction focuses on cultivating and sustaining interoceptive awareness (Kabat-Zinn 1990).

From a functional perspective, the ability to maintain interoceptive awareness is a good candidate marker of mindfulness training because it affords a set of strong predictions in terms of patterns of neural recruitment. Because interoception is a sensory rather than conceptual process, one can hypothesize the involvement of primary interoceptive cortices, theorized to reside in the posterior insula through a spinothalamocortical pathway was proposed to carry sympathetic afferents that signal the physiological condition of all tissues of the body (Craig 2002). The integration of interoception into present moment awareness by contrast is hypothesized to depend upon the anterior insula (Craig 2009), which has been associated with switching of attentional states, acting as a conductor of sorts for determining which mental processes occupy conscious awareness (Menon and Uddin 2010). Mindful, non-judgmental interoceptive attention is theorized to lead to well-being because it allows practitioners to understand the transitory nature of visceral experience, and to generalize this understanding to recognize the transitory nature of thoughts which might otherwise be viewed as objective truth (Silananda 2002). It should be noted that interoceptive attention alone is not an absolute good when paired with habitual, catastrophic appraisals that are defining characteristics of anxiety and post-traumatic stress disorders. However, when paired with attitudes of equanimity and acceptance of experience, the transitory signals provided by the body can provide a rich attentional environment for understanding one's habitual patterns of cognition and reactions.

Using this neuroscience model, my colleagues and I have attempted to test the role of interoception in MT. In testing the model we had a number of concerns. First, it was important to demonstrate that the voluntary direction of attention towards interoceptive signals such as the breath would reliably engage this interoceptive network, particularly within the insula, putative primary interoceptive cortex. In one recent study, we were able to demonstrate that relative to attention to external visual targets, attention to the breath activated the posterior insula, even in participants with no meditation experience (Farb et al. 2013a). This finding corroborated participants' reports that respiratory sensation was immediately accessible as a target of attention. However, because even non-meditators were able to recruit posterior insula activity during interoception, MT did not modulate the degree of recruitment in primary interoceptive cortex. Thus the neuroscientific data contradicts a theory of MT as increasing the raw input of interoceptive signal from the body. How such information is then processed and integrated into contextual awareness is, however, a different story: our second region of interest, the anterior insula, demonstrated altered activity in the MT group (Farb et al. 2012). In untrained participants, exteroception led to significantly greater anterior insula recruitment than interoception, suggesting that despite ease of interoceptive sensation, access to such sensation was not considered motivationally relevant. In the MT group however, there was no difference in interoception and exteroception-related recruitment, and the interaction between group and attention condition was significant. These results suggest that MT increases access of interoceptive information from the posterior insula to the anterior insula, integrating momentary changes in body sensation within the broader information context afforded to consciousness.

The finding of increased integration of body sensation into the prefrontal attention system via the anterior insula is consistent with Buddhist theoretical accounts of developing awareness of momentary arising and passing of body sensation (Silananda 2002). At the same time, this work advances our knowledge of brain plasticity, demonstrating that interoceptive attention can be a source of neuroplasticity much as has been demonstrated for the external senses. Further research may corroborate these findings through cheaper behavioral measures of breath or body awareness. Research may then attempt to extend these theories by linking indices of interoceptive awareness to changes to more general attention direction and stability, attention effects which have already been linked to meditation practice (Jha et al. 2007; MacLean et al. 2010; Tang et al. 2007). In this way, a mechanistic model linking specific practices to broad changes in the attentional system begins to emerge. With this rung of the 'research ladder' in place, investigation may then progress to further steps along the theoretical path, such as the realization of a lack of self underlying the continuous arising and passing of sensation and thought (Sayadaw 2006).

## The Cultivation of Deeper Meaning

Findings of altered interoceptive recruitment following MT are encouraging, but serve as only a candidate link in an explanatory chain for how meditation promotes well-being and relieves suffering. An initial criticism of findings that describe altered perception or attention is that they are short-lasting. For every improved capacity linked to MT, it behooves us to consider how enduring and ingrained such changes are: have participants simply learned an isolated attentional technique, or has a deeper shift in values or the nature of self-reference occurred? The possibility of a change in skill without deeper, more meaningful change may even be a likely outcome of secular MT that has been excised from a broader cultural tradition: from the Buddhist psychological perspective, meditative practice alone is insufficient to promote enduring well-being; instead, there is a broader context of lifestyle factors which may prove to be necessary for one to undergo positive self-transformation. Furthermore, even if sensory capacity is chronically altered through training, the link between such changes and altered regulatory capacity remains unproven. A skeptic could argue that training perceptual capacity is one thing, but to alter the basic principles of self-reference underlying stress reactivity is a more monumental task, and ultimately unsustainable. Echoing such sentiments, the rapper Jay-Z famously remarked:

> Check out my swag' yo, I walk like a ballplayer
> No matter where you go, you are what you are player
> And you can try to change but that's just the top layer
> Man, you was who you was 'fore you got here (2003)

The challenge for contemporary contemplative science is to document changes that extend beyond this "top layer" of attention change, and determine whether such changes alter the habitual patterns of reactivity to emotional challenge, improving the adaptiveness of such responses.

One way to address the issue of altered emotion regulation is to identify trait-like training-related changes to both the attention system and practitioner skill in emotion regulation. For example, a recent longitudinal study examined the effects of a 3-month meditation retreat on a number of attentional and emotional outcome

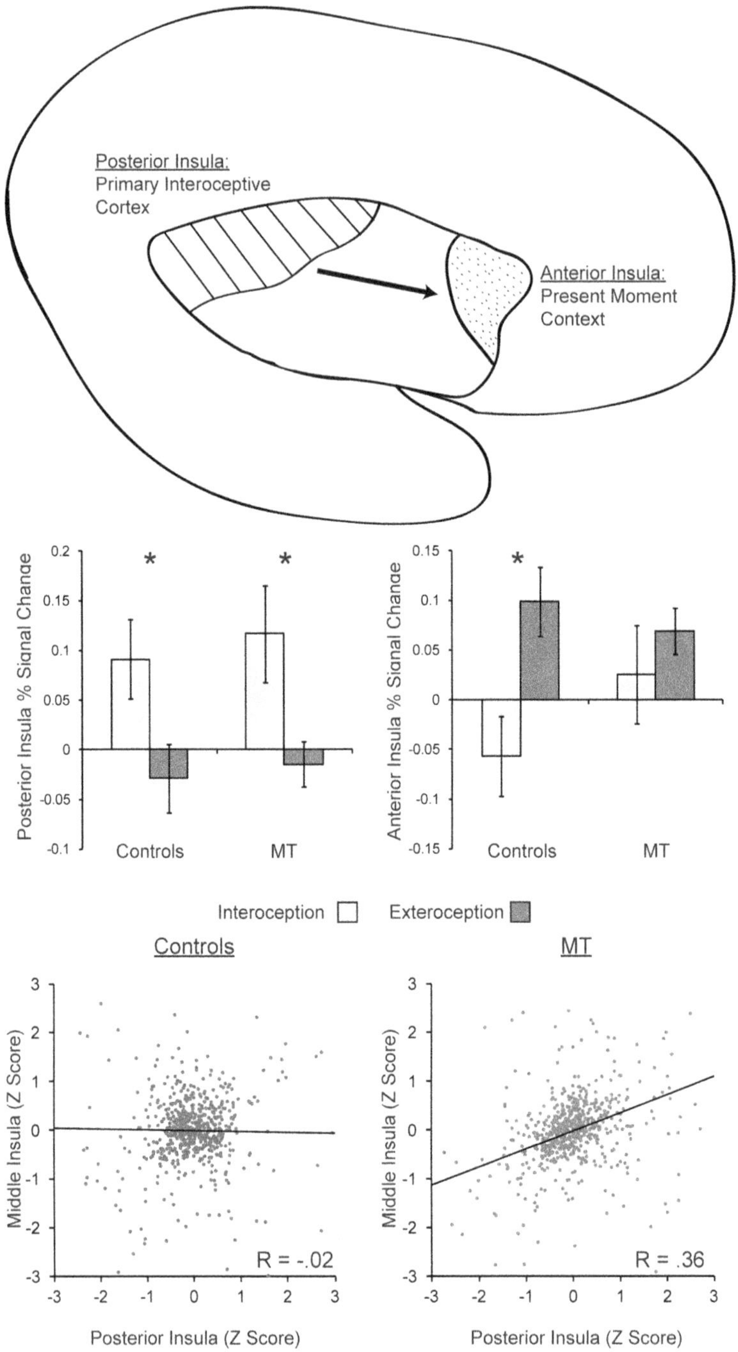


**Fig. 1** Summary of neural evidence for how mindfulness meditation training (MT) affects interoceptive attention. *Top panel*: training is theorized to promote propagation of interoceptive signal from primary interoceptive cortex in the posterior insula to motivational and contextual awareness

measures (Sahdra et al. 2011). Training-related improvements in response inhibition on an extended attention task were linked to improvements in adaptive function, a composite score based upon personality and emotion regulation measures. A wait-listed control group did not show such improvements with repeated testing, but when this control group subsequently participated in their own 3 month retreat, their improvements on the attention task were again associated with improvements in adaptive function. While such a study design cannot rule out expectation effects contributing to improvements in both attention and well-being, it does provide suggestive evidence that training-related changes to the attention system have wider downstream consequences for mental health. It should be noted that this study did not measure interoception, but instead sustained visual attention over a 30 min period, investigating a potentially distinct mechanism of meditation training, i.e. sustained attention rather than body awareness. Nevertheless, the functional approach to meditation research is well exemplified by this work.

With respect to interoception as a mechanism for change, a parallel set of findings has begun to emerge. In the Farb et al. (2013a) study described above, connectivity analyses suggested that posterior insula connectivity with the middle insula increased in a trait-like manner, irrespective of attention condition (Fig. 1). Similarly, while the control group increased posterior-anterior insula connectivity as a function of engaging interoceptive attention, the MT group constantly demonstrated this elevated level of connectivity, during both interoceptive and exteroceptive attention. Together, these findings suggest the trait-like elevation of interoceptive signal propagation towards the prefrontal cortices following MT. In a related study, interoceptive signal in the right insula following emotional challenges was greater in an MT than control group, and this activity was associated with lower depression inventory scores, corroborating the notion that interoceptive access is important for adaptive regulation following an emotional challenge (Farb et al. 2010). Neither of these studies employed a longitudinal design, making it premature to conclude that MT led to increased insula connectivity and activity following emotional challenge. In both the attention and interoception studies, the role of expectation is not experimentally controlled through an active control group, so these inferences remain speculative in nature, though they are consistent with the idea of adaptive MT-related improvements to specific attentional capacities.

Accounts such as these are promising in showing that measuring changes to specific mental capacities may support an empirical account of MT. These findings

**Fig. 1** (continued) in the anterior inusla. *Middle panel*: bar graphs contrasting interoceptive and exteroceptive attention in the posterior and anterior insula. The interaction between group and attention condition was significant in the anterior insula but not posterior insula, suggesting the anterior insula is a better activation-based marker of MT-related change. Error bars represent standard errors, and asterisks represent significant differences at the $p<.05$ level. *Bottom panel*: connectivity analysis plots between the posterior and middle insula for Control and MT groups. Points on the scatter graph are whole time-course data for a representative participant from each group, revealing positive anterior connectivity from the posterior insula in the MT but not control group. These findings were originally presented in (Farb et al. 2013b)

are, however, only low rungs on a theoretical ladder connecting meditation practices to the cultivation of deeper meaning and positive personal transformation. There are many intermediate steps between changes in these capacities and broad mindfulness benefits, such as the weakening of self-interest, the extinction of unhealthy patterns of emotional reactivity, or the cultivation of positive emotional states. Ideally, a complete theory of meditation training would even link proximal functional changes associated with MT to the deep metaphysical insights described in Buddhist texts, such as the awareness of the emptiness underlying the perceptual world, or the universal nature of human suffering and its resolution (Trungpa and Lief 2009). Such a complete theory may also need to include other training practices that are complementary to the ego-dissolution associated with mindfulness. For example, when asked how to best know, whether someone is progressing in meditation, the Dalai Lama did not reference changes in the sensory capacity, but rather asserted that "kindness" is the true marker of advanced meditative practice, suggesting that the inclusion of loving-kindness practices and compassionate behavior may also be important aspects in the model. The field of potential inquiry is indeed broad, and a great deal of work remains in addressing this arc of self-improvement through meditation.

## Concluding Remarks

Despite the challenges of the work still to come, we may consider that a scientific model of meditation need not be complete to effect cultural change. A successful account of MT is already emerging and may be realizable in the short term: such an account may succeed by relating changes in specific perceptual or attentional capacities with reported changes positive changes in personal values, feelings of well-being, and altruistic behavior. If these benefits continue to be documented and the mechanisms for generating them refined, such success will hopefully attract sufficient interest from the broader community so as to affect public policy in funding MT programs and providing additional public exposure for such training's beneficial effects. This increased exposure can in turn generate greater interest in the deeper mechanisms of meditative practice, promoting reinforcing cycles of scientific inquiry, institutional commitment and public exposure. From a research perspective, the priority in this cycle is to continue to tackle empirically testable theories of training-related capacity change. Validation and refutation of such hypotheses are equally valuable in informing implementation of mindfulness practices, and in generating public understanding of how such practices function.

We stand at a crossroads in the future of meditation practice in the West. The decline of religious authority and rise of secular, liberal society has left a normative gap in how best to exercise our culturally-inherited freedom. Undoubtedly most people desire to live a life that is satisfying and meaningful, and it is becoming increasingly clear that reliance on technology and consumption alone cannot provide such satisfaction. There is a growing willingness to explore techniques which can provide such meaning, yet given our emphasis on individual freedom, Westerners are justifiably skeptical of dogmatic practices, particularly ones which address issues of identity and personal values. Just as religion was once a source of moral authority that provided opportunities for social unity, science now provides a dominant mode of discourse for assessing the value of unfamiliar techniques and ideas. Contemplative science therefore presents one potential avenue to address skepticism of technologies for self-change, but only if clear and convincing evidence emerges for why MT practices are worth the investment. While this reliance on science may seem ironic given meditation technique's dedication towards the realization of knowledge through direct experience, it is important to consider that the motivations for pursuing such direct experience are inevitably going to be culturally inherited. Contemplative science has the opportunity to provide a source of culturally integrated acceptance of MT's potential. Through functional approaches to investigating MT practices, contemplative research suggests that MT improves capacities for interoception, sustained attention, and emotion regulation. Whether such capacity changes can be robustly linked to improvements in well-being and personal meaning will define the next decade of meditation research, and help to determine contemplative practice's relevance in Western society.

## References


Bikhchandani, S., D. Hirshleifer, and I. Welch. 1992. A theory of fads, fashion, custom and cultural change as informational cascades. *Journal of Political Economy* 100(5): 992–1026.

Brefczynski-Lewis, J.A., A. Lutz, H.S. Schaefer, D.B. Levinson, and R.J. Davidson. 2007. Neural correlates of attentional expertise in long-term meditation practitioners. *Proceedings of the National Academy of Sciences of the United States of America* 104(27): 11483–11488. doi:10.1073/pnas.0606552104.

Chiesa, A., P. Brambilla, and A. Serretti. 2010. Functional neural correlates of mindfulness meditations in comparison with psychotherapy, pharmacotherapy and placebo effect. Is there a link? *Acta Neuropsychiatrica* 22: 104–117.

Craig, A.D. 2002. How do you feel? Interoception: The sense of the physiological condition of the body. *Nature Reviews Neuroscience* 3(8): 655–666. doi:10.1038/nrn894.

Craig, A.D. 2009. How do you feel–now? The anterior insula and human awareness. *Nature Reviews Neuroscience* 10(1): 59–70. doi:10.1038/nrn2555.

Farb, N.A.S. 2012. Mind your expectations: Exploring the roles of suggestions and intention in mindfulness training. *Journal of Mind-Body Regulation* 2(1): 27–42.

Farb, N.A., A.K. Anderson, H. Mayberg, J. Bean, D. McKeon, and Z.V. Segal. 2010. Minding one's emotions: Mindfulness training alters the neural expression of sadness. *Emotion* 10(1): 25–33. doi:10.1037/a0017151.

Farb, N.A., Z.V. Segal, and A.K. Anderson. 2013a. Attentional modulation of primary interoceptive and exteroceptive cortices. *Cerebral Cortex* 23(1): 114–126. doi:10.1093/cercor/bhr385.

Farb, N.A., Z.V. Segal, and A.K. Anderson. 2013b. Mindfulness meditation training alters cortical representations of interoceptive attention. *Social Cognitive and Affective Neuroscience* 8(1): 15–26. doi:10.1093/scan/nss066.

Farb, N.A.S., H.A. Chapman, and A. Anderson. 2013c. Emotions: Form follows function. *Current Opinion in Neurobiology*. doi:10.1016/j.conb.2013.01.015.
Fox, K.C., P. Zakarauskas, M. Dixon, M. Ellamil, E. Thompson, and K. Christoff. 2012. Meditation experience predicts introspective accuracy. *PLoS One* 7(9): e45370. doi:10.1371/journal.pone.0045370.
Gibson, Eleanor J. 1953. Improvement in perceptual judgments as a function of controlled practice or training. *Psychological Bulletin* 50(6): 401–431. doi:10.1037/h0055517.
Goleman, D.J. 2004. *Destructive emotions: A scientific dialogue with the Dalai Lama*. New York: Bantam Dell.
Hölzel, B.K., S.W. Lazar, T. Gard, Z. Schuman-Olivier, D.R. Vago, and U. Ott. 2011. How does mindfulness meditation work? Proposing mechanisms of action from a conceptual and neutral perspective. *Perspectives on Psychological Science* 6(6): 537–559.
Hunt, L. 2003. *Kung fu cult masters*. London: Wallflower Press.
Jay, Z. 2003. *Public service announcement*. The Black Album. New York: Roc-A-Fella Records.
Jha, A.P., J. Krompinger, and M.J. Baime. 2007. Mindfulness training modifies subsystems of attention. *Cognitive, Affective, & Behavioral Neuroscience* 7(2): 109–119.
Kabat-Zinn, J. 1982. An outpatient program in behavioral medicine for chronic pain patients based on the practice of mindfulness meditation: Theoretical considerations and preliminary results. *General Hospital Psychiatry* 4(1): 33–47.
Kabat-Zinn, J. 1990. *Full catastrophe living: Using the wisdom of your body and mind to face stress, pain and illness*. New York: Delacorte.
Kabat-Zinn, J. 2003. Mindfulness-based interventions in context: Past, present, and future. *Clinical Psychology: Science and Practice* 10(2): 144–156.
Khalsa, S.S., D. Rudrauf, A.R. Damasio, R.J. Davidson, A. Lutz, and D. Tranel. 2008. Interoceptive awareness in experienced meditators. *Psychophysiology* 45(4): 671–677. doi:10.1111/j.1469-8986.2008.00666.
Kroll, U. 1974. *The healing potential of transcendental meditation*. Atlanta: John Knox Press.
MacLean, K.A., E. Ferrer, S.R. Aichele, D.A. Bridwell, A.P. Zanesco, T.L. Jacobs, et al. 2010. Intensive meditation training improves perceptual discrimination and sustained attention. *Psychological Science* 21(6): 829–839. doi:10.1177/0956797610371339.
McCown, D., D. Reibel, and M.S. Micozzi. 2010. *Teaching mindfulness: A practical guide for clinicians and educators*. New York/London: Springer.
Menon, V., and L.Q. Uddin. 2010. Saliency, switching, attention and control: A network model of insula function. *Brain Structure & Function* 214(5–6): 655–667. doi:10.1007/s00429-010-0262-0.
Miles, B. 2004. *Hippie*. New York: Sterling.
Penfield, W., and E. Boldrey. 1937. Somatic motor and sensory representation in the cerebral cortex of man as studied by electrical stimulation. *Brain* 60: 389–443.
Popper, K.R. 1962. *Conjectures and refutations; the growth of scientific knowledge*. New York: Basic Books.
Sahdra, B.K., K.A. MacLean, E. Ferrer, P.R. Shaver, E.L. Rosenberg, T.L. Jacobs, et al. 2011. Enhanced response inhibition during intensive meditation training predicts improvements in self-reported adaptive socioemotional functioning. *Emotion* 11(2): 299–312. doi:10.1037/a0022764.
Sasaki, Y., J.E. Nanez, and T. Watanabe. 2010. Advances in visual perceptual learning and plasticity. *Nature Reviews Neuroscience* 11(1): 53–60. doi:10.1038/nrn2737 10.1038.
Sayadaw, M. 2006. *Practical insight meditation: Basic and progressive stages*. Kandy: Buddhist Publication Society.
Schmidt, S. 2011. Mindfulness in East and West- Is it the Same? In *Neuroscience, consciousness and spirituality: Studies in neuroscience, consciousness and spirituality*, vol. 1, ed. H. Walach, S. Schmidt, and W.B. Jonas, 23–28. Dordrecht/New York: Springer.
Shapiro, D.H. 1992. A preliminary study of long-term meditators: Goals, effects, religious orientation, cognitions. *The Journal of Transpersonal Psychology* 24(1): 23–39.
Shapiro, S.L., L.E. Carlson, J.A. Astin, and B. Freedman. 2006. Mechanisms of mindfulness. *Journal of Clinical Psychology* 62(3): 373–386. doi:10.1002/jclp.20237.

Silananda, S.U. 2002. *The four foundations of mindfulness*. Boston: Wisdom Publications.
Tang, Y.Y., Y. Ma, J. Wang, Y. Fan, S. Feng, Q. Lu, et al. 2007. Short-term meditation training improves attention and self-regulation. *Proceedings of the National Academy of Sciences of the United States of America* 104(43): 17152–17156. doi:10.1073/0707678104 10.1073/pnas.
Tashi, T., and G. McDougall. 2009. *Emptiness*, The foundation of Buddhist thought, vol. 5. Boston: Wisdom Publications.
Trungpa, C., and J.L. Lief. 2009. *The truth of suffering and the path of liberation*, 1st ed. Boston: Shambhala: Distributed in the United States by Random House.

# Spiritual Phenomena as Public Goods: Exploring Meditation Beyond the Standard Model

**Torkel Falkenberg**

**Abstract** There are many academic and philosophical entry points available to us in describing the enigmatic effects of meditation. The standard model of enquiry logically starts with a neurobiological explanatory mode, describing anatomical and physiological correlates to meditation. For example, meditation practice can be correlated positively with cortical thickness in several brain regions. The standard model is based on cause and effect relationships, which are essentially linear and that any phenomena must be ultimately explained as physical manifestations rooted in matter. Such a viewpoint has been described as *scientism,* a philosophy that purports to define that the world 'really is' reducable to matter. Here knowledge derived from sociological, psychological, anthropological, hermeneutical, philosophical or theological perspectives is mainly irrelevant. However, increasingly a diversity of qualitative and quantitative research methods have been called for in building evidence-informed medicine, whereby the thinking underpinning *scientism* is being challenged to its very core. For meditation research beyond the standard model, standardised as well as explorative statistical procedures, including multivariate regression analyses and principal component analysis, may assist in analysing differences and exploring correlations between a multitude of different qualitative and quantitative meditation outcomes. Such outcomes may include effects regarding mortality and morbidity, telomerase activity, sense of coherence (SOC), and even address possible non-local phenomena such as synchronicity and spirituality. Results from integral meta-theoretical enquires may tell us if meditation practice link subjective markers of spirituality, synchronicity and SOC with objective markers of telomerase-mediated genetic integrity and morbidity and mortality. Such results together with health economy data may inform evidence-informed decision making and healthcare reform globally in relation to spiritual phenomena as public goods.

T. Falkenberg (✉)
I C – The Integrative Care Science Center, Järna, Sweden
e-mail: torkel.falkenberg@integrativecare.se

S. Schmidt and H. Walach (eds.), *Meditation – Neuroscientific Approaches and Philosophical Implications*, Studies in Neuroscience, Consciousness and Spirituality 2,
DOI 10.1007/978-3-319-01634-4_15, 

## The Philosophy of Science and the Standard Model

There are many academic and philosophical entry points available to us in describing the enigmatic effects of meditation. Conventional avenues of enquiry logically start with a neurobiological explanatory mode, describing anatomical and physiological correlates to meditation in an attempt to proxy meditation purely in its material dimension.

Such an approach would be coherent with the common understanding of cause and effect relationships, which are essentially and most of the time linear. This approach to enquiry stems from the unprecedented successes of approaches founded in materialism. Since the seventeenth century, mechanistic and materialistic explanations have revolutionized mankind's ability to isolate, measure and weight complex phenomena and account for them within a theoretical framework which is communicable without subjectivity and bias; at least that has been the theory and the intention.

Hence, a molecular approach to the effects of meditation is deemed by many as having the best potential for success in conveying mechanisms of action acceptable to rational man. From such a perspective, the obscure historical trajectory of meditation practice and theory derived from ancient cosmologies such as Tibetan Buddhism is to be rejected as mainly pre-rational phenomena. In such a view, the fact that effects of meditation can be evidenced by neurobiological events is not seen as the result of insight into true and real phenomena, but rather the result of random cultural experiments leading to meditation practices whose survival and success is purely a matter of a kind of Darwinian selection and cultural habit in which the fittest practices are retained. In short, the standard model of logical reasoning can account for everything that has been stipulated about meditation.

Any other explanation deviating from such logical and deductive reasoning would be explained as devoid of validity, similar to, for example, theories attempting to explain Traditional Chinese Medicine by energy flow in meridians. It is this scientific philosophy and reasoning which has revolutionized our world, or as Bertrand Russell puts it: "Everything that distinguishes the modern world from earlier centuries is attributable to science, which achieved its most spectacular triumphs in the seventeenth century" (Russell 1961).

The Copernican revolution involved a major thrust of self-confidence whereby man was empowered to break free from the dogmas and politics of religion, as interpreted by the Church and its allies for millennia. By way of an increasingly sophisticated understanding of the body and its functions alongside with theories underpinning a separation of the body from the ancient trinity (body, soul and spirit), a new model of man as machine devoid of soul and spirit was the result of those developments.

Had it not been for the French Revolution and the aim for *equality, liberty* and *fraternity*, the mechanistic model would probably have had less impact. The French Revolution was able to jumpstart the materialistic evolution, without accounting for *fraternity* and *equality* as qualities that need to be based on moral and ethical values beyond the material ones, if liberty is to be realized.

Ironically, some of the heads that were chopped off during the French Revolution belonged to key representatives of the *Gnostic* traditions, which at that time opposed the dogma of the Church with a countervailing view of the trinity itself embedded in concepts derived from ancient traditions of Alchemy. Notably, Alchemical principles are the building blocks in many ancient medical cosmologies ranging from Asian to European traditional medical systems. As far as the Gnostic movement was concerned, the trinity was never intended to be institutionalized and dogmatically governed by authorities. Rather, the opportunity lies in every man and woman learning to explore, handle and integrate the trinity in a way that harmonizes with the core ideals of the French Revolution.

## Science and Scientism as a Departure for Meditation Research

The Nobel prizes in medicine or physiology, physics and chemistry are expressions of a society that predominantly rewards a materialistic inquiry and thereby supports a worldview that values above all else quantitative analysis and deductive reasoning. This approach has been immensely successful in enhancing prosperity, reducing mortality and morbidity and has led to unprecedented global increases in life expectancy.

Hence, it comes as no surprise that this forceful avenue of discovery – leading to rocket science, nuclear bombs, central heating and iPhones – has been capitalized on by the market economy. This implicit and explicit authority, conveyed by industry, medicine and mechanisms with patentability as the common goal, has seriously undermined any other attempts for enquiry and explanation beyond the materialistic perspective. This forceful joint venture, now for the first time in mankind, has the power and ability to explain the earth on the one hand, and to irreversibly destroy the earth as we know it, on the other, if the power of science is not balanced by commonly shared values.

By using conventional quantitative, deductive scientific approaches in investigating meditation, quite a few interesting results have been described. For example, meditation practice can be correlated positively with cortical thickness in several brain regions corresponding to the cingulo-fronto-parietal attention networks (Grant et al. 2013). Additional ample evidence suggests that meditative practice activates specific cortical areas and improves attention (Jha et al. 2010) (see also the chapters by Austin, Farb and Esch in this volume). Notably, recent research by the Nobel Prize laureate Elisabeth Blackburn and co-workers even show that meditation positively influences genetic integrity by increasing telomerase activity subsequent to meditation practice (Jacobs et al. 2011).

Hence, in explaining meditation we might well proceed along this successful avenue of enquiry, being considerate of the fact that the landmarks of quantitative science have had an unforeseen and dramatic side effect. By mainstream society omitting phenomena that we are currently unable to describe by linear and quantitative analysis, a vital, fundamental aspect of life has eclipsed scientific and rational scrutiny. Materialism insists that all phenomena must be ultimately explained as physical manifestations rooted in matter. If this is not seen as possible, the phenomena themselves are ignored, and are ridiculed as being pseudoscientific or, worse still, unimportant. Such a viewpoint has been described as *Scientism,* a philosophy that purports to define what the world 'really is' solely by the current methods of scientific inquiry (Loughlin et al. 2012). It adopts what the philosopher Thomas Nagel called 'an epistemological criterion of reality', defining what is real as that which can be discovered by certain quite specific methods of investigation. As a consequence, all features of experience not revealed by those methods are deemed 'subjective' in a way that suggests they are either not real, or lie beyond the scope of meaningful rational inquiry and thereby, by default, unimportant and irrelevant for consideration. This devalues capacities that are in fact essential components of good reasoning and moral practice. Ultimately, the implications of scientism for statements of value undermine judgements essential for science itself to have a sound basis. Scientism has implications, therefore, for ontology, epistemology and also for which claims we can assert as objective truths about the world, and as a consequence also for morality.

## The Middle Ages and Scientism

If we analyse the implications of this approach, we must conclude that we need to embrace science but reject scientism if we are to avoid stifling virtuous practice, and in order to develop richer conceptions of human reasoning. To reverse this trend, we need to find, or revive, a conceptual framework that replaces the impersonal, overly technical approaches towards evaluating meditation reason and practice. Alternatives to scientism are possible where the full range of sources of knowledge that form our intellectual heritage – including insights from the humanities, ethics and philosophy – are combined to explore meditation.

The restrictions of scientism can also be remedied by the academic revolutionaries of quantum theory, which amazingly enough are able to evidence fundamental phenomena that go beyond linear causality. By including entanglement, non-locality and scientific theorems like *Bell's Theorem* (Shimony 2012) new models of reality and new ways of conceptualizing the world are possible.

What if meditation is also a spiritual phenomenon? What if we could describe spiritual phenomena as real, but whose reality lies in a dimension beyond what can be conceived materialistically, a form of goods that are, in principle, available to everyone and hence public goods? What if they are even common resources essential for humanity but not subject to corruption or sales profits, because they lie outside of economic purviews, not least because they are simply not patentable? Would such a theoretical scenario stand a chance of success in today's secular world, given that it would challenge most of the values and methods we hold dear and which for most people seem essential to their happiness, and their release from misery and suffering? Who – in the age of the *web* and rockets to Mars – will take the risk of challenging our maybe unconscious and deep seated cultural aversion to times when the keys to the spiritual "goods" were in the hands of totalitarian despots, with very little mindfulness, self-knowledge and understanding as in the late period of the Middle Ages. These dreadful times were characterized by hardship and calamities, such as famine, plague, and war, which decimated the population of Western Europe. In the 4 years from 1347 through 1350, the Black Death killed approximately a third of the European population. Controversy, heresy, and schisms within the Church paralleled the warfare between states, the civil war, and peasant revolts occurring in the kingdoms.

It is important to acknowledge that the period before this, the High Middle Ages, was known for intellectual life marked by scholasticism and the founding of universities as well as the construction of Gothic cathedrals which was one of the outstanding artistic achievements of this particular time period. These cathedrals, were often dedicated to Mary and hence named *Notre Dame* to emphasize the importance of the feminine principle within general religion and the conceptual framework of Alchemy.

Despite such achievements, most people fearfully refer to the long period of the Middle Ages ranging from eighth to the sixteenth century, with the circumstances typical of the late Middle Ages (fourteenth to sixteenth century). This historical paradigm shift, from a flourishing intellectual and culturally outstanding Europe, to a continent marked by calamities, corruption, inquisition and death, resulted in a fear driven development of mankind towards individual empowerment and freedom. This development has been accompanied with outstanding materialistic inventions but also an increasing aversion against anything explicitly religious or spiritual in contemporary society. Acknowledging these historical traumas is a prerequisite to fully explore and understand meditation beyond the notion of scientism. Unless we can consciously work with addressing such fears and historical wounds, we will not be able to advance our society towards fully taking on research into complex phenomena related to dimensions beyond the body. Most decision makers, funders, and societies will not reward such fishing expeditions into uncertainty given the historical disappointments. Hence, we seem bound to explore meditation from a scattered materialistic perspective, and where knowledge derived from sociological, psychological, anthropological, hermeneutical, philosophical or theological perspectives will have a minor impact on its narrative. Neurobiological cause-effect thinking will dominate the picture. Possibly, some views derived from quantum physics may be allowed as a footnote or curiosity by mainstream medical science, and possibly also as an affordable insurance policy in the unlikely event that future findings might retrospectively endorse the footnote.

## Challenges Beyond the Standard Model

So here we are, apparently having arrived with a rather pessimistic outlook implying that truthful insight into all the realms of meditation, spiritual as well as physical, is not really attainable given the present circumstances. However, increasingly a diversity of research methods have been called for in building evidence-informed medicine, whereby the thinking underpinning scientism is being challenged to its very core (Rawlins 2008). Here, qualitative and quantitative methods are envisaged as complementary and findings are triangulated in order to build the best foundation for informed decision-making. Despite this emerging trend among health technology assessment boards, the prevailing notion of what constitutes excellent science among most medical researchers today still clearly leans towards the ideas of scientism.

Given this situation, the greatest challenge for a paradigm shift of magnitude in science probably requires proving the existence of spiritual dimensions that cannot be reduced to material phenomena beyond doubt in a reproducible fashion, including repetition by independent researchers. However, this is in itself not a likely scenario for events, which are probably characterized by non-locality and entanglement. The psychologist Carl Gustav Jung and physicist Wolfgang Pauli collaborated in an attempt to describe phenomena influenced by meaning and only describable as non-linear events, and referred to them as *synchronicity*, but they were not very successful in conveying what that actually meant even in psychological terms. This does not come as a surprise given the insurmountable challenges that Jung and Pauli faced in their time. These were challenges that no one before had attempted to investigate academically: the deep aspects of individual and collective psychology. Parallel attempts by Freud and the like were not as bold, which may explain why Jung's ground breaking work has never really surfaced into mainstream psychology and society. In fact, most individuals proposing alternative views on spirituality – ranging from Giordano Bruno to Jung and Rudolf Steiner – have been banished in one way or another by representatives of the prevailing doctrine. In retrospect, some of their intelligent contributions may actually provide trustworthy and indispensable domains of theory in building a comprehensive basis for understanding meditation and spirituality, alongside with e.g., the theories of quantum physics. As a matter of fact, quite a few researchers are now trying to merge complementary domains of ideas to achieve a theory that accounts for the spiritual realm in a rational way and potentially testable by science. Leaning on the suggestions and potential theories of these individuals, some notably co-authors of this very book, the time might be ripe to explore meditation from a trans-rational and integral meta-theoretical perspective. Integral means comprehensive, inclusive, non-marginalizing, embracing. Integral approaches to any field attempt to be exactly that: to include as many perspectives, styles, and methodologies as possible within a coherent view of the topic. In a certain sense, integral approaches are "meta-paradigms", or ways to draw together an already existing number of separate paradigms into an interrelated network of approaches that are mutually enriching (Visser 2003).

A number of bold experiments have been able to show significant effects that can so far not be accounted for by the standard model. Here, experiments suggestive of non-locality and entanglement outside strictly physical contexts and a large body of anecdotal evidence suggesting non-locality on an individual level may be included. The latter category of phenomena involves an enormous disadvantage in that they are generally non-reproducible. To conventional scientific thinking, this is in itself a criterion of pseudo-science and may in no way be accounted for as contributing to our knowledge base. This very fact is the Achilles heel of most "out-of-the-box" research including qualitative research, and explains why such results are often discarded. And it is here that this disagreement becomes the tipping point between excellence and garbage science.

This points to one problem: The methodological prerequisite of experimental science presupposes the separability and replicability of phenomena at will. If parts of our reality do not conform to this presupposition, then by default they will eschew any scientific scrutiny. What if the phenomena of spirituality and non-locality are exactly of this kind?

Thus, non-reproducibility challenges the essence of experimental science and means that any deviation from this core tenet makes findings either untrue, or would mean that the standard model is incomplete. Hence, if Copernicus had problems in arguing that the sun was at the centre of the solar system rather than the earth, anyone may imagine what might happen if we were to reject the standard model of scientific enquiry as impaired, given all the success stories it has yielded, and how painful this would be to the totality of contemporary society, which has built its foundation on this prerequisite. Looking on the positive side, surely any credible and successful attempt to broaden the standard model will almost automatically change mankind in the same way as Copernicus did.

How might this be achieved, and what role might meditation play here, apart from being the study object itself? Exploring phenomena beyond the materialistic viewpoint may very well require an enquiry into, as yet, non-standard methods in science, as the standard methods determine to a great extent what we are able to detect. Hence, we suggest the idea that we must use something exceptional to detect the exceptional, otherwise it will merely be seen as a type of noise or obscure event. We may recall the exons in DNA, which for decades were regarded as "trash", instead of something we have good reason to believe is important but which we do not have the means to understand at present.

## Exceptional Ways of Inquiry into the Unknown

The greatest chance of reforming the standard model comes from findings obtained by means of new technologies, like the telescope for Galileo that made it possible to extend the reasoning and ideas of Copernicus. These were so to say hand-in-glove inventions essential for the resulting material landmarks. Similarly, it is fair to ask, what type of technology or method could guide us into researching the spiritual effects of meditation? Technology may play a role here, but probably not in the front seat. Rather, a method more aligned with the expected result, is likely to be more relevant. Perhaps the "telescope" in this case is meditation and systematic introspection itself, whereby meditation can be used to enquire about meditation – a contemplative enquiry of the mind as described by Arthur Zajonc (2009).

This was the standard way of Hindu and Buddhist psychology for centuries (see the contribution by Puta and Sedlmeier in this volume). Here meditation is used to confront the depths and heights, the moral and spiritual realities that underlie all things.

In addition, non-local phenomena, or synchronicity if one prefers, cannot reasonably be captured by the standard model, because of the very fact that it moves outside the model but with implications for effects determined in the standard model. Hence, it is likely that there is a relationship between a possible spiritual realm and what the standard model can measure on a general level, but which is not reproducible on an individual level, as is to be expected. Remember that although individual quantum events such as the radioactive decay of a single nucleus cannot be predicted, the half-life of a large aggregate can be measured and even understood statistically. The trick is to understand how we might experiment with such probability relationships in this new arena. One prerequisite for such a relationship is that synchronicity – despite it being an unpredictable phenomenon not bound to the confines of the material world – is hypothetically crucial for life itself. If so we can ask, given testing conditions that are non-artificial and spiritually meaningful to the individual, could one reveal the true nature of this relationship? How can this be experimented on, when experiments per se are created?

Well, the phenomenon of spirituality is probably best tested under real-life conditions which are not tampered with, but where a systematic approach may uncover effects which are not likely to be the result of the standard model. Hence, qualitative and quantitative inquiries in natural environment, without experimental manipulations will be necessary.

## An Experimental Framework

Given all the previous attempts to describe meditation, ranging from in-depth interview findings to experimental studies into anatomical and genomic correlates, it might be the time to put the puzzle together. Standardised as well as exploratory statistical procedures, including multivariate regression analyses and principal component analysis, can assist in analysing differences and exploring correlations between a multitude of different qualitative and quantitative domains and outcomes. Findings from qualitative explorations of meditation may be aggregated through meta-synthesis and can be included in statistical meta-evaluation procedures, together with data from various domains of quantitative enquiry. Aggregates of domains and outcomes may very well include effects of meditation on an individual level regarding mortality and morbidity, telomerase activity, levels of sense of coherence (SOC), synchronicity and spirituality.

An explanatory hypothesis could be that meditation is linked to higher levels of spirituality, leading to greater SOC, which affects telomerase-mediated genetic integrity over time with implications for morbidity and mortality. How meditation might be related to spirituality, SOC and genomic regulation are challenges that can be addressed and where the results can be used to describe an evidence-based comprehensive theoretical framework of meditation effects.

Hence, by means of a combination of the highest quality of scientific design, multidisciplinary as well as multi-method enquiry and innovative statistical approaches, we may be able to explore meditation effects beyond the standard scientific model. Results of such enquiries may inform decision-making and healthcare reform globally. A careful approach will help ensure that findings related to meditation and spiritual matters are done in a manner that can win respect, given the strength of the research design and the quality of the anticipated results.

Most likely, phenomena with implications beyond the linear rationalist world view, can not be convincingly studied on a single rational level, but rather on a trans-rational and integral meta-theoretical level as discussed previously.

## Epilogue

Historically, the mystical traditions of all peoples have treasured meditation's illumination of consciousness and the insights that derive from contemplative experience. Texts concerned with such experiences can be found in every culture and in every faith tradition (Zajonc 2009). Despite its acknowledged contributions to cultural enrichments over centuries, the scientific undertakings linked to exploring meditation on a spiritual level may seem to pose insurmountable challenges. But, let's imagine for a moment that meditation indeed could be studied systematically and its effects proven beyond doubt even on a spiritual level. What would the consequences be? Would such paradigm shifting create a tipping point for political unrest globally, or would it quantum leap mankind into fuller integration?

Irrespective of such speculations, the very fact that mindfulness practices, such as Mindfulness Based Cognitive Therapy (MBCT) and Mindfulness Based Stress Reduction (MBSR), are seen as stripped of spiritual and religious values, is undoubtedly contributing to their exponential increase in health care utilisation worldwide. Like a Trojan horse, the putative inherent link between spiritual phenomena and meditation practice is hence manifesting itself in mainstream secular society, mainly unintentionally. This time without the historical politics and stewardship of religious doctrine, but now rather giving every practitioner individually the opportunity to discover for herself what the interior telescope of meditation practice may reveal and reward.

**Acknowledgments** I am very grateful for the reviewing and editing of versions of the text by Bo Ahrenfelt, Katarina Falkenberg, David Finer, Walter Osika, Stefan Schmidt, Harald Walach and Arthur Zajonc.

## References

Grant, J.A., E.G. Duerden, J. Courtemanche, M. Cherkasova, G.H. Duncan, and P. Rainville. 2013. Cortical thickness, mental absorption and meditative practice: Possible implications for disorders of attention. *Biological Psychology* 92(2): 275–281.

Jacobs, T.L., E.S. Epel, J. Lin, E.H. Blackburn, O.M. Wolkowitz, D.A. Bridwell, et al. 2011. Intensive meditation training, immune cell telomerase activity, and psychological mediators. *Psychoneuroendocrinology* 36(5): 664–681.

Jha, A.P., E.A. Stanley, A. Kiyonaga, L. Wong, and L. Gelfand. 2010. Examining the protective effects of mindfulness training on working memory capacity and affective experience. *Emotion* 10(1): 54–64.

Loughlin, M., Lewith, G., and Falkenberg, T. 2012. Science, practice and mythology: A definition and examination of the implications of scientism in medicine. *Health Care Analysis* 21(2): 130–145.

Rawlins, M. 2008. *The Harveian oration of 2008*. London: Royal College of Physicians.

Russell, B. 1961. *History of western philosophy*. London: Routledge.

Shimony, A. 2012. "Bell's Theorem". In *The stanford encyclopedia of philosophy* (Winter 2012 Edition), ed. Edward N. Zalta. http://plato.stanford.edu/archives/win2012/entries/bell-theorem/. http://plato.stanford.edu/entries/bell-theorem/

Visser, F. 2003. Wilber, K. Foreword. In *Thought as passion*, ed. F. Visser, xii–xiii. Albany: SUNY Press.

Zajonc, A. 2009. *Meditation as contemplative inquiry: When knowing becomes love*. Massachusetts: Lindisfarne Books.

# Does Meditation Give Us Unique Insight into Ultimate Reality? The Ethical Aim of Buddhism

**Hoyt L. Edge**

**Abstract** The first part of the paper focuses on the narrow philosophical question of whether one can know ultimate reality through meditation. I argue that such knowledge is not possible, that experience is always mediated. Interpreting the Buddha naturalistically, I argue that his aim was less about knowledge and more about providing insight into how one can live a flourishing life. In the second section of the paper, I discuss ways in which a philosophical approach (broadly understood) may benefit a team engaging in neurological studies of meditation. Such an approach might help in linguistic analysis, as well as in understanding the cross-cultural context of traditional meditation traditions. I further suggest that mindfulness might make one more open to scientific discovery. Finally, I propose that the study of meditation will both benefit and be benefited by theories of extended cognition.

## Introduction

Meditation grew up within spiritual traditions of the East and the West, and when Western philosophers approach the topic of meditation within these wider contexts, the road is fraught with dangers. I mention three here. There is first the problem of cross-cultural understanding. Significant work has been accomplished over the past decades in teasing out major cognitive differences between these two general cultural traditions. Second, there is the issue of simply understanding the millennial-old religious traditions in which multiple voices have spoken, sometimes in contradictory utterances. And, third, there is the problem of a Western

H.L. Edge (✉)
Department of Philosophy and Religion, Rollins College, Winter Park, FL, USA
e-mail: hedge@rollins.edu

S. Schmidt and H. Walach (eds.), *Meditation – Neuroscientific Approaches and Philosophical Implications*, Studies in Neuroscience, Consciousness and Spirituality 2,
DOI 10.1007/978-3-319-01634-4_16, 

philosopher trying to understand the philosophical import when religious practitioners make assertions about what they know through meditative experiences. This is especially true for someone like me who has had a long-term interest in meditation, religious traditions, and mysticism, and has even practiced meditation on an elementary level, but I would never claim to be an expert in any of these practices; indeed, these traditions and the writings about them are so vast that they are overwhelming to someone who has not spent years studying and practicing them. Nevertheless, I will examine only one claim, that certain meditative experiences give practitioners knowledge of ultimate reality, an unmediated access to reality. I will examine several ways in which we can understand this claim. This discussion will raise only a narrow philosophical issue in light of the vast literature. Although my conclusion about this claim for knowledge of ultimate reality is negative, that we cannot have such knowledge, we can gain important insights that can be applied to more general questions of meditation and neuroscience. In a second section of the paper, I will take the conclusions from the first section and use them to give us some further insight into meditation and into its relationship with neuroscience.

## Epistemological Claims Made in Meditation

Claims to know ultimate reality are usually attributed to what is often referred to as a mystical experience, which sometimes occurs spontaneously but is usually the result of deep meditative practice. There are a number of characterizations of mystical experience,[1] but the one I want to focus on is its noetic quality, as James (1902) has pointed out, that the experience is both a feeling and a knowing, and what is known is often purported to be ultimate reality. This claim is often found in the popular literature, but it is also found in more scholarly works and in the claims of those who have had a mystical experience.

When looking at the popular literature, claims abound that assert the meditator has direct insight into the nature of things. Meditation, it is asserted, is a process where one can experience ultimate reality; it is "seeing things as they really are" (Collins 1982, p. 115). In a description of insight meditation, we see the assertion that it is a meditation that enables a person "to see the real nature of external things" (London Buddhist Vihara).[2] We can find another example in the statement that meditation can cultivate the "penetrative mental faculty which perceives and understands the realities the way they really are" (Meditation).[3]

[1] See, for instance Ellwood (1980), James (1902), Stace (1972), Underhill (1967).

[2] This is part of a longer discussion to be found at: http://www.londonbuddhistVihara.org/med_aware/pmedinsight.htm/

[3] This section "The Buddha as Healer", entitled Meditation, is part of a larger discussion found at: http://www.buddhanet.net/cmdsg/Gettings8.htm. In fairness, I want to note that the discussion begins with a different focus on the aim of meditation, one that aims primarily at mental development. I will return to this aim later in the paper.

But one does not find such claims only in the popular literature. In a classic philosophical analysis of mysticism, W. T. Stace (1972) divides mystical experiences into two types, an extrovertive type and an introvertive type. Both types, however, seem to claim epistemological objectivity in terms of reality. This sense of objectivity is not derived from a argument but it is asserted to be constitutive of the experience itself. "An inner light assures him of this" (p. 134).

While Ellwood (1980) is arguing for understanding mysticism in its context and within its special spiritual tradition, he nevertheless affirms that mystics "claim to contact ... transhistorical reality" (p. 19).

In her scholarly analysis, Underhill (1967) describes mysticism in two doctrines: Absolute Truth speaks to the real self, and "only in so far as the self is real can it hope to *know* reality" (p. 23). Quoting Coventry Patmore, she affirms that "'mysticism' is the science of *ultimates*,... the science of self-evident Reality, which cannot be 'reasoned about,' because it is the object of pure reason or perception" (p. 25).

Some mystics assert that what one knows is the Absolute (The Godhead, or outer reality) and others focus on knowing the true self. In either case, the view is that one is in contact with ultimate reality.

## Precautions About These Claims

We must be careful when accepting these claims. Let me make two preliminary remarks before I focus on the issue of whether it is possible to know ultimate reality through an unmediated perception of the world in meditation.

The first precaution focuses on a problem of cross-cultural understanding. Much of the literature concerning mysticism derives from the East, and particularly it comes from experiences when using Buddhist meditation, and most of my remarks will be directed toward and derived from this religious tradition. In addition to longstanding anthropological reports of a radically different understanding of the world throughout cultures, there have been a growing number of experiments that point to significant cognitive differences between Eastern and Western cultures. Nisbett (2003), for instance, argues that those two represent two kinds of cultures, Collectivist and Individualist, respectively, with different understandings of the self. They also use different "habits of mind" (45) that have different assumptions about stability and change, ways of organizing the world, patterns of attention and perception, and ways of categorizing objects in the world. Experiments show that these differences are so deep that the two cultures even perceive differently when looking at a scene. Nisbett and Miyamoto (2005) have developed evidence that there is a difference in perception between Japanese and American participants in their studies. Cognitive differences in the more holistic thought of Easterners as opposed to the more individualistic thought of Westerners bleeds over into one's perceptual experience. Japanese tend to view objects in their context, not focusing on the individual objects, while Americans focus on the individual primary object in a scene. And to mention one more example, Weinberg et al. (2008) found cross-cultural differences when examining basic epistemological problems. These cross-cultural differences seem to be deep and fundamental; Western thinkers unfamiliar with these cross-cultural differences may not fully understand what is being asserted by someone not from a Western tradition.

A second precaution that I would like to point out deals with language. We are all too familiar with the difficulty of translating some words from one language to another. But this problem can be exacerbated by what Rosemont (1991) has called "concept clusters." He introduces this concept within a discussion of two fundamentally different views of the person in Confucianism and the West, and points out that there are two concept clusters, systematic ways of thinking about Western "rights-bearing individuals" and Eastern "role-bearing persons," that reflect individualist versus holistic orientations. Even when we use the same words, such as "morality," "person," and "autonomy," the concepts may vary systematically between the two cultures; these terms need to be understood as part of a cluster of ideas that have different meanings and relationships to each other in the two cultures.

Even within a culture, however, concepts can change meaning, either adding or subtracting meaning over time. An example would be the use of "gender." Fifty years ago, the primary usage of the term referred to a linguistic use, for instance in German in referring to masculine or feminine or neutral nouns, or in a biological use. Only later did the practice of using the word in a sociological sense develop and become the primary use of the term. Another example comes from when I started teaching over 45 years ago. At that time I asked (before teaching a section of the course on Descartes) whether students thought that the statement, "the thought of fried chicken is a brain process," made sense to them or if they thought it was as logically incoherent as saying "a square is a circle." At this time, 23 of 25 students thought the statement was logically incoherent because thinking is a mental event and not a physical event. On the other hand, beginning two decades ago, the voting was reversed, with most people wondering why I would even ask such a question because it was so obvious that thoughts are nothing but brain processes to these recent students. So our popular understanding of such a fundamental concept like "thinking" has changed dramatically over several decades.

But there are also concepts used in science that may be ambiguous. Take the concept of "representation," for instance. This idea gained a philosophical foothold through Cartesian dualism, in which consciousness was supposed to represent the world in veridical perception. Descartes literally thought that what we saw – what was contained in consciousness – was a mental re-presentation of an external physical object in consciousness. Thus, the concept of representation entails a dualist understanding of mind and matter. Yet, today, the term is used widely in psychology and in philosophy by those people who eschew dualism. Presumably, it is believed that one can use the term without evoking its historically dualist view. One needs to ask whether the term has dramatically changed its meaning or whether using the term implicitly but thoughtlessly plays upon a dualist view while putatively rejecting it. This is not a question of idle speculation, but it has real implications, especially for the question I wish to turn to now, whether meditators can have an unmediated experience, a representation, of ultimate reality.

## Experience Is Mediated

Given our ordinary understanding of representation, certainly coming out of the Cartesian tradition, and seemingly in its more modern psychological usage, mystical claims seem to assert that they re-present reality, that the meditator has direct insight into reality, into the way things really are. However, I want to argue that there is good reason to doubt these claims. And, I do not think that there are any differences in this regard between a literal perception of the world and an internal perception, an in-sight or introspection, because the arguments I propose should hold for both varieties of cognition. My argument is that experience always is mediated. Let me turn to that discussion now.

As mentioned above, Nisbett and Miyamoto (2005) produced evidence that Japanese and Americans see different things when viewing a scene, with Japanese viewing objects within their context, while Americans focus on the individual primary object in a scene. So, these cultural differences seem to mediate the perception. We "see through" these cultural filters.

However, one might present a counter-argument that perception is actually a two-step process. This view asserts that the first step is a passive reception of the perceptual experience, and the second step is an instantaneous interpretation of this experience that gives rise to the conscious experience. Hence, the argument goes, while the initial experience is unmediated, the conscious experience derived from the second step is mediated through one's worldview, to say it in a general way. One might argue that meditation allows one to engage in the first step, blocking the second step in which the experience is mediated through one's conceptual apparatus, and thus the meditative experience can yield an unmediated experience of reality, the real nature of things.

Such an approach is wrong, I believe, for several reasons. First, this approach is a version of traditional foundationalism in philosophy in which one asserts that there is an ultimate foundation to knowing, that if one peeled away the interpretative processes, one would arrive at a piece of foundational knowledge on which one could test various interpretations to get to the truth of the matter. Such a foundationalism has been criticized on a number of grounds. A quite colorful approach was taken by the nineteenth century existentialist, Nietzsche (1960), who, when asserting that God is dead, was arguing that neither God nor any thing or any process could serve as a foundation to knowledge.

More recently and directed at the question of whether science can provide a foundation, N. R. Hanson (1958) argues against perception being a two-step process, one in which the first step perception is viewed similar to a camera taking a photo, which is subsequently interpreted in the second step. He says: "Instantaneous interpretation hails from the Limbo that produced unsensed sensibilia, unconscious inference, [and] incorrigible statements.... These are ideas which philosophers force on the world to preserve some pet epistemological or metaphysical theory" (p. 10). Employing the experimental evidence from Gestalt psychology, he examines closely what it means to see. He says that when we look at a reversible perspective figure, such as a Necker cube, differing perceptions are not due to interpretation imposed upon a sensation. The sensations are the same; if two people are asked to draw what they see, they will draw the same thing. "Seeing is not only the having of a visual experience; it is also the way in which the visual experience is had," (p. 15) and this is due to differing conceptual organizations in the experience. The context and background knowledge are necessary for the particular organization, but one does not first have a bare sensation, a raw feel, and then interpret it; rather, what one sees is conceptually organized. Perception is thus "theory-laden."[4] These arguments provided background for Thomas Kuhn (1962) to propose in his ground-breaking, *The Structures of Scientific Revolutions*, that our scientific theories are mediated by our scientific paradigms. Although others have argued that scientific progress does not proceed in exactly this way, the basic idea that scientific observation has to be understood as being mediated by theory is generally accepted. We see (and understand) through our knowledge; seeing is not a two-step process of raw sensation that is subsequently interpreted.

Phenomenology is also an approach to experience that rejects experience as a two-step process. For Merleau-Ponty (1962), to experience is to interpret. Any experience is already interpreted. Or, as Heidegger (1963) puts it, we are engaged in projects in our daily lives as Being-in-the-world, and any notion of experience as a purely passive activity does not accord with how we experience the world.

## Meditation and Perception

With this background, we can ask the question whether an experience in meditation can yield knowledge of ultimate reality, as some people have claimed. I think the answer must be negative. All experience, whether internal or external perception, must be mediated. We come to any experience with background knowledge and assumptions, and we experience through this tacit knowledge.[5] Meditation does not offer us a foundation of special knowledge. It does not provide an unmediated experience in which we can know ultimate reality. My conclusion is based on the argument that experience is not a two-step process, where there is the Given, as Rorty (1979) expresses it, in the first step. The notion of a Given is a myth. All of our experience is mediated through our cognitive systems; we come to all experience with a background. Even the act of meditation is part of a larger project that we have chosen in our lives. There is always a cognitive context for our experience.

But, mystics may say, there is a noetic quality to the experience, as James has pointed out. It feels like one knows when one is experiencing. They assert that there is a certainty that is attached to the mystical experience (and only the mystical experience). The mystical experience is unique in this regard. It is the perfect case of knowing when experiencing.

[4] Hanson distinguishes between "seeing as" and "seeing that," but these are distinctions made in seeing-talk, "in our seeing concept" (p. 21), but they are logical distinctions, not different steps in knowing.

[5] For an additional argument, see Polanyi (1966).

I can only respond as a philosopher that there is a distinction between subjective certitude and objective certainty. I have no doubt, after reading descriptions of mystical experiences, that the experience comes with subjective certitude. But I can only try to understand it using the best arguments available. I cannot take any one's word on this issue. Even if the experience in every culture and in every time is described the same way, as Stace (1972) has asserted, it can still be illusory. Given the strong arguments against the Myth of the Given, simple assertions to the contrary cannot be accepted. And it is also the case, as Ellwood (1980) has argued, that "no such thing as mysticism in the abstract exists, only mystics of particular traditions" (p. 19). My own experience with mystical states confirm the interpretive aspect of these experiences.

Several decades ago, when I was studying the mystical experience, I employed several techniques, including an Altered States of Consciousness Induction Device,[6] to produce experiences in several participants that seemed to mimic traditional mystical experiences. One of the participants described his experience as "turning inside out" and "becoming the universe," a state that he described as being without any feeling. This was a student who had had no knowledge of the mystical literature. Another student, who was enrolled in a colleague's class on mysticism, described a much different experience. He writes,

> I was at the source of awareness, enlightenment, and existence, manifested in a form of energy linking all objects animate and inanimate.... I realized for the first time, in the sense of direct contact and exposure, the existence of a life-force of energy that was manifested in every human being and form.... I had mystical union and at-one-ment with a life-force... I was surrounded by meaning and freed from the despair of meaninglessness, guilt, and time.

And he even quoted a passage from a book on mysticism about having a truth not about anything particular but a knowledge of the whole. The differences in these two descriptions reflect different tacit knowledge.

However, one might propose another counter argument here. It might be said that what happens in meditation is that we rid ourselves completely of all concepts and we arrive at a state of pure awareness which is not mediated. I will leave aside the possibility of such a state,[7] and simply say that, even if such an experience took place, it is not an epistemological moment. The meditator has not gained any knowledge but simply was aware. Of course, one might reflect on that experience, which may be ineffable, and try to describe the experience with the available concepts, but then the person must interpret. For the person herself to make sense of the experience, she must interpret it, but in doing so, we are back to the level of the conceptual. Put simply, pure awareness cannot guarantee the truth of a statement

---

[6] This is a device that induces an altered state in a participant, developed originally by Bob Masters and Jean Houston. One stands in a cradle that is hung by a rope so that it swings freely; the participant is blindfolded and, in my case, had headphones playing white noise, so that the combination of loss of spatial orientation and minor sensory deprivation puts one into an altered state which can then be guided by the experimenter. I used the device only a short time before turning to a more analytic approach to the mystical state.

[7] I believe my arguments against experience being a two-step process apply here.

about an experience describing it. So, even if we can make sense of having an unmediated awareness, and I must admit that I cannot, one could not even know that one has had it, much less describe it or think that one has confronted ultimate reality. Hence, mysticism cannot provide us with unmediated knowledge of ultimate reality.

## Another View of the Mystical Experience

D. T. Suzuki (1972) offers a related way of thinking about the mystical experience when he says, "The ultimate reality, whatever name you may give to it – the Self, the Mind, the Absolute, or God – is really a something, or a nothing which is altogether beyond the grasp of the thinkable thinking agent" (pp. 10–11). In this context Suzuki is not only arguing against logical and dualistic thinking, but he is also saying that our conceptual categories are inadequate to re-present the experience. The words "ultimate reality" are not appropriate to describe the experience. One reason for this conclusion is that thinking discriminates, and the experience does not seem to be about anything that is discriminable. Thus, we can conclude that the mystic's standpoint is not conceptual. The point I want to make here is that there is an argument within the Zen tradition of rejecting the idea that the mystical experience should be described as experiencing ultimate reality.[8] That kind of language is not adequate to the task. Suzuki is suggesting that the function of meditation is not to engage in a kind of peek show, that one does not simply to look at, or gaze at, some sort of reality. Suzuki asserts, "You must not try to catch the self or ultimate reality in the way you perceive the star, a tree or a table before you… Zen is therefore not a ware on sale in the public market" (1972, p. 18). Rather, he suggests, there is another task that is the aim of meditation, which I will discuss below.

It is important to emphasize that the conclusion that meditation does not give us unmediated access to reality does not in my mind lessen the value or important of the insights gained in meditation. Indeed, I suspect that placing an emphasis on such an unmediated cognitive experience sidetracks us from what appears to be the more important, perhaps the traditional, aim of the meditative experience. I believe that this conclusion is supported by a number of statements, particularly about Buddhism, made by practitioners of Buddhist meditation as well as others who have analyzed the experience. But let me make one more point before we turn to consider more closely what seems to be said to be the aim of meditation within the spiritual traditions.

Stace (1972), you will recall, made a distinction between the Extrovertive mystical experience and the Introvertive mystical experience. When one makes the claim that

[8] To be fair to Suzuki, it should be pointed out that he also sometimes uses language that seems to say that one experiences reality. He says, for instance, "The Zen master had taken hold of the real thing" (1972, p. 3). But, I think the direction of his thought is to reject these attempts in favor of another approach.

one can experience ultimate reality, it suggests that we may be referring to the Extrovertive experience. But, if one cannot talk about meditation grasping ultimate reality in this sense, perhaps we can talk about experiencing in the Introvertive Experience the true nature of the Self. However, I do not think that such an approach can be successful, taken at face value. We encounter the same issues when having an in-sight as we do when we have an "out-sight," if I may use that term. Internal experience is just as mediated as an external experience. Nevertheless, the turn toward self is an important one and is worth pursuing. In fact, we have good evidence to think that the meditative experience is aimed at self-knowledge, although we do not have to think that it is knowledge of some ultimate Self, even if it is a constructed self, as the Buddhists assert.

## What Might Be the Spiritual Aim of Meditation?

Once again we can turn to Suzuki who said that "Zen purposes to discipline the mind, itself, to make it its own master, through an insight into its proper nature. This getting into the real nature of one's own mind or soul is the fundamental object of Zen Buddhism" (1977, p. 40). I don't think we have to enter the argument, for instance, about whether the Atman is the Brahman (in Hindu thought) or whether we need to deny that there is an individual self or there is only the Universal Self. Again, I think that these metaphysical claims fall victim to the previous arguments that there is no unmediated experience. But the meditative experience certainly does give us knowledge about the self and about how we experience the world in our ordinary modes of experience. We learn on the one hand that we are liable to experience the world in an attached way – immediately judging, rejecting, accepting, avoiding, and grasping – and it is possible to control these immediate reactions. On the other hand, we can experience the impermanent nature of the self, its constantly changing nature.

Owen Flanagan (2011, p. 95) makes this latter point when he says:

> When properly interpreted, Buddhists believe that there are persons, and the talk of persons themselves is harmless so long as we recognize that *person* and *self* refer to something, a pattern that is conventionally useful but that does not name anything "ultimate" or "really real." ….

We are familiar with the tradition of Western philosophers discussing the nature of the self. There are philosophers who have argued for an existing mind/thing with the essential property of thinking; the Cartesian *cogito (res cogitans)* is certainly an example of this approach. But a large number of thinkers have rejected such a characterization of the self as an entity, David Hume (1988) being an example. Further, phenomenology has argued that we can closely examine our experience to see what kind of self we experience, and generally they argue that we find no enduring Cartesian entity. As the Sanskrit scholar Gombrich (2009) explains, "The third hallmark is very often mistranslated (sometimes by me, too, in the past) has 'not having a self or essence'. That is indeed how later Buddhists came to interpret it, but that was not its original meaning…. [I]t became the one-word expression of the Buddha's anti-essentialism" (p. 70). So, one can argue that there is a self, but that there is no self with an essence. There is no self with a permanent, immutable core. Everything, including the self, seems impermanent and changing.

Therefore, the meditative experience can be a window into a closer examination of how we experience something that we want to call a self and to describe its nature. Gombrich has argued that "[t]he Buddha rejected 'being' as a reified category; for him there is no such *thing* as 'existence'. He likewise rejected the concept of consciousness that went with it: just as being was a process, not a thing, so was consciousness. In fact, consciousness was a process we all experience and one which he analyzed" (2009, p. 69). Similar to Husserl (1973), who argued that we ought to bracket questions of existence and simply describe experience, so the Buddha argued that we should pay attention to experience and not worry about questions of being. As Gombrich (2009) puts it: "The Buddha rejected all questions of the type 'Does *x* exist?' He rephrased it: 'Can we experience *x*?'" (p. 73).

I am open to going even further to suggest that the meditative experience can offer us a unique experience of the self, perhaps an understanding of self that is hard to experience in any other way, given the special character of meditation. A number of commentators on Buddhism have talked about the experience of the impermanence of the self, the fleeting nature of our monkey mind, and the absence of any enduring self. The claim does not have to be about ultimate reality nor does one need to argue that this experience is totally unmediated for it to be a significant and important insight about us as humans.

Let me give a short summary of where we are in the chapter. Having examined and rejected the claims that the aim of meditation is to engage in epistemology and metaphysics about ultimate reality, we noticed that there was a distinct focus on the self but also rejected the idea that we were engaged in epistemological claims about any ultimate self. It is time to follow up on what strikes me (and others) as the driving force of Buddhist meditation. Rather than engaging in epistemology or metaphysics, the Buddha was more engaged in an ethical exploration of how we should live our lives if we are interested in flourishing as humans. When one is confronted with questions about existents, one can affirm them or deny them, or simply set aside the question. I believe that the Buddha simply put them aside.

## The Buddha as Healer

While not wanting in any way to disparage the world's great religions, all of which have had wonderful meditative and mystical traditions, nevertheless as a philosopher I want to try to understand the basic aims and motivations of these religions and their religious founders. As Flanagan (2011) sought to understand Buddhism from a naturalized frame of reference, I am seeking to do the same. In this respect, I am impressed by the voice of the former Tibetan (Gelug) and Korean (Zen) monk,

Stephen Batchelor (1997), in his book *Buddhism Without Beliefs*,[9] at least on this point. Batchelor points out that the Buddha used the language of medicine, offering a prescription against suffering. Much as a doctor would recommend aspirin or a particular medicine if a patient came to her seeking relief, so the Buddha offered his prescription for overcoming suffering (or the impermanence of life) in the Four Noble Truths. These statements were not statements to be believed in, Batchelor argues, but prescriptions to be tried. Essentially, as a physician, the Buddha was saying: "This worked for me; I think it will work for you." Experience it for yourself and see if it works. Gombrich (2009, p. 161) also argued for this position, saying:

> Again and again the Buddha emphasized that his goal as a teacher was entirely pragmatic. His followers came to know him as the great physician; the *Dharma* was the medicine he prescribed… Though there is no canonical evidence for this interpretation, modern scholars have plausibly argued that the formulation of the Four Noble Truths follows the medical idiom of the time: first the disease is diagnosed, then its origin or cause is established, then it is accordingly stated what a cure would consist of, and finally the treatment to achieve that cure is prescribed. The Buddha described himself as the surgeon who removes the arrow of craving.

And Suzuki (1977) explains, "Zen has nothing to teach us in the way of intellectual analysis; nor has it any set doctrines which are imposed on its followers for acceptance" (p. 38).

My own introduction to this view took place decades ago when a visiting Buddhist initiate came to our campus. She led us daily in meditation and talked about Buddhism, and one day after the meditation session, I began a discussion with her in an attempt to better understand what she was saying. I spoke about the ultimate Oneness of all things and the metaphysical implications of such a worldview, and she stopped me to tell a story. Early in her training, she said, one of her tasks had been to serve tea daily to the head of the monastery. The Roshi was very specific in how the tea should be served, so every day she brought in a pot of tea, waited a certain time and then poured four cups of tea. The Roshi put two of the cups on the opposite side of the table and slowly drank the other two. Fascinated by this ritual, she finally got up the courage to ask the Roshi why he was doing that – was he symbolically offering tea to the spirits? What metaphysical meaning lay behind the two cups across from him. The Roshi smiled and said, "I only want to drink two cups of tea." The point of her story became obvious. The point of Buddhism was not metaphysics, and we too often get tangled up in metaphysics and forget what Buddhism is really about.

Of course, Buddhism has subsequently been developed as a religious system with a set of over-beliefs, as William James (1902) calls them. Just as one should not mistake the map for the territory, so one should not mistake the beliefs built up over generations as the aim of the Buddha. Of course, the Buddha was not interested in physical healing as much as mental healing, or more generally, we can see his intent in his talks with disciples as suggesting what can be thought of as a path to a flourishing life. As such, his aim is not epistemological but moral.

---

[9] As well as his newer *Confessions of a Buddhist atheist* (2010).

Flanagan (2011) is correct, I think, when he says the Buddha was "offering advice, philosophical therapy, about how best to live a good and meaningful life as a person" (p. 94). If we take this naturalized view of Buddhism – that is, if we strip it of its religious tradition that developed after the Buddha – I believe that the aim of meditation is to help humans achieve a flourishing life, and a path to this life is through mindfulness,[10] a calm and non-judgmental awareness of things.

Mindfulness is at the heart of Buddhism, but most often we in the West think of mindfulness as a specific form of meditation. In turn we think of meditation as practiced in a sitting posture, but strictly speaking, we need to distinguish the practice of mindfulness from the practice of what we normally think of as meditation. Mindfulness is an attitude toward life and not restricted to sitting postures or a number of other practices developed in the West. Transcendental Meditation and other forms of focus meditation probably introduced most Westerners to meditation and these meditations are usually practiced in a sitting position, so it is easy to see why meditation and a sitting posture are conflated in the West. However, sitting postures are not regularly practiced in several Buddhist countries in Asia[11]; nevertheless mindfulness is central to their practices. It is better if we begin to think of meditation in a broader sense that encompasses all forms of mindfulness practice, as I will mention later.

If the aim of the Buddha's sayings is focused toward producing human flourishing, what are his suggestions? First, the distinction between happiness and flourishing needs to be made. Philosophers for centuries have been discussing happiness as the aim of life. Aristotle (1962) said that it is the only end that is done for itself and is self-sufficient. However, he distinguished the flourishing life from the life of pleasure, and he introduced the term *eudaimonia*, which has often been translated as happiness, but is better understood as the flourishing life. The flourishing life is one that is most fully human. Aristotle's view of the flourishing life contains aspects that are tied to his cultural context, and for the purposes of this paper we do not need to consider his list of virtues. Rather, we need only note the ancient Greek view that, just as good health consists of a fully functioning body, so the flourishing life is brought about by the full functioning of the whole person. In the same way, the Buddha was diagnosing the human condition and suggesting ways to live a healthy, flourishing life, how best to live a good and meaningful life: one of serenity, self-contentment and well-being.

First, the Buddha offers advice on what is not the way to a flourishing life, a life (solely) of the intellect (which is good advice for a philosopher). Buddhists speak of the Great Death of our ordinary life, which is too often composed of judgments and pronouncements. We should put aside the intellectual life to try to experience life without intellectual judgment. Suzuki (1972) describes it: "Slay, they would say, with one stroke this meddling intellect, and throw it to the dogs" (p. 75). My teacher decades ago said it less challengingly to me with her story about the Roshi and his

[10] I am using mindfulness in a general way referring to Eastern approaches. See Schmidt (2011) for an analysis of the differences in mindfulness in the East and West.

[11] See Flanagan (2011, p. 106).

tea. Suzuki (1972) says, "When the Zen master gives you any answer, you must remember that his standpoint is not at all conceptual. It is always deeply rooted in his innermost experience itself.... Intellectually or logically, his utterances are unintelligible and nonsensical" (p. 4). In meditation, we give ourselves up to something that is not conceptual. Mystics always say that the experience is ineffable, and this is because it is not a conceptual experience. As such, it is unthinkable and inexpressible. Zen points the way, and in this experience, we learn something about the flourishing life. As Suzuki (1977) puts it, "The discipline of Zen consists in opening the mental eye in order to look into the very reason of existence" (p. 40).[12]

## Mindfulness

I have argued that the aim of the Buddha was not primarily an epistemological one, as trying to understand ultimate reality, but I am not denying that knowledge is one of the things that the Buddha shared with his disciples. Buddhism is an in-depth examination of the psychology of self; it offers an extraordinary taxonomy of various states of mind and types of experiences. With this knowledge, we will be better able to control our meditation. If we understand the various states and types of consciousness, we will better able to achieve the attentive state that is called for in meditation, realizing that many experiences are mere distractions or beside the point. We can recognize and understand destructive states of mind and eliminate them. If others have experienced various levels of consciousness as they have progressed into deeper, more attentive states, and share such a taxonomy, then we may be better able to find the way to an attentive mind.

Gombrich is insightful when he points out that the historical context of early Buddhism was dramatically different from our own. What little schooling took place outside of the monasteries for men was essentially training for the job that their father had, as well as training for the women directed toward the hearth. We take for granted the benefit of our schooling, which, in addition to basic intellectual skills, we also learn to focus by reading or solving a math problem. We learn to interact, to become sensitive to other people and to the environment. At the initial level, mindfulness taught basic skills of attentiveness and focus, something that is a bi-product of our education today. "Today we tend to over-interpret what was meant at one level by awareness and concentration" (2009, p. 172). Of course, the Buddha aimed at enlightenment, not simply at developing focus, but such "mental training must initially have been... a basis for moral and intellectual understanding" (2009, p. 172).

As such, training in mindfulness was taking a moral position. Modern philosophy has stressed a rational approach to ethics in utilitarianism and in Kantian deontology – that one calculates the greatest amount of happiness for the greatest number of people, or that one tries to universalize the maxim of our action to see if an action is a moral one. But, traditional ethics did not consist primarily in such

[12] So, at this point, you may want to stop reading this chapter and just sit!

rational activity. Aristotle, for instance, focuses on character and the virtues appropriate for a good person. The aim was to be a good person, not create rules for good conduct. It is in this sense that the Buddha was offering an ethics, an understanding of how we can *live* a good and fulfilling life. Mindfulness is a way that we can become less egoistical and recognize that we are responsible for ourselves.

The notion of character is prescriptive, not descriptive. To describe a good character and the attendant virtues is not simply to describe a personality trait; it is rather to say that this is how we *ought* to live. So, the prescription for the mindful life is not basically an attempt to merely describe the self or reality. Rather, meditation can be used for moral insight and training. As the physician offers you a prescription for better health, so the Buddha was prescribing a better kind of life. Wisdom is needed for this life, but so is mindfulness because an attentive and open life is a good life. It is a life in which we can flourish.

This is one reason why practitioners describe the meditative experience as a kind of "coming home." It feels like they have arrived at a natural place, a natural way to walk in the world. "Zen purposes to discipline the mind itself, to make it its own master, through an insight into its proper nature. This getting into the real nature of one's own mind or soul is the fundamental object of Zen Buddhism" (Suzuki 1977, p. 40). In arriving at a *satori* experience, we achieve not simply a new viewpoint but a new way of living, a way that had been obstructed by our own over-intellectualization. "All of your mental activities will now be working to a different key, which will be more satisfying, more peaceful, and fuller of joy than anything you ever experienced before. The tone of life will be altered. There is something rejuvenating in … .Zen" (Suzuki 1977, p. 97). Amid the impermanence of life, we can achieve a sense of serenity, peace, and contentment. And this is true enlightenment.

I can make the point in a slightly different way. One of the important ideas in Buddhism is the assertion of no-self, the denial of a permanent self. Believing that such a permanent self exists gives rise to illusion, grasping, and ultimately suffering. The literature on this idea is vast, and I do not want to enter into that discussion, but simply ask one question: why is the view of no-self so important in Buddhism? On the one hand, one can assert that one is engaging in metaphysics; having made contact with reality as it really is, one is simply describing it. In other words, in saying that one experiences no-self one is engaging in philosophy, in talking about the ultimate constituents of "what is." This is the approach I have been criticizing. Or, on the other hand, as I have been suggesting, one can answer the question by saying that the Buddha was trying to cure suffering. "The Buddha did not see himself as a philosopher constructing an ethic of argumentation but as a healer concerned to cure the suffering of mankind, from which he himself had recovered" (Collins 1982, p. 117). When one achieves enlightenment, one is relieved of suffering and the illusion of a permanent self (or world); one can engage in the "flow" of living, with greater clarity and spontaneity and with less "baggage."

In other words, if it is craving that causes suffering, and if we are able to have an experience in meditation without craving, then one is no longer experiencing suffering (or discontent, or disharmony, or emptiness). Without craving and its fear of losing that which is craved, I believe the Buddha says that we will experience generosity and loving-kindness. In turn, in the experience of being healed from suffering as well as the loving-kindness that results from it, we have wisdom; acting out of this state of loving-kindness, without suffering or discontent, we will live with a sense of contentment and joy. It will be a life of personal flourishing.

The Westerner, on the other hand, will think that there is a rational approach to living correctly, asking if one's behavior can be universalized in a Kantian fashion or if it can derived from a utilitarian calculus. In other words, one monitors behavior by comparing it to an objective standard external to the living of life. Moral thought in the major theories of Western philosophy is rationalistic; one uses reason to figure out what the objective standards are and in turn how one should live, and having settled on the correct course of action, one employs will to act in this way, and the standard reason for failing to act this way is either that we have not used reason appropriately or we have experienced a weakness of will. Juxtapose the experience of living in this objective way in which one is comparing one's actions with an objective standard to the experience in meditation of seeming to "come home," of living without grasping or craving. And if we carry that sense into everyday living, I think this is what Suzuki means when he says that life has a different tone. I think Morris (1994) was correct when he said, "The Buddha… was… in search of a psychological state that would give the person a sense of peace and tranquility in an otherwise turbulent and changing world" (p. 54).

## Meditation and Neuroscience

Let us turn to a discussion of meditation and neuroscience. Others in the book may be discussing the physiological correlates of meditation (see the e.g. the chapters by Clausen et al., Hasenkamp, Hinterberger and by Trautwein et al. in this volume) and what we can learn from these. But, I will leave that topic to the experts, and focus on more philosophical questions.

It may be worth pointing out that both neuroscience and the philosophical approach I have been taking have two things in common. First, both attempt an objective, external understanding of states of consciousness. Second, both take a naturalizing approach to these states. I will later make a suggestion that neuroscience can make progress toward mediating the split between the internal and the external.

One difference between philosophy and neuroscience is important to point out, however. If I am correct that the aim of the Buddha was for us to live flourishing lives, neuroscience cannot tell us what makes for a flourishing life; this is primarily the job of philosophy. Neuroscience cannot tell us what *eudaimonia* is. However, there is no doubt that a neurological investigation of the meditative state can contribute uniquely to our understanding of meditation. As a non-reductionist, I do not believe that neurology fully explains meditation, and certainly it does not explain away moral aspect of meditation, but neurology not only can give us a unique understanding of meditation, but it offers us insight into the meditative processes and thus could actually help facilitate the meditative process. Additionally I will argue that philosophical inquiries into meditation may also benefit neuroscience.

Let us look more closely at the connection between the flourishing life and neuroscience. How to live one's live is basically a normative question; it is laying out what sort of life would be good for humans to live. It tries to explain how humans can be fulfilled and what sort of life will lead to that fulfillment. It is assumed that such a life fulfills the highest human good. For Aristotle, it is a life lived out of virtues; which virtues one ought to possess is a philosophical question (and I am speaking of philosophy broadly here, so that it includes religion and other areas). Most Westerners today would question whether all of the virtues that Aristotle recommended would be ones we would want to follow, and surely a list of virtues in Europe would differ from those in Tibet. But, many of them on the list would overlap, and one can make a good argument that the Buddha was talking about some of these fundamental ones.[13] While the neurosciences and other sciences may give us empirical data concerning how people think and make decisions and what the neurological correlates are, they are not equipped to answer the prescriptive question about which life constitutes the flourishing life and which virtues humans should live by, although they surely can give us the sort of data that would make us rethink our philosophical answers. And there is no question that scientists can also engage in such philosophical discussion, but when they do so, they are moving beyond the narrower confines of their empirical work.

## Philosophy and Neuroscience

The topic of meditation is, of course, complicated. In the first part of the paper, one of the questions I focused on was the narrow one of evaluating claims that one can know ultimate reality in the mystical experience. In addition to the complexity of any meditative discipline, there is the additional problem of the numerous kinds of meditation. Not only are the instructions on how to meditate different, but different practices claim to have different outcomes. Therefore, it is difficult to talk about meditation in general. While I often talk about Zen meditation in the first part of the paper, there are a number of Zen practices, in addition to there being a plethora of other practices. For instance, Lutz, Dunne, and Davidson (2007) discuss a number of forms in their paper on meditation and the neuroscience of consciousness. Among others, they talk about the Recollection of the Buddha, Lovingkindness, Tantric forms, Focused Attention, Vipassana Realization of Selflessness, Open Presence, and Non-Referential Compassion. Some of these forms focus attention, while others don't. Some try to manipulate energies, and some don't. Some try to produce an emotional state, and some don't. And there are a myriad of other differences, and these authors do not consider more physical forms of meditation such as Tai Chi or Yoga's use of asanas.

[13] Certainly virtues such as compassion, altruistic joy, equanimity, truthfulness, patience, and vigor would make both lists.

Not only are there different styles, but there are different stages of the depth of meditation, which often take years of practice to reach, and the styles aim at different depths of consciousness. As Austin (2006) has pointed out, such complexity poses a challenge for neuroscientific research. If we additionally consider the different subjective experiences of meditators, as well as their differing reports, even if we assume the meditators are able to adequately describe their experiences, it is easy to see how complex neuroscientific research on meditation can become.

However, although the complexity introduces specific challenges, it also offers some opportunities. Typically, there is a rich history found in the main meditative approaches, and these traditions often characterize the states of consciousness that are aimed at with good precision. Each form of the various meditations aims at a particular goal, and the researcher is thereby able to investigate those particular states of consciousness. Thus, depending on the practice, there seems to be a predictable effect on the state of consciousness, and often the bodily state. Therefore, the researcher is able to examine the subjective reports of the meditator, and if they follow the expected outcomes, is able to map the neurophysiological correlates of those states.

Furthermore, the traditions argue that there are stages in the process and that these stages succeed each other gradually over time with practice. Thus it is possible for the researcher to gain a much more finely tuned set of correlations with the progressing stages.

We should not forget the practical implications for such research. When differences of brain waves were discovered to correlate with different states of consciousness, the idea of biofeedback was initiated to great practical effect. Using simple machines that gave different forms of feedback about one's physiology, one was able to begin to learn how to control a number of bodily states. From learning simply to relax to learning to control heart rhythm, the practical value of feedback as a way to achieve physiological goals has become mainstream. In the same way, beginning to correlate different levels of deepening stages of meditation, for instance in a Zen meditation, might have great practical value in terms of feedback, allowing a meditator to make significantly better progress in meditation, and perhaps with health benefits, as well.

I do not want to imply that we should seek a magic bullet to achieve *satori* in three easy steps. If I am right about the Buddha being interested in our living a flourishing life, the higher stages of meditation cannot be achieved easily. Simply having a profound experience does not guarantee that one's life will be changed and that one will automatically develop the attendant virtues. The point is not simply having an experience, but it is rather the kind of life that one develops after all the hard work of achieving the special experience. This is the reason why a number of people argue that, even if an experience has the external markings of a mystical experience – such as in an experience of unity – such an experience should not be classified as mystical if it does not change one's life. The aim of the Buddha was to say that we should be a certain kind of person and live a certain kind of life, not be a spectator in experience.

So, my argument that the results of experiments in neuroscience can have practical value to a meditator is not to suggest that we can develop a technology of achieving mystical states. Rather, I want to propose that such a technology might be used in conjunction with traditional training practices to help a meditator recognize ways to deepen her experience and continue a disciplined approach to meditation, leading to a flourishing life.

## The Team Approach to Research

I have mentioned above several complexities involved in research. Let me add to that list, because we find additional complexities in the fact that these practices are typically described within spiritual traditions, which arose in particular cultural contexts. These forms of meditation subsequently have spread cross-culturally. Given all of these complexities, it seems to me that such research would benefit from a cross-disciplinary team in which philosophical perspectives (in the broad sense of this term) are included along with a number of others. Further, it would also be extremely helpful if at least one member of the team is a practicing meditator, and probably even a practitioner of the particular form of meditation that was being currently researched. This may sometimes be the neuroscientist, as in the case with James Austin, or it could be another member of the team. Such a person would be invaluable in planning the experiment and helping to interpret the phenomenological reports of the participants. A bit later, I will discuss what may be a more radical advantage.

## Help in Terminology

A philosophical perspective can help in a number of ways, and since philosophers are often trained in analysis of language, a philosophical approach might help in discerning appropriate terminology. Lutz, Dunne and Davidson point out that "…we have yet to standardize the English lexicon of technical terms for the analysis of meditation (2007, p. 507)." To be helpful for research, terminology needs to be naturalized from what is most often a religious context. A particular problem, of course, derives from the well-known characterization of mystical experiences as ineffable. Nevertheless, mystics still proceed to describe their experience, sometimes asserting that they have direct experience of some truth, or sometimes they introduce metaphoric language. All religions overlay religious experience with interpretive language and beliefs, so it is important to examine what is central about the experience and what is interpretation in order to develop a naturalized terminology helpful for research.

As mentioned above, these meditation practices have typically evolved out of particular spiritual traditions that have been developed over time and within certain cultural contexts. Often the language is transferred to another culture and repeated by practitioners because that is the language that they have become familiar with in being introduced to the spiritual tradition and the meditation. Such language needs to be analyzed. One must take into account central goals of particular meditative practices, understand its cultural context and history, and attempt to develop a naturalized understanding that can be useful for empirical research.

The problems may come not only from spiritual traditions, but they can derive from academic traditions just as easily. Let me give an example by referring to a point I made earlier. It is typical in cognitive psychology, for instance, to refer to representations that are found in consciousness, and it is correlations with these representations that neuroscientists are attempting to map out. But since this term arose in a dualistic context, it gives the impression of it being a mental object that re-presents an object in the physical world in perception (i.e., in consciousness). We naturally tend to think of a representation as a mental thing (even when we call it a symbol). However, it is not clear a naturalized conception of consciousness would describe consciousness in the language of objects. We might want to use more dynamic language to talk about consciousness (or even eschew the term "consciousness" and talk about "consciousing"). For instance, if you talk about having "representings" rather than having *a* representation, for instance, a different view of how to think about consciousness arises, one incidentally that seems closer to the dynamic experience described as "impermanence" in Buddhist meditation. Such an active, dynamic approach and analysis would yield a different vocabulary and potentially different experimental questions than using vocabulary whose original context is dualistic and assumes a kind of entity ontology, asserting that the world is composed of things rather than processes. Such questions about terminology are fundamental for both philosophical psychology and for neuroscience, and they can be more fully analyzed and discussed with a philosophical approach, in a broad understanding of this term, within a multi-disciplinary team. Our language contains implicit assumptions, and we need to become clear about what they are.

Among others, phenomenologists have argued that when we reflect on our experience, we typically interpret it using concepts that do not adequately describe the experience. Phenomenology has developed a rich vocabulary and descriptions of experience, which are alternatives to the dominant approach. In general, these approaches attempt to eschew implicit dualistic descriptions of our experience, and they reject an entity ontology. Rejecting the Cartesian question of whether we can really know that the external world exists, they describe our experiences as dynamic and grounded within the world. Heidegger (1963) proposes that we are Beings-in-the-world, engaged in projects as a woodworker engaging intently on a project in his workshop. Merleau-Ponty (1962) argues perceptively that the body is the primary site of knowing in the world, not consciousness, and his naturalizing phenomenology emphasizes that knowing is a function of our sensory-motor functions. While this approach has had influence in cognitive psychology, it might be interesting to begin to interpret the meditative experience through this perspective.

However, a research team should not intentionally employ a pre-existing interpretive framework but should encourage the participants to engage in a kind of phenomenology in which they describe their experiences as carefully as possible. Only under those circumstances will researchers make progress in correlating physiological states with participants' reports.

## What Can We Know?

In the first part of the chapter, I examined the question of whether we know ultimate reality through meditation. I argued that we cannot, but furthermore that this does not seem to have been the aim of the Buddha, but rather it was to encourage a flourishing life. Although I gave a negative answer to the possibility of grasping ultimate reality, I suggested that we nevertheless gain important knowledge in meditation.

Many Zen practices, as well as other practices, such as a mantra meditation, attempt to focus attention on one object. Zen practices encourage non-judgmental attention; in such a state, one does not name the object or make a judgment about it. Austin (2006) describes it as a state "relatively free from discursive mythological interventions including metaphors, similes, analogies, and related symbols" (p. 231). Suzuki (1977) says that the standpoint of the Zen master "is not at all conceptual" (4). He argues that Zen attempts to get beyond logical, dualistic thinking in which the intellect "must give itself up" (p. 11), and he writes that "language with all its logicality has its limit in dealing with experience which takes place in the realm of unthinkables" (p. 12). He further points out that "we are too much of slaves to words and logic" (1977, p. 61). Zen practice, therefore, is one that produces non-judgmental and non-linguistic attention, and thus attempts to by-pass the natural tendency to name and judge in experience.

Neurophysiologically, Zen training encourages mediators to experience more and more times in which "their left hemispheric language functions are diminished. The training enables word-thoughts to fade into the background" (Austin 2006, p. 231). One of the things that tends to happen in such states is that, without the typical linguistic and dualistic naming of objects, one begins to experience the object as less and less of an "other," and the sense of separation of the self and the object begin to disappear. One senses a kind of merging; one can imagine that as the object begins to fade from attention that one can experience a sense of merging with no object of focus. In such a case, it might feel like a merging with a reality that is larger than the individual self, as the individual self and individual object of attention have faded into the background, and hence one has a sense of uniting with some kind of reality. This seems to be Deshmukh's (2006) point when he says, "An attentional act creates the duality of the attended-attendee, resulting in a specific conscious experience. Disengaging attention to the limit results in a resting attentiveness with global, nonspecific … mental-awareness" (p. 286).

But let me focus on the point about giving up language and the dualistic intellect. Although, as I argued before, I do not think that there is a totally unmediated experience, nevertheless when one focuses on giving up judgment, it seems to be the case that one could *reduce* the level of mediation. The more one practices non-judgmental attention, say in mindfulness meditation, it seems reasonable that one might develop greater cognitive flexibility and reduced cognitive rigidity (Greenberg et al. 2012), and thus one might sense decreasing levels of mediation. The traditional conceptual binds would have been loosened. In such a state, with reduced conceptual filtering, it may be that one could emerge from the state with new insights, with new connection among ideas. This is one of the reasons that it is important to have an experienced meditator on the team; he could not only describe in sharper fashion the experience and therefore the research could make more detailed correlations with specific states of consciousing, but also such experiences might open up creative thinking.

Philosophers of science have made a distinction between the logic of discovery and the logic of justification, and they have focused on the latter, the logic of justification (whether theories can be justified). They have had little to say about the logic of discovery because the creative process of discovery was thought to be mainly outside of the realm of control. However, it might be the case that the training in non-judgmental attention might be employed as an aspect of scientific training. As Kuhn (1962) pointed out, formal training in science aims at inculcating a particular understanding of how things work in the scientific discipline, as well as its assumptions. In a certain way, it therefore trains one to ignore anomalies and work within the confines of the current understanding of how things work. Its aim is to create filters through which the scientist observes the world and theorizes about it. What I am suggesting is that training in non-judgmental meditation might be employed to encourage the type of seeing in which anomalies are more readily noticed and considered, and alternative approaches to standard practice might be suggested.

Charles Tart (1975) put forth a rather radical suggestion a number of years ago about an idea that he developed after having experienced a Rolfing session. The suggestion seems analogous to the process I mentioned above concerning non-judgmental processes (although I think he actually judged that the Rolfing was quite painful). He argued that one could develop state specific sciences. Since each altered state of consciousness has its unique characteristics, to fully understand that state Tart suggested that we train observers or scientists to enter the altered state and make observations. They might be trained to share these observations within the altered state and see what consistency they have among their descriptions. Additionally, Tart speculated that these scientists might be able to theorize within these altered states and develop new insights and understandings of the specific states. I am not sure that meditators in states of non-judgmental attention could share their insights; nevertheless, they might engage in the sharing of descriptions and theorizing immediately following the meditative experience, even if they are using dualistic language at this point. I am not suggesting that such state specific sciences are possible, but if a scientist is trained in non-judgmental attention, one could develop a greater ability to look at data non-judgmentally, without seeing it as much through the lens of the current paradigm.

At the very least, it might be helpful to have as many of the team of researchers trained in non-judgmental meditation and having reached a certain level of practice.

## Are We Atomic Individuals or Are We Connected?

At the conclusion of this chapter, I cannot address in any adequate way the topic of what humans are like, or what has been the subject of much theorizing, what human nature is.[14] This is a complicated topic with far-reaching implications, and I am humbled by the insight into this topic given over the millennia by religious figures, philosophers, psychologists, and other social and natural scientists. All I want to do is to make one potential addition to this vast literature, one that derives from my own and others' experience in meditation.

It has always struck me as ironic that meditation seems like such an isolated, individualistic enterprise. It focuses so much on the individual and her state of consciousness that it may seem almost self-indulgent. Yet the sense of self that one derives from the experience is just the opposite. The more profound experiences in meditation are ones in which the individual self shrinks in focus and in importance, and there may be the sense that one is intimately connected to something larger. Rather than experiencing oneself as an isolated, atomic self, one comes away from the experience with the sense of a related, connected self. In the more profound experiences, this seems to be what is described as feeling connected with nature or with the ultimate.

In my own and others' research on cross-cultural understandings of the self, we find a distinction between what has been called Western individualism and non-Western collectivism. This distinction has been described in many ways. Schweder and Bourne (1984) make the distinction between the egocentric and the sociocentric self, and Markus and Kitayama (1991) talk about the independent vs. the interdependent self. Marsella (1985) describes the Western individuated self as "separate, detached, and self-sufficient," while the non-Western self is "extended to include a wide variety of significant others" (p. 209). The basic distinction is that the self in the West is thought of as atomic, independent, and unconnected to others and the world, a separate entity. On the other hand, peoples outside of the Western tradition think of and experience the self as more extended and connected to others and even to the land (consider the Australian Aboriginal totemic view of themselves).

While I am mindful of distinctions not only among these descriptions of self, as well as differences between these and the unitive experience of mystical insight, the point I want to make is that the idea of self that comes through meditation is more similar to the non-Western sense of self than the individualistic descriptions that are traditional in the West. It is true that phenomenologists have argued that we do not ordinarily experience ourselves as atomic beings, but this has been

[14] Or whether there exists a human nature; the Buddhists seem to deny it.

a minority position in Western thought. On the other hand, the profound experience of the loss of self in meditative experience, and certainly the unitive experience of mystical experiences, leads us singularly to say that our more fundamental experience of ourselves is not as an atomic being but as a self that is related and connected. I believe that this experience may have implications for the following contemporary topic.

An interesting discussion in contemporary philosophy of mind and cognitive science tries to understand how best to characterize human cognition. The more classical individualist approaches have been challenged by an argument for extended cognition. Rather than viewing cognition as a function solely of the brain, the hypothesis of extended cognition argues that the brain readily uses the external environments in cognition, such as a notebook or a cell phone.[15] Individual cognition, it is argued, takes place in an extended process, so that these external objects become part of the cognizing mind. Mason Cash (forthcoming) has argued that these views of extended cognition still seem implicitly to incorporate an individualistic conception of mind, and so he argues for the view that we need to substitute the traditional Western view of individual autonomy with the concept of relational autonomy, and we can thus think of cognition as extended into the social milieu of cultural and institutional practices. Thus, any cognition can be counted as mine if I can be held responsible. He suggests we can turn to the much discussed feminist notion of relational autonomy to understand this alternative view of cognition. Feminism has widely discussed a more relational understanding of the self.

We can also turn, it seems to me, to the cross-cultural literature such as the ones I have cited above to understand a more relational view of the self and of autonomy. But, we might also turn to the literature on meditation, especially the literature of the great Asian religions to think about a view of the self that is basically relational as opposed to atomic. I would argue, therefore, that the meditative experience is precisely one that gives a profound sense of an extended self and of extended cognition, and the meditative experience might have important contributions to our understanding of extended cognition.

Therefore, I would recommend to a team of researchers investigating the neuroscience of meditation that it should not only continue the traditional theorizing based on consciousness as stemming from the brain, but I would recommend that the team should put the research into the context of extended cognition. Not only might the results of research on meditation have important implications for the theory of cognition as extended, but, also, it might be the case that the idea of extended cognition might have implications for research. The meditation research that I am aware of seems to have implicitly incorporated the individualistic view of cognition into the experimental setting. This research looks at the brains of meditators and tries to discern correlates of the various meditative states. But, it seems to me that the brain is one organ in a physical system, extending at least to the whole body, if not beyond. Given our understanding of the brain and of meditation, even the more

[15] For instance, see Clark (1997, 2008).

individualistic approach of considering only brain function is a tremendously complicated task, and it might be that this is precisely where research must begin. But, I'd place my bet on the view that the cognitive process is extended at least throughout the body, and ultimately, a better understanding of the physiological correlates of meditation will occur when we begin to understand how the whole physical system is studied.

## Conclusion

I have argued that meditation cannot give us unmediated access to ultimate reality, but I concluded that the aim of the Buddha was to lead us to a flourishing life; the Buddha was interested in a good life, not epistemology. Meditation is one factor in helping one live such a life. As such, we should learn as much as we can how to successfully meditate, and neuroscience can facilitate that process in noting more objective measurements of various states of consciousness and levels of consciousness that could benefit a practitioner in progressing through specific levels of meditation toward a specific chosen state of consciousness. To achieve a successful research program, it would be advantageous in various ways to include a philosophical approach in a multi-disciplinary team.

## References


Aristotle. 1962. *Nichomachian ethics*. Trans. Oswald, Martin. Indianapolis: The Bobbs-Merrill Co.

Austin, J.H. 2006. *Zen-brain reflections: Reviewing recent developments in meditation and states of consciousness*. Boston: MIT Press.

Batchelor, S. 1997. *Buddhism without beliefs*. New York: Riverhead Books.

Batchelor, S. 2010. *Confessions of a Buddhist atheist*. New York: Spiegel & Grau.

Cash, M. forthcoming. Cognition without borders: "Third wave" socially distributed cognition and relational autonomy. *Cognitive Systems Research*. http://dx.doi.org/10.1016/j.cogsys.2013.03.007

Clark, A. 1997. *Being here: Putting brain, body and world together again*. Cambridge, MA: MIT Press. A Bradford Book.

Clark, A. 2008. *Supersizing the mind*. New York: Oxford University Press.

Collins, S. 1982. *Selfless persons: Imagery and thought in Theravada Buddhism*. Cambridge: Cambridge University Press.

Deshmukt, V.D. 2006. Neurscience of meditation. *TSW Holistic Health and Medicine* 1: 275–289. doi:10.1100/tswhhm.2006.244.

Ellwood Jr., R.S. 1980. *Mysticism and religion*. Englewood Cliffs: Prentice-Hall.

Flanagan, O. 2011. *The Bodhisatva's brain*. Cambridge, MA: MIT Press.

Gombrich, R. 2009. *What the Buddha thought*. London: Equinox Publishing.

Greenberg, J., K. Reiner, and N. Meiran. 2012. "Mind the trap": Mindfulness practice reduces cognitive rigidity. *PLoS ONE* 7(5): e36206. doi:10.1371/journal.pone.0036206.

Hanson, R.N. 1958. *Patterns of discovery*. New York: Cambridge University Press.

Heidegger, M. 1963. *Sein und Zeit*. Tübingen: Max Niemeyer Verlag.

Hume, D. 1988. *An enquiry concerning human understanding*. LaSalle: Open Court.

Husserl, E. 1973. *Logical investigations*. Trans. J.N. Findlay. London: Routledge.
James, W. 1902. *The varieties of religious experience: A study in human nature*. London: Longmans, Green.
Kuhn, T. 1962. *The structure of scientific revolutions*. Chicago: University of Chicago Press.
London Buddhist Vihara. http://www.londonbuddhistvihara.org/med_aware/pmedinsight.htm. Accessed 10 Oct 2012.
Lutz, A., J.D. Dunne, and R.J. Davidson. 2007. Meditation and the neuroscience of consciousness: an introduction. In *Cambridge handbook of consciousness*, ed. P. Zelazo, M. Moscovitch, and E. Thompson, 499–554. Cambridge/New York: Cambridge University Press.
Markus, H.R., and S. Kitayama. 1991. Culture and the self: Implications for cognition, emotion, and motivation. *Psychological Review* 98: 224–253.
Marsella, A. 1985. A category of the human mind: The notion of the person; the notion of the self. In *Culture and self: Asian and American perspectives*, ed. Anthony Marsella, George DeVos, and Francis Hsu, 281–308. New York: Travistock Publiscations.
Meditation. http://www.buddhanet.net/cmdsg/getting8.htm. Accessed 10 Oct 2012.
Merleau-Ponty, M. 1962. *Phenomenology of perception*. Trans. Colin Smith. New York: Humanities Press.
Morris, B. 1994. *Anthropology of the self*. Boulder: Pluto Press.
Nietzsche, F. 1960. *Joyful wisdom*. Trans. Common, Thomas. New York: Fredrick Ungar.
Nisbett, R. 2003. *The geography of thought*. New York: The Free Press.
Nisbett, R.E., and Y. Miyamoto. 2005. The influence of culture: Holistic versus analytic perception. *Trends in Cognitive Science* 9: 467–473.
Polanyi, M. 1966. *The tacit dimension*. London: Routledge.
Rorty, R. 1979. *Philosophy and the mirror of nature*. Princeton: Cambridge University Press.
Rosemont Jr., H. 1991. Rights-bearing individuals and role-bearing persons. In *Rules, rituals, and responsibilities*, ed. Mary I. Bockover, 71–101. LaSalle: Open Court.
Schmidt, S. 2011. Mindfulness in east and west – is it the same? In *Neuroscience, consciousness and spirituality*, ed. H. Walach, S. Schmidt, and W.B. Jonas, 23–38. New York: Springer.
Stace, W.T. 1972. *Mysticism and philosophy*. London: Macmillan.
Suzuki, D.T. 1972. *What is Zen*. New York: Harper & Row.
Suzuki, D.T. 1977. *An introduction to Zen Buddhism*. New York: Grove Press.
Sweder, R.A., and E.J. Bourne. 1984. Does the concept of the person vary cross-culturally? In *Culture theory: Essays on mind, self, and emotion*, ed. R.A. Schweder and R.A. Levine, 158–199. New York: Cambridge University Press.
Tart, C.T. 1975. Science, states of consciousness, and spiritual experiences: The need for state-specific sciences. In *Transpersonal psychologies*, ed. C. Tart, 11–58. New York: Harper & Row.
Underhill, E. 1967. *Mysticism: A study in the nature and development of man's spiritual consciousness*. London: Methuen.
Weinberg, J., S. Nichols, and S. Stich. 2008. Normativity and epistemic intuitions. In *Experimental philosophy*, ed. J. Knobe and S. Nichols, 17–47. New York: Oxford University Press.

# God or Ultimate Reality in Theory and Practice: A Philosophical Analysis

**Anne L.C. Runehov**

**Abstract** The present chapter explores how human experiences, including experiences of God or Ultimate Reality should be understood in relation to reality. It is suggested that experiencing is the sine qua non of human existence. It is argued that human beings cannot not experience. Experiences are real in the sense that they have causal effects on the brain, and the cultural-religious-personal environment in which human beings are embedded. Also a distinction is made between concepts, conceptions and conceiving. In order to answer the question how human experiences can be justified, two principles or criteria are adapted (1) The experience should have de facto evidence and (2) it should have effective evidence. In order to answer the question how such experiences should be understood during the course of interdisciplinary research, four main types of naturalism are analyzed, ontological, methodological, epistemological naturalism and supernaturalism. The result of the analyses suggests that a minimalist coherent ontological naturalism or an extended or flexible interferential ontological naturalism should be adapted. Finally, the problem of the gap between descriptive and normative claims is considered.

## Introduction

Saint Francis of Assisi experienced many years a struggle between two discrepant ideals of life. He could not decide upon whether to live the life of the world or to obey the persisting call of the spirit. One day when he was walking in the country side, he passed a little forgotten church (the church of S. Damiano). He went in to pray. As he fell down before the Crucifix, he experienced being 'smitten by unwanted visitations [and] found himself another man than he who had gone

A.L.C. Runehov, Th.D., MPhil (✉)
Department of Systematic Theology, Copenhagen University, Regensburg, Germany
e-mail: aru@teol.ku.dk

S. Schmidt and H. Walach (eds.), *Meditation – Neuroscientific Approaches and Philosophical Implications*, Studies in Neuroscience, Consciousness and Spirituality 2,
DOI 10.1007/978-3-319-01634-4_17, 

in' (Underhill 1912, p. 218). Still being moved by what happened, he experienced the painted picture of Christ Crucified speaking to him, calling his name: '"Francis" it said, "go, repair my House, the which as thou seest is falling into decay"' (Underhill 1912, p. 218). Francis obeyed. His entire worldview had changed for ever; there was no more struggle as to which type of life to choose. There were no more hesitations and no more uncertainties.

Another example of such fascinating experience is the conversion of Saint Catherine of Genoa. It is said that Catherina possessed a religious nature already from childhood but suffered a life of loneliness and was often depressed, not least due to an unhappy marriage. At the age of 26 at the occasion of a Catholic festival, her sister, being a nun, asked her to make a confession. Not convinced at all, Catherine obeyed and went to see the confessor. As she knelt before him, her whole world changed. As described by Evelyn Underhill, 'she received in her heart the wound of the unmeasured Love of God, with so clear a vision of her own misery and her faults, and of the goodness of God, that she almost fell upon the ground' (Underhill 1912, p. 219). She experienced infinite love. Her experience did not end there but continued for many hours afterwards. Also Madame Guyon, having longed to become a nun since she was 12 but similarly to Catherine had been pushed into an unhappy marriage, suddenly experienced 'a *profound wound*, which was full of delight and of love–a wound so sweet that [she] desired that it might never heal' (Underhill 1912, p. 223). Also here the experience changed her way of living completely.

There are plenty of similar descriptions of life-changing experiences. Evelyn Underhill's magnificent book addresses only the subject of Christian mysticism but contains 600 pages of exciting stories. One finds such stories across religions as well as in non-religious worldviews. The experiences seem to come spontaneously or are obtained by way of long term meditation, yoga and prayer, for example. They are not only found in history but also occur today. Consider the following two examples of transforming experiences as they were told to me by one of my informants whom I called Grace:

> One Sunday, when I was about eight years old, I was sitting in the living room on my own, reading the children's Bible. It was natural for me to do so, because I am the daughter of a minister, and I liked the Bible stories very much. Suddenly, a kind of white smoke came out of the page I was reading. Being the daughter of a minister and having heard many Biblical stories, I was not afraid at all but curiously watched what happened. The white smoke turned into figures that smiled at me and left. I remember this experience as something that really happened, and have no doubt about the reality of the spiritual figures. (Runehov 2007, p. 122)

She continued:

> About a year later, on an occasion when I was lying in bed making shadows on the wall with my hands (forming a duck etc.), I suddenly saw that one of my hands burned. There were flames coming from the palm of my hand and my fingers. Again, being the daughter of a minister and having heard so many Biblical stories about fire and since my hand did not hurt at all; I was not really afraid and watched what happened. The fire disappeared as quickly as it came, leaving no marks of course. Later in life, when I became an icon-painter, I remembered my burning hand and believe that this was a Divine sign that my hands would become important instruments for the religious task I would be given. (Runehov 2007, pp. 122–123)

Indeed, such experiences seem to be immune to the increasing secularism and atheism of today. Regardless of the way in which they are obtained, regardless of the cultural-religious and personal settings of the experiencer, these experiences have lasting effects on the lives of those undergoing them. The conversion can be phrased as – once I was blind but now I see; I was unconscious but now I am aware; I was asleep but now I am awake, I was lost but now I am found. The past is forgotten, all things are made new. Furthermore, suddenly one seems to know what to do, one has a purpose with and in one's life. Even the most hard-skinned atheist, neuro-psychologist Michael Persinger who maintains that such experiences are illusory also admits that '[m]ultiple variants of sensed presence often precede mystical and religious experiences that are frequently followed by sudden, permanent changes in the self-concept' (Persinger 1993, p. 915).

How can we understand these experiences, which some refer to as experiences of God or Ultimate Reality, other as experiences of a higher state of consciousness and Albert Einstein as a 'cosmic religious feeling' (Einstein [1931] 2006, p. 28).

## What It Is to Be a Human Being

Philosophical explanations of what it is to be a human being or what is the *sine qua non* of human beings come in different variations. I will only name a few. According to philosopher Jean-Paul Sartre, to be a human being is to never be what one really is but to always be what one is not. Hence he seems to argue that the *sine qua non* of human beings is always to have to choose (which is a freedom without being free). Sartre maintains that human beings cannot not choose. In the act of choosing, the person is alone. Her choices are absolute, unique and individual (1992). Nineteenth century philosopher Arthur Schopenhauer saw this *sine qua non* to be the will, a will that is in need of development from being a *will for one-self* to becoming a *will in itself* (from will to WILL). He argues that our body and our will are one. The will gives a-priori knowledge about the body while the body provides aposteriori knowledge about the will (1992, p. 168). For philosopher Ludwig Wittgenstein, the *sine qua non* of being human is to possess the capacity of language, which enables a variation of language games (1973).

However, also scientific explanations of what is the most important feature of being human come in different variations. For example, some neuroscientists and evolutionary biologists would argue that the uniqueness of human beings is (the evolution of) their neocortex. Related to this, some cognitive scientists argue that it is especially the cognitive capacities generated by the neocortex what constitute human beings. Physicist would argue that since, at the end of the day, everything can be explained by physics, a human being is ultimately constituted by the properties and relations, actions and interactions of particles and fields or whatever basic entities physics deals with (Lockwood 1989, p. 20).

Within religious contexts, it is the soul or spirit that is regarded to be the essence of a human being. Contemporary philosopher of religion Linda Trinkaus Zagzebski argues that it is personhood (rather than rationality) that is the quality of humankind that distinguishes them from other instances of nature and furthermore, that the personhood of human beings is the most significant way in which humans are like God (Zagzebski 2004, p. 203).

I agree with Sartre that one of the quandaries of human existence is frequently having to choose. I also agree with Schopenhauer that the will is a powerful instrument in human life. For instance, the relationship between God or Ultimate Reality and humans is seen as a "union of feeling and will" (Burns 2002, p. 15). I agree that the neocortex and its further developments are significant for distinguishing between humans and other species. I also agree with the philosophical idea of the uniqueness of the human mind and the uniqueness of personhood. However, all serve, in one way or another what it actually is to be human, namely to be an *experiencer*. A la Sartre I argue, *human beings cannot not experience*. Accordingly I suggest a pragmatic way of thinking concerning reality. As anthropologist and theologian Lluis Oviedo writes,

> Putting the problem in pragmatic terms helps to avoid some shortcomings of a more ideological stance, and helps to escape a too precipitate scientific closure of the subject. Even in its minimal version, the pragmatic account ensures at least a possibility of gaining some access to the religious conscious experience and its ability to influence real life and relationships; even it can generate the illusion of a 'black box'. (Oviedo 2007, p. 128)

I am not denying that God or Ultimate Reality may exist independent from human experiences; rather, I am emphasizing Ultimate Reality as experienced by human beings; as such, it is a reality that can be investigated similar to the reality of other human experiences.

In my opinion then, body and mind (soul or spirit) are the tools for making a human being what she is. The body is without doubt the sustainer or sometimes the destroyer (as for example due to brain damage or heart failure); and similarly, the mind contains the drives human beings need (for good and worse). Neither by itself is what human beings are, not even in combination. The *sine qua non* of (human) existence or being (sein) is to be an experiencer.[1] Experiences are what constitute our reality (Walach and Runehov 2010, p. 158). We experience all the time, from the foetus stage to dying. In doing so we transcend different levels of reality but all of them are part of our reality. Some examples of such transcendent experiences are from dream- or coma-state to waken-state, from experiencing nature to experiencing experiencing nature. Human beings transcend one dream level with yet another dream level when dreaming that they are dreaming. Humans transcend a level of reality also when they recover from illness or leave a period of sorrow or stress behind. Finally, when they experience unity in Ultimate Reality, leaving distinctness in ordinary reality behind, they have transcended levels of reality.

However, postulating that our experiences are real suggests that our experiences have causal effects on our brain, mind and the world. This implies that there is mutual causality between the subjective to the intersubjective and to the objective worlds. Let me illustrate my point further.

---

[1] In a previous article I made a distinction between being as sein and being as ein Wesen – a being. (Runehov 2006, p. 62). I put human between brackets because to be an experiencer does not only apply to human beings.

## The Realness of Experiences

The philosophical debate that concerns the realness of experiences has a long history. At the opposite sides of the debate there are those arguing that our experiences are real because they are experiences of something that exists out there (independent of us). Differently put, our experiences are *true* representations of the outer reality, the object or subject of experience is exactly as we experience it. This kind of realism is described by Hillary Putnam as, 'there is exactly one true and complete description of "the way the world is"' (Putnam 1981, p. 49). Those defending such a view of reality are advocates of *naïve or metaphysical realism.* On the other side of the scale we find those defending *metaphysical anti-realism.* According to this view there is no reality independent of our experiences at all. Between these two opposite versions of realism we find critical -, internal, pragmatic - and non-metaphysical realism. Even though it certainly is interesting to compare the different versions of realism, it is not what I have in mind here. For the time being it is not important to investigate whether or not the object of experience *exists.* Rather, what matters is that since human beings cannot not experience, it is more fruitful to try to say something about the realness of these experiences as such. Hence, what is more interesting is to look at the *concept* and *conception* of some item of reality but also how it is *conceived.*

For example, the concept of water is $H_2O$ and represents the chemical properties and necessities for water to be water. However, we do not conceive of $H_2O$. The conception of water then represents the ability to apply the concept water in a correct way so that it is possible to determine whether or not one is referring to water.[2] For example, a conception of water could be that it boils at a temperature of 100 °C. Conceiving water concerns how water is experienced, for example as warm, cold, wet, colourless, etc. We have concepts of God and Ultimate Reality. These concepts differ. Indeed, the concept of the Christian theistic God is quite different form the Buddhist concept of Nirvana. However, the conception of God and Ultimate Reality makes it possible to determine that we are referring, be it in different ways, to a reality that is partly or wholly transcendent in relation to ordinary reality. These concepts and conceptions would not have been created if such reality had not been experienced and if human beings had not made the effort to make these experiences intelligible. Hence, conceiving precedes concepts and conceptions and at the same time, it also transcends them. The phenomenon of colours provides a splendid example of such transcendence.

Colours are real from the perspective of human perception. Physicists, relying on their methodological enquiries, maintain that colours as such do not *exist* (Ward 2005, p. 207).[3] What they mean is that they cannot detect colours or the sources of colours other than through human experiences (colours as conceived).[4] Hence,

[2] The difference between concept and conception derives from philosopher Herrmann (2008).

[3] One could perhaps argue in a similar way concerning time.

[4] This also counts for gravitation and what is hypothesized as the graviton.

because we conceive colours as we do we can establish their conception and concept (the physical preconditions, light absorption, reflection, or emission spectra). This example also shows that, in the course of our life, we 'discover that there is sometimes, perhaps often, a difference between the way that the world appears to us and the way that it really is' (Ward 2005, p. 207). Nevertheless, the wonder of colours is not only that we experience a variety of colours; we also create out of our subjective knowledge of the experience of colours. Indeed, we transfer our subjective reality of colours to an objective (or at least intersubjective) reality of colours, as, for example, in arts and architecture. Hence, how we conceive of the world has an impact on reality (Walach and Runehov 2010, p. 159).

Also the following example of experiences that have an impact on outer reality is the experience of good or bad. Those who often experience good events will most probably project the good onto their environment, while individuals who have a serious lack of such experiences are more likely to project their bad experiences into the world (Walach and Runehov 2010, p. 158).[5] When we have a nightmare, for example, we are scared "for real", our body starts sweating (triggers a reaction from our nervous system) and we try to find our way out of the situation (triggers mental reasoning). We turn around in our bed. When we dream that somebody hurts us we might cry or become angry - not in a fictive sense but factually. If we wake up at that very moment, we realise that we actually have been crying (our cheek or pillow is wet) or we discover that we are still shedding tears! It is not our brain that makes us have these dreams, neither is it our mind. Rather our dreams are reactions of what we have consciously or unconsciously *experienced* earlier. Previous experiences trigger new ones at another level of reality, thereby triggering adequate reactions in our brain and mind.

Whatever behaviour comes from whatever experience, it will bring about new experiences for us and others, shaping new behaviours shaping new experiences. Human beings and other animals also, cannot exist without experiences. A difference between human beings and animals is that humans have the ability to "analyse" their experiences, which, as a matter of fact, is nothing but yet another experience at another level of reality, a level of reality perhaps not (yet) reachable for other animals. Human experiencers want to understand why they experience. They want to know whether others experience as they do. They compare experiences. Human beings experience that they are experiencing, they experience that they want to experience, they experience the experiences of others, and so on. It is also here ontological and epistemological questions come in. Ontological questions require an answer as to whether our experiences are merely real for us, intersubjectively real or ontologically real. Epistemological questions concern how we can know that our experiences are reliable and which conclusions we ought to draw from them. These are questions only raised by the human race.

Because of the increasing possibilities for humans to experience (environmental, cultural, personal), and the projection of these into the natural world, the human

[5] I do not believe I have to go into this any further - our own experiences and what we experience through media channels have sufficiently clarified and convinced us of this matter of fact.

brain developed and continues to develop accordingly, something that is necessary in order to sustain the increasing experiencing (mutual causation). It is this that entailed the development of the neocortex. For instance, one capacity humans gained through the development of the neocortex is known as the capacity of *Theory of Mind*. Philosophically, *Theory of Mind* commonly refers to the capacity to attribute mental states; i.e. beliefs, intents, desires, pretending, knowledge, etc., to oneself and others and to understand that others may have beliefs, desires and intentions that are different from one's own. Philosophically, we attribute mental states to ourselves and others by way of analogical inference. Roughly explained, the analogical inference principle is the idea that other human beings are "very like me". In the language of logic, we induce the other from ourselves. Simply put, x observes y, y is like x, hence x understands y. This capacity, it is said, is crucial for intersubjective communication. It is also crucial, so it is argued, for the development of the capacity of empathy. While low-level empathy comprising selfish or biological altruism, emotional contagion and compassion may be evolutionary innate (in humans and non-humans), high-level empathy comprising sympathy, selfless altruism and agape have emerged during the course of evolution.

Humans as non-humans are *experiencers*. What we call life is a chain of experiences and it has consequences for the world and all life in it.

For instance, we had a wonderful cat called Gizmo who unfortunately passed away 2 years ago. He could really show me that he experienced my stress and did not appreciate it at all. On one of these occasions I was working on an article and felt stressed, the deadline was approaching and still the article did not make sense to me. In one of our rooms we had installed a corner for meditation. Gizmo, apparently not liking the situation, jumped on my computer, and on my knees but since I did not pay enough attention to his call he started meowing with an intensity that gave me no choice than to follow him. He went into the meditation room, forced me to sit down and meditate. When I did as he asked, he started spinning and I relaxed. Besides the fact that my experience of stress affected my cat, this anecdote shows that intersubjective communication is not only established between humans.

## Justifying Experiences

Thus far, I have argued that to be a (human) being is to be an experiencer. It is now time to investigate the status of reality of the experiences. In other words, can they be justified?

Since human beings are embedded in a natural-social-cultural-personal reality, they experience different levels of one reality constituting their reality, which includes Ultimate Reality. Because of the embededness of reality in a natural-social-cultural environment (through conceptualisation of the experienced realities), the reality experienced is shared with others (including animals and perhaps vegetation). This means that even though there are discrepancies between experiences, human experiences are not relative.

However, some may ask whether it is, after all, not necessary in order for an experience to be justified that the experience is based on something independent of the experience. In my view, this something independent of the experience need not to be something *out there*. What is needed is that there is something that challenges us, something that offers resistance (Herrmann 2004). The challenging something may be a thing, an event or a mental or physical process. To return to the experience of colours, it is not the colours that challenge us, rather it is the preconditions as explained by physics together with the human neurological conditions of perception that challenge us to experience colours. In any case, the experience will be an expression of a relation between the experiencer and the experienced. Still, the question of justification is relevant, not least when the subject matter of the experience is not part of the material-physical world. One way to go is suggested by Keith Yandell.

Yandell divides experiences into those having de facto *evidence* and those having *effective evidence* (Yandell 1993, p. 44). The idea is that an experience $x$ is de facto evidence for a claim $c$ if the experience $x$ meets the relevant conditions (those conditions making the experience legitimated) and an experience $x$ is effective evidence if it is both de facto evidence and accepted as such by the experiencer. Roughly, suppose that Marie experiences a sweet scent of roses $x$. She looks around her and indeed notices flowers, which she defines as roses in the back of the garden. Hence, her experience is de facto evidence for her experiencing the scent of roses. However, she might be too distant from the flowers to be able to distinguish roses from other flowers. When she approaches the flowers, she recognises them as roses, which gives her experiences of the scent of roses both a de facto and effective evidence. On the other hand, suppose she approaches the flowers but does not recognise the flowers as roses. The flowers are roses but a type Mary does not yet know about. In this case, her experience of the scent of roses will remain de facto evidence (it are roses) but will not become effective evidence for her claim (Runehov 2007). Suppose now that two friends are with her who have the necessary knowledge. To them, the same experience will have both de facto and effective evidence. The experience of the friends is justified.

Relating this to experiences of God or Ultimate Reality, the experiences are justified if they have de facto evidence and effective evidence. An experience of God or Ultimate Reality $x$ is de facto evidence for a claim $c$ if the experience $x$ meets the relevant conditions for $x$ to be $x$. These conditions are found in the doctrines and soteriology of the religions (the concepts and conceptions) and their continuities and in the descriptions of mystics, philosophers and others having undergone the experiences themselves or studied them profoundly. The argument of continuity is important because it leaves room for experiences that emerge from living in multicultural environments causing the borderlines between different religious traditions to be less articulated. The openness in this parameter is that ultimate experiences are not restricted to the ones known today by the world's religious traditions. However, the experiences also need to fulfil the criteria of effective evidence. The examples presented in the introduction are examples of experiences having both de facto and effective evidence for the experiencer. However, similarly as in the example of Mary not recognising the roses as roses, it might be the case that a person does not recognise her experience as an experience of Ultimate Reality or God. In this case, as long as the person does not recognise her experience as an experience of Ultimate Reality, and regardless of the amount of effective evidence available through history, her experience of Ultimate *x* can not be said to be justified as an experience of Ultimate Reality. Indeed, sometimes it takes years before a person realises what she has experienced and sometimes the person simply refuses to recognise the experience as such.

In conclusion, human beings are experiencers and as such transcend different levels of reality, including Ultimate Reality. They conceive different types of reality during the course of their lives. They also share their experiences and create concepts and conceptions, which makes it possible to determine what the experiences are about. The advanced human brain sustains the richness and variety of human experiencing but also, the increasing richness and variety of human experiencing has causal effects on the brain. These facts have consequences for the cultural-religious-personal environment human beings are embedded in, for good and for worse. Experiences of Ultimate Reality of God are real in the same sense as are all human experiences. This is so because all experiences are subjective experiences but receive their de facto and effective evidence in intersubjectivity.

## Studying Experiences of God or Ultimate Reality Interdisciplinarily

Experiences of Ultimate Reality remained on the operation table of philosophers and theologians for a long time. Today, the situation has changed. Not least perhaps due to the advanced technologies contemporary neuroscientists have at their disposal, the subject matters of religious experiences as well as of other subjective experiences and of consciousness have taken the attention of neuroscience. However, often these experiences are explained in terms of something else, something that belongs to the natural world. One of the reasons is that due to post-Cartesian dualism, enlightenment and post-modernism, we have come to perceive ourselves as minds on the one hand and as bodies on the other. Furthermore, due to the enlightenment and post-modernism we have come to divide the whole human enterprise into public-primary and private-secondary domains. A consequence is that the physical ‘body’ has gained public-primary interest while the ‘mind’ has become, if not a part of the body (brain), then a property of the personal-subjective realm. Hence, in order for experiences of Ultimate Reality or God to become a subject matter of science, they need to become *part of the body*. The *supernatural* needs to become *natural.*

The scope of the present section is to analyse some types of naturalism and highlight the advantages as well as the shortcomings of these naturalistic views of reality. Is it plausible to endorse a naturalistic view of Ultimate Reality or should we keep to a supernaturalistic apprehension when studying experiences of God or Ultimate reality interdisciplinary?

One problem is that different neuroscientists who perform research into experiences of Ultimate Reality have different interpretations of the results of their inquiries. Indeed, Nina Azari, Eugene d'Aquili, Thilo Hinterberger, Andrew Newberg, Michael Persinger, Rüdiger Seitz and Harald Walach, for instance, have different opinions of what is at stake. Persinger wants us to believe that neuroscience has shown that religious experiences are illusory. Azari, Seitz and Newberg want to understand the neurocorrelations of religious experiences and behaviour. However, where Azari and Seitz merely want to understand the neurological underpinnings, Newberg and d'Aquili also want to understand the religious experiences and behaviour in their own right. Hinterberger and Walach go even further. On the basis of high resolution EEG and fMRI experiments with advanced meditators from all over the world, accept the religious reality as reality in its own right and seek to give it a rightful place in human life. Furthermore, they also emphasize the importance of such experiences for human health. Nevertheless, neuroscience, as other branches of the natural sciences, suggests a naturalistic view of reality. Does this mean that science and religion are at odds?

## Naturalism and Philosophy

How did we come to endorse a naturalistic view of the world? David Papineau and Owen Flanagan maintain that the term naturalism has no very precise meaning or lacks a single determinate meaning (Papineau 2007; Flanagan 2006, pp. 432–452). Nevertheless, the term naturalism was coined 400 years ago, and denoted "a view of the world, and of [human] relation to it, in which only the operation of natural (as opposed to supernatural or spiritual) laws and forces is admitted or assumed" (Flanagan 2006, p. 432).[6] In modern times, the term is associated with the twentieth century philosophers John Dewey, Ernest Nagel, Sidney Hook and Roy Wood Sellars who wanted to ally philosophy more closely to science (Papineau 2007). Their definition, however, does not differ remarkably from the initial seventeenth century one. Naturalism came to denote that "Reality is *exhausted* by nature, containing nothing supernatural, [hence] the scientific method should be used to investigate *all areas of reality*, including the human spirit" (Papineau 2007).[7]

Philosopher Mikael Stenmark would call such a type of naturalism comprehensive scientism which signifies that science can and probably will solve all or almost all our genuine problems. The natural sciences are able to answer *all our empirical, theoretical, practical, ethical and existential questions*. This type of scientism, he argues, embraces all other types of scientism, including ontological scientism (Stenmark 2001, pp. 15–16). In other words, comprehensive scientism is the strongest version of scientism.

---

[6] I changed "man's" into "human's".

[7] My italics.

The observant eye will have noticed that naturalism is in itself a normative position since it concerns the terms in which reality ought to be understood. From the fact that there is no consensus of how reality should be understood, it follows that we have a variety of types of naturalism to choose from. From what we shall call the pure types of naturalism, ontological, methodological and epistemological naturalism as well as supernaturalism, a wide variety of new types have emerged.

## Ontological Naturalism

Some advocates of *ontological naturalism* argue that mental phenomena are equal to natural phenomena as described by the natural sciences. They will argue that mental phenomena are *reducible* to natural items because otherwise, they would not have qualities such as "special location, causal interaction with physical things" (Schmitt 2005, p. 343). Mental phenomena are physical/material phenomena, $\forall x$ $(P(x) \rightarrow Q(x))$, where P denotes mental states and Q all that is physical or material. Let us call this variety of naturalism, *comprehensive ontological naturalism* that implies that there is nothing over and above the physical and material.

Advocates of such a view are for instance, Patricia and Paul Churchland, Daniel Dennett, Michael Persinger, Richard Dawkins and others. For instance, Persinger maintains that religious experiences are generated by neural activity and hence, he exhaustively explains religious experiences in terms of brain activity. In other words, there is nothing more to add to the picture. Reality P can be entirely explained in terms of *nature* Q. Hence if P then Q, where Q embodies all that is matter and physical and P may embody all that is matter, physical subjective and spiritual. Hence, $\forall x(P(x) \rightarrow Q(x))$, all P are (reducible to) Q. Persinger argued that if we could understand exactly which neural activity causes epileptic seizure experiences we would know which neural activity generates religious ones. In other words, the brain is the only *cause* of subjective experiences.

Apparently there is something wrong with assuming $\forall x\ (P(x) \rightarrow Q(x))$. To give an example, few, I believe, would argue that music is nothing but sound waves, i.e., air that vibrates in different sequences and with different amplitudes, because music can only occur when these sound waves meet a listener who interprets the sounds and comprehends them, according to his or her cultural, biological, and personal background, as harmonies and disharmonies (Runehov 2007, p. 212). Secondly, neither scientists nor scholars have any means to account for reality as a whole. They may be able to account for particulars of reality (for example biological or historical facts), but from this it does not logically follow that they are able to account for reality as a whole. Hence, from $\exists x\ (P(x) \rightarrow Q(x))$ it does not follow that $\forall x\ (P(x) \rightarrow Q(x))$. Differently argued, it does not follow that because some religious experiences are in fact related to some brain disorder that all religious experiences are, as Persinger and others argue. Consequently, comprehensive naturalism is not a fruitful view of reality to be endorsed.

Another variety of ontological naturalism is advocated by philosopher Samuel Guttenplan, who has a more flexible view on naturalism when he says that naturalism should be seen within the frame of some *realm*. Everything that exists and takes place in that realm consists of the empirically accessible features of the world (Guttenplan 2001, p. 449). The realms may be the *laws* and/or *theories* of the natural sciences. Guttenplan's view on naturalism can be formulated as it is within a specific realm R that $\forall x\ (P(x) \rightarrow Q(x))$. I will refer to this kind of naturalism as *minimalist ontological naturalism* (Walach and Runehov 2010, p. 159).[8]

Yet other defenders of *ontological naturalism* argue that there is no distinctive metaphysical area (realm) of inquiry. There is only one natural order and it *comprises* all of reality. Nothing can be explained independently of the natural order. However, which theory is the best one to explain the natural order is not determined (Jacobs 2008). Hence, $\forall x\ (P(x) \vee Q(x))$.[9] Let us call this version of ontological naturalism *extended or flexible ontological naturalism* (Walach and Runehov 2010, p. 159).[10]

Philosopher of Religion Willem Drees, amongst others, advocates an ontological naturalistic position of the type I called extended or flexible ontological naturalism. He argues that "[t]he natural world is the whole of reality that we know of and interact with; no supernatural or spiritual realm *distinct* from the natural world shows up *within* our natural world, not even in the mental life of humans" (Drees 1996, p. 2; Rottschaefer 2001, p. 414; Griffin 2000, p. 66).[11] If I have understood Drees correctly, he means that the mind also belongs to the natural order as a whole. Such ontological naturalism need not be reductive. Actually it allows a degree of holism and may include both religious and moral dimensions (Drees 2003, p. 595). In *Religion and Science in Context* Willem Drees goes even further and maintains that such a form of reduction

> is not eliminaton; rather, it is integration of these phenomena into our multifaceted picture of the world. Such a reductionism is a form of holism as it reveals the pervasive coherence of reality. (Drees 2010, p. 88)

Hence, this does not mean that everything there is can be explained in natural terms. There may be processes and events which the natural sciences, by way of their methods cannot depict, for example, the phenomenology of colours. What neuroscientists are able to show is how a correlation between the brain and a colour perception is established. As Drees argues, science suggests a naturalistic view of reality but it has its limits. Extended or flexible ontological naturalists need not

---

[8] What I refer to as minimalist ontological naturalism should not be confused with Griffin's idea that naturalism can be understood in a minimal or maximal sense (Griffin 2000) 11–12. What I mean by minimalist naturalism is a naturalism useful within a specific realm or scope of inquiry.

[9] Where "v" stands for "inclusive v".

[10] What I refer to as minimalist ontological naturalism should not be confused with Griffin's idea that naturalism can be understood in a minimal or maximal sense (Griffin 2000, pp. 11–12). What I mean by minimalist naturalism is a naturalism useful within a specific realm or scope of inquiry.

[11] Rottschaefer: my italics. Drees calls his ontological naturalism supernaturalistic religious naturalism. Furthermore, Drees defends the view of a physically closed universe.

deny that colours are real. Furthermore within the realm of neuroscientific research, the question whether colours exist independently of experience or not is not an important issue. Similarly, it is not essential for neuroscientists to question whether the object of a religious experience exists independently of experience or not. What is important is to know how such experiences are possible and why they have the quality they have as described by the experiencer, affective and or cognitive.

To summarise, we have established three types of ontological naturalism:

1. Comprehensive ontological naturalism stating that $\forall x\ (P(x) \rightarrow Q(x))$,
2. Minimalist ontological naturalism stating that **within a certain realm R**, $\forall x\ (P(x) \rightarrow Q(x))$,[12]
3. Extended or flexible ontological naturalism, where $\forall x\ (P(x) \vee Q(x))$.

From what has been said above as well as from the assumption that experiences, being the essence of what it is to be a human being, are real, I argue that in order to give justice to the experiences studied from an interdisciplinary perspective, endorsing the minimalist or the extended/flexible ontological naturalistic view of reality is most fruitful.

## Methodological Naturalism

We have arrived at the second type of naturalism to be analysed on its validity, namely methodological naturalism. This type of naturalism emphasises that the methods used in the natural sciences are the ideal methods and hence it is maintained that the methods of the social sciences and philosophy ought to be modelled on the methods of natural sciences (Schmitt 2005, p. 343). Methodological naturalists, perhaps starting with Isaac Newton, seek to apply scientific methods in non-scientific disciplines because of the success that the natural scientific methods have shown in the enterprise to describe the world. The assumption is that if the methods of natural sciences were applied to the social sciences and philosophy, success would automatically follow. Hence, methodological naturalism is not about what does or what does not exist; rather it is about applying the best possible method to understand the world. Because scientists and scholars apply different methods of investigation, endorsing methodological naturalism is not a good research strategy for performing interdisciplinary reserach. Indeed, from what has been argued above, and particularly in relation to the reality of experiences and the need for justification on their rationality of the subjective descriptions of the experiences, research methods ought not to be restricted.

[12] The difference between 1. and 2. is, while 1. says that all P (apples) are Q (green); 2. says that it is only within the realm R (Uppsala) that all P (apples) are Q (green). This is not the same as saying that "some P are Q", because this would say that there are apples that are green here and there (there is no specific realm where all apples are green).

Methodological naturalism is not the same as methodological reductionism. Methodological reductionism can be understood in different ways. Firstly, it can be understood as every explanation ought to be continually reduced to the very simplest possible explanation. Secondly, it may be apprehended as a research strategy for analysing things to study, such as cells, in terms of their parts, such as, macromolecules, as well as for applying successful theories in one area, such as Darwinian evolution, to other areas, such as sociology or religion. Thirdly, methodological reductionism may aim to reduce a whole into a specific set of parts for a scientific purpose without claiming that this is all there is to that whole. An illustration of such reductionism would be to explain pain in terms of its neurological correlations without the claim for exhaustiveness, nor is there claim for a mutual method.

Indeed, the different disciplines involved should use their own specific methods of research, because it leaves open the possibility of comparing and eventually developing different (reduced) explanations. Furthermore, this kind of reduction may actually capture important (or new) elements that belong, for example, to an experience of Ultimate Reality, which are valuable for the overall understanding of religious experience. For example, experiences of God or Ultimate Reality may be explained as a feeling of union (Rudolf Otto); feeling of total dependency (Friedrich Schleiermacher); feeling of insight (William James); sense of the heart (Jonathan Edwards) and a total deafferentation of both the left and right posterior superior parietal lobes (Andrew Newberg and Eugene d'Aquili 2001).

## Epistemological Naturalism

The view that epistemology should be naturalized derives from Willard Van Orman Quine. The idea was to have an objective approach to epistemology by way of using the empirical sciences to investigate human epistemic activity (Goldman 1999, p. 598). This is because epistemological naturalism dictates that human knowledge is a natural phenomenon and hence should be studied in the same way as any other aspect of nature. Quine tries to convince us that philosophy *is* a science, only one of a more general character (Quine 1960, pp. 3–4). Philosophy is concerned (amongst other things) with epistemology, hence epistemology should be analysed in a scientific manner. This has implied that a-priori knowledge is denied by those defending this position. Our beliefs must be "well-formed and reliable hypotheses", formulated as precisely as possible in order to become subject to (empirical) experiential tests.[13]

The epistemological question "what can we know" is not unimportant, but it is a philosophical question. However, it becomes important in the interface between (neuro)science and philosophy (of religion). This means that epistemological

[13]Tomas Hançil as quoted by Spurway, N. "Evolutionary Epistemology", forthcoming in *The Encyclopedia of Science and Religions,* Springer, 2010/2011.

naturalism can be useful in some interdisciplinary research but not in all. Our purpose is however to look for a naturalism that may encompass several disciplines.

## Supernaturalism

The last type of naturalism to take into consideration is known as supernaturalism. Supernaturalism does not belong to the category of non-naturalism; rather supernaturalism takes an anti-naturalistic stand concerning questions about God and Ultimate Reality. By supernaturalism is meant "the invocation of an agent or force which somehow stands outside the familiar natural world and so whose doings cannot be understood as part of it" (Flanagan 2006, p. 432). Supernaturalism, in my opinion, suffers from similar objections as comprehensive naturalism. Let us recall Owen Flanagan's three components of supernaturalism:

1. There exists some kind of Supernatural Reality *outside* the natural world.[14]
2. Super Reality acts upon the natural world [from outside].
3. The basis of belief in Super Reality and its acts is invisible, indiscoverable, and cannot be inferred by way of any epistemological methods (Flanagan 2006, p. 432).

There are obviously theologians who would defend all three premises. For instance, a consequence of accepting these premises together with the wish to find proof for the existence of God entails that part of Alvin Plantinga's philosophical endeavour has been to try to give proof for the *possibility* of Ultimate Reality by way of modal logics. Modal logics are different from Aristotelian logic in the sense that they include, besides the logical notions of *true* and *false*, the logical notion of *possibility*. And since possibility cannot be proven true or false, it remains open. However, in contemporary theology the first premise that some kind of Supernatural Reality exists *outside* the natural world is no longer uncritically accepted. Rather, such a dualistic theistic view is strongly debated, at least within contemporary Christian theology. Furthermore, the increasing inter-religious debate makes it difficult to argue for premise two as well. Thus, in order for supernaturalism to fit into the program, Flanagan's three premises would at least need to be rephrased, perhaps as follows:

1. There exists some kind of *Ultimate* Reality (UR) *within* the natural world but UR is not *reducible* to the natural world.
2. UR acts upon the natural world *from within*.
3. The basis of belief in UR and its acts remains ineffable but its marks or signs are not (Arthur Peacocke, Philip Clayton and John Polkinghorne, amongst other argue for such a view though by different theories).

[14] My italics.

Some would argue that such a revised supernaturalism is no longer a supernaturalism. Perhaps they are right in some sense; perhaps such supernaturalism rather refers to what Willem Drees calls supernatural religious naturalism.

To round off, supernaturalism, as defined by Flanagan and others, is not suitable for our purpose, firstly because as with comprehensive ontological naturalism, it is difficult to receive interdisciplinary as well as interreligious consensus for this view. Secondly, since human beings are part of nature and are experiencers of that nature, also experiences of Ultimate Reality come natural to them.

## The Problems to Solve

I believe we have now reached the stage of evaluation which allows us to revisit the problem put forward: what type of naturalism is to be preferred. However, there is yet another problem in need for a solution, namely the problem of the gap between descriptive and normative claims.

## Adequate Type(s) of Naturalism

In order to do justice to interdisciplinary studies on the subject matter of experiences of God or Ultimate Reality, we need a naturalism that can be approached from those who line up with different types of realism. What I mean is that agreement over whether or not reality is as human beings perceive it is not really essential. It is entirely possible to suppose that there is a reality out there and this reality is or is not as human beings perceive it. Such questions are not the scope of our investigation. We are not concerned with particles, planets or the vegetable kingdom. What is essential is the reality of the experiences. As mentioned earlier, physicists tell us that colours do not exist outside our subjective perception of them - there are no colours in nature such as human beings experience them. However, colours are real.

The type of naturalism that is needed is monistic but without reducing the mental to the material. The most adequate type of naturalism discussed in the previous section is ontological naturalism. However, not all three varieties of ontological naturalism states above are likely to be adequate. For the reasons given above, comprehensive ontological naturalism is not an adequate approach. On the other hand, minimalist and extended or flexible ontological naturalism are worth further investigation.

## Minimalist Ontological Naturalism

Minimalist ontological naturalism is a type of naturalism seen from within the frame of some realm. Everything that exists and takes place in that realm consists of the empirically accessible features of the world. The realm may be the laws and/or theories of the natural sciences. This type of ontological naturalism was formulated as follows, within a certain realm, $\forall x\ (P(x) \rightarrow Q(x))$. In the course of interdisciplinary research on experiences of Ultimate Reality, the realm of these experiences is limited to those experiences that are scientifically possible to investigate, e.g. leaving out the ones obtained spontaneously like, for example, those mentioned in the introduction. An agreement needs to be made that everything there is within a certain realm (experiences of Ultimate Reality in this case) is related to one network of realities that includes both the physical and the mental and hence can be justified by way of reference to some part(s) in that network. In other words, minimalist ontological naturalism needs to be coherent.

Extended or flexible ontological naturalism states that there is only one natural order and that this comprises all of reality. Nothing can be explained independently of the natural order. However, which theory is the best one to explain the natural order is not determined. This type of naturalism was formulated as $\forall x\ (P(x) \vee Q(x))$. However, even if this type of ontological naturalism seems at first sight to be more open than the former one, it also needs some adjustment. With regard to the possibility of novel religious experiences as argued above, and with regard to the continuing evolution of the human brain, attention should be paid to the fact that there are aspects which have yet to be explored or which have not yet been manifested. Therefore I suggest that the principle of interference should be added. As such, *extended or flexible interferential ontological naturalism* provides a fruitful alternative to minimalist coherent ontological naturalism.[15]

## Descriptive and Normative Claims

There is a common understanding that scientists describe the world in terms of facts only while, for example theologians describe the world only in terms of values. To put it differently, science concerns the *is* of the world while theology concerns the *ought* of the same world. Willem Drees, however, argues that theology is both descriptive and normative. Theology has both a cosmology and an axiology, i.e. a theory of values (Drees 2010, pp. 76–77). The problem to be solved is how to bring the two disciplines (or other of the like) together? A neurophysiologist will give a descriptive explanation of, for example, a certain religious experience, accounting for the neurocorrelates involved. A theologian, on the other hand, will give a normative explanation of the same religious experience, stating that in order to be that type of religious experience the experience ought to be of a certain kind, as described in the Scriptures, doctrines and soteriology of the religions and its continuities. Even though the theologian will seldom equate religious experience

[15] Interference: as in Physics where the term means: The variation of wave amplitude that occurs when waves of the same or different frequency come together. In other words, the meeting of two waves (sound, light, etc.) which reinforce or neutralise each other according to their relative phases on meeting (depending on whether they are in or out of phase).

with neural activity, he or she does not need to integrate some Ultimate Reality which is completely independent of the natural order (Clayton and Peacock 2004). Theologians need not affirm dualism. As a matter of fact, many contemporary theologians do not.

However, is science always free from normative claims? If we, as a matter of example, look at how neuroscientist describe the brain, is their statement free from norms? Indeed, is not a normal functional brain described by neuroscience a normative claim? Is this norm not the basis upon which they further describe deviations from this norm in terms of below normal or above normal brain activity? Isn't it this basis that allows neuroscientists to describe what happens in the brain during some specific meditational state of mind? In this sense, isn't it also the case that, for instance, theology has norms for how an experience of God or Ultimate Reality should be and measures the experiences as it is conceived against these norms? Clearly, experiences of God or Ultimate Reality may be at least partly based on normative learning but are they also normative when practised?

In my view, neither neuroscientific, philosophical nor theological research on religious experiences can stand aloof from the complex natural-cultural-personal environment in which the subject matter of research is involved. Indeed, descriptive neuroscientific accounts of experiences of God or Ultimate Reality are similar to "naturalistic accounts of our beliefs, feelings, and actions, [...] subject to the same forms of adjudication as accounts" of what we ought to experience and how we ought to experience.[16] Neuroscientific inquiry does not separate itself from practical inquiry, because neuroscientific inquiry concerns the neuro-correlations of *religion in practise* (experiences through behaviour/practice). Similarly, theological inquiry, even though it relies on the *oughts* and *shoulds* of experiences of God or Ultimate Reality, can also not be removed from practical inquiry. Seen as such, and within the realm of interdisciplinary research, theology needs to approach *religion in practise* in a similar way that neuroscience approaches the subject matter.

It seems that religious practice needs to be measured against the natural-social-personal features of the environment in which the religious practitioner is embedded. Hence, the descriptive and normative claims of religious experiences merge in practice, leaving open the question of their rationality for both neuroscientists and theologians. The gap between descriptive and normative claims need not be problematic. We need not be concerned with the issue of whether Ultimate Reality or God *exists independently of our experiences;* nor need it concern us how the experiences or behaviour ought to be vis à vis a certain religion; what is essential is whether the experiences in practise can be rationally supported by the disciplines involved in the research. Let me end this section with the words of James Ashbrook:

> One way to recover the power of God [the sacred] may be to recover primary sensory processing [the neural]. As one experiences the sensory richness of reality – apart from words – one experiences the reality of God. (Ashbrook 1996, p. 556)

[16] Inspired by Ward (2005, pp. 210–211).

And a different but related argument pointing to the necessity of interdisciplinarity that comes from Malcolm Jeeves:

> There is a steadily accumulating body of evidence indicating a link between mind-brain states and spiritual awareness. It indicates that it would be wrong to believe that spirituality in some way stands apart and separate from our human embodiment. (Jeeves 2006, p. 71)

## Conclusion

Religious experiences are part of human experiences, which were said to be real because of the postulation that being experiencers is the sin qua non of human existence. Human beings cannot not experience. Experiences are real in the sense that they have causal effect on the brain, and the cultural-religious-personal environment in which human beings are embedded. This implies that human experiences are not relativistic in scope. Four main types of naturalism were analysed according to their validity for interdisciplinary research on experiences of God and Ultimate Reality; ontological, methodological, epistemological naturalism and supernaturalism. To be fruitful for interdisciplinary research, the principles of coherence and interference were added. This resulted in minimalist coherent ontological naturalism and extended or flexible interferential ontological naturalism. The problem of the gap between descriptive and normative claims was solved by focusing on the difference between religious experiences in theory and religious experiences in practise. What is at stake is that experiences of Ultimate Reality as practised need to be investigated on their rationality by the different disciplines involved in the process of investigation.

**Acknowledgments** The present has been made possible with the supported by the Copenhagen University Star Research Program "Naturalism and Christian Semantics" at the Copenhagen University.

## References

Ashbrook, J.B. 1996. Interfacing religion and the neurosciences: A review of twenty-five years of exploration and reflection. *Zygon: Journal of Religion and Science* 31(4): 545–582.

Burns, Charlene P.E. 2002. *Divine becoming: Rethinking Jesus and incarnation*. Minneapolis: Ausburg Fortress Press.

Clayton, P., and A. Peacock (eds.). 2004. *In whom we live and move and have our being*. Cambridge/Michigan: Eerdmans Publishers Co.

Drees, W.B. 1996. *Religion, science, and naturalism*. Cambridge: Cambridge University Press.

Drees, W.B. 2003. Naturalism. In *The encyclopedia of science and religion*, ed. W. Van Huyssteen, 593–597. New York: Macmillan Library Reference.

Drees, W.B. 2010. *Religion and science in context: A guide to the debates*. London/New York: Routledge.

Einstein, A. [1931] 2006. *The world as I see it*. New York: Citadel Press Books.

Flanagan, O. 2006. The varieties of naturalism. In *The Oxford handbook of religion and science*, ed. P. Clayton and Z. Simpson, 432–452. Oxford: Oxford University Press.
Goldman, A.I. 1999. Naturalism. In *The cambridge dictionary of philosophy*, ed. R. Audi, 596–599. Cambridge: Cambridge University Press.
Griffin, D.R. 2000. *Religion and scientific naturalism: Overcoming the conflicts*. Albany: SUNY Press.
Guttenplan, S. (ed.). 2001. *A companion to the philosophy of mind*. Oxford: Blackwell.
Herrmann, E. 2004. *Religion, reality and a good life: A philosophical approach to religion*. Tübingen: Mohr Siebeck.
Herrmann, E. 2008. On the distinction between the concept of God and conceptions of God. *International Journal of Philosophy of Religion* 64: 63–73.
Jacobs J. 2008. Naturalism. Resource document. *The internet encyclopedia of philosophy*. www.iep.utm.edu/naturali/. Accessed 15 June 2010.
Jeeves, M. 2006. *Human nature: Reflection on the integration of psychology and christianity*. West Conshohocken: Templeton Foundation Press.
Lockwood, M. 1989. *Mind, brain & the quantum. The compund 'I'*. Cambridge, MA/Oxford: Blackwell.
Newberg, A.B., and E. d'Aquili. 2001. *Why God won't go away*. New York: Ballantine.
Oviedo, L. 2007. Has theology anything to suggest to consciousness studies? In *Herausforderungen und Grenzen wissenshaftlicher Modelle in Naturwissenschaft und Theologie*, ed. F. Vogelsang and H. Meisinger, 125–140. Bonn: Evangelische Akademie im Rheinland.
Papineau, D. 2007. Naturalism. Resourse document. *Stanford encyclopedia of philosophy*. http://plato.stanford.edu/entries/naturalism. Accessed 15 June 2010.
Persinger, M. 1993. Vectorial cerebral hemisphericity as differential sources for the sensed presence, mystical experiences and religious conversions. *Perceptual and Motor Skill* 76: 915–930.
Putnam, H. 1981. *Reason, truth and history*. Cambridge: Cambridge University Press.
Quine, W.V.O. 1960. *Word and object*. Cambridge, MA: M.I.T. Press.
Rottschaefer, W.A. 2001. Emerging naturalism: William Rottschaefer and Willem Drees in converstation: How to make Naturalism Safe for Supernaturalism: An evalutation of Willem Drees's Supernaturalistic Naturalism. *Zygon: Journal of Religion and Science* 36(3): 407–453.
Runehov, A.L.C. 2006. A being or to be? Philosophical thoughts about future research on neuroscience and religions and the need for interdisciplinarity. *European Journal of Science and Theology* 2(1): 55–66.
Runehov, A.L.C. 2007. *Sacred or neural? The potential of neuroscience to explain religious experience*. Götingen: Vandenhoeck & Ruprecht.
Sartre, J.P. [1943] 1992. *Varet och intet*. Göteborg: Korpen.
Schmitt, F.F. 2005. Naturalism. In *A companion to metaphysics*, ed. J. Kim and E. Sosa, 343–345. Malden/Oxford/Victoria: Blackwell.
Schopenhauer, A. [1818] 1992. *Världen som vilja och föreställning*. Nora: Nya Doxa
Stenmark, M. 2001. *Scientism: Science, ethics and religion*. Hants: Ashgate Science and Religion Series.
Underhill, E. 1912. *Mysticism, a study in the nature and development of man's spiritual consciousness*. New York: E.P. Dutton.
Walach, H., and A.L.C. Runehov. 2010. The epistemological status of transpersonal psychology: The data-base argument revisited. *Journal of Consciousness Studies* 17(1–2): 145–165.
Ward, A. 2005. Defending ethical naturalism: The roles of cognitive science and pragmatism. *Zygon: Journal of Religion and Science* 40(1): 201–220.
Wittgenstein, L. 1973. *Philosophical investigations*. New Jersey: Prentice Hall.
Yandell, K. 1993. *The epistemology of religious experience*. Cambridge, UK: Cambridge University Press.
Zagzebksi, L.T. 1998. *Virtues of the mind: An inquiry into the nature of virtue and the ethical foundations of knowledge*. Cambridge/New York: Cambridge University Press.
Zagzebksi, L.T. 2004. *Divine motivation theory*. Cambridge/New York: Cambridge University Press.

# The Concept of Tri-Guna: A Working Model

**Maika Puta and Peter Sedlmeier**

**Abstract** This literature review presents an overview of an ancient Indian personality system that shows promise for playing an important role in the applied research on well-being and spirituality: the concept of *tri-guna*. The core proposition of this concept is that the psyche consists of three energies ("*gunas*") called *sattva, rajas* and *tamas*. They are said to be present in everyone in different degrees, explaining differences not only in behavior but also in well-being and spirituality. It is assumed that a dominance of *sattva* is favorable for well-being. In the first part of this chapter, we provide a summary of indicators for the three *gunas*, extracted from the available literature, and present empirical findings. The indicators are given separately for cognition, emotion, motivation, social and physical factors, the environment and behavior in general. In the second part we discuss interventions that are claimed to increase *sattva* and thereby further well-being. This review can be used as a theoretical basis for a more systematic empirical examination of the concept.

## Introduction

As defined by the WHO (1948) "Health is a state of complete physical, mental and social well-being and not merely the absence of disease or infirmity." This definition indicates the necessity of finding interventions and conditions that lead to well-being. In the past decades the focus on well-being has increased in psychological research and various approaches have contributed to the understanding of well-being. Antonovsky (1979, as quoted in Faltermeier 2005) was one of the first to formulate a theory with the orientation on salutogenesis, that is, on the conditions that lead

M. Puta (✉) • P. Sedlmeier
Chemnitz University of Technology, Chemnitz, Germany
e-mail: maika.puta@web.de; peter.sedlmeier@psychologie.tu-chemnitz.de

S. Schmidt and H. Walach (eds.), *Meditation – Neuroscientific Approaches and Philosophical Implications*, Studies in Neuroscience, Consciousness and Spirituality 2,
DOI 10.1007/978-3-319-01634-4_18, 

to health (not to disease). Meanwhile, research on well-being and happiness is becoming main-stream, as the involvement of several prominent researchers indicates. For instance, Diener and colleagues have focused on investigating subjective well-being (e.g., Diener and Ryan 2009) and Ryff and Singer (2008) formulated a model of fundamental aspects of psychological well-being which are self-acceptance, positive relationships with others, personal growth, purpose in life, environmental mastery and autonomy. According to another theoretical approach by Seligman et al. (2006) the components of happiness are positive emotions, engagement and meaning in life. Moreover, several other groups of researchers have worked on respective models and interventions (Cloninger 2006; Fava and Ruini 2009; Fordyce 1977; Frisch 2009; Gallagher et al. 2009; Seligman et al. 2005).

Besides the approaches that have emerged from the western academic research listed above, there is another group of theoretical approaches which gives elaborate answers to the question as to how well-being can be increased: the psychological theories stemming from ancient Indian theorizing (for an overview see Cornelissen et al. 2011; Rao et al. 2008). Although, so far, this "re-discovered" Indian psychology has not received much attention from western academic psychologists, the potential gain of knowledge from the Indian approach to psychology (termed "Indian psychology" in the following) is enormous thus justifying research in this field (Sedlmeier 2006; Sedlmeier et al. 2012). Indian psychology consists of different approaches, of which most are connected through their scriptural sources. Interestingly, this tradition has empirical features and it is often claimed that its theories rest on observation and experience (Feuerstein 2001; Sedlmeier 2011). The scriptural sources of Indian psychology are the Vedas (lit. "knowledge") and their corollaries, a collection of ancient Indian scriptures and their interpretations and explanations.

Indian psychology contains, as a core element, a model of psychological health, the concept of *tri-guna* which can potentially be investigated by the sturdy methods of academic psychology. This could be fruitful for many reasons. Firstly, to date, different aspects of well-being are usually investigated separately and integrative approaches are rarely found. In contrast, the concept of *tri-guna* integrates several concepts that are also dealt with in academic psychology, ranging from subjective well-being, spirituality, virtues, and character strengths to the fundamental aspects of psychological well-being. Secondly, health psychology is in need of a model for measuring both psychological disease and well-being. The concept of *tri-guna* provides such a model. Thirdly, the concept of *tri-guna* could provide a theoretical background for interventions from India that are said to increase well-being which could complement interventions presently investigated academically and are mainly informed by western culture. Moreover, considering the significance of health and well-being, every potential source of knowledge should be explored. Indian health psychology, based on the concept of *tri-guna*, provides a vast potential source of useful knowledge; and it is likely to contribute to the knowledge on well-being, as other Indian traditions have successfully done before. Examples of interventions which were developed from traditional Indian knowledge that are now receiving the attention of academic psychologists are mindfulness meditation and loving-kindness meditation (Eberth and Sedlmeier 2012; Fredrickson 2008).

As Murthy and Kumar (2007) have pointed out, so far, investigators have given only partial descriptions of the *tri-guna* concept, often taking only one or a few scriptural sources into account. This leads to differing theoretical conceptions, ways of measuring the concept and problems of construct validation. Murthy and Kumar (2007, p. 110) propose that "[…] what requires further study is how ancient seers and sages primarily conceived *tri-guna* in psychological terms […]." In this spirit, the purpose of this chapter is to review the literature on *tri-guna* and to offer a comprehensive description of the concept which may serve as a basis for further empirical work in finding a complete and accurate construct definition.

This chapter first gives a general description of the *tri-guna* concept as it is described in traditional Indian psychological thought and then focuses on discussing theoretical assumptions and empirical findings concerning the indicators of the manifestation of the *gunas* in different variables relevant to individual experience and behavior. In the second part, possible interventions to increase well-being on the basis of the *tri-guna* concept are explored.

## The Concept of Tri-Guna and Its Relevance to Well-Being

The following delineation of the *tri-guna* concept is meant to be seen as a working model. This working model was constructed by collecting the various perspectives on the concept from Indian literature. Due to different translations and interpretations of the concept there are a few inconsistencies between these perspectives. We will discuss them and try to resolve them whenever possible.

### *The Context of the Tri-Guna Concept*

The concept of *tri-guna* can be applied to different contexts. Most frequently psychologists examine it from the perspective of personality psychology (e.g. Das 1991; Marutham et al. 1998; Wolf 1998). However, it has also been applied to the contexts of mental health (e.g. Lakshmi Bhai et al. 1975; Singh 1971; Wolf & Abell 2003) and occupational psychology (e.g. Elankumaran 2004; Kaur and Sinha 1992; Sharma 1999). Previous research on the *tri-guna* concept in the context of mental health focused on measuring the *gunas* (Singh 1971), investigating their relationships to mental disturbances (Archana Das and Venu Gopal 2009; Lakshmi Bhai et al. 1975; Stempel et al. 2006), and assessing the effect of *hatha*-yoga (Deshpande et al. 2008; Khema et al. 2011; Rani and Rani 2009) and mantra meditation (Wolf and Abell 2003) on the *gunas*. In this chapter, the concept of *tri-guna* is investigated from the perspective of health psychology.

In classical Indian thought the concept of *tri-guna* is part of a larger philosophical approach, termed *Sankhya*. On the most general level of analyzing reality, *Sankhya* distinguishes between matter and spirit. This philosophy is not equivalent to other dualistic philosophies which distinguish between the physical and mental reality (Larson 1983). This is because *Sankhya* draws the line of two entirely different realities not between physical and mental, but between material and spiritual. Physical and mental substances are both seen as material: in other words, the psyche (or mind) and the body are both material (however, the mind is said to consist of a finer form of matter than the body). Apart from the mind and body there is a third component, "the *purusha*" (soul or *atman*), it is said to consist of spirit or pure consciousness, and to emanate awareness. According to *Sankhya*, the soul is what gives life and consciousness to the psyche and gross physical body. The goal of interventions based on the *Sankhya* philosophy is to become fully aware of the soul by mental practices (Bryant 2009). This is seen as a continuous process and the more it progresses, the more well-being is claimed to increase. Why well-being increases by becoming fully aware of the soul is explained differently according to different conceptions of the nature of the soul. Relevant descriptions are found in the classical *Sankhya* philosophy and another, later traditional philosophical approach, the *Vedanta* philosophy. The *Sankhya* tradition simply describes the experience of deep soul-awareness as being free from suffering, whereas the *Vedanta* tradition postulates that the soul is blissful (*Vedanta Sutras* I.1.13, as quoted by Bryant 2009). In other words, according to *Sankhyan* descriptions, by becoming aware of the soul suffering is annihilated. The *Vedanta* tradition additionally claims that the soul emanates happiness which penetrates the psyche to a stronger degree, the more the soul is realized. This idea is also represented by modern descriptions of Ayurveda (which build on the *tri-guna* concept): the happiness and well-being which the soul emanates penetrate the psyche to different degrees and the capability of the psyche to experience this well-being is increased by strengthening *sattva*, one of the components of *tri-guna*, described below (Verma 1997).

## A General Description of the Gunas and Their Mechanisms

Turning to the role of the *gunas* in the *Sankhya* philosophy, it states that matter consists of three components or attributes ("*gunas*") called *sattva*, *rajas* and *tamas*. The term *guna* has a multitude of meanings, and the concept of *tri-guna* is interpreted differently by different authors. After a dictionary study and a textual and interpretative analysis of the term *guna*, Murthy and Kumar (2007) conclude that originally the three *gunas* are conceived of as three different *components* of matter and not different *qualities* of one same substance (although the term *guna* may refer to quality in other philosophical contexts). Larson and Bhattacharya (1987) also depict the basic *Sankhya* philosophy like this – the *gunas* as being constituents (not qualities) of one continuous process of transformation. *Sattva* is commonly translated as goodness, intelligibility or lucidity; *rajas* as activity or passion and *tamas* as inertia, ignorance or darkness. An analysis of the terms can be found in Murthy and Kumar (2007).

Concerning the coactions of the *gunas*, a primary scripture about the *Sankhya* philosophy, the *Sankhya Karika* (Ganganatha 1896), describes them to "[…] mutually subdue, and support, and produce each other and consort together […]." On the one hand, the *gunas* subdue each other, which means a *guna* will only dominate when the two others are suppressed, on the other hand they support and produce each other and consort together, meaning they depend on each other and never exist or act separately. This is an important axiom of the theory which when applied to the context of the psyche means that even though one *guna* may be dominant, the other two will still take an effect, albeit to a lesser degree. The *gunas* increase and decrease in dependence on each other, and to outer and inner stimulation. They are said to be the basic components of all matter and therefore present in everything in different proportions. Anything that is predominantly composed of or dominated by a particular *guna* is a stimulus for that *guna.* An event, thought or behavior which is mainly *sattvic* for example, strengthens and stimulates *sattva guna* in somebody who thinks or behaves accordingly. In other words, acting in a *sattvic* way like helping someone, will increase *sattva* in other areas also, leading to e.g. positive emotions like appreciation. According to the *Mahabharata* (Ganguli 2005) an excess of ill-luck is a stimulus for *tama guna*, leading to disgrace, error and stupefaction. These are then stimuli for further qualities appertaining to *tamas*. Such an accumulation of *tamas* can be interrupted by strengthening *rajas* or *sattva.* The *Mahabharata* (Ganguli 2005) gives more information concerning the mechanisms of *tri-guna.* It states that when *tamas* is restrained, *rajas* increases in a greater proportion, and when *rajas* is restrained, *sattva* increases in a greater proportion. Principally any one or two *guna(s)* can be dominant; however it is also possible that all three are present in equal proportions.[1]

## *The Gunas in the Context of Psychology*

Just as the manifestation of the gross external world is said to be based on the *gunas,* also all psychological phenomena are seen as the results of the interaction of the *gunas* (Bryant 2009). Hence, different psychological conditions can be explained through proportional mixtures of the *gunas*. The *gunas* fulfill different functions

[1] For readers who know Ayurveda it might be interesting to read a few words about the connection between the *gunas* and the *doshas*. The *gunas* are said to be the psychological correlates of the three *doshas* (*vata, pitta* and *kapha*) of Ayurveda (Shilpa and Murthy 2011). It is described that the *doshas* affect the body, whereas the *gunas* take effect on the psyche (Bhardwaj 2003): In Indian medicine the body is mainly discussed in relation to the *doshas* and the psyche mainly in relation to the *gunas*. The *gunas* are seen as a more basic and subtle component of matter (which includes the psyche) than the *doshas*. *Sankhya* gives a detailed description of the evolution of matter, starting from the *gunas*. To depict this here would go beyond the scope of this chapter. However, it is relevant that the *doshas* are seen as composed of the *mahabhutas* (gross elements: space, wind, fire, water and earth), which are based on an evolution of the *tri-gunas* (Bryant 2009; Larson and Bhattacharya 1987). The *doshas* and *gunas* are said to interact with each other, and for a complete characterization of personality according to Indian psychology it is fruitful to consider both (Shilpa and Murthy 2011). Since the emphasis of our work is on examining psychological interventions to increase well-being from Indian psychology, and the interventions described in connection with the *gunas* are more numerous and psychological in nature, we chose to focus on the *gunas*.

complementing each other. In relation to gross matter, the function of *sattva* is described to be maintenance, *rajas* is the principle of activity and *tamas* provides the substance of nature and is the cause of destruction or decomposition. In the context of the psyche (= subtle matter) *sattva* provides for intellectual clarity, reflective discernment (Larson and Bhattacharya 1987) and serves well-being (Deussen and Strauss 1906). When the psyche is governed by *sattva* the light of the soul is reflected in it, thus *sattva* is described as pure or illuminating. *Rajas* provides the capacity for change and spontaneous desire (Larson and Bhattacharya 1987), desire being its essence (Ganguli 2005) and *tamas* is the basis for continually perceiving the surrounding world (Larson and Bhattacharya 1987). In general, a dominance of *sattva* is said to lead to balance and stability; a dominance of *rajas* to change, activity and imbalance and a dominance of *tamas* to inertia and dullness (Frawley 1997). A relative dominance of *rajas* and *tamas* is seen as a cause for dysfunctional conditions (Balodhi 2005; Kapur 2008). A psyche which is penetrated by the energy of the soul to a great extent is *sattvic*, or calm and balanced. The energy of the soul penetrates the psyche less, if it is overly passionate, and active (= *rajasic*), and even less if it is dull, confused and dejected (= *tamasic*). Of the three, *tamas* is the least favorable dominant *guna, rajas* is the intermediate *guna* and most favourable is *sattva* (Deussen and Strauss 1906; *Manu-Samhita* as quoted in Knapp 2010; Prabhupada 1987a; Tagare 1976a).

According to this concept, well-being is dependent on the extent of *sattva guna* that is dominant in the experience and behavior of a person. Being reminded of the axiom that the *gunas* depend on each other and never act separately, this does not mean that in order to increase well-being *rajas* and *tamas* should be totally nullified, as this is not possible. People are influenced by all the *gunas* and will show characteristics of all three (Kaur and Sinha 1992), or as Mohan and Sandhu (1986, p. 49) put it "Every person is a triad." *Rajas* and *tamas* have their roles to play, *rajas* as a force energizing action and *tamas* as a state of exhaustion, signalizing the need for rest and inactivity. In the context of increasing well-being the goal lies in strengthening *sattva* so that it becomes predominant. Predominant *sattva* not only enhances well-being, but also strengthens the individual's invulnerability to disorders (Kapur 2008). Although *sattva* is not simply a balance between *rajas* and *tamas,* one of its functions is to harmonize and balance these two *gunas* so that their energy can be used as necessary (Braud 2008).

The concept of *tri-guna* is said to be a concept of human traits *and* states (Paranjpe and Rao 2008; Sitamma 2005). This is not uncommon: mindfulness, for example, can also be viewed as state or trait (Prazak et al. 2012). Both personality and psychological states are explained through the dominating *guna(s).* The proportions of *sattva, rajas* and *tamas* present in the psyche are said to change continuously depending on outer circumstances like time and place, but also inner circumstances like thoughts (Bhardwaj 2003). This may seem contradictory, if a personality trait is defined as a *lasting* attribute in which people differ (Asendorpf 2004). However, as Fisseni (2003) points out, in the concept of personality it is assumed that different traits, processes, states and actions of a person form a *relatively constant but dynamic* whole. So, one could interpret the trait-proportion of the *tri-guna* concept to be the "relatively constant" and the state-proportion to be the "dynamic" part of the whole. The relatively constant combination of *gunas* accounts for the different personality types described in Ayurveda. The state-proportions form the "dynamic" part of the whole, meaning that a particular combination of the *gunas* will be relatively constant in an individual within which there is continuous change. The *Mahabharata* (Deussen and Strauss 1906), for example, describes various qualities of the respective *gunas* to *emerge occasionally*. The focus of this paper lies on the changeable aspects of the *gunas* in a person. This calls for a detailed description of the functioning of the *gunas* in different aspects of the psyche, in which changes can be observed (see next paragraph). The dynamic change of the *gunas* is well described in Goswami and Adhikari (1988c, p. 10): "A good person is sometimes torn by passion, and a passionate person sometimes wants to give up everything and rest. An ignorant person may sometimes become disgusted with his depraved life, and a passionate person may sometimes indulge in bad habits in the mode of ignorance."

Interestingly, Kumar (2007) compares the *guna* concept to a compass leading the way to wholesome living. No matter "where" one is, the "*guna* compass" sets the example of a *sattvic* lifestyle and can thereby guide a person through crises or despair into the direction of well-being and harmony. In the same line Theodor (2010) portrays the *gunas* as three different "universal paths" on which people choose to travel. It is this understanding of the *gunas* being relevant to different life-styles, which is crucial to health psychology. Of course the *tri-guna* concept can be used to assess personality types and make predictions of behavior according to the type, but Kumar (2007) and Theodor (2010) argue that the *tri-guna* concept provides for much more. It describes life-styles, motives and behavior people choose and the results they attain by following them. Most importantly, it can help to improve well-being by depicting a life-style which leads to well-being, the *sattvic* life. The path of *sattva* may seem somewhat boring at the beginning, as Theodor describes, but will eventually lead to well-being. This reminds one/us of endeavors for a healthy life-style which are not attractive in the short-term, but will gradually lead to increased health and well-being in the long-term. The path of *rajas* first appears attractive but in the long-term it only leads to exhaustion and distress. And the path of *tamas* is a path of carelessness which ultimately results in self-destruction. Clearly, life-styles governed by the influences of *rajas* and *tamas* refer to unhealthy living which are at first comfortable or exciting but eventually lead to a ruined health and decreased well-being. Also Murthy and Kumar (2007) propose that the core traits of each *guna* are related to the results of *sattvic, rajasic* or *tamasic* actions – as the experiences of purity (well-being), suffering or ignorance respectively. This understanding puts more emphasis on choices and life-styles than on traits.

In summary, relevant keys to the construct definition of the *tri-guna* concept are the axioms that: (a) matter (including the psyche) constantly changes and this process of transformation has three constituents, (b) these three constituents act together and depend on each other, (c) the constituents struggle for dominance and when one (or two) become(s) dominant the influence of the other(s) decreases, and (d) there is no state in which one of them is completely absent.

The scriptural sources of Indian psychology give detailed descriptions of the influence of the dominance of a particular *guna* on the psyche, and also depict how *sattva* can be strengthened, so that it becomes the dominating *guna*. Thus, these descriptions give the necessary details to diagnose the manifestation of the *gunas* and to intervene so that well-being can be increased. In the following paragraph these symptoms of the manifestation of the different *gunas* are addressed in detail, in order to provide a working model for the empirical investigation of the *guna-concept*. It is of crucial importance for us to make the model we will work with transparent, so that it can become open to scrutiny. It is an attempt to provide a comprehensive description of the *guna*-concept that integrates many of the partial descriptions which are the basis of work in this field at present. Some of the components of the model have been tested empirically and these results will also be discussed in the following.

## The Influence of the Gunas on Cognition, Emotion, Motivation and Other Variables

Although the *gunas* manifest in many aspects of human behavior and experience, their long-term effects (especially the effect of the dominant *guna*) are often summed up by one main term for each *guna* according to different authors. The different long-term manifestations of *sattva* are summarized by the terms pleasure, satisfaction and joy; the variety of *rajas* is summarized by the terms suffering, frustration and pain and the manifestations of *tamas* are summarized as bewilderment, indifference and oppression. Although dominant *rajas* and *tamas* can cause short-term positive effects, these one-word-summaries of the *gunas* only capture the negative effects because the focus lies on the long-term effects. *Rajas* and *tamas* also have necessary and beneficial effects (e.g. facilitating action (*rajas*) and rest (*tamas*)) in combination with dominant *sattva* (as discussed above, the *gunas* work together). The detrimental effects of excessive *rajas* and *tamas* are nullified when *sattva* is strengthened. In the traditional literature there are only a few hints of the positive functions of *rajas* and *tamas* when they are subdominant.

Again, it is evident that the effects of dominant *sattva* are most favorable for well-being and the question arises, how *sattva* can be increased to augment well-being. In order to answer this question it is necessary to know the detailed manifestations of the *gunas* in the psyche and other relevant variables. Such a description and classification can help in two ways: (a) to develop a comprehensive scale to diagnose the predominance of the *gunas* and (b) to set goals for interventions that further *sattvic* qualities and behavior. What follows in the next section of this chapter are descriptions of the indicators for the respective *gunas*.

Westhoff et al. (2007) *behavioral equation* was used as a foundation to structure the many aspects of human experience and behavior in which the *gunas* are manifested. Originally, the behavioral equation was developed in order to structure job demands that are important for assessing occupational aptitude. The equation states that human behavior is a function of a variety of factors and their interactions (*I*), and summarizes these factors into six groups. It provides a useful tool for systematizing the factors that are important for the prediction, explanation and modification of individual behavior also outside of the occupational context. The equation is: *behavior = fI (cognitive, emotional, motivational and social variables, variables of the organism and the environment)*. Westhoff (2005) comprehensively describes the constructs belonging to these groups of factors which have proven to be of value in the context of assessing occupational aptitude. This description was used as a foundation for structuring the different factors relevant to the *tri-guna* concept, however due to the different context, some factors were added (e.g. "way of dealing with success and/or failure"), while for a few factors relevant in the occupational context no scriptural descriptions for the manifestation of the *gunas* were found (e.g. "way of dealing with emotional challenges" or "artistic talent").

The indicators of the respective *guna* were taken from traditional Indian texts, as they should present the concept closest to the original formulation. Because *Sankhya* is one of the oldest in classical Indian philosophical systems, some of its basic ideas can also be found in other philosophies. This applies especially to the *guna*-concept. To increase the search domain, we also searched through texts that are not purely *Sankhyan* but that have picked up on the *guna*-concept and integrated it into their philosophy. Information given in purports to the texts was also added to the compilation. Inconsistent information that was given in the different texts was treated equally and added to the compilation. (The working-model character of our compilation once again becomes evident here.) Scriptures taken into consideration are the *Maitri-Upanishad* (Hume 1962, as quoted by Suneetha and Srikrishna 2009), *Maitrayana-Upanishad* (Michel 2006), the *Caraka Samhita* (Sharma 1994) and other texts on *Ayurveda* (Buhrman 1997, 1998, 2005; Frawley 1997; Frawley and Summerfield Kozak 2001; Ranade 1994; Verma 1997), the *Yoga Sutras* (Bryant 2009), the *Bhagavad-gita* (Bhanu Swami 2003; Maheshwar 1978; Prabhupada 1989; Theodor 2010), which is part of the *Mahabharata,* the *Mahabharata* itself (Deussen and Strauss 1906; Ganguli 2005), the *Sankhya Karika* (Colebrooke and Wilson 1837; Ganganatha 1896) and other scriptures on *Sankhya* (summarized by Larson and Bhattacharya 1987), and the *Bhagavata Purana* (Goswami and Adhikari 1988a, b, c; Prabhupada 1987a, b, c, d, e; Tagare 1976a, b, 1978). Some details from two modern descriptions of *guna* indicators (Gupta 1977, as quoted by Mohan and Sandhu 1986; Kumar 2007) were also added. However, information from these two recent attempts, not only to explain the *guna* concept but also to find indicators of the *gunas* in contemporary society inconsistent with or totally different from the traditional interpretations, was not added.

When reading the indicators listed in the following tables, it might seem that each scripture proposes different indicators for the respective *guna* for a specific variable. However, this is not the case; there is a great overlap in the information given by the different texts. In order to avoid redundancy in the tables, these common indicators were not mentioned more than once. Furthermore, the profiles might look like extreme descriptions of people that rarely exist. And in fact – they are rare, as in any case for the extremes of any quality. The difference of the influence of the three *gunas* will not be as grave in most people, as portrayed below. This characterization of the extremes is only for the purpose of singling out the specific indicators of the respective *gunas*. Moreover, it might be surprising that the *Mahabharata* (Deussen and Strauss 1906) repeatedly describes symptoms of *sattva* as being "aimless", as for instance "aimless cognition" or "aimless cult", or that it describes "inactivity" to be a quality of *sattva*. It does not fit into the other details of the concept to assume a person with dominant *sattva* would act aimlessly. However, this translation can be understood to indicate goal-oriented action without a desire for the fruits of the activity. Such activity is considered "aimless", "disinterested action" or "inactivity" in the Vedic worldview as it does not create karma (reactions to actions) (see Theodor 2010, p. 51 for further details).

## Cognitive Variables

Before considering the manifestations of the *gunas* in different cognitive variables (Table 1), it is important to note that what is understood as "knowledge" in *Sankhya* is not the same as the western understanding of knowledge. In *Sankhya, vidya* (knowledge) mainly refers to (direct or intuitive) knowledge of reality, that is, of the difference between matter and spirit – recall that body and mind are seen as matter and the soul as spirit. Thereafter, a person with this kind of knowledge discriminates between these two and a person devoid of this knowledge does not, or rather mistakes the non-self (body & mind) as the self, ignoring the real nature of the self as the spirit soul (Palsane et al. 2002).

What follows in the next three paragraphs is a summary of the cognitive characteristics according to the respective *guna* and relevant empirical findings connected to these theoretical assumptions.

When *sattva* is the dominating *guna* the organs involved in perception are expected to be quiescent and receptive, allowing for accurate perception. People with dominant *sattva* are intrinsically motivated to learn and have a good memory. They have a concept of the difference between the temporary body and the eternal spirit soul. They see the eternal spirit in others and live with a feeling of relatedness to others on the spiritual level. When *sattva* dominates, thinking is rational and calm, and the individual appraises his or her perception realistically. Furthermore, the individual has a high self-efficacy and broods and worries seldom. People with dominant *sattva* have a high stress tolerance, as they think positively and are able to perceive surroundings, own emotions, thoughts and behavior without expecting them to be different, or wanting to change or control them. With predominating *sattva* a person gives attention to his or her inner world of spirituality, emotions and thoughts. Furthermore, a person with dominating *sattva* is more likely to be above average in intelligence. Concerning "creativity" Kumar (2007) proposes that it is energized by *sattva* while Bryant (2009) attributes it to *rajas*. A person with prevailing *sattva* has an efficient working style. He or she considers the consequences of his or her actions and keeps up the enthusiasm until a task is completed.

**Table 1** Symptoms of the *Gunas* in cognitive variables

| | Sattva | Rajas | Tamas |
|---|---|---|---|
| Perception | Organs capable of perception (Larson and Bhattacharya 1987) | "Inability of the perceiving senses to disentangle themselves from mundane objects" (Goswami and Adhikari 1988c, p. 570) | Inert organs, incapable of perception (Larson and Bhattacharya 1987) |
| | Quiescent senses (Tagare 1978) | | Stupor (Tagare 1978) |
| | Seeing, hearing, tasting correctly (Prabhupada 1989) | | |
| | Receptive mind (Frawley and Summerfield Kozak 2001) | | |
| Learning | Elucidation, studying gratuitously (Deussen and Strauss 1906) | | |
| Memory | Remembrance (Deussen and Strauss 1906) | Fluctuating memory (Ranade 1994) | Forgetfulness (Deussen and Strauss 1906) |
| | Good memory (Ranade 1994) | | Bad memory (Ranade 1994) |
| | | | Loss of memory (Ganguli 2005) |
| Knowledge | Spiritual knowledge, knowledge of the atman [the soul] (Tagare 1978) | Inconsistent knowledge (Maheshwar 1978) | Ignorance (Larson and Bhattacharya 1987) |
| | Truth, knowledge itself (Larson and Bhattacharya 1987) | "Scientific knowledge of the material body", "knowledge based on duality" (Goswami and Adhikari 1988c, p. 575) | Knowledge concerned with the comfort of the body (Prabhupada 1989) |
| | More knowledge and wisdom than others, seeing the spirit soul present in all entities (Prabhupada 1989) | Considering the material body to be the self (Prabhupada 1989) | Seeing one activity as everything, minute and meager, not aimed at the truth, not based on a well-founded cause (Theodor 2010) |
| | Knowledge, seeing one reality in all beings (Theodor 2010) | Seeing a variegated reality (Theodor 2010) | |

(continued)

**Table 1** (continued)

| | Sattva | Rajas | Tamas |
|---|---|---|---|
| Thinking | Intellect is luminous (Colebrooke and Wilson 1837) | Worries, shamelessness, doubts, thinking of past, present and future possibilities (Deussen and Strauss 1906) | Dullness of mind (Colebrooke and Wilson 1837) |
| | Freedom from worries and doubts, calm mental behavior, awareness of one's autonomy, purity, aimless cognition, non-delusion, enlightening insight, wisdom (Deussen and Strauss 1906) | Inability to see things the way they are (Prabhupada 1987a) | Not discriminating, lack of judgment, rigidity, in confusion, mental darkness, immaturity, seeing truth in untruth, criticizing (Deussen and Strauss 1906) |
| | Rational (Gupta 1977, as quoted by Mohan and Sandhu 1986) | Unsteady perplexity (Goswami and Adhikari 1988c) | Misunderstanding or non-understanding, obstinate in stupidity, not looking for cause and effect, absorption in one routine, "lazy or dully obstinate way of looking at things which has no eye for the real nature of the world" (Maheshwar 1978, p. 228) |
| | Discrimination (Tagare 1978) | Uncontrollable, racing thoughts (Buhrman 1997) | "Faculty of understanding fails" (Tagare 1978, p. 2079) |
| | Purity, clearness (Larson and Bhattacharya 1987) | Strong opinions, inconsistent in problem solving, blaming others for problems (Frawley 1997) | Insanity, dream, veiling, covering, delusion (Larson and Bhattacharya 1987) |
| | Seeing things the way they are (Prabhupada 1987a) | Hectic, nervousness (Verma 1997) | Unable to tell right from wrong, "distortion of the intelligence because of too much activity" (Goswami and Adhikari 1988c, p. 570) |
| | "Careful study of the past and future" (Goswami and Adhikari 1988c, p. 560 f.) | Reliving painful events in one's mind, or playing out solutions to them (Buhrman 1998) | Inability to understand the things as they are, foolishness, illusion, madness (Prabhupada 1989) |

| | | | |
|---|---|---|---|
| | Ability to understand reality (Prabhupada 1989) | Regret (Ganguli 2005) | Negativity (Buhrman 1997) |
| | | | Ignorance about own problems, not dealing with own problems, "allow negative influences to dominate them" (Frawley 1997, p. 35) |
| | Clarity, seeing the good in everything, life is seen as a learning experience (Frawley 1997) | Looking for improvement of things (Kumar 2007) | Inability to comprehend, partial and erroneous comprehension, ignorance, illiberality, censuring good acts, unripe judgment , want of discrimination, stupid reasoning (Ganguli 2005) |
| | Liberality (Ganguli 2005) | | Stubbornness (Theodor 2010) |
| | "Maintaining a balance andharmony of opposites […] recognition and understanding of things as they are, without judgment or desire to alter" (Kumar 2007, p. 22), finding miracles in the ordinary, "[…] trusting the process of the universe and believing that things will work out." (Kumar 2007, p. 23) | | |
| | Lucidity (Bryant 2009) | | |
| Concentration | | Scattered, fickle minded (Larson and Bhattacharya 1987) | Inattention (Maheshwar 1978) |
| | | Distraction, inability to concentrate (Buhrman 1997) | Inability to concentrate (Goswami and Adhikari 1988c) |
| | | Absentmindedness (Michel 2006) | |

(continued)

**Table 1** (continued)

| | Sattva | Rajas | Tamas |
|---|---|---|---|
| Focus of attention | Internal focus (Frawley and Summerfield Kozak 2001) | External focus (Frawley and Summerfield Kozak 2001) | Mind fixed on sense objects without higher awareness (Goswami and Adhikari 1988c) |
| | Living in the moment, the present (Kumar 2007) | Worldly affairs (Ganguli 2005) | Mind locked into physical existence, overly preoccupied with own sensory pleasure and pain (Buhrman 1998) |
| | | Focus on the future (Kumar 2007) | External focus, "blindness to the internal world of consciousness" (Frawley and Summerfield Kozak 2001, p. 15) |
| | | | Focus on the past (Kumar 2007) |
| Intelligence | Intelligence itself (Larson and Bhattacharya 1987) | Averagely intelligent (Ranade 1994) | Unintellectual (Gupta 1977, as quoted by Mohan and Sandhu 1986) |
| | Very intelligent (Ranade 1994) | | Unintelligent (Ranade 1994) |
| Creativity | Creativity (Kumar 2007) | Creative activity (Bryant 2009) | |
| Style of working/acting | Efficiency, endurance (Deussen and Strauss 1906) | Unjust, over endeavor for goals, laxness, endurance, heroism, boldness, fight, passion (Deussen and Strauss 1906) | Unregulated actions, lack of persistence (Deussen and Strauss 1906) |
| | Characterized by serenity (Tagare 1976a) | Characterized by vehemence (Tagare 1976a) | Characterized by dullness (Tagare 1976a) |
| | Full of foresight (Gupta 1977, as quoted by Mohan and Sandhu 1986) | Blinded by personal desires, audacity, experiences struggle (Goswami and Adhikari 1988c) | Lacking foresight (Gupta 1977, as quoted by Mohan and Sandhu 1986) |
| | Free of attachment (Goswami and Adhikari 1988c) | Harmful, passionate (Theodor 2010) | Negligence of error (Maheshwar 1978) |
| | Regulated, without attachment, love, hatred or desire for the fruitive results, determination, enthusiasm till the work is completed (Prabhupada 1989) | | Unregulated, without principle (Prabhupada 1989) |
| | Free from false ego and attachment (Bhanu Swami 2003) | | "The end justifies the means" (Kumar 2007, p. 21) |
| | Concerned about purity of means, work done in small steps, search for long-lasting, simple benefits, non-violent means (Kumar 2007) | | Procrastination, undisciplined (Theodor 2010) |

Some of these descriptions have been confirmed empirically. Sitamma (2005) reports that studies have found that *sattva* correlates positively with intelligence, short-term memory and concentration, and negatively with neuroticism. In 2003 J. A. Hopkins investigated correlations between the five factors of the NEO-PI-R and the *gunas* measured with the Vedic Personality Inventory (Wolf 1998). The manuscript has not been published yet but since the results are of high relevance we will consider them. These results confirm the negative correlation between *sattva* and neuroticism ($r=-.41$, $p<.001$). Furthermore, *sattva* correlated with openness to experience ($r=.23$, $p<.01$). Kaur and Sinha (1992) found correlations between *sattva* and personal effectiveness ($r=.24$, $p<.01$) and self-actualizing behavior ($r=.31$, $p<.01$). Daftuar and Anjuli (1997) demonstrated that *sattva* does not correlate with occupational stress. Furthermore, *sattva* correlated negatively with the subscales attention problems ($r=-.23$, $p>.05$) and thought problems ($r=-.2$, $p>.05$) of the Adult Self Report (Archana Das and Venu Gopal 2009). Other consistent correlations (Hopkins) are correlations with subscales of the NEO-PI-R which indicate that *sattvic* people appreciate art and beauty ($r=.43$, $p<.001$), are intellectually curious ($r=.42$, $p<.001$), believe in their efficacy ($r=.52$, $p<.001$), are personally organized ($r=.41$, $p<.001$), think before speaking or acting ($r=.31$, $p<.01$) and are generally not very vulnerable to stress ($r=-.74$, $p<.001$). Contrary to what might be expected for *sattva* Hopkins found neither a significant correlation between *sattva* and the subscale measuring openness to emotions and feelings, nor between *sattva* and the subscale measuring openness to rethinking one's values.

The organs involved in perception of a person with predominating *rajas* are strained to the full limit and over agitated, making it impossible to perceive subtleties. The memory is mediocre, or fluctuates. A person with prevailing *rajas* has no concept of a spiritual commonality of all beings, and rather only conceives of differences and commonalities on the basis of material characteristics. When *rajas* dominates, thoughts are uncontrollable and rotate around worries. Immediately associated to perception are thoughts of how to change reality, which leads to a low stress-tolerance. Persons with dominant *rajas* have difficulty concentrating, are averagely intelligent and give little conscious attention to spirituality, emotions and thoughts.

According to studies Sitamma (2005) refers to, *rajas* correlates negatively with self-actualization. An inconsistent result is that *rajas* did not correlate with occupational stress in one study (Daftuar and Anjuli 1997). Hopkins found a negative correlation between *rajas* and openness to experience ($r=-.28$, $p<.001$) and Archana Das and Venu Gopal (2009) report a correlation between *rajas* and attention problems ($r=.2$, $p>.05$).

If *tamas* prevails perception is said to become inaccurate, as will be the case after the consumption of alcohol or other drugs. The memory of people with dominant *tamas* is defective. They are not interested in the truth and rarely question or correct their concepts. When *tamas* is predominant an individual has difficulties in understanding circumstances and facts. Thoughts focus on negative consequences, and the individual ignores his or her problems instead of trying to solve them. He or she has difficulties concentrating, rarely gives attention to spirituality, emotions and thoughts and is most likely to be of under-average intelligence.

Sitamma (2005) reports studies that found negative correlations between *tamas* and short-term memory, intelligence, concentration and positive correlations between *tamas* and psychoticism, neuroticism and self-actualization. Also Stempel et al. (2006) found a correlation between *tamas* and psychoticism ($r=.3$, $p<.05$). Kaur and Sinha (1992) found negative correlations between *tamas* and personal effectiveness ($r=-.12$, $p<.05$) and self-actualizing behavior ($r=-.21$, $p<.01$). In Daftuar and Anjuli's study (1997) dominant *tamas* correlated very highly with occupational stress ($r=.97$, $p<.01$). Furthermore, *tamas* seems to correlate positively with the subscales thought problems ($r=.26$, $p<.01$; Archana Das and Venu Gopal 2009) and attention problems ($r=.64$, $p<.01$; Archana Das and Venu Gopal 2009) of the Adult Self Report and negatively with openness to experience ($r=-.19$, $p<.05$; Hopkins). Contrary to the theoretical assumptions, Hopkins found no significant negative correlation between *tamas* and the subscale of the NEO-PI-R measuring openness to rethinking one's values.

Correlations between different relevant NEO-PI-R subscales and *rajas* and *tamas* respectively were very similar in Hopkins' study. Probably this is because the factors *rajas* and *tamas* correlate positively in the Vedic Personality Inventory (Wolf 1998) which was used in this study. Both *rajas* and *tamas* correlated negatively with the tendency to appreciate art and beauty (*rajas:* $r=-.3$, $p<.01$; *tamas:* $r=-.4$, $p<.001$), to be intellectually curious (*rajas:* $r=-.31$, $p<.01$; *tamas:* $r=-.37$, $p<.001$), to believe in one's efficacy (*rajas:* $r=-.35$, $p<.01$; *tamas:* $r=-.5$, $p<.001$), to be personally organized (*rajas:* $r=-.32$, $p<.01$; *tamas:* $r=-.36$, $p<.01$) and to think before speaking or acting (*rajas:* $r=-.25$, $p<.05$; *tamas:* $r=-.27$, $p<.05$). Both *rajas* and *tamas* correlated positively with a vulnerability to stress (*rajas:* $r=.52$, $p<.001$; *tamas:* $r=.71$, $p<.001$).

## Emotional Variables

The following table (Table 2) lists the symptoms of dominating *sattva, rajas* and *tamas* in emotional variables. Although it also has cognitive components, the variable "satisfaction and happiness" is discussed together with the other emotional variables. For the emotional variables there is a considerable overlap between the symptoms of *rajas* and *tamas*, for example anger, fear, despondence, grief and capriciousness are symptoms of *rajas* and *tamas*. However, whereas *rajas* may lead to some kind of pleasure or love, *tamas* is not said to lead to any positive emotions. Furthermore, the negative emotions are generally described in more severe terms when listed as symptoms of *tamas*. Hopkins' study shows some empirical evidence for this: positive correlations between the neuroticism subscales and *rajas* and *tamas* respectively were strongest for *tamas*.

There is also a slight overlap between *sattva* and *rajas* for emotions such as love and pleasure. However, the love experienced by persons influenced by *sattva* is more global, as it rests on a vision of seeing one reality in all beings (Theodor 2010). *Rajasic* love is based on distinguishing between beings on the basis of material

**Table 2** Symptoms of the *Gunas* in emotional variables

| | Sattva | Rajas | Tamas |
|---|---|---|---|
| Frame of mind, emotional state | Love, pleasure, delight, cheerfulness, joyfulness, satisfaction, delight, sense of well-being, peace of mind, calmness of the heart, courage (Deussen and Strauss 1906) | Anguish, grief, anger, fear, despondence, impatience, pain, hatred, hankering, dissatisfaction, disgust, remorse, love, pleasure (Deussen and Strauss 1906) | Anger, fear, despondence, grief, timidity, sullenness, cowardice, bewilderment (Deussen and Strauss 1906) |
| | Agreeable, renunciation, contentment, purity, compassion, lightness, pleasantness, affection, tranquility (Larson and Bhattacharya 1987) | Disagreeable, frustration, anxiety, misery (Larson and Bhattacharya 1987) | Oppressive, depression, misery, sorrowfulness (Larson and Bhattacharya 1987) |
| | Fearlessness, detachment from the mind (Goswami and Adhikari 1988c) | Agitation (Prabhupada 1987b) | Violent hatred, delusion, unhappiness, depression (Goswami and Adhikari 1988c) |
| | Happiness (Prabhupada 1989) | Sorrow (Ganguli 2005) | Madness, morose (Prabhupada 1989) |
| | Love uniting all things (Frawley 1997) | Restlessness, passion, attachment (Bryant 2009) | Hopelessness (Buhrman 2005) |
| | Gladness (Ganguli 2005) | Affected by sorrow and joy (Theodor 2010) | Gloom, cheerlessness (Ganguli 2005) |
| | Detachment (Bryant 2009) | | Despair (Michel 2006) |
| | | | Dejected (Theodor 2010) |
| Emotional stability | Tranquility, balance, gravity, even-tempered, firmness, collectedness, balance, composure (Deussen and Strauss 1906) | Capriciousness (Colebrooke and Wilson 1837) | Proneness to lose one's temper (Goswami and Adhikari 1988a) |
| | Forbearance (Tagare 1978) | Impetuousness (Deussen and Strauss 1906) | Capriciousness (Michel 2006) |
| | Peace of mind (Frawley 1997) | | |
| | Tolerance, equanimity (Goswami and Adhikari 1988a) | Proneness to lose one's temper (Goswami and Adhikari 1988a) | |
| | | "Rash eagerness to fight" (Goswami and Adhikari 1988c, p. 561) | |
| | | Agitated mind (Frawley 1997) | |
| | | Enjoyment of happiness, endurance of sorrow (Ganguli 2005) | |
| | | Fickleness (Michel 2006) | |
| | | Restlessness, passionate (Theodor 2010) | |

(continued)

**Table 2** (continued)

| | Sattva | Rajas | Tamas |
|---|---|---|---|
| Way of dealing with feelings | Fitting expression of emotions (Ranade 1994) | Exaggerated expression of emotions (Ranade 1994) | |
| Satisfaction & happiness | Not-lamenting, satisfaction, sense of well-being (Deussen and Strauss 1906) | Dissatisfaction (Deussen and Strauss 1906) | Lamentation (Goswami and Adhikari 1988c) |
| | "Satisfaction in the bliss of the Self" (Tagare 1978, p. 2077) | Insatiability (Hume 1962, as quoted by Suneetha and Srikrishna 2009) | Discontent (Ganguli 2005) |
| | "Satisfied in any condition of life" (Prabhupada 1987d, p. 137) | Joy which turns into disappointment, fatigue, satiety, disgust, suffering, sin (Maheshwar 1978) | Seeking pleasure by giving pain to others (Kumar 2007) |
| | Lasting contentment & happiness (Frawley 1997) | Experiencing struggle (Goswami and Adhikari 1988c) | Happiness based on illusion, the pleasure derived from indolence, negligence and sleep (Theodor 2010) |
| | Happiness based on long-term results, springing from the tranquility of the soul and heart (Theodor 2010) | Seeking for happiness in external things (Frawley and Summerfield Kozak 2001) | |
| | | Happiness based on immediate satisfaction, which later turns into suffering, springing from sense-gratification (Theodor 2010) | |

characteristics (Theodor 2010). This leads to liking and disliking. Also, the positive *rajasic* emotions (e.g. love, affection) are directed to worldly objects (Ganguli 2005). In general the overlaps between *sattva* and *rajas,* and *rajas* and *tamas* support the assumption that in some ways *rajas* is the intermediate *guna* between *sattva* and *tamas.*

To sum up, when *sattva* dominates, a person experiences a variety of positive emotions like happiness, joy and love. A person with dominant *sattva* is said to be emotionally stable and to express emotions adequately. *Sattvic* persons are satisfied with their life, and rarely lament. *Sattvic* happiness is steady and independent from external circumstances, as it comes from the tranquility of the psyche which allows the energy of the soul to unfold itself.

There is empirical evidence for the connection between *sattva* and positive emotions: *sattva* was found to correlate positively with the extraversion subscale "positive emotions" of the NEO-PI-R ($r = .56$, $p < .001$; Hopkins). *Sattva* also correlated negatively with the subscales "anxiety" ($r = -.6$, $p < .001$), "angry hostility" ($r = -.69$, $p < .001$) and "depression" ($r = -.75$, $p < .001$). Stempel et al. (2006) found that *sattva* correlated negatively with the subscale measuring phobic anxiety ($r = -.44$, $p < .01$). Inconsistently with the theoretical assumptions, *sattva* did not correlate with job satisfaction (Sharma 1999). Sharma argues this could be because of the emphasis on material benefits in the working world, instead of facilitating self-actualization, which should be a priority for persons with dominant *sattva.* Khema et al. (2011) found positive correlations between *sattva* and emotional intelligence (pre: $r = .25$, $p < .01$, post: $r = .34$, $p < .001$).

When *rajas* dominates, the individual experiences mainly negative emotions and seldom some joy and pleasure. Characteristic for *rajas* are frustration, dissatisfaction and restlessness stemming from unfulfilled desires. Also, the person is emotionally unstable and expresses his or her emotions in an exaggerated manner. Persons with dominant *rajas* enjoy pleasure which disappoints them in the long term. Consistent with these descriptions is that Sharma (1999) found that *rajas* correlates negatively with job satisfaction ($r = -.29$, $p < .05$) and Hopkins reports a positive correlation with neuroticism ($r = .58$, $p < .001$). *Rajas* correlated with anxiety ($r = .29$, $p < .05$) in Stempel et al.'s study (2006) and the Adult Self Report subscale anxious/depressed ($r = .2$, $p > .05$) in Archana Das and Venu Gopal's (2009) study.

When *tamas* dominates, a person is described to experience negative emotions like despair, depression and timidity. Like for *rajas* a person with dominating *tamas* is said to be emotionally instable. *Tamasic* persons are happiest when they do not have to do anything constructive, and mostly they are unhappy and lament a lot. Inconsistent with this is that *tamas* did not correlate negatively with job satisfaction in Sharma's study (1999). Like *rajas,* also *tamas* correlated positively with neuroticism ($r = .72$, $p < .001$) in Hopkins' study. Khema et al. (2011) report negative correlations between *tamas* and emotional intelligence (pre: $r = -.28$, $p < .01$, post: $r = -.24$, $p < .05$). Additionally, *tamas* correlated positively with the subscales measuring depression ($r = .31$, $p < .05$), anxiety ($r = .43$, $p < .01$) and phobic anxiety ($r = .33$, $p < .05$) of the Brief Symptom Inventory (Stempel et al. 2006). Archana Das and Venu Gopal (2009) found correlations between *tamas* and the Adult Self Report subscale measuring anxiousness/depression ($r = .49$, $p < .01$).

For both *rajas* and *tamas* Hopkins found correlations with NEO-PI-R subscales relevant to emotional variables. *Rajas* and *tamas* correlated positively with the subscales "anxiety" (*rajas:* $r=.32$, $p<.01$; *tamas:* $r=.65$, $p<.001$), "angry hostility" (*rajas:* $r=.55$, $p<.001$; *tamas:* $r=.59$, $p<.001$) and "depression" (*rajas:* $r=.44$, $p<.001$; *tamas:* $r=.77$, $p<.001$) and negatively with the subscale "positive emotions" (*rajas:* $r=-.33$, $p<.01$; *tamas:* $r=-.59$, $p<.001$).

## Motivational Variables

The next table (Table 3) lists the influence of the *gunas* on motivational variables. The variable "spirituality" is discussed together with the other motivational variables, although it also has other components. In short, persons with dominant *sattva* are motivated by duty, persons with dominant *rajas* are motivated by desires for personal gain and persons with dominant *tamas* are characterized by a lack of motivation.

In summary, people with dominant *sattva* are said to have a strong motive to fulfill their duties in society. They do not harbor ulterior motives when helping others. Also, they can easily coordinate sub-functions of the psyche (emotion, cognition etc.) to attain their goals, in other words they are disciplined and determined. Another feature of a *sattvic* person is that he or she has few (or no) material desires or goals, and lives according to moral principles. *Sattvic* people are spiritually inclined and if they worship God they do so only for spiritual reasons and not in order to fulfill material desires. A person with prevailing *sattva* tends to make wise choices, harbors few or no doubts, and is able to stay enthusiastic independently from success or failure. He or she strives for a balance between activity and regeneration.

Correlations (Hopkins) between *sattva* and the NEO-PI-R subscale "self-discipline" ($r=.6$, $p<.001$) and "impulsiveness" ($r=-.58$, $p<.001$) confirm the assumption between *sattva* and self-discipline. The balanced use of energy of a person with dominant *sattva* may enable them to be more active than *rajasic* and *tamasic* persons: Hopkins' study shows that *sattva* correlated positively ($r=.28$, $p<.05$) with the "activity" extraversion subscale of the NEO-PI-R, whereas the correlation was negative for *tamas* ($r=-.29$, $p<.05$) and not significant for *rajas*. Also, *sattva* correlated with the scale conscientiousness ($r=.53$, $p<.001$). Daftuar and Anjuli (1997) found a correlation between *sattva* and organizational commitment ($r=.53$, $p<.05$). Another consistent result is that persons who regularly recite the *Bhagavad-gita*, which focuses on spiritual topics, have significantly higher scores on *sattva* than persons who do not recite the *Gita* (Anjana and Raju 2001). This is further confirmed by a correlation between *sattva* and Underwood and Teresi's (2002) Daily Spiritual Experience Scale ($r=.41$, $p<.01$; Stempel et al. 2006). Also Bhal and Debnath (2006) report a positive relationship between *sattva* and spirituality ($r=.34$, $p<.01$) measured with the scale developed by MacDonald (2000); however, *sattva* did not correlate significantly with all subscales. Additionally, Bhal and Debnath (2006) found a positive relationship between *sattva* and an idealistic ethical orientation ($r=.26$, $p<.01$).

**Table 3** Symptoms of the *Gunas* in motivational variables

| | Sattva | Rajas | Tamas |
|---|---|---|---|
| Motives | Purity, no ill motives (Deussen and Strauss 1906) | Ill motives, satisfying lust (Deussen and Strauss 1906) | Ill motives (Deussen and Strauss 1906) |
| | To be engaged for the sake of God, not expecting material benefits (Prabhupada 1989) | Hard work for fortune and prestige (Goswami and Adhikari 1988c) | Foolishness: things which are impossible to achieve (Prabhupada 1989) |
| | Tranquility (Ganguli 2005) | To fulfill desires of oneself, family and society, acting out of pride or to get respect (Prabhupada 1989) | "Disinclination toward constructive activity" (Bryant 2009, p. xlix) |
| | Dharma, duty, performing actions for their own sake, expecting no fruit in return, above sensual repulsion or attraction, determined to sustain health (Theodor 2010) | "Seek power over others" (Frawley 1997, p. 35) | Obscure notions, aim to harm others, self-torture, determined not to let go of sorrow, intoxication, sleep, fear and dejection (Theodor 2010) |
| | | Social achievements, completion of tasks, fame, action (Buhrman 2005) | |
| | | Earthly motives (Ganguli 2005) | |
| | | Concentrated on achievement, outcome, success, "concerned with achieving the ends" (Kumar 2007, p. 21) | |
| | | To gain fruits of action like success or honor; hypocrisy, ostentation, satisfaction of one's desires, determined to make money and enjoy (Theodor 2010) | |
| Volition | Discipline, practicing asceticism gratuitously (Deussen and Strauss 1906) | Non-renunciation, addiction to lust and pain, impatience (Deussen and Strauss 1906) | Lack of self-control, impatience (Deussen and Strauss 1906) |
| | Mind & sense control (Tagare 1978)) | Unsteadfastness (Hume 1962, as quoted by Suneetha and Srikrishna 2009) | Choosing the path of least resistance (Maheshwar 1978) |
| | Self-control (Goswami and Adhikari 1988c) | Uncontrollable hankering, determination to strive for personal gain (Prabhupada 1989) | "Determination which cannot go beyond dreaming, fearfulness, lamentation, moroseness and illusion" (Prabhupada 1989, p. 822) |
| | Unbreakable determination which controls mind and senses (Prabhupada 1989) | Restlessness, hankering, determination for enjoyment (Theodor 2010) | Lacking will-force (Buhrman 2005) |
| | Determination to avoid harmful but momentarily pleasing things (Ranade 1994) | | Laziness, undisciplined (Theodor 2010) |
| | Bodily "gates" are controlled, determination to maintain one's psycho-physical functions, self-restraint, austerity, determination (Theodor 2010) | | |

(continued)

**Table 3** (continued)

| | Sattva | Rajas | Tamas |
|---|---|---|---|
| Desires | Equanimity, non-lustfulness, free of desires, strives for tranquility, renouncement (Deussen and Strauss 1906) | Desirousness, lustfulness, ambitions, greed, desire to control, desire for "war", lust, the inclination to acquire things (Deussen and Strauss 1906) | Desirousness, lustfulness, cravings, greed (Deussen and Strauss 1906) |
| | Without undue desires (Gupta 1977, as quoted by Mohan and Sandhu 1986) | "Lust after sensual pleasures", insatiable greed (Tagare 1978) | Indulging in false hopes (Goswami and Adhikari 1988c) |
| | Non-covetousness, senses quiescent (Tagare 1978) | Sexual desire (Larson and Bhattacharya 1987) | Cupidity (Ganguli 2005) |
| | No attraction or repulsion, not desiring the fruits of one's actions (Theodor 2010) | Material desire (Goswami and Adhikari 1988c) | |
| | | Hankering for sense gratification, desire for honor, material riches and a family, greed, no desires for a higher elevation (Prabhupada 1989) | |
| | | Cupidity (Ganguli 2005) | |
| | | Greedy for the fruits of one's actions (Theodor 2010) | |
| Interests | | Inclination towards the mundane (Deussen and Strauss 1906) | Disinterest, taking pleasure in eating and similar activities, taking pleasure in good smells, clothes, amenities, lying and sitting, sleeping during the day, gloating over destruction (Deussen and Strauss 1906) |
| | | | Disinterest in the eternal spirit (Prabhupada 1989) |
| Values | Knowing what should be done and what not (Prabhupada 1989) | Ego and desire misrepresent the truth to serve their own purpose (Maheshwar 1978) | Devoid of morality (Gupta 1977, as quoted by Mohan and Sandhu 1986) |
| | Valuing silence, stillness, caring for nature (Kumar 2007) | Not being able to distinguish between irreligion and religion and what should be done and what not (Prabhupada 1989) | Considering wrong to be right (Prabhupada 1989) |
| | | Valuing speed, extravagance, prestige, power, technical solutions, comfort, convenience, paying lip service to justice and fairness, value nature only in regard to its usefulness to humans (Kumar 2007) | |

| | | | |
|---|---|---|---|
| Goals | Accepts status quo, no strong ambition (Ranade 1994) | Fame & wealth (Tagare 1978) | No ambition (Ranade 1994) |
| | Seeking end of suffering, enlightenment for others (Kumar 2007) | Concerned to make the world more and more materially comfortable (Prabhupada 1989) | Disregard toward own ability to accomplish a goal (Theodor 2010) |
| | | Seeking enlightenment for oneself (Kumar 2007) | |
| Beliefs | Faith, aimless cult (Deussen and Strauss 1906) | Difficult to understand or appreciate God (Prabhupada 1987d) | Faithlessness, ignorant faith (Deussen and Strauss 1906) |
| | Possibility of understanding bhakti-yoga (Prabhupada 1987d) | "Seeking heavenly blessings for prosperity" (Tagare 1978, p. 2077) | Religious neglect, atheism (Hume 1962, as quoted by Suneetha and Srikrishna 2009) |
| | Favorable guna for spirituality (Goswami and Adhikari 1988b) | Faith in fruitive work (Goswami and Adhikari 1988c) | Difficult to understand or appreciate God (Prabhupada 1987d) |
| | Faith in spiritual life (Goswami and Adhikari 1988c) | Worship of God to get a material benefit (Prabhupada 1989) | Faith in irreligious actions (Goswami and Adhikari 1988c) |
| | Worship of God out of duty (Prabhupada 1989) | Incredulousness (Ganguli 2005) | Reluctant to advance in spiritual understanding, worship without regard for scripture (Prabhupada 1989) |
| | Religious piety (Theodor 2010) | Wandering from one spiritual practice to another (Kumar 2007) | |
| Spirituality | Senses detached from matter, detachment from material mind (Goswami and Adhikari 1988c) | Manipulative, ambitious, egoistic, passionate (Frawley and Summerfield Kozak 2001) | Harmful, unaware, deceptive, perverted, criminal (Frawley and Summerfield Kozak 2001) |
| | Enlightened, spiritual, loving, compassionate (Frawley and Summerfield Kozak 2001) | | |
| | Emancipation from attachments, indifference, renunciation, freedom from the idea of "I" and "mine" (Ganguli 2005) | | |
| Expectations | Freedom from expectations (Ganguli 2005) | | False expectations (Goswami and Adhikari 1988c) |

(continued)

**Table 3** (continued)

| | Sattva | Rajas | Tamas |
|---|---|---|---|
| Choice behavior | Freedom from doubts (Deussen and Strauss 1906) | Inconsideration, delusion, cockiness, doubts, rashness (Deussen and Strauss 1906) | Inconsideration, delusion, not discriminating, indecisiveness, rashness, thoughtlessness (Deussen and Strauss 1906) |
| | Right decisions (Ranade 1994) | False decisions (Ranade 1994) | False decisions (Ranade 1994) |
| | Certainty (Ganguli 2005) | Hesistancy, doubt (Ganguli 2005) | Not knowing what to do, heedlessness (Ganguli 2005) |
| | Wisdom, discrimination (Bryant 2009) | | Delusion (Bryant 2009) |
| | | | Not considering future consequences or loss (Theodor 2010) |
| Way of dealing with success and/or failure | Maintaining enthusiasm in success and failure (Theodor 2010) | Happy with success, distressed by failure (Prabhupada 1989) | |
| Activity | Inactivity, tirelessness (Deussen and Strauss 1906) | Slackness, aversion, activity itself, strain (Deussen and Strauss 1906) | Lethargy, sleep, slackness, aversion, exhaustion, rigidity, overactivity, sluggishness (Deussen and Strauss 1906) |
| | Striving for balance (Frawley 1997) | Impulse to activity, urge to work (Maheshwar 1978) | Incapacity, negligence of action (Maheshwar 1978) |
| | Enthusiasm, absence of laziness (Ganguli 2005) | Rigidity, perspiration (Larson and Bhattacharya 1987) | Sleep, inactivity, sloth, extreme idleness, sleepiness (Larson and Bhattacharya 1987) |
| | | Too much activity (Goswami and Adhikari 1988c) | Chronic fatigue (Goswami and Adhikari 1988c) |
| | | Hard work, intense endeavor (Prabhupada 1989) | Indolence, sleep, laziness, no endeavor although capacity is there (Prabhupada 1989) |
| | | Hyperactivity (Buhrman 1997) | Lack of motivation (Buhrman 1997) |
| | | Excessive activity, danger of burnout (Frawley 1997) | Lack of will to live (Buhrman 1998) |
| | | Obsessively active (Buhrman 2005) | Attached to inactivity (Buhrman 2005) |
| | | Want of tranquility, fatigue, toil (Ganguli 2005) | Lethargy, stillness, sleep manifest (Bryant 2009) |
| | | Energetic endeavor, power (Bryant 2009) | Indolence (Theodor 2010) |
| | | Great effort, excessive endeavours (Theodor 2010) | |

A person with prevailing *rajas* is said to be mainly motivated to fulfill his or her personal desires and to attain prestige and power. Due to strong hankerings he or she sometimes has difficulties keeping up his or her determination and discipline. *Rajasic* people have strong desires for material pleasure and in its pursue; they have difficulties acting according to moral principles and will neglect them if they compromise their enjoyment. They want to be wealthy and famous. If they dedicate themselves to spirituality or religion they do so in order to attain material benefits from it. They are rash in decision-making and often have doubts. *Rajasic* people are overly active and do not consider a moderate use of their energy.

Hopkins reports a negative correlation between *rajas* and conscientiousness ($r=-.3$, $p<.001$). Anjana and Raju's (2001) comparison shows that non-reciters of the *Bhagavad-gita* have higher scores on *rajas* than *Bhagavad-gita* reciters, indicating that persons with *rajas* are not particularly interested in spirituality. Stempel et al. (2006) report similar findings: *rajas* correlated negatively with daily spiritual experiences ($r=-.41$, $p<.01$). Bhal and Debnath (2006) found no significant correlation between *rajas* and spirituality. Elankumaran (2004) compared the amount of job involvement of persons with dominant *sattva, rajas* or *tamas* and found that persons with dominant *sattva* and *rajas* had a significantly higher level of job involvement than persons with dominant *tamas*.

People with dominant *tamas* are expected to have difficulties motivating themselves for constructive activities. If *tamas* is very strong they will consciously harm others. Although their life-style does not benefit them they leave things as they are. Dominant *tamas* leads to a lack in self-control, high expectations and strong desires. *Tamasic* people do not care for ethics and are generally not interested in spirituality. They are indecisive and do not think of the consequences of their decisions. Concerning their level of activity *tamasic* persons are lazy and have a strong lack of energy and enthusiasm.

As for *rajas*, empirical results hint at the spiritual indifference of persons with dominant *tamas:* non-reciters of the *Bhagavad-gita* have higher scores on *tamas* than *Bhagavad-gita* reciters (Anjana and Raju 2001) and *tamas* correlates negatively with daily spiritual experiences ($r=-.46$, $p<.01$; Stempel et al. 2006). In addition, Bhal and Debnath (2006) found a negative correlation between *tamas* and an idealistic ethical orientation ($r=-.25$, $p<.01$). Hopkins reports a negative correlation between *tamas* and conscientiousness ($r=-.58$, $p<.001$). Both *rajas* and *tamas* correlated negatively with the NEO-PI-R subscale "self-discipline" (*rajas:* $r=-.42$, $p<.001$; *tamas:* $r=-.57$, $p<.001$) and positively with the subscale "impulsiveness" (*rajas:* $r=.35$, $p<.01$; *tamas:* $r=.6$, $p<.001$), although both correlations are stronger for *tamas*.

## Social Variables

What follows is a list of how the *gunas* manifest in social variables (Table 4). What remains inconclusive here is that the *Mahabharata* (Deussen and Strauss 1906; Ganguli 2005)

**Table 4** Symptoms of the *Gunas* in social variables

| | Sattva | Rajas | Tamas |
|---|---|---|---|
| Attitudes toward oneself | Self-less, not boasting, pride (Deussen and Strauss 1906) | Pride, egoism, selfishness (Deussen and Strauss 1906) | Pride, overestimating one's competences and knowledge (Deussen and Strauss 1906) |
| | Complex free (Gupta 1977, as quoted by Mohan and Sandhu 1986) | Arrogance (Tagare 1978) | Vanity (Larson and Bhattacharya 1987) |
| | "[..] may not consider themselves sattvic or claim to be sattvic." (Kumar 2007, p. 23) | Dishonesty, vanity (Larson and Bhattacharya 1987) | "Do not like to be responsible for their lives" (Frawley 1997, p. 35) |
| | Freedom from egoism, confidence, humility (Ganguli 2005) | Being fond of hearing oneself praised, false pride, "considering oneself different and better than others… advertising one's own prowess" (Goswami and Adhikari 1988c, p. 561) | Presumption of knowledge in ignorance, haughtiness, boastful assertions of performance when there has been no performance (Ganguli 2005) |
| | Freedom from self-absorption (Theodor 2010) | Boastfulness of speech, desire of concealment, ostentation, haughtiness (Ganguli 2005) | |
| | | Exaggerated ego-notion (Theodor 2010) | |
| Behavior & attitudes toward others | Leadership, faith, non-damaging, gentleness, straight-forwardness, non-enviousness, interest for others, compassion for all beings, benevolence, generosity, not wishing ill to anyone, not boasting, placability, chastity, giving and accepting gratuitously (Deussen and Strauss 1906) | Arrogance, non-compassion, defamation, termagancy, selfishness, inhospitableness, hostility, cruelty, robbery, conflict, harshness, leadership, to argue, violence, conflict, quarrel, envy, gossiping, egoism, protection, imprisonment, verbal abuse, pointing out the weaknesses in others, wrong charity, criticizing, praising, superiority, rape, caring, obedience, service, support, preferential treatment, belittlement, honoring, gossip (Deussen and Strauss 1906) | Arrogance, unfairness, mean actions, meanness, unfriendliness, egoism (Deussen and Strauss 1906) |
| | Hospitable, equally favorably disposed to all (Gupta 1977, as quoted by Mohan and Sandhu 1986) | Secretiveness (Hume 1962, as quoted by Suneetha and Srikrishna 2009) | Jealousy (Hume 1962, as quoted by Suneetha and Srikrishna 2009) |
| | Giving for the sake of giving (Maheshwar 1978) | Ridiculing others (Tagare 1978) | Insolent contempt for others, especially superiors (Maheshwar 1978) |

| | | | |
|---|---|---|---|
| | Charity, modesty (Tagare 1978) | Blame, dishonesty, murder, malice, envy, hatred, animosity, jealousy, rigidity, wickedness, deception, disgrace (Larson and Bhattacharya 1987) | Quarrelsomeness (Tagare 1978) |
| | Agreeableness, kindness, honesty, modesty, propriety, compassion, straight-forwardness, respect, forgiveness (Larson and Bhattacharya 1987) | "Seek power over others" (Frawley 1997, p. 35) | Infidelity, crookedness (Larson and Bhattacharya 1987) |
| | Not expecting anything in return (Prabhupada 1989) | Expecting something in return (Prabhupada 1989) | Insensitivity, allow others to dominate them (Frawley 1997) |
| | Absence of cruelty, compassion to all creatures (Ganguli 2005) | Vindictiveness, ostentatious charity, exhibition of manliness, disposition to revile, injuring others, quarrel, false speech (Ganguli 2005) | Cruelty, hostility (to all creatures), incapacity of association, frequently speaking ill of others (Ganguli 2005) |
| | Appreciative, affirmative (Kumar 2007) | Deviousness, cantankerousness (Michel 2006) | Malevolence (Michel 2006) |
| | Tolerance, honesty (Theodor 2010) | Impress, make an impact (Kumar 2007) | Attacking to defend oneself (Kumar 2007) |
| | | | Disregard to injury to others (Theodor 2010) |
| Style of communication | Dialogue, based on trust, mutual understanding and exploration, with open heart and mind, reaches compromise, "There is no fixed position, no dogma, no desire to convert; rather there is a desire to reach a stage which is respectful to all sides and honours the intrinsic qualities of every position, making dialogue a conversation among equals." (Kumar 2007, p. 26) | Diplomacy, concealed fixed position, concealed self-interest, desire to convince and win through argument or bribery, avoiding confrontation and breakdown of communication, ego is massaged, "[…] everyone is made to feel important and the red carpet is rolled out." (Kumar 2007, p. 26) | Monologue, fear of losing, blind faith in own correctness, threatening, insulting, denigrating, imposing brute force, justifying errors, never recognizing errors, difficulty with apologizing (Kumar 2007) |

(continued)

**Table 4** (continued)

| | Sattva | Rajas | Tamas |
|---|---|---|---|
| Norms & reaction to violating them | Having a sense of shame, good conduct (Deussen and Strauss 1906) | Shamelessness, lack of righteousness, theft, damage (Deussen and Strauss 1906) | Unfairness, nihilism (Deussen and Strauss 1906) |
| | Avoiding unbecoming acts (Tagare 1978) | Policy heedlessness (Ganguli 2005) | Violation of rules of conduct (Ganguli 2005) |
| | Embarrassment at improper action (Goswami and Adhikari 1988c) | | Shamelessness (Michel 2006) |
| | Righteousness (Ganguli 2005) | | |
| Duty/ies | Dedication to one's duties, unabbreviated and gratuitous fulfillment of one's duties (Deussen and Strauss 1906) | Non-performance out of bodily discomfort (Prabhupada 1989) | Distaste for duties (Deussen and Strauss 1906) |
| | Devoted to religious, social and moral duties (Gupta 1977, as quoted by Mohan and Sandhu 1986) | Non-performance out of fear of suffering (Theodor 2010) | Not acting out of attachment to inaction (Maheshwar 1978) |
| | No doubts in one's activities, "neither hateful of inauspicious work nor attached to auspicious work" (Prabhupada 1989, p. 803) | | Non-performance out of illusion (Theodor 2010) |
| | Performance for its own sake without interest in the fruits (Theodor 2010) | | |
| Behavior of others | Experiences love, praise, honor and help from others (Gupta 1977, as quoted by Mohan and Sandhu 1986) | | |

lists pride as an indicator for all three *gunas* and doesn't offer any further explanation; therefore the reasons for this remain unclear. It is reasonable to assume that, according to this source, the pride will differ in its quality, but how it differs is not addressed.

All in all, people with dominant *sattva* are expected to have an unperturbed, realistic view of themselves. They are said to be complex free, humble, confident and not self-absorbed. They have a well-meaning attitude toward others, which shows itself in straight-forwardness, kindness, generosity and affection. *Sattvic* people feel committed to laws and conventions and react with shame if they do something which contradicts laws or their ideals. They are dedicated to their duties and experience support from others.

Sharma (1999) examined the relation between self-concept and the *gunas* and found that *sattva* correlated with a positive view of oneself ($r=.56, p<.01$). Stempel et al. (2006) report a negative correlation between *sattva* and feelings of personal inferiority and inadequacy ($r=-.29, p<.05$). *Sattvic* people's kind attitude towards others is supported by a positive correlation between *sattva* and agreeableness ($r=.49, p<.001$; Hopkins) and the extraversion subscale "warmth" of the NEO-PI-R ($r=.51, p<.001$; Hopkins), which measures friendliness towards and interest in others. Further NEO-PI-R subscale correlations (Hopkins) indicate that *sattvic* people are not socially anxious ($r=-.56, p<.001$), are dutiful ($r=.59, p<.001$), assertive ($r=.36, p<.01$), concerned for others' welfare ($r=.59, p<.001$), tender minded toward others ($r=.42, p<.001$), compliant ($r=.33, p<.01$), straightforward ($r=.46, p<.001$); and that they believe in the sincerity of others ($r=.65, p<.001$). In Archana Das and Venu Gopal (2009) *sattva* correlated negatively with the subscales rule-breaking ($r=-.33, p>.05$) and aggressive behavior ($r=-.19, p>.05$) of the Adult Self Report. Kaur and Sinha (1992) found that *sattva* correlates with a higher leader-member exchange score ($r=.14, p<.05$), which indicates the closeness of the relationship between an employee and supervisor. Furthermore, they found a positive correlation between *sattva* and an orientation towards the protestant work ethic ($r=.21, p<.01$). Sitamma (2005) summarizes that *sattva* correlates positively with introversion and negatively with extraversion. Conflicting with this, *sattva* correlated positively with extraversion ($r=.23, p<.01$) in Hopkins' study.

When *rajas* dominates, people are said to be proud and arrogant. They may support others, if it contributes to their goals; otherwise they are critical and harsh toward others. They have friends, but only as long as they can benefit materially from the relationship. *Rajasic* people do not feel very committed to laws and conventions and are not ashamed if they break them. They try to avoid unpleasant experiences and become neglectful in performing their duties if that would require undergoing personal discomfort. In Sharma's study *rajas* was also correlated with a positive view of oneself ($r=.45, p<.01$). Hopkins reports a negative correlation between *rajas* and the NEO-PI-R subscale "modesty" ($r=-.25, p<.05$). *Rajas* correlated with intrusive behavior ($r=.21, p>.05$) in Archana Das and Venu Gopal (2009). Again, there are conflicting results concerning extraversion: According to Sitamma's (2005) summary, *rajas* correlates positively with extraversion. Hopkins, however, found that *rajas* correlates negatively with extraversion ($r=-.24, p<.01$) and agreeableness ($r=-.5, p<.001$).

A person with dominant *tamas* is also described as proud and boastful, but to a greater degree than the *rajasic* person. *Tamasic* persons lie about their accomplishments to receive honor. They are insensitive and cruel and do not consider if their behavior injures others. Like people with dominant *rajas* they violate laws and conventions and are not ashamed of it. They have a strong dislike for performing their duties because they are rather inactive. This is supported by the negative correlation between *tamas* and an orientation towards protestant work ethic ($r=-.2$, $p<.01$) reported by Kaur and Sinha (1992). Hopkins reports negative correlations between *tamas* and extraversion ($r=-.37$, $p<.001$) and agreeableness ($r=-.46$, $p<.001$) and the NEO-PI-R subscale "assertiveness" ($r=-.42$, $p<.001$). Archana Das and Venu Gopal (2009) found correlations between *tamas* and the Adult Self Report subscales withdrawn ($r=.41$, $p<.01$), aggressive behavior ($r=.56$, $p<.01$), rule-breaking ($r=.56$, $p<.01$) and intrusive behavior ($r=.44$, $p<.01$). In addition, *tamas* correlated with the Brief Symptom Inventory subscale measuring feelings of personal inferiority and inadequacy ($r=.31$, $p<.05$; Stempel et al. 2006).

*Rajas* and *tamas* again had very similar correlations to the subscales of the NEO-PI-R in this variable group in Hopkins' study. *Rajasic* and *tamasic* people were more socially anxious (*rajas:* $r=.36$, $p<.01$; *tamas:* $r=.55$, $p<.001$), less friendly and interested in others (*rajas:* $r=-.4$, $p<.001$; *tamas:* $r=-.45$, $p<.001$), trusted less in the sincerity of others (*rajas:* $r=-.55$, $p<.001$; *tamas:* $r=-.53$, $p<.001$), were less straightforward (*rajas:* $r=-.41$, $p<.01$; *tamas:* $r=-.36$, $p<.001$), less altruistic (*rajas:* $r=-.46$, $p<.001$; *tamas:* $r=-.52$, $p<.001$), less sympathetic toward others (*rajas:* $r=-.37$, $p<.01$; *tamas:* $r=-.33$, $p<.01$) and less dutiful (*rajas:* $r=-.51$, $p<.001$; *tamas:* $r=-.47$, $p<.001$).

This concludes the delineation of the indicators for the *gunas* in the psychological factor groups. Two studies have investigated the relationship between the *gunas* and psychological illness in general which confirm the assumption that *rajas* or *tamas* are related to psychological illness. Lakshmi Bhai et al. (1975) compared people with a clinically relevant mental disorder to those without a diagnosis and found that the former had significantly higher scores on *rajas* and *tamas*. Stempel et al. (2006) found significant correlations between the global severity index of the Brief Symptom Inventory and *rajas* ($r=.33$, $p<.05$) and *tamas* ($r=.51$, $p<.01$) respectively (the negative correlation with *sattva* was not significant ($r=-.25$, $p>.05$).

## Physical Variables

In the following table (Table 5) the symptoms of the *gunas* in physical variables is listed. People with dominant *sattva* are clearly most healthy.

With regards to variables of the organism strong *sattva* is indicated by acute senses, a calm nervous system and a feeling of having a light body. Health or well-being is a symptom of *sattva*. When *sattva* dominates in sleep one feels refreshed upon awakening. Rani and Rani (2009) investigated the relationships between the *gunas* and physiological measures (heart rate, galvanic skin response, blood

**Table 5** Symptoms of the *Gunas* in physical variables

| | Sattva | Rajas | Tamas |
|---|---|---|---|
| Physical variables | Frame is light, senses acute (Colebrooke and Wilson 1837) | Beautiful features (Deussen and Strauss 1906) | Body is sluggish and heavy, senses obtuse (Colebrooke and Wilson 1837) |
| | Light limbs (Larson and Bhattacharya 1987) | Self-adornment, ostentation (Frawley and Summerfield Kozak 2001) | Lack of flexibility (Deussen and Strauss 1906) |
| | Senses quickened, nervous being calm, filled with ease and clarity (Maheshwar 1978) | | Insensibility, ugliness, sense of limbs being heavy (Larson and Bhattacharya 1987) |
| | "Generates a sense of lightness in the limbs and clarity of the senses." (Larson and Bhattacharya 1987, p. 215) | | Obesity, sloppy appearance (Frawley and Summerfield Kozak 2001) |
| Health | Vitality (Colebrooke and Wilson 1837) | Pain (Deussen and Strauss 1906) | Low energy (Frawley and Summerfield Kozak 2001) |
| | Health itself, sense of well-being (Deussen and Strauss 1906) | "Unhealthy condition of the working physical organs" (Goswami and Adhikari 1988c, p. 570) | Lassitude (Ganguli 2005) |
| | Hereditary good health (Ranade 1994) | Disequilibrium (Frawley 1997) | |
| | Harmony, balance (Frawley 1997) | Indigestion, high blood pressure (Verma 1997) | |
| | | Nervousness, irregular breathing (Buhrman 2005) | |
| | | Heart attacks, cancer, auto-immune diseases (Chopra in Kumar 2007) | |
| | | Power (Bryant 2009) | |
| Sleep | Refreshing, upon awakening one feels lucid (Bryant 2009) | Insomnia (Buhrman 1997) | To sleep more than necessary (Prabhupada 1989) |
| | | Disturbances (Verma 1997) | Upon awakening one feels tired and sluggish (Bryant 2009) |
| | | Restless, upon awakening one feels confused (Bryant 2009) | |

pressure and lung functions). They did not find many significant correlations. For *sattva* there was a significant negative correlation with diastolic blood pressure ($r = -.35$, $p < .05$).

A person with dominant *rajas* is said to care more for short-term pleasures and looks than for health. It is suggested that too much *rajas* leads to various diseases like insomnia, heart attacks, auto-immune diseases or cancer. When *rajas* dominates in sleep, it becomes restless and upon awakening one feels confused. *Rajas* has been found to correlate with systolic ($r = .41$, $p < .05$) and diastolic blood pressure ($r = .58$, $p < .01$; Rani and Rani 2009).

*Tamasic* persons are expected to have an inflexible, heavy body and suffer from a lack of energy. When *tamas* dominates sleep one feels tired upon awakening. Archana Das and Venu Gopal (2009) report a correlation between *tamas* and somatic problems ($r = .45$, $p < .01$).

A few researchers investigated gender differences, but the results are inconsistent. Uma et al. (1971, as quoted by Sitamma 2005) report that women had higher scores on *tamas* than men. Mohan and Sandhu (1986) found that women had significantly higher scores on *sattva* and lower scores on *rajas* than men. Marutham (1992, as quoted by Sitamma 2005) found that on average the dominant *guna* for men was *rajas*, whereas the average dominant *guna* in women was *sattva*. Stempel et al. (2006) and Bhal and Debnath (2006) found no correlation between any of the *gunas* and gender. In the sample of Stempel et al. (2006) age was positively correlated with *sattva* ($r = .41$, $p < .01$) and negatively with *rajas* ($r = -.41$, $p < .01$) and *tamas* ($r = -.46$, $p < .01$).

## *Variables of the Environment*

Table 6 shows variables of the environment according to the influences of the *gunas* (Westhoff (2005) also groups the financial situation amongst these variables). After the lengthy description of the influence of the *gunas* on different psychological and physical variables, this factor group may be surprising. What does the particular strength of each *guna* in a person have to do with the person's environment? As detailed above, the *gunas* are said to be the fundamental components of all matter, therefore they not only comprise the psyche but also the environment. It is claimed that people with the respective dominating *guna* are attracted to be in places where the same *guna* dominates. Therefore, this categorization of different environments where humans reside may be useful indicators for the diagnosis of the *gunas*.

To sum up, people with dominant *sattva* like to reside in bright and simple surroundings. They are not overly concerned with their possessions and are satisfied with whatever they can attain without over endeavoring. People with dominant *rajas* like to live in the hustle of big cities and accumulate a lot of property. When *tamas* prevails, people are attracted to, according to the Indian texts, "casinos and brothels".

**Table 6** Living conditions categorized according to the *Gunas*

| | Sattva | Rajas | Tamas |
|---|---|---|---|
| Environment | Brightness (Deussen and Strauss 1906) | City life (Prabhupada 1987d) | Gambling houses (Tagare 1978) |
| | Abundance (Larson and Bhattacharya 1987) | | Brothels, drinking houses (Prabhupada 1987d) |
| | Living in the forest (Prabhupada 1987d) | | |
| | Simplicity (Prabhupada 1989) | | |
| Property | "Food" obtained without difficulty (Goswami and Adhikari 1988c) | As much as possible (Prabhupada 1989) | Poverty (Larson and Bhattacharya 1987) |
| | Lifestyle doesn't demand much of the natural resources, life is about being not having (Kumar 2007) | Don't mind waste, excessive demand on natural resources (Kumar 2007) | |

## *Behavior*

The following table (Table 7) lists indicators of the *gunas* for general behavioral categories. Speaking in terms of the behavioral equation, this table discusses the outcome of the different factor groups, that is, behavior itself. Although "competences and virtues" could also be grouped with other variables, we chose to list them here in order to do justice to the different components that play a role in displaying a virtue or competence.

In general the behavior of a person with dominant *sattva* is described to lead to positive results like joy and success. People with prevailing *sattva* have a healthy lifestyle and prefer food that increases health. *Sattvic* people are highly virtuous. *Rajas* energizes actions for worldly purposes with temporary outcomes. Persons with dominant *rajas* find it difficult to care for their health, although they do it partially, and they prefer unhealthy food. *Tamasic* people do not care for their health at all; they do not exercise and consume too much alcohol and other drugs. They eat unhealthy food that is not freshly prepared. *Rajasic* and *tamasic* persons are not particularly virtuous.

All in all, the *guna* profiles are mostly coherent in themselves and distinct from another. In some cases the difference between the indicators for *rajas* and *tamas* are subtle. For example both, people with dominant *rajas* and *tamas,* tend to neglect their duties; however they do so for different reasons. *Rajasic* persons neglect their duties because they want to avoid discomfort, whereas *tamasic* persons are additionally overly fond of being inactive.

Although there are inconsistencies, all in all, the first empirical findings support the *tri-guna* concept: there is a relatively high congruence between theoretical assumptions and empirical evidence. Some or all of the inconsistent findings may result from the differing and partly suboptimal operationalizations of the *gunas* in the different studies.

**Table 7** Symptoms of the *Gunas* in behavior

| | Sattva | Rajas | Tamas |
|---|---|---|---|
| Actions | Purity, aimless actions and endeavors, fulfilling vows (Deussen and Strauss 1906) | Unjust, over-endeavor for aims, purchase and sales, politics, all actions based on desire, heroism (Deussen and Strauss 1906) | Mean/bad actions, excessive talking (Deussen and Strauss 1906) |
| | | Engagement in actions which bear fruits (Prabhupada 1989) | Whimsically, without purpose (Prabhupada 1989) |
| | | Acts with earthly purposes (Ganguli 2005) | Acts springing from wrath, fraught with cupidity, vileness of behavior (Ganguli 2005) |
| | | Discrepancy between words and deeds (Kumar 2007) | |
| Results of actions | Non-damaging (Deussen and Strauss 1906) | Temporary, unstable, misery (Prabhupada 1989) | Foolishness, good neither for oneself, nor for others (Prabhupada 1989) |
| | Pure (Prabhupada 1989) | Uncertain (Bhanu Swami 2003) | Vain, uselessness (Ganguli 2005) |
| | Success (Ganguli 2005) | Lead to suffering, harmful (Theodor 2010) | Result in foolishness (Theodor 2010) |
| | Lead to spotless results, joy, end of suffering (Theodor 2010) | | |
| Health behavior | Good lifestyle, self-care (Deussen and Strauss 1906) | Follows healthy lifestyle if convinced (Ranade 1994) | Insatiable when eating and drinking, unregulated eating (Deussen and Strauss 1906) |
| | Cleanliness (Gupta 1977, as quoted by Mohan and Sandhu 1986) | No desire to take the time to recuperate (Frawley 1997) | Intoxication (Larson & Bhattacharya, 1987) |
| | Determination to avoid harmful but momentarily pleasing things, being conscious of one's limits (Ranade 1994) | Harsh exercise (Frawley and Summerfield Kozak 2001) | Taking shelter in intoxication (Prabhupada 1989) |
| | Everything which is calming and pure (Deussen and Strauss 1906) | Difficulty calming down mentally and physically (Buhrman 2005) | Everything which is unclear, incomprehensible and incognizable (Deussen and Strauss 1906) |
| | Takes care of him-/herself (Frawley 1997) | Overeating (Michel 2006) | Careless in health (Ranade 1994) |
| | Gentle exercise (Frawley and Summerfield Kozak 2001) | Practices which begin like nectar, but gradually resemble poison (Theodor 2010) | Lack of exercise (Frawley and Summerfield Kozak 2001) |
| | Determination to sustain mind, life air and senses (through yoga), practices which start like poison, but gradually resemble nectar (Theodor 2010) | | Not wanting to abandon intoxication, fear, dejection, sleep, sorrow (Theodor 2010) |

| | | | |
|---|---|---|---|
| Sleeping behavior | "Rising early in the morning" (Prabhupada 1987e) | Dreams (Goswami and Adhikari 1988c) | Sleeping during the day (Deussen and Strauss 1906) |
| | Wakefulness (Goswami and Adhikari 1988c) | | Rising from bed late, staying awake at night (Goswami and Adhikari 1988b) |
| | | | Deep sleep (Goswami and Adhikari 1988c) |
| Preferred food | Juicy, fatty, wholesome (Prabhupada 1989) | Too bitter, reduce mucus in stomach (Prabhupada 1989) | Prepared long before consumption (>3 h), decomposed, bad odor (Prabhupada 1989) |
| | Increases strength of will, juicy, tasty, mild, with oil, long lasting effect, beneficial to organs, rice, grains, milk, wheat, sugar (Bhanu Swami 2003) | Produces pain while eating, afterwards sorrow and at last disease (Bhanu Swami 2003) | Putrid, stale, tasteless, leftovers; meat, eggs, fish, alcoholic drinks (Theodor 2010) |
| | Increases satisfaction, health, strength, lifespan and happiness; tasty, substantial, mild and pleasant; vegetables, fruits, fresh milk, beans (Theodor 2010) | Causes suffering, disease and misery; spicy, scorching, burning, sour, salty, too hot, pungent, too dry; fried foods, mixed nuts, sorts of cheese (Theodor 2010) | |
| Competences & virtues | Truthfulness, innocence, gentleness, cleanliness, not boasting with detachment (Deussen and Strauss 1906) | Impatience, greed, harshness, deception, cruelty (Deussen and Strauss 1906) | Helplessness, miserliness, impatience (Deussen and Strauss 1906) |
| | Mercy, renunciation, contentment, truthfulness (Tagare 1978) | Dishonesty (Larson and Bhattacharya 1987) | Intolerance, miserliness, deceitfulness, hypocrisy (Tagare 1978) |
| | Kindness, honesty, modesty, patience, compassion, simplicity, sweetness, forbearance, endurance, mildness (Larson and Bhattacharya 1987) | Forbidden acts, cleverness, dexterity (Ganguli 2005) | Crookedness, miserliness (Larson and Bhattacharya 1987) |
| | Generosity, simplicity, virtue (Goswami and Adhikari 1988c) | | Wickedness, unrighteousness, illiberality, absence of faith, evilness, crookedness, senselessness, unforgiving (Ganguli 2005) |
| | Loyalty (Frawley and Summerfield Kozak 2001) | | |
| | Righteousness, sincerity, liberality, faith, nobility, confidence, forgiveness (Ganguli 2005) | | |

After this detailed characterization of the indicators for *sattva, rajas* and *tamas* respectively, we would like to remind the reader of the context in which we are investigating the concept of *tri-guna*: how can this contribute to well-being? By now it is obvious that according to the *tri-guna* concept, thoughts, feelings and behavior are influenced by three types of energies. All three are necessary for healthy functioning, however to different degrees. If *rajas* and *tamas* get too strong, health deteriorates. Thus, in order to increase well-being, it is necessary to find interventions which increase *sattva,* and decrease excessive *rajas* and *tamas.* As the different correlates of the *gunas* show, there is reason to hypothesize that a relative dominance of *sattva* could lead to well-being. Sitamma (2005) remarks that it is necessary to search for methods which will increase *sattva* and to describe the characteristics of such interventions. The aim of the next paragraph is to review such possible interventions. The classical Indian texts which discuss the *tri-guna* concept mention many possible interventions. Knowledge of possible interventions permits testing their effectiveness, alone or in packages.

## Interventions to Increase Sattva Guna and Decrease Rajas and Tamas

The general mechanism of increasing the influence of a particular *guna* is to align one's behavior (whenever possible) with the action modes of this *guna* that are listed in detail above. It is claimed that by cultivating a particular *guna* (internally and externally), one causes it to manifest/increase within oneself. In order to increase *sattva* means that "One can strengthen the mode of goodness by cultivation of those things that are already situated in goodness […]. […] Thus, by choosing food, attitudes, work, recreation, etc., strictly in the mode of goodness, one will become situated in that mode." (Goswami and Adhikari 1988c, p. 4).

Therefore, the first step in increasing *sattva* would be to acquire knowledge of *sattvic* behavior and then try to adjust one's behavior to this. Or as Theodor (2010, p. 10) puts it in his introduction to the *Bhagavad-gita,* "For this purpose, various characteristics of the *gunas'* bondage are delineated, and these enable a process of self-examination or self-study. Consequently, one is able to change one's habits for the purpose of raising oneself on this ladder of the *gunas*." Theodor uses the word "bondage" for the action modes of the *gunas,* since the influence of the *gunas* is so strong that they force a particular mode of behavior, unless the individual actively tries to change his or her life-style.

This relatively simple mechanism of increasing the influence of *sattva* explains why some of the *de*scriptions of *sattvic* behavior exactly correspond to *pre*scriptions for increasing it. For example the process of yoga given by Patanjali recommends practicing celibacy and truthfulness for increasing *sattva*, whereas the *Bhagavad-gita* lists these two qualities when describing the behavior of a *sattvic* person. Another example for this principle is that foods mentioned as preferred food of persons with dominating *sattva* are the same as foods recommended for a *sattvic* diet.

There are some differences in opinion concerning the question of adopting *sattvic* behavior as being the best course of action for everyone, or whether they should act differently according to their dominating *guna*. For example, the *Bhagavata Purana* (Tagare 1976b, p. 988) recommends, "He should vanquish *rajas* and *tamas* by *sattva*, […]." indicating that it is possible to increase *sattva* as a first step for anyone. Frawley (1997), however, suggests that people with dominating *tamas* should first increase the influence of *rajas* and then go on to develop *sattvic* qualities, while people with dominating *rajas* may immediately start to cultivate *sattva*. These recommendations can be harmonized by assuming that it is easier for *tamasic* persons to first develop *rajasic* than *sattvic* traits, although it is not impossible to immediately decrease the influence of *tamas* by strengthening *sattva*.

Once *sattva* is the dominant *guna*, or more prominent in an individual, the individual will naturally be more inclined to *sattvic* things, thoughts and behavior, all of which increase strength, prolong life, and further satisfaction and well-being (Maheshwar 1978). A *sattvic* lifestyle becomes, so to say, natural and spontaneous (Bryant 2009). Thus, this development in a way reinforces itself and it becomes easier to keep *sattva* prominent. As Bryant (2009, p. 141) describes, "[…], once the mind becomes stilled, its *sattvic* nature can manifest, as a result of which the qualities of *sattva*, insight and lucidity, also gradually manifest. These qualities, in turn, start to pervade all aspects of a practitioner's life […]." This reminds one of the notion in positive psychology that well-being develops in a positive up-ward spiral, one positive experience, behavior or trait furthering another.

In addition to the information given in the description of *sattvic, rajasic* and *tamasic* thinking and behavior (Tables 1, 2, 3, 4, 5, 6, and 7) one can also find specific procedures and interventions meant to increase *sattva guna*. What follows is a review of possible interventions given in the traditional literature, which could be tested empirically with regards to their effectiveness.

As a general practice to increase the influence of *sattva*, it is recommended to practice yoga (Ranade 1994). In the *Mahabharata* (Ganguli 2005, Santi Parva, Section CCXVII) it is even stated, "Whatever acts are destructive of Passion [*rajas*] and Darkness [*tamas*] constitute yoga in respect of its real character." suggesting that an essential purpose of yoga is to reduce *rajas* and *tamas*. Of course, this is a broad domain, as there are many kinds of yoga. For example many of the commentators to the *Yoga Sutras* of Patanjali remark that increasing *sattva* and decreasing the influence of *rajas* and *tamas* is a goal of Patanjali's yoga system (Braud 2008; Bryant 2009), and the *Bhagavat Purana* states that *sattva* can be maximized by the practice of *bhakti*-yoga (Tagare 1976a).

However, increasing *sattva* is not actually the ultimate aim of any yoga process; it is mostly considered as an important helpful and intermediate step to attaining the ultimate goal (which is different from process to process, ranging from attaining liberation by freeing the *purusha* from its misidentification with matter, to additionally developing love of God). As Bryant (2009, p. 130) describes, "Cultivating the higher qualities of *sattva* is a continuous and constant requirement of the *yogic* path and spills over into all aspects of life's affairs and social interactions." Different yoga schools give recommendations on how to increase *sattva*. Generally, a yoga system might be considered as a package of interventions to increase *sattva*. In order to organize the practices by the following categorization, recommendations of different yoga processes like those of the *yamas* and *niyamas* have been isolated from their "package" and integrated in the following list of possible interventions. The next section discusses specific interventions which are claimed to enhance *sattva guna*.

## Meditation

Meditation is depicted as an important activity to strengthen *sattva*. It is the heart of Patanjali's yoga process, "defined as keeping the mind fixed on any particular object of choice without distraction." (Bryant 2009, p. lvii). Apart from this, *Ayurveda* recommends concentration exercises (Verma 1997), singing of mantras and meditating daily to strengthen *sattva* and attain/keep mental peace (Ranade 1994). Wolf and Abell (2003) examined the effects of *maha*-mantra meditation on the *gunas,* and found that it significantly increased *sattva* and decreased *tamas*.

## Spiritual Awareness

As described above, according to *Sankhya* philosophy the light of the soul is said to reflect stronger in a *sattvic* psyche. This means a *sattvic* psyche is more capable of connecting to the spiritual source of energy, joy and love of the soul. This gives *sattvic* people the energy to act the way they do. The capability of the psyche to let the energy of the soul flow can be increased by practicing spiritual awareness.

Strengthening this awareness of the difference between matter and spirit is said to increase *sattva* (Ranade 1994) and it provides the mindset required for other practices as well (e.g. that of nonviolence). *Tamas* can be decreased by knowledge of the nature of the soul (Nilakantan 1989). In practice this means frequently reminding oneself of the eternality of the self/the soul and the impermanence of the body and mind, and adjusting one's behavior to this awareness. Wolf (2008) advises us to cultivate spirituality in order to strengthen *sattva*. This kind of practice requires believing in the eternality of the self and seems to be reasonable only for people who believe in this.

## Monitoring, Controlling and Restraining Oneself

An indispensable principle of a *sattvic* lifestyle is observing, controlling and restraining the activity of senses, mind and speech. This is one of the main functions of *sattva,* as Frawley (1999, p. 27) puts it, "Sattva is the neutral or balancing force, harmonizing the positive and negative, which oversees and observes." The *Maitrayana Upanishad* (Michel 2006) declares: *sattva* is attained by *tapas* (restraint). To increase *sattva,* balance, and to avoid acting myopically it is necessary to avoid being carried away by the forces of *rajas* and *tamas* (Tagare 1978). The *Caraka Samhita* (Sharma 1994) has a section in which it describes health behavior, which according to Ranade (1994), strengthens the *sattvic* qualities of the mind. In this section, the *Caraka Samhita* also prescribes that one should be self-controlled. The *Bhagavad-gita* defines the restraint of senses, mind and speech in three verses. Sri Aurobindo summarizes these verses in relation to the *gunas*: "Here comes in all that quiets or disciplines the rajasic and egoistic nature and all that replaces it by the happy and tranquil principle of good and virtue. This is the askesis of the sattwic Dharma so highly prized in the system of ancient Indian culture. Its greater culmination will be a high purity of the reason and will, an equal soul, a deep peace and calm, a wide sympathy and preparation of oneness, a reflection of the inner soul's divine gladness in the mind, life and body." (Maheshwar 1978, p. 217).

Patanjali lists *tapas,* austerity or restraint, as the first necessary practice of *kriya-yoga,* an action centered approach to the goal of *yoga* (chapter "Towards an Epistemology of Inner Experience", first *sutra*). *Kriya*-yoga is especially relevant for people with dominant *rajas* or *tamas.* Bryant (2009, p. 170) explains "*Tapas* means the control of the senses – controlling the quantity, quality, and regularity of one's food intake, for example; the quality of what one listens to or reads or talks about – in other words, the "*sattvicizing*" of one's sensual engagements."

Vijnanabhiksu, a commentator to the *Yoga Sutras* of Patanjali, emphasizes that restraint should be of a gentle nature that doesn't weaken the mind or body (Bryant 2009). This is important to note, otherwise the point of restraint – to increase the influence of *sattva* – is missed. The following three paragraphs discuss specific methods for restraining the senses, mind and speech.

## Control of the Senses/Body

The *Bhagavad-gita* (Theodor 2010, p. 127) positively defines restraint of the body as "Worship of the gods, of Brahmins, of teachers and wise men as well as purity, honesty, celibacy and nonviolence, are said to be the austerity of the body." This clearly corresponds to and overlaps with recommendations in regard to social behavior and developing virtues. Celibacy is also an important part of Patanjali's yoga system (the fourth *yama*), as it helps cultivating spiritual awareness in opposition to engaging with the temporary body (Bryant 2009). Furthermore, purity or external cleanliness is also recommended by Patanjali (*niyama* one), which Vyasa defines as cleaning the body and eating pure food (Bryant 2009). With regards to eating, one can find lists of *sattvic* types of food in the *Bhagavad-gita* and texts on Ayurveda (e.g. Deshpande 2003; Frawley 2000). Ayurveda also gives prescriptions on how to eat in a *sattvic* way. Especially persons with dominating *tamas* should strictly avoid overeating (Buhrman 1998). In addition, eating can be regulated by fasting (Jawa 2002).

The *Bhagavat Purana* (Goswami and Adhikari 1988c, p. 6) lists items which are especially relevant to *sattvic* sensual engagements: "According to the quality of religious scriptures, water, one's association with one's children or with people in general, the particular place, the time, activities, birth, meditation, chanting of mantras, and purificatory rituals, the modes of nature become differently prominent." Sridhara Swami (as quoted in Tagare 1978) gives details on how these items are *sattvic,* for example a *sattvic* time is the early morning, best suited for meditation.

In order to *sattvicize* body and mind and to reduce *rajas* and *tamas,* Patanjali (Bryant 2009) and Verma (1997) recommend practicing *asanas* of yoga (exercises for the body) and *pranayama* (breathing exercises). Ranade (1994) recommends regular exercise in general. Furthermore, the *Caraka Samhita* (Sharma 1994) prescribes that one should stop exercising before one is fatigued. Deshpande et al. (2008) investigated the effects of an 8-week practice of *asanas, pranayama* and meditation on the *gunas,* compared to an 8 week program of physical exercise. Both the yoga group and the physical exercise group showed a significant increase in *sattva guna*, although the effect size was greater in the yoga group. *Rajas* and *tamas* only decreased significantly in the physical exercise group. Rani and Rani (2009) compared *guna* scores of short-term (6 months to 2 years) and long-term (2–5 years) *hatha*-yoga practitioners from the same yoga center. As expected, the long-term practitioners had a significantly higher score on *sattva* and significantly lower scores on *rajas* and *tamas* than the short-term practitioners. Khema et al. (2011) found that *rajas* and *tamas* decreased significantly after 4 weeks of an "integrated approach to yoga therapy (IAYT)" intervention which included *asanas*, *pranayama,* singing, meditation and lectures on stress-management, the four yoga paths and diet. The mean *sattva* score increased slightly, but not significantly.

Jawa (2002) describes techniques for strengthening mental health from the *Bhagavad-gita* and mentions two strategies for controlling the senses: restraining the senses and withdrawing the senses. Restraining the senses means to refrain from the activity of the sense(s) for a certain period of time, for example fasting, or not speaking. Withdrawing the senses refers to mentally withdrawing from sensory input/information by withdrawing one's attention from them. The latter strategy is especially important in situations in which certain sensory input needs to be tolerated.

## Mind Control

For increasing *sattva* in the mind the *Caraka Samhita* (Sharma 1994) recommends being optimistic, cheerful, devoted to knowledge and calm, and giving up habitual criticism and continued grief. One should not (regularly) remember one's "scandals" and not be exhilarated by success, or depressed by failure.

The definition of restraint of the mind of the *Bhagavad-gita* is: "Peace of mind, gentleness, silence, self-control and purification of one's existence are said to be the austerity of the mind." (Theodor 2010, p. 127).

With regards to controlling and restraining the mind or rather thoughts, Patanjali's *yamas* and *niyamas* are crucial, for without practicing them the mind will not manifest *sattva*. All of the *yamas* and *niyamas* have an internal aspect to their practice relevant to the mind. Nonviolence (the first *yama*) for example includes giving up hatred and malice, the second *yama* (truth) is explained by Vyasa to entail keeping one's thoughts in correspondence to facts that can be perceived, inferred or which have been testified by others (Bryant 2009). Refrainment from stealing (*yama* three) entails giving up all desires to steal, and ultimately even material desires themselves. Celibacy (*yama* four) requires giving up the tendency to see members of the opposite sex as objects of desire; and renouncing possessions (*yama* five) helps to cultivate a spirit of detachment from temporary matter. Hariharananda points out that "Hoarding wealth without sharing it is sheer selfishness […]." (Bryant 2009, p. 247). Here we find another practice which quiets the egoistic and *rajasic* tendencies.

*Niyama* one – cleanliness – not only refers to external cleanliness, but according to Vyasa (Bryant 2009) also to internal cleanliness. Internal cleanliness means to free oneself from "contamination" in the mind, which for example consists of pride, hatred, jealousy, attachment and vanity, according to the commentators on the *Yoga Sutras*. Bhoja Raja mentions that the mind is cleaned by cultivating a friendly attitude towards all creatures (Bryant 2009). *Niyama* two – contentment – "[…] manifests as disinterest in accumulating more than one's immediate needs in life […] True happiness comes from contentment with whatever one has, not with thinking that one will be happy when one gets all that one desires. Even if there is some lack, says Sankara, one thinks, "It is enough." Or, as Hariharananda puts it, to avoid injury from thorns, one only has to wear one pair of shoes – one doesn't need to cover the entire earth with leather!" (Bryant 2009, p. 253). Also Verma (1997) addresses the necessity of giving up greed and dissatisfaction. This leads to mental peace by removing the motor of the *rajasic* tendency for over-endeavor and hankering. The third *niyama* (austerity) also has a mental component, that of practicing tolerance in the face of physical discomfort. Patanjali also requires a theistic practice with his fifth *niyama* – devotion to God, which Vyasa interprets as offering all activities to God, without a desire for enjoying the fruits of this offering. Of course, in the context of modern times, this recommendation as part of a *sattvic* lifestyle can only be an optional part of any intervention, which is likely to be adopted or rather deepened by theists. Also Frawley (1997, p. 42) recommends spiritual practice to "[…] heal our relationship with God or the inner Self." Moreover, Frawley (1997) advises learning from suffering by considering it as life's arrangement for further spiritual growth and seeing the good in everything. Deussen and Strauss (1906) mention that a person with dominant *sattva* is aware of his or her autonomy, this indicates that increasing self-efficacy will enhance *sattva*. Kumar (2007) recommends keeping the mind focused on the present 80 % of the time, thinking of the future 15 % and of the past only 5 % of the time. Also, he advises developing a respectful, reverential worldview that recognizes the intrinsic value of life, understanding things the way they are and developing a non-judgmental attitude of accepting reality. Wolf suggests cultivating the mindset of satisfaction thinking that "I don't need to do or have anything to experience satisfaction, aliveness, courage, and clarity – because these qualities are who I am." (Wolf 2008, p. 16), since according to the *guna* concept it is the nature of the soul to be satisfied, and this nature can be uncovered by increasing *sattva*. This is based on a worldview that doing and having things we cherish follow being (in *sattva*), rather than worldviews which require having and doing certain things in order to experience satisfaction.

## Speech Control

In regard to speaking, the *Caraka Samhita* (Sharma 1994, p. 56) prescribes "[…] one should speak useful, measured, sweet and meaningful words; […]". Also, one should not lie or speak about the faults of others.

Similarly, in the *Bhagavad-gita* restraint of speech is defined as "Words that do not agitate others, and are truthful, pleasant and beneficial, as well as constant recitation of scriptures are said to be the austerity of speech." (Theodor 2010, p. 127).

Recommendations for *sattvic* speaking can also be derived from Patanjali's *yamas* and *niyamas.* According to Patanjali and his commentators, a person striving for *sattva* should avoid using harsh words, speak the truth, avoid gossip and repeat *om* or devotional mantras. Verma (1997) advises to speak only as much as necessary and to be exact in one's choice of words. If one's work demands that one speak a lot, Verma recommends balancing this by remaining silent in the evening.

### *Behavior Towards Others*

A very important aspect of developing *sattva* concerns the social behavior. In addition to the description of *sattvic* behavior toward others in Table 4, this paragraph lists concrete recommendations that are given separately from the description of *sattva guna* for ways to increase *sattva* by *sattvic* social behavior. These aspects might be of special importance.

The most important prescription is the first *yama* or prescription for a yogic lifestyle – that of remaining nonviolent towards all creatures, which, strictly speaking, also calls for a vegetarian diet (Bryant 2009). It also entails being compassionate and empathetic toward others and is based on the practice of spiritual awareness, which means to consider every creature equally in regard to the spiritual self. The third *yama,* refraining from any kind of theft, also falls into the category of social behavior. Frawley (1997) advises letting go of individual sorrow, opening oneself to others' suffering and surrendering one's motivation and drive to the greater good by service and charity. Furthermore, *Ayurveda* offers elaborate recommendations of friendliness and ethical behavior to enhance *sattva:* one should be straightforward, attentive, generous, kind and gentle (Ranade 1994). One should donate, entertain guests, be devoted to superiors and teachers, committed to compassion and friendship, and not start quarrels or create fear in others (Sharma 1994). Also "[…] one should behave like kith and kin to all living beings, pacify the angry, console the frightened, help the poor, be truthful, peaceful, tolerant of others' harsh words [...]" (Sharma 1994, p. 57). The *Yoga Sutras* (Bryant 2009) suggest different kinds of behavior for different people: the practicing yogi should befriend happy people, be compassionate towards those in distress, joyful towards pious persons and equanimous towards the impious. Bryant (2009, p. 129) explains, "By thus removing these traits of envy, desire to inflict harm, and intolerance, which are characteristics of *rajas* and *tamas*, the *sattva* natural to the mind can manifest."

## Virtues

Developing virtues is described as a method to increase *sattva.* The virtues listed in the column for *sattva* in Table 7 could constitute developmental goals for individuals wanting to further *sattva*. These are truthfulness, innocence, gentleness, cleanliness, not boasting with detachment (Deussen and Strauss 1906), mercy, renunciation, contentment (Tagare 1978), patience, compassion, simplicity, sweetness, forbearance, endurance, mildness (Larson and Bhattacharya 1987), generosity, simplicity (Goswami and Adhikari 1988c), loyalty (Frawley and Summerfield Kozak 2001), righteousness, sincerity, liberality, faith, nobility, confidence and forgiveness (Ganguli 2005). Special recommendations as prescriptions for enhancing *sattva* could be found for truthfulness (Bryant 2009; Ranade 1994) and patience (Ranade 1994).

Swami (2005, p. 65) describes a way to develop *sattvic* virtues, which he calls "the principle of acting to attain a quality". He explains that by practicing virtuous, *sattvic* behavior, one automatically comes under the influence of *sattva guna.* This reflects the general mechanism of strengthening a particular *guna* mentioned above.

## Style of Working

In one's work one should balance activity (energized by *rajas*) and inactivity (energized by *tamas*) (Verma 1997). Excessive activity should be given up altogether (Verma 1997). Furthermore, one should not try to accomplish too many things at once, not be too pedantic, not take up a task out of anger, joy or without inquiring about it and not procrastinate (*Caraka Samhita*, as quoted in Rhyner 2003). It is advised to avoid hardships, not to work when one is sick, not to overburden oneself, and to act according to time and place (Ranade 1994).

## Routines

Routines are frequently recommended in Ayurvedic literatures as a tool for enhancing *sattva* (Ranade 1994; Verma 1997). Especially the course of the day should be guided by a routine. One should sleep regularly and not too little or too much

(Ranade 1994) and rise early in the morning (Prabhupada 1987e). Late night hours are governed by *tamas* and should be used to sleep, whereas the morning hours are *sattvic* and should be utilized for spiritual development (Wolf 2008). Verma (1997) also recommends rituals for new beginnings in life.

## Conclusion

In summary, according to the *tri-guna* concept, well-being is influenced by three energies called *sattva, rajas* and *tamas*. Illness and psychological disturbances result from an excess of *rajas* and *tamas,* whereas a strong *sattva* leads to well-being. When *sattva* is strong a person is calm, relaxed and has a high stress-tolerance. He or she often experiences positive emotions, is enthusiastic and disciplined. People with strong *sattva* have a well-meaning attitude toward others and perform their duties reliably. They have a healthy life-style and are satisfied with their life. People with an excess of *rajas* have a low stress-tolerance and worry a lot. They are often frustrated, restless and have strong desires for prestige and wealth. Their life-style is unhealthy and they do not consider moderating the use of their energy. They have difficulties relaxing and take pleasure in actions which bring short-term relief but are harmful in the long-term. People with an overly strong *tamas* focus on the negative and do not try to solve their problems. They mostly experience negative emotions and have difficulty finding motivation for constructive activity. They are undisciplined, lack in energy and are insensitive towards others. *Tamasic* people neglect their health the most, and are dissatisfied with their lives.

The *tri-guna* concept also provides for ways to make life more positive and enjoyable, by making *sattva* the dominant *guna*. In short, activities which increase *sattva* and decrease *rajas* and *tamas* are meditation, reflecting on the difference between matter and spirit, restraining the activity of senses, mind and speech, developing certain virtues such as honesty and patience, as well as a well-meaning attitude and behavior toward others. Furthermore, it is helpful to take care of balancing activity and inactivity in one's style when working and resting, and to implement routines during the course of the day.

As this literature review shows, the concept of *tri-guna* has been elaborated on extensively in different ancient Indian scriptures. The context in which *tri-guna* has been developed has, of course, implications on its postulated effects, some of which might sound quite strange and outdated to contemporary readers. However, if one abstracts from such context specific pecularities, the concept encompasses details for diagnosing the dominating *guna(s)*, and statements on how to intervene so that well-being can be enhanced. As already hinted at above, many of the concepts and relationships described in the chapter might carry with them contextual properties (stemming from ancient India) that might not be necessary for yielding the results suggested in the scriptural sources analyzed here. However, the questions of whether the suggested effects (and maybe others) can be produced in a systematic way, and which parts of the proposed interventions really work, can and should be examined in a rigorous scientific way. This paper forms the theoretical basis for investigating these postulations empirically.

## References


Anjana, R., and Raju, S. 2001. *Personality of Bhagavad Gita reciters: A comparative study.* National Seminar on Psychology in India: Past, present and future. http://www.infinityfoundation.com/mandala/images/Souvenir-Internet1.pdf. Accessed 19 Sept 2012.

Antonovsky, A. 1979. *Health, stress and coping*. London: Jossey-Bass.

Archana Das, G.M., and D.V. Venu Gopal. 2009. Trigunas and psychological problems. *Journal of Indian Psychology* 27(1 & 2): 47–52.

Asendorpf, J.B. 2004. *Psychologie der Persönlichkeit*. Heidelberg: Springer.

Balodhi, J.P. 2005. Classical Indian Approaches to Psychological Dysfunction. In *Towards a spiritual psychology – Essays in Indian psychology*, ed. K. Ramakrishna Rao and S.B. Marwaha, 337–344. New Dehli: Samvad India Foundation.

Bhal, K.T., and N. Debnath. 2006. Conceptualizing and measuring Gunas: Predictors of workplace ethics of Indian professionals. *International Journal of Cross Cultural Management* 6(2): 169–188.

Bhanu Swami. 2003. *Sarartha Varsini – Srimad Bhagavad-Gita with commentary by Srila Visvanatha Cakravarti Thakura*. Chennai: Parampara.

Bhardwaj, A. 2003. Stellenwert und Bedeutung der Prakrti. In *Ayurveda -Basislehrbuch*, ed. S. Ranade, C. Hosius, and J. Heckmann, 85–89. München: Urban & Fischer.

Braud, W.G. 2008. Patanjali Yoga and Siddhis: Their Relevance to Parapsychological Theory and Research. In *Handbook of Indian psychology*, ed. K. Ramakrishna Rao, A.C. Paranjpe, and A.K. Dalal, 217–243. New Delhi: Foundation Books.

Bryant, E.F. 2009. *The yoga sutras of Patanjali*. New York: North Point Press.

Buhrman, S. 1997. Ayurvedic psychology and psychiatric approaches to the treatment of common affective disorders. *The Protocol Journal of Botanical Medicine* 2(1): 1–8.

Buhrman, S. 1998. Leaving depression behind: The yogic way out. *Yoga International* 40: 26–33.

Buhrman, S. 2005. What is the mind, how does it create depression, and what can we do about it? *Light on Ayurveda Journal* 3(4): 13–16.

Cloninger, C.R. 2006. The science of well-being: an integrated approach to mental health and its disorders. *World Psychiatry* 5(2): 71–76.

Colebrooke, H.T., and H.H. Wilson. 1837. *The Sankhya Karika*. London: A. J. Valpy.

Cornelissen, R.M.M., G. Misra, and S. Varma. 2011. *Foundations of Indian psychology*, Concepts and theories, vol. I. Delhi: Longman/Pearson.

Daftuar, C.N., and Anjuli. 1997. Occupational stress, organizational commitment and job involvement in Sattva, Rajas and Tamas personality types. *Journal of Indian Psychology* 15: 44–52.

Das, R.C. 1991. Standardization of the Gita inventory of personality. *Journal of Indian Psychology* 9: 47–58.

Deshpande, R. 2003. Ernährungstherapie und diätetische Maßnahmen. In *Ayurveda -Basislehrbuch*, ed. S. Ranade, C. Hosius, and J. Heckmann, 127–148. München: Urban & Fischer.

Deshpande, S., Nagendra, H. R., and Raghuram, N. 2008. A randomized control trial of the effect of yoga on Gunas (personality) and Health in normal healthy volunteers. *International Journal of Yoga* 1:2–10. http://www.ijoy.org.in/text.asp?2009/2/1/13/43287. Accessed 11 Jan 2011.

Deussen, P., and O. Strauss. 1906. *Vier Philosophische Texte des Mahabharatam: Sanatsujata-Parvan, Bhagavadgita, Mokshadharma, Anugita*. Leipzig: Brockhaus.

Diener, Ed., and K. Ryan. 2009. Subjective well-being: A general overview. *South African Journal of Psychology* 39(4): 391–406.
Eberth, J., and P. Sedlmeier. 2012. The effects of mindfulness meditation: A meta-analysis. *Mindfulness* 3: 174–189. doi:10.1007/s12671-012-0101-x.
Elankumaran, S. 2004. Personality, organizational climate and job involvement: An empirical study. *Journal of Human Values* 10(2): 117–130.
Faltermeier, T. 2005. *Gesundheitspsychologie*. Stuttgart: Kohlhammer.
Fava, G.A., and C. Ruini. 2009. Well-Being Therapy. In *The encyclopedia of positive psychology*, ed. S.J. Lopez, 1034–1036. Malden: Blackwell.
Feuerstein, G. 2001. *The yoga tradition: Its history, literature, philosophy and practice*. Prescott: Hohm Press.
Fisseni, H.J. 2003. *Persönlichkeitspsychologie*. Göttingen: Hogrefe.
Fordyce, M.W. 1977. Development of a program to increase personal happiness. *Journal of Counseling Psychology* 24(6): 511–521.
Frawley, D. 1997. *Ayurveda and the mind: The healing of consciousness*. Wisconsin: Lotus Press.
Frawley, D. 1999. *Yoga and Ayurveda – Self healing and self realization*. Wisconsin: Lotus Press.
Frawley, D. 2000. *Ayurvedic healing*, 2nd Revised and Enlarged Edition. Wisconsin: Lotus Press.
Frawley, D., and S. Summerfield Kozak. 2001. *Yoga for your type – An Ayurvedic approach to your Asana practice*. Wisconsin: Lotus Press.
Fredrickson, B.L. 2008. Promoting positive affect. In *The science of subjective well-being*, ed. M. Eid and R.J. Larsen, 449–468. New York: Guilford Press.
Frisch, M.B. 2009. Quality of Life Therapy and Coaching (QOLTC). In *The encyclopedia of positive psychology*, ed. S.J. Lopez, 824–826. Malden: Blackwell.
Gallagher, M.W., S.J. Lopez, and K.J. Preacher. 2009. The hierarchical structure of well-being. *Journal of Personality* 77: 1025–1049.
Ganganatha, J. 1896. *Tattva-Kaumudi (Sankhya) of Vachaspati Misra*. Bombay: Tookaram Tatya.
Ganguli, K.M. 2005. *Mahabharata of Krishna-Dwaipayana Vyasa*. http://www.sacred-texts.com/hin/maha/index.htm. Accessed 8 Mar 2012.
Goswami, H., and G. Adhikari. 1988a. *Srimad Bhagavatam. Tenth Canto – Part Two*. Los Angeles: Bhaktivedanta Book Trust.
Goswami, H., and G. Adhikari. 1988b. *Srimad Bhagavatam. Eleventh Canto – Part One*. Los Angeles: Bhaktivedanta Book Trust.
Goswami, H., and G. Adhikari. 1988c. *Srimad Bhagavatam. Eleventh Canto – Part Two*. Los Angeles: Bhaktivedanta Book Trust.
Gupta, S.K. 1977. *Madhusudan Saraswathi on the Bhagavad Gita*. Delhi: Motilal Banarsidass, Indological Publishers & Booksellers.
Hume, D. 1962. *The principles Upanishada*. Delhi: Motilal Banarsidass.
Jawa, S. 2002. Bhagavad Gita and techniques of mental health. In *Perspectives on indigenous psychology*, ed. G. Misra and A.K. Mohanty, 262–280. New Delhi: Concept Publishing Company.
Kapur, M. 2008. Psychological theories and practices in Ayurveda. In *Handbook of Indian psychology*, ed. K. Ramakrishna Rao, A.C. Paranjpe, and A.K. Dalal, 299–310. New Delhi: Foundation Books.
Kaur, P., and A.K. Sinha. 1992. Dimensions of Guna in organizational setting. *Vikalpa* 17(3): 27–32.
Khema, S.S., N.H. Ramarao, and A. Hankey. 2011. Effect of integral yoga on psychological and health variables and their correlations. *International Journal of Yoga* 4(2): 93–99.
Knapp, S. 2010. *The secret teachings of the Vedas – Answers to the mysteries of life*, New Revised & Expanded Edition. Mumbai: Jaico Publishing House.
Kumar, S. 2007. *Spiritual compass – The three qualities of life*. Devon: Green Books.
Lakshmi Bhai, A.J., H.N. Murthy, and S.V. Nagalakshmi. 1975. Rajas and Tamas in psychological disturbances. *Indian Journal of Clinical Psychology* 2: 135–138.
Larson, G.J. 1983. An eccentric ghost in the machine: Formal and quantitative aspects of the Samkhya-Yoga dualism. *Philosophy East and West* 33(3): 219–233.

Larson, G.J., and R.S. Bhattacharya. 1987. *Encyclopedia of Indian philosophies: Sankhya a dualist tradition in Indian philosophy*. New Jersey: Princeton University Press.
MacDonald, D.A. 2000. Spirituality: Description, measurement, and relation to the five factor model of personality. *Journal of Personality* 68(1): 153–197.
Maheshwar. 1978. *Bhagavad Gita – In the light of Sri Aurobindo*. Pondicherry: Sri Aurobindo Ashram Trust.
Marutham, P., J.P. Balodhi, and H. Mishra. 1998. Satva, Rajas, Tamas (SRT) inventory. *NIMHANS Journal* 16(1): 15–19.
Marutham, P. 1992. *Sattva, Rajas and Tamas Factors among college students*. Unpublished M. Phil dissertation, Banglore, NIMHANS.
Michel, P. 2006. *Upanishaden – Die Geheimlehre des Veda*. Wiesbaden: Marix Verlag.
Mohan, V., and S. Sandhu. 1986. Development of scale to measure Sattvic, Rajasic and Tamasic Guna. *Journal of Indian Academy of Applied Psychology* 12: 46–52.
Murthy, P.K., and S.K.K. Kumar. 2007. The concept of Triguna – A critical analysis and synthesis. *Psychological Studies* 52: 103–113.
Nilakantan, R. 1989. *Gitas in the Mahabharata and the Puranas*. Delhi: Nag Publishers.
Palsane, M.N., S.N. Bhavsar, R.P. Goswami, and G.W. Evans. 2002. The Concept of Stress in the Indian Tradition. In *Perspectives on Indigenous Psychology*, ed. G. Misra and A.K. Mohanty, 249–261. New Delhi: Concept Publishing Company.
Paranjpe, A.C., and K.R. Rao. 2008. Psychology in the Advaita Vedanta. In *Handbook of Indian psychology*, ed. K. Ramakrishna Rao, A.C. Paranjpe, and A.K. Dalal, 253–285. New Delhi: Foundation Books.
Prabhupada, A.C.B.S. 1987a. *Srimad Bhagavatam. First Canto*. Los Angeles: Bhaktivedanta Book Trust.
Prabhupada, A.C.B.S. 1987b. *Srimad Bhagavatam. Second Canto*. Los Angeles: Bhaktivedanta Book Trust.
Prabhupada, A.C.B.S. 1987c. *Srimad Bhagavatam. Third Canto – Part Two*. Los Angeles: Bhaktivedanta Book Trust.
Prabhupada, A.C.B.S. 1987d. *Srimad Bhagavatam. Fourth Canto – Part Two*. Los Angeles: Bhaktivedanta Book Trust.
Prabhupada, A.C.B.S. 1987e. *Srimad Bhagavatam. Tenth Canto – Part One*. Los Angeles: Bhaktivedanta Book Trust.
Prabhupada, A.C.B.S. 1989. *Bhagavad-gita as it is*, Secondth ed. Los Angeles: Bhaktivedanta Book Trust.
Prazak, M., J. Critelli, L. Martin, V. Miranda, M. Purdum, and C. Powers. 2012. Mindfulness and its role in physical and psychological health. *Applied Psychology: Health and Well-Being* 4(1): 91–105.
Ranade, S. 1994. *Ayurveda – Wesen und Methodik*. Heidelberg: Haug.
Rani, N.J., and M.U. Rani. 2009. Cumulative impact of yoga practice on Trigunas and on their autonomic nervous system correlates. *Journal of Indian Psychology* 27(1&2): 1–8.
Rao, K.R., A.C. Paranjpe, and A.K. Dalal (eds.). 2008. *Handbook of Indian psychology*. New Delhi: Cambridge University Press.
Rhyner, H. 2003. Allgemeine therapeutische Prinzipien. In *Ayurveda – Basislehrbuch*, ed. S. Ranade, C. Hosius, and J. Heckmann, 60–84. München: Urban & Fischer.
Ryff, C., and B.H. Singer. 2008. Know thyself and become what you are: A Eudaimonic approach to psychological well-being. *Journal of Happiness Studies* 9: 13–39.
Sedlmeier, P. 2006. Ancient Indian psychology: Can it offer anything to academic psychology? In *Perspectives on cognition: A Festschrift in honor of Manfred Wettler*, ed. R. Rapp, P. Sedlmeier, and G. Zunker-Rapp, 199–214. Lengerich: Pabst.
Sedlmeier, P. 2011. Indian psychology and the scientific method. In *Foundations of Indian psychology*, Concepts and theories, vol. 1, ed. R.M.M. Cornelissen, G. Misra, and S. Varma, 253–287. Delhi: Longman/Pearson.
Sedlmeier, P., J. Eberth, M. Schwarz, D. Zimmermann, F. Haarig, S. Jaeger, et al. 2012. The psychological effects of meditation: A meta-analysis. *Psychological Bulletin*. doi:10.1037/a0028168.

Seligman, M., T. Steen, N. Park, and C. Peterson. 2005. Positive psychology progress: Empirical validation of interventions. *American Psychologist* 60(5): 410–421.
Seligman, M., T. Rashid, and A. Parks. 2006. Positive psychotherapy. *American Psychologist* 61: 774–788.
Sharma, P.V. 1994. *Caraka Samhita*, vol. I. Delhi: Chaukhambha Orientalia.
Sharma, R. 1999. Self-concept and job satisfaction in Sattva, Rajas and Tamas personalities. *Journal of Indian Psychology* 17(2): 9–17.
Shilpa, S., and C.G.V. Murthy. 2011. Understanding personality from Ayurvedic perspective for psychological assessment: A case. *An International Quarterly Journal of Research in Ayurveda* 32(1): 12–18.
Singh, R. 1971. An inventory from Mahabharata. *Indian Journal of Psychiatry* 13: 149–61.
Sitamma, M. 2005. Trigunas: A Review of Empirical Studies. In *Towards a spiritual psychology – Essays in Indian psychology*, ed. K. Ramakrishna Rao and S.B. Marwaha, 262–276. New Dehli: Samvad India Foundation.
Stempel, H.S., S.E. Cheston, J.M. Greer, and C.K. Gillespie. 2006. Further exploration of the Vedic Personality Inventory: Validity, reliability and generalizability. *Psychological Reports* 98(1): 261–273.
Suneetha, S., and C.H. Srikrishna. 2009. Triguna personality theory: Classical representation and modern research. *Journal of Indian Psychology* 27(1 & 2): 35–46.
Swami, S. 2005. *The Gayatri book*. Schöna: Vasati Publishers.
Tagare, G.V. 1976a. *The Bhavata-Purana. Part 1*. Delhi: Motilal Banarsidass.
Tagare, G.V. 1976b. *The Bhavata-Purana. Part 3*. Delhi: Motilal Banarsidass.
Tagare, G.V. 1978. *The Bhavata-Purana. Part 5*. Delhi: Motilal Banarsidass.
Theodor, I. 2010. *Exploring the Bhagavad Gita – Philosophy, structure and meaning*. Ashgate: Surrey.
Underwood, L.G., and J.A. Teresi. 2002. The daily spiritual experience scale: Development, theoretical description, reliability, exploratory factor analysis, and preliminary construct validity using health-related data. *Annals of Behavioral Medicine* 24(1): 22–23.
Uma, K., Y.S. Lakshmi, and E.G. Parameswaran. 1971. Construction of a personality inventory based on doctrine of three gunas. *Reasearch Bulletin* 6: 49–58.
Verma, V. 1997. *Ayurveda – Gesund und erfolgreich im Alltag und Beruf*. Neuhausen: Urania Verlags AG.
Westhoff, K. 2005. Konstrukte und Operationalisierungen. In *Grundwissen für die berufsbezogene Eignungsbeurteilung nach DIN 33430*, ed. K. Westhoff, L.J. Hellfritsch, L.F. Hornke, K.D. Kubinger, F. Lang, H. Moosbrugger, A. Püschel, and G. Reimann, 128–142. Lengerich: Pabst.
Westhoff, K., C. Hagemeister, and A. Strobel. 2007. Decision-aiding in the process of psychological assessment. *Psychology Science* 49(3): 271–285.
WHO. 1948. *Constitution of the World Health Organization*. http://apps.who.int/gb/bd/PDF/bd47/EN/constitution-en.pdf. Accessed 11 July 2011.
Wolf, D.B. 1998. The Vedic Personality Inventory: A study of the Gunas. *Journal of Indian Psychology* 16(1): 26–43.
Wolf, D.B. 2008. *Relationships that work – the power of conscious living*. San Rafael: Mandala.
Wolf, D.B., and N. Abell. 2003. Examining the effects of meditation techniques on psychosocial functioning. *Research on Social Work Practice* 13(1): 27–42.

# Meditation: A Link to Spirituality and Health. A Novel Approach to a Human Consciousness Field Experiment

**Eduard van Wijk, John Ackerman, and Roeland van Wijk**

**Abstract** Spirituality is an experience that, in large part, comes from within. It is an attitude wherein one intuits the self as being a tiny part of a much bigger whole. Transcendental meditation (TM) helps individuals to deal better with stress as proven by a variety of biologic markers. The physiologic biomarkers, reactive oxygen species production (ROS), ultra-weak photon emission (UPE), electroencephalographic (EEG) data in conjunction with TM can not only begin to document the presence or absence of health, but also the degree of synchronization existing between systems of the body and mind. Much of the data that correlated with the above can demonstrate a reduction of ROS production which seems to facilitate a healthier biologic pathway.

## Background

A growing body of basic and clinical scientific evidence supports the longstanding hypothesis that mind-body effects might be evoked by various spiritual and religious practices. In 2001, the authors initiated a research program to study the role of spirituality in health care based on the ideas: (a) that everything is connected, All-in-One, and (b) then can be experienced within a milieu of wholeness in mind,

E. van Wijk (✉) • R. van Wijk
Meluna Research, Geldermalsen, The Netherlands

Sino Dutch Centre for Preventive and Personalized Medicine,
Leiden University, Leiden, The Netherlands
e-mail: epa.vanwijk@lacdr.leidenuniv.nl; roeland_van_wijk@meluna.nl

J. Ackerman
Cottage Hospital, Santa Barbara, CA, USA
e-mail: johnmackerman@gmail.com

S. Schmidt and H. Walach (eds.), *Meditation – Neuroscientific Approaches and Philosophical Implications*, Studies in Neuroscience, Consciousness and Spirituality 2,
DOI 10.1007/978-3-319-01634-4_19, 

body, spirit, community and environment. By what physiologic mechanisms do spirituality, prayer or meditation facilitate healing and health? Twenty first century physiologic technologies are beginning to piece together some answers regarding intention and non-intention, consciousness and unconsciousness, fields and healing.

Meditation can evoke mind-body effects. Meditative styles can be classified into two types – mindfulness and concentrative meditation – depending on how the attention processes are directed. Mindfulness practices involve allowing any thoughts, feelings, or sensations to arise while maintaining a specific attentional stance. Examples include Zen, Vipassana and the Western adaptation of mindfulness meditation. Concentrative meditation techniques involve focusing on specific mental or sensory activity: a repeated sound, an imagined image or specific body sensations such as the breath. Examples include forms of yogic meditation and the Buddhist Samatha meditation.

Transcendental meditation (TM) fits somewhat within the concentrative forms due to studies at practice centers focusing on the repetition of a mantra. However, the method places a primary emphasis on absence of concentrative effort and the development of a witnessing, thought-free "transcendental awareness." (Mahesh Yogi 1963; Travis et al. 2002). The experience in Transcendental Meditation (TM) is called the field of "absolute consciousness" (Being) (Maharish Mahesh Yoga 1975, 1986). Meditation in a group setting can facilitate, in almost everyone, the experience of "absolute consciousness". It is thought to eliminate disharmony and bring both the surroundings and the inner experience of an individual to the level of an "All-in-One" state of Being.

Meditation, even without spirituality, can have positive impact on physiologic parameter as measured for instance by the electroencephalogram (EEG), and molecular indicators as blood constituents that are related to reduced stress and health. Meditation may also have a preventive role by reducing human oxidative status, in particular the level of reactive oxygen species (ROS). Convincing evidence supports the role of ROS interactions with DNA, lipids and the aggregation of proteins that may, ultimately, result in development of chronic disease such as cancer, atherosclerosis and a number of brain pathologies (see for review: Van Wijk et al. 2008b).

Historically, during many different times, states of harmony or disharmony envelop a community or society. Based on the understanding of consciousness in the Vedic tradition, absolute (transcendental) consciousness has been speculatively equated with the unified field proposed by quantum field theory, which underlies all matter and energy in the universe (Dillbeck et al. 1981; Hagelin 1987, 1989; Orme-Johnson 2003; Orme-Johnson et al. 1988). The inference from the theory is that a shared feeling of harmony or disharmony inside the group may "radiate", without intention, to other people outside the group. The goal of this chapter is to address a way to document this.

If the activation of transcendental consciousness gives rise to beneficial biochemical and physiologic changes, then one might predict that individuals outside the group might also exhibit similar changes to those observed in the group of TM practitioners. A few studies have been performed regarding the influence of group meditation on subjects that were non-meditating and not involved in the practice of meditation (Pugh et al 1988). The non-meditating subjects in these types of recording studies are named "witness" since they may register the field's state without being the direct object of attention and intention of humans.

The Dutch study is progressing to utilize human subjects to "witness" changes in the field with the emphasis on physiologic parameters that are meaningful for health. In the following paragraphs, information is detailed regarding the concept of health which may lead to the selection of meaningful physiologic parameters (section "Defining Physiologic Biomarker Parameters Regarding Future Documentation of Health: EEG, Neuro-endocrine, Biochemical Stress Parameters, Reactive Oxygen Species (ROS) and Ultra-weak Photon Emission (UPE)").

## Defining Physiologic Biomarker Parameters Regarding Future Documentation of Health: EEG, Neuro-endocrine, Biochemical Stress Parameters, Reactive Oxygen Species (ROS) and Ultra-weak Photon Emission (UPE)

### *Neuro-endocrine Stress Parameters*

In the human experiments, health was related to the concept of stress-related neuro-endocrine changes. Investigations regarding stress-related neuro-endocrine changes began with the work of Selye (1936, 1956). Stressors are defined as "physiologic or psychologic perturbations that throw us out of homeostatic balance". This is followed by a stress-response defined as, "the set of neural and endocrine adaptations that help us to re-establish homeostasis" (Sapolsky 1992). Usually, the stress-response involves activation of the hypothalamic-pituitary-adrenocortical (HPA) axis resulting in increased secretion of the glucocorticoid cortisol. This is often referred to as the principal stress hormone. While increases in cortisol are often protective and adaptive in the short run, persistent activation of the HPA axis can cause prolonged elevation of cortisol that is damaging to health. Extreme, prolonged or chronic intermittent stress is known to cause long-lasting changes in the HPA axis and other systems of neuro-endocrine regulation resulting in detrimental health and maladaptive behaviors. Many of the latter can be traced to excess levels of cortisol or other glucocorticoids (McEwen 1998; Sapolsky 1992; Seeman et al. 1997).

The biochemical stress parameters have been utilized in studies regarding the influence of meditation. Meditation may combat the problem of stress; an ultimate goal would be to restore optimal functioning to neuro-endocrine mechanisms that maintain homeostasis. Programs that reduce or reverse the effects of stress produce a variety of benefits for physical and mental health in those individuals (e.g. Alexander et al. 1996; Castillo-Richmond et al. 2000; Jevning et al. 1992; Orme-Johnson and Walton 1998). Several studies support the hypothesis that long-term practice of TM can reverse enduring negative effects of stress on neuro-endocrine regulation that include reduction of cortisol secretion and other biochemical indicators of stress (Jevning et al. 1978; MacLean et al. 1997; Walton et al. 1995). Increased excretion of 5-hydroxyindoleacetic acid (5-HIAA; a main metabolite of serotonin) has also been reported in individuals practicing TM (Bujatti and Riederer 1976; Walton et al. 1995). These increases were found to be correlated with reductions in anger, anxiety, aggression and other negative emotions (Walton and Pugh 1995). Cross-sectional comparison of TM practitioners and non-meditating controls found that the ratio of 5-HIAA to cortisol demonstrated a greater difference between the groups than were found for 5-HIAA or cortisol secretion alone (Walton and Pugh 1995). The biochemical changes described above have been linked to decreased health problems such as hypertension, heart disease (Walton et al. 1995; Alexander et al. 1996; Senders et al. 2012; Bhasin et al. 2013), reduced substance abuse and improvement in the rehabilitation of criminal offenders (Walton and Levitsky 1994, 2003; Himelstein 2011).

## *Other Physiologic Parameters*

Collaborations between members of meditative traditions and neuroscientists have begun to distil the range of neurophysiologic changes from long-term contemplative practice (Goleman 2003; Mason et al. 1997; Rapgay et al. 2000; Travis et al. 2004). A current review of meditation state and trait indicates considerable discrepancy among results although EEG meditation studies have produced a reasonable consistency (Cahn and Polich 2006). Electroencephalography (EEG) is one of the most frequently used objective techniques to study the human brain and its relationship to consciousness and meditation.

EEG, which measures the patterns of brain wave activity, provides an overall measure of electrical activity of specific regions of the brain. The findings from EEG meditation studies have demonstrated that alpha power increases are often observed when meditators are evaluated during meditating compared with control conditions and this band is stronger at rest in meditators compared with nonmeditator controls. It suggests that both state and trait alpha changes emerge during meditation practice. In addition, theta power increases for meditative practice have been widely reported. Another interesting field refers to EEG coherence (i.e., the squared cross-correlation between EEG power from two scalp locations within a frequency band). It indexes the functional covariation of activity among different cortical areas. Increased alpha–theta range coherence among recording sites has been observed intra- and interhemispherically for state effects during meditation (Aftanas and Golocheikine 2001; Gaylord et al. 1989; Travis 2001; Travis and Pearson 1999; Travis and Wallace 1999; see also the chapter by Esch in this volume).

The impact of meditation can be observed directly utilizing the EEG. However, to find a physiologic connection between meditation and health vs. disease, further steps are required. This issue will now be further developed.

## *Reactive Oxygen Species (ROS) and Health*

In 1954 Gerschman and Gilbert proposed that most of the damaging effects of elevated oxygen concentrations in living organisms might be attributed to the formation of free radicals (Gerschman 1981). Aerobic life is connected with the continuous production of free radicals, particularly radical oxygen species (ROS) that might be dangerous for the living organism. The reactive species attack biomolecules producing alterations in DNA, proteins and lipids and have been implicated in the pathogenesis of age-related pathologies (reviewed in Van Wijk et al. 2008b).

In 1956, Harman proposed the "free radical theory of aging" which suggested that free radical damage of cellular macromolecules is responsible for the aging process (Harman 1956). However, this idea did not capture the interest of many biologists and clinicians until the discovery in 1969 of the enzyme, superoxide dismutase (SOD) with the function of catalytically removing a specific free radical. In the 1970s and 1980s, many scientists, unfamiliar with free radicals, regarded the field as highly specialized or irrelevant to mainstream biology, biochemistry and medicine. In fact, however, it is just the opposite.

Reactive oxygen species are physiologically controlled and released in an orderly fashion to help avoid damage to vital components. To maintain cell and tissue integrity, the antioxidant system maintains a "check and balance" over their production. One such balancing system is the repertoire of antioxidant protection that includes antioxidants, such as vitamins C or E, protective enzymes, such as gluthatione, coenzymes, such as enzyme Q, and regenerating pathways.

Damage to DNA, lipid and protein structure may not be long-lasting. In cells, repair systems are present for each of these types of damage. As is, perhaps, not surprising for a molecule so important to an organism's health and survival, a variety of mechanisms have evolved for repairing DNA. Hundreds of proteins are involved in repair processes as part of excision repair pathways correcting DNA defects. Lipid structures are easier to repair. The fluid nature of the membrane is a critical feature that, rather than rigidly locking lipid components in place, allows the lipid components to constantly move both within the membrane and in their exchange processes between membrane and cytoplasmic locations.

The third type of damage impacting proteins seems the most complex. The defense tier regarding age-related pathologies addresses the faulty assembly and aggregation of proteins. It is activated when the above two major tiers of cellular defense are insufficient. It is hypothesized that cellular protein quality dyscontrol is the underlying common denominator of these diseases (Dobson 2001). The protein quality control system that prevents chronic disease development and shields the organism from the impact of free radical damage is based on the role of Heat Shock Stress Proteins (HSP's). The HSP's or stress proteins are also classified by their function as "chaperone proteins". They form complexes with protein and other cellular structures in an effort to prevent deleterious interactions between proteins. Each protein in a cell has its own intrinsic propensity to spontaneously unfold and misfold. A continuous flux of toxic, misfolded proteins can be spontaneously formed during the lifetime of a cell. The chaperones disentangle stable dysfunctional aggregated proteins by unfolding and refolding in order re-stabilize them into "re-educated and born again" native, functional cellular proteins (Shtilerman et al. 1999; Ben-zvi and Goloubinoff 2001).

During extreme situations such as oxidative stress, chaperone systems become overloaded by toxic protein forms (Westerheide and Morimoto 2005). In addition, as we age, fewer molecular chaperones are produced and we may also react poorly to environmental stress (Hinault and Goloubinoff 2006). Simultaneously, irreversibly damaged proteins accumulate (Heydari et al. 2000; Soti and Csermely 2003). In addition to their general cytotoxic effect, irreversibly damaged proteins can inhibit the activity of the remaining minority of functional chaperones and proteases. The misfolded proteins and aggregates that are hallmarks of degenerative disorders have been, in some cases, identified and classified.

## *Ultra-weak Photon Emission (UPE) and Reactive Oxygen Species (ROS)*

Ultra-weak photon emission is a non-invasive, quantitative technique that registers light emanating from biochemical interactions that produce reactive oxygen species (ROS). The quantity of light production is directly correlated with ROS production. Many techniques are available to measure ROS and changes in oxidative status. Most of them utilize ROS products. These measurements require tissue-invasive techniques and are not available for non-invasive, on line recording (Gutteridge and Halliwell 1990).

Many years ago, UPE was accepted by the scientific world as a method of validating the production of the electronically-excited states in biological systems particularly, the oxygen dependent chain reactions involving ROS (McElroy and Seliger 1962, 1963; Tarusov et al. 1961, 1962; Baremboim et al. 1969; Seliger and McElroy 1960). Early research on low-level chemiluminescence was largely unnoticed. However, reports on light emission during lipid peroxidation, both in isolated microsomes and during other oxidative reactions, revived the interest in chemiluminescence and suggested its use as a tool for the investigation of the radical reactions under physiologic conditions (Nakano et al. 1975; Sugioka and Nakano 1976; Hamman et al. 1977).

The most important aspect of organ chemiluminescence is that it provides, on a non-invasive, non-destructive basis, a signal of oxidative metabolism that is readily and continuously detectable. In that respect, chemiluminescence has been favored compared to other indirect, invasive biochemical assays (Chance et al. 1979; Cadenas et al. 1981; Riely et al. 1974; Dillard et al. 1977; Okuda et al. 1992).

## *Ultra-weak Photon Emission (UPE) and Humans*

Since 2002 the authors standardized protocols for (a) the management of human subjects in UPE measurement protocols, and (b) technical requirements for the recording of dynamic and steady state UPE characteristics. In the initial studies, subjects were recorded in a specially designed dark room. They were positioned on a bed under a hanging, computerized, moveable photomultiplier system which almost touched targeted anatomic locations (Van Wijk and Van Wijk 2004, 2005a, b, 2006). The photomultiplier tube has a 5 cm opening for recording anatomic surface areas. Highly reliable data documented by wavelength analysis of UPE from different anatomic locations revealed spontaneous emissions mainly in the 470–570 nm range. This range corresponds to the emission range of specific electron-excited states.

From 2003–2006, the overhead system systematically registered UPE from 60 male and 25 female healthy subjects. A "common" human body emission pattern became apparent. From subject to subject, anatomic emission distribution percentage did not differ (Van Wijk et al. 2006a, b, c, 2008a). In collaboration with Dr. M. Kobayashi (Tohoku Institute of Technology, Sendai, Japan) the unique and highly sensitive charge-coupled device (CCD) imaging system was utilized to independently confirm anatomic differences within subjects (Van Wijk et al. 2006a, b, c, 2008a). However, subjects do differ in photon strength (intensity) from identical anatomic areas.

## *Ultra-weak Photon Emission (UPE) in Meditation*

The general goal of biological systems during oxidative stress is to attenuate ROS production and mediated reactions. A role of meditation has been suggested in three studies utilizing serum lipid peroxide levels to estimate oxidative stress. Data have demonstrated that serum lipid peroxide levels have actually decreased in long-time practitioners of TM (Schneider et al. 1998), Zen meditation (Kim et al. 2005) and yoga (Yadav et al. 2005). The effects of the meditative life style were confirmed utilizing ultra-weak photon emission. The data demonstrated that the intensity of UPE compared to controls was 27 % lower for those practicing TM and 17 % lower for those practicing a form of meditation other than TM (OTM) (Van Wijk et al. 2006b, 2008a). UPE data have also validated that the anatomic percentage emission patterns of the TM, OTM and control groups were similar.

UPE was also studied in experienced meditators immediately before and after meditation. Besides intensity, the influence of meditation on the fluctuations of photon signals was estimated to study the fractal properties of the signal. Fractals have attracted wide attention in mathematics and physical sciences and they are now being used in biology and medicine (Teich 1989; Bassingthwaighte et al. 1994; Kobayashi et al. 1998; Gebber et al. 1999; Lewis et al. 2001; Van Wijk et al. 2005, 2010, 2011; Herman et al. 2009; Werner 2010). In essence, fractals represent objects or processes whose small pieces resemble the whole. When a non-fractal process is magnified by studying its fluctuation characteristics over longer time periods, no new features are revealed. In contrast, as a fractal process is magnified, ever finer features are revealed. Thus the fluctuation characteristics depend on the time resolution used to make the measurement. This analysis utilizes the photon number distribution in consecutive time windows, or time dependent Fano factor. The Fano factor is equal to the variance divided by the mean of the number of events, in this case photon counts, in the time windows. This type of analysis has been regularly used to study fractal properties of a phenomenon. It was utilized before by Teich (1989), Lewis et al. (2001) and Gebber et al. (1999) to describe the properties of action potentials. F(T) was used by Vekaria (2003), Kobayashi and colleagues (1998) and Van Wijk and colleagues (2005, 2010, 2011) to evaluate UPE data. Photon emission in the pre-meditation period was characterized by an increase in the Fano factor values when window sizes rose above 6 s (Van Wijk et al. 2005). This means that before meditation the count number distribution differs from a Gaussian distribution and that the distributions became rougher as the length of the time window was increased. This roughness arises from long-range correlations of photon events. We determined that calculations for the Fano factor curve are correct and truly reflect a fractal process by constructing simulated data sets in which we used sets of randomized data. The situation after meditation was different. The increase in Fano factor with window size in post-meditation never rose above the level of the surrogate data set (Van Wijk et al. 2005). Fractality of pre- and post-meditation UPE were significantly different for the subjects ($p<0.05$). The data led to the conclusion that meditation leads to a decrease in UPE fractal properties. The decrease in correlation, i.e., photon clustering has been discussed in relation to the habitual response in classical conditioning and the role of contextual stimuli in the "associative strength" in classical conditioning (Van Wijk et al. 1997a, b). It is in favor of the opinion that meditation lowers the associative strength with respect to conditioned responses that are non-functional in adaptation. The system tends towards an increased balance, a view that corresponds with the observations that long-term practice initiates a substantial decrease of free radical levels and thus UPE.

## Towards a Human Witness Model: Hypothesis and Testing Technical Possibilities

### *Human Subject as a Witness: A Perspective for Future Experimental Setting*

Information has pointed to the Human Witness Model based on the neuro-endocrine stress parameters. Increased cortisol secretion and decreased levels of 5-HIAA, have been studied in human "witness" subjects in relationship to the group practice of TM. A time series quasi-experiment demonstrated the influence of a TM group practice program on overnight 5-HIAA excretion in different samples of non-meditators that were not present at the group meditation site (Pugh et al. 1988). These findings were consistent with another previous time series study that examined 5-HIAA excretion in TM-group participants (Löliger 1990). A long-term quasi-time series experiment of 77 days focused on the influence of changes in group size regarding biochemical stress parameters. An increase in the day-to-day change in the size of the afternoon TM group session was a significant predictor of the decreased urinary cortisol excretion that same night. Such increases in the size of the group were also a significant predictor of subsequent increases in 5-HIAA excretion as well as in the 5-HIAA-to-cortisol ratio.

## *Hypothesis*

The data suggest meditation to produce the influence of pure consciousness in the surrounding environment and improve the surrounding's capability to experience harmony and state of health. Accordingly, the hypothesis can be erected that a human subject is an information-processing system induced by the surrounding's "field" of potential information without the surrounding field paying direct attention to the human witness. Validating this Human witness hypothesis might be accomplished by placing a human witness adjacent to a group of highly experienced TM meditators at the same time that the group begins to meditate. It is essential that the witness not be aware of the beginning and end of the group meditation and that the witness already be, for some time, in a baseline emission recording mode.

## *Technical Facilities: A Photomultiplier Device*

The device is based on research data documenting that common anatomic photon emission percentage does exist. Theoretically, this means that the registration of only one or a few anatomic locations are sufficient to characterize photon emission properties of a subject. Registration of photon characteristics of the hands documented definite advantages to other locations. The hands are excellent to use for UPE registration since they exhibit relatively high photon strength culminating in a relatively high signal to noise ratio. They do have relationships for UPE parameters that correlate with other anatomic locations and can be easily recorded in the sitting position (Van Wijk et al. 2006a, b, c, 2008a).

This photomultiplier device was designed to facilitate the simultaneous recording of both hands of a subject. This prototype is ambulatory (50×35×20 cm; weight 25 kg) and is compatible with any laboratory and clinical practice for physiologic registration in combination with other equipment. The central part of the prototype device is a metal box internally divided into a left and right chamber. Each chamber has an opening for a subject to place one hand into that chamber. The inserted arm can be covered with a flexible light-tight tube of material that is fixed around the arm. A photomultiplier tube (25 cm) can be positioned on top of each chamber. The photomultiplier tubes have a spectral sensitivity range of 200–650 nm that facilitates recording specific ROS-related processes. It is placed in an 18 °C climate in a dimly lit room. The temperature in the registration chambers is maintained constant to facilitate low electronic background noise.

## *Improvement of the Mediation – EEG – ROS – UPE – Health Connection: Cross Correlation (Cross Coherence) in Humans Between the Electro-cortical and the ROS-systems; Tuning with Oscillating Processes*

The photomultiplier device can facilitate the simultaneous recording of UPE and EEG. Studies on the fractal properties of UPE have suggested fluctuations which appear when periods of more than 5 s were taken into account. Data also suggested that these fluctuations were influenced by meditation. Although these fluctuations are not always easy to discriminate from the overall fluctuations in UPE signals, they may show cross coherence with long-term fluctuations in EEG signals.

Studies with human subjects have demonstrated two types of coherence: coherence in this context is defined as synchronized interactions between rhythmic oscillations of various physiologic systems. This synchronization can exist between identical parts of the same system (auto coherence). Examples include neuronal coherence, muscular coherence, cardiac coherence and sympathetic coherence. The other type is cross coherence, synchronization between different systems in the same person such as cortico-muscular coherence, cardio-respiratory synchronization, somato-sympathetic synchronization and heart-brain synchronization. The combination of both types of synchronizations may be an important organizing factor that facilitates optimal functioning in a living system.

It is anticipated that these data can provide a better understanding of health and meditation when fluctuations in the ROS system, as recorded by UPE, were included and combined with fluctuations in brain states.

In one study, the prototype was utilized with only one photomultiplier tube to study cross-correlation between fluctuations of photon emission intensity and alpha activity. Subjects sat in front of the device with the right hand inserted into the dark chamber. The subject was connected simultaneously to an EEG system that recorded electro-cortical activity at P3 and P4. Both physiologic technologies recorded simultaneously, synchronously and continuously, both documenting fluctuations in time. Photon emission of the right hand was registered with 50 ms dwell times; mean photon strength was computed for 5 s periods. EEG was registered with a sample rate of 125 Hz per channel and a FFT frequency spectrum that was computed for each 5 s. The electro-cortical and ROS cross- coherence was detected in subjects with a sufficiently high mean photon emission (Van Wijk et al. 2008c). The analysis of time periods of different durations suggested that periods with a high degree of cross-coherence alternate with periods without detectable cross-coherence.

## *Application in Field Studies*

The registration of both EEG and ROS activity of the human witness makes it possible to document baseline values for EEG, UPE and EEG-UPE cross-coherence compared to changes recorded in the witnesses while the group is meditating.

Human on-line recordings of diverse system fluctuations plus data analysis of coherence, such as the relationship between electro-cortical activities and the system of ROS-mediated reactions, may be a highly sensitive model that can estimate one's state of health. The authors propose to utilize this model in a final study of the "field" emanating from transcendental consciousness and its relationship to All-in-One interconnectedness.

## Conclusion and Perspective

This chapter focuses on meditation research in order to demonstrate its role in helping to create a healing environment. Other prior physiologic studies utilizing the electroencephalogram and blood constituents have highlighted the fact that meditation can have a positive impact on stress. This chapter addresses "Reactive oxygen species (ROS) in both health and disease" providing evidence that the role of ROS interactions with DNA, lipids and certain aggregations of proteins may, ultimately, result in the development of chronic disease (i.e., cancer, atherosclerosis, and a number of brain pathologies). The general goal of biological systems during oxidative stress is to attenuate ROS production and its mediated reactions. A role of meditation (vis-à-vis TM, Zen, or yoga) in reducing ROS levels has been suggested in studies utilizing serum lipid peroxide levels to estimate oxidative stress.

Ultra-weak photon emission (UPE) is a non-invasive, quantitative technique that records light emitted from human skin emanating from certain specific biochemical interactions that produce reactive oxygen species (ROS). The positive health results of a meditative life style were confirmed utilizing the recording of UPE. The use of the "meditation/EEG/ROS(UPE)/health" connection has been recently improved vis-à-vis the estimation of the cross correlation between the fluctuations of the electro/cortical and the ROS systems. The photomultiplier device can facilitate the simultaneous recording of UPE and EEG. Human on-line recordings of system fluctuations plus the relationship between electro/cortical activities and the system of ROS mediated reactions may be a highly sensitive model that can estimate a person's state of health.

Such experimental methodology and technology is now available to test the hypothesis that a shared feeling of harmony inside of a group meditation may "radiate" to other people outside the group. One hypothesis, then, might be that individuals close to but outside the group might also exhibit harmony similar to those members inside the group. The people both inside and outside of the meditation group are also expected to improve regarding physiologic health parameters. Validating this "human witness" hypothesis might be accomplished by placing a human "witness" adjacent to a group of highly experienced meditators and record the "meditation/EEG/UPE" connection simultaneously once the group members begin to meditate. The simultaneous recording of both EEG and ROS of the human witness outside of group facilitates the possibility to document baseline values for EEG, UPE and EEG/UPE cross-coherence compared to changes recorded of the members (witnesses) meditating in the group.

The perspective of the authors is to utilize this experimental research model in order to do a study of a "field" which might emanate from a group "transcendental consciousness" that has been speculatively equated with a unified field (as for instance in the Vedic tradition) underlying all matter and energy.

**Acknowledgments** This work was supported by the Samueli Institute for Information Biology (grant number: 2 EA-0000047), the BIAL foundation (grant numbers: 81/04; 71/06) and Het Johan Borgman Fonds.

## References

Aftanas, L.I., and S.A. Golocheikine. 2001. Human anterior and frontal midline theta and lower alpha reflect emotionally positive state and internalized attention: High-resolution EEG investigation of meditation. *Neuroscience Letters* 310: 57–60.

Alexander, C.N., R.H. Schneider, F. Staggers, W. Sheppard, B.M. Clayborne, M. Rainforth, et al. 1996. Trial of stress reduction for hypertension in older African American: II. Sex and risk subgroup analysis. *Hypertension* 28: 228–237.

Baremboim, G.M., A.N. Domannski, and K.K. Taroverov. 1969. *Luminescence of biopolymers and cells*. New York: Plenum.

Bassingthwaighte, J.B., L.S. Liebovitsch, and B.J. West. 1994. *Fractal physiology*. New York: Oxford University Press.

Ben-zvi, A.P., and P. Goloubinoff. 2001. Mechanisms of disaggregation and refolding of stable protein aggregation by molecular chaperones. *Journal of Structural Biology* 135: 84–93.

Bhasin, M.K., J.A. Dusek, B.H. Chang, M.G. Joseph, J.W. Denninger, G.L. Fricchione, et al. 2013. Relaxation response induces temporal transcriptome changes in energy metabolism, insulin secretion and inflammatory pathways. *PLoS One* 8(5): e62817.

Bujatti, M., and P. Riederer. 1976. Serotonin, noradrenaline, dopamine metabolites in transcendental meditation technique. *Journal of Neural Transmission* 39: 257–267.

Cadenas, E., H. Wefers, and H. Sies. 1981. Low-level chemiluminescence of isolated hepatocytes. *European Journal of Biochemistry* 119: 531–536.

Cahn, B.R., and J. Polich. 2006. Meditation states and traits: EEG, ERP, and neuroimaging studies. *Psychological Bulletin* 132: 180–211.

Castillo-Richmond, A., R.H. Schneider, C.N. Alexander, R. Cook, H. Myers, S. Nidich, et al. 2000. Effects of stress reductions on carotid atherosclerosis in hypertensive African Americans. *Stroke* 31: 568–573.

Chance, B., H. Sies, and A. Boveris. 1979. Hydroperoxide metabolism in mammalian organs. *Physiological Reviews* 59: 527–605.

Dillard, C.J., E.E. Dumelin, and A.L. Tappel. 1977. Effect of dietary vitamin E on expiration of pentane and ethane by the rat. *Lipids* 12: 109–114.

Dillbeck, M.C., D.W. Orme-Johnson, and R.K. Wallace. 1981. Frontal EEG coherence, H-reflex recovery, concept learning, and the TM-Sidhi program. *International Journal of Neuroscience* 15: 151–157.

Dobson, C.M. 2001. Protein folding and its links with human disease. *Biochemical Society Symposia* 68: 1–26.

Gaylord, C., D. Orme-Johnson, and F. Travis. 1989. The effects of the transcendental meditation technique and progressive muscle relaxation on EEG coherence, stress reactivity, and mental health in black adults. *International Journal of Neuroscience* 46: 77–86.

Gebber, G.L., S. Zhong, C. Lewis, and S.M. Barman. 1999. Differential patterns of spinal sympathetic outflow involving a 10-Hz rhythm. *Jornal of Neurophysiology* 89: 833–840.

Gerschman, R. 1981. Historical introduction to the "free radical theory" of oxygen toxicity. In *Oxygen and living processes, an interdisciplinary approach*, ed. D.L. Gilbert. New York: Springer.

Goleman, D.J. 2003. *Destructive emotions: How can we overcome them? A scientific dialogue with the Dalai Lama*. New York: Bantam Books.
Gutteridge, J.M.C., and B. Halliwell. 1990. The measurement and mechanisms of lipid peroxidation in biological systems. *Trends in Biochemical Sciences* 15: 129–135.
Hagelin, J.S. 1987. Is consciousness the unified field? A field theorist's perspective. *Modern Science and Vedic Science* 1: 29–87.
Hagelin, J.S. 1989. Restructuring physics from its foundation in the light of Maharishi's Vedic Science. *Modern Science and Vedic Science* 3: 3–72.
Hamman, J.P., D.R. Gorby, and H.H. Seliger. 1977. A new type of biological chemiluminescence: The microsomal chemiluminescence of benzo[a]pyrene arises from the diol epoxide product of the 7, 8-dihydrodiol. *Biochemical and Biophysical Research Communications* 75: 793–798.
Harman, D. 1956. Aging, a theory based on free radical and radiation chemistry. *Journal of Gerontology* 11: 298–305.
Herman, P., L. Kocsis, and A. Eke. 2009. Fractal characterization of complexity in dynamic signals: Application to cerebral hemodynamics. *Methods in Molecular Biology* 489: 23–40.
Heydari, A.R., S. You, R. Takahashi, A. Gutsmann-Conrad, K.D. Sarge, and A. Richardson. 2000. Age-related alterations in the activation of heat shock transcription factor 1 in rat hepatocytes. *Experimental Cell Research* 256: 83–93.
Himelstein, S. 2011. Meditation research: The state of the art in correctional settings. *International Journal of Offender Therapy and Comparative Criminology* 55: 646–661.
Hinault, M.P., and P. Goloubinoff. 2006. Molecular chaperones and proteases: Cellular fold-controlling factors of proteins in neurodegenerative diseases and aging. *Journal of Molecular Neuroscience* 55: 74–78.
Jevning, R., A.F. Wilson, and J.M. Davidson. 1978. Adrenocortical activity during meditation. *Hormones and Behavior* 10: 54–60.
Jevning, R., R.K. Wallace, and M. Beldeback. 1992. The physiology of meditation: A review. A wakeful hypometabolic integrated response. *Neuroscience and Biobehavioral Reviews* 16: 415–424.
Kim, D.H., Y.S. Moon, H.S. Kim, J.S. Jung, H.M. Park, and H.W. Suh. 2005. Effect of Zen Meditation on serum nitric oxide activity and lipid peroxidation. *Progress in Neuro-Psychopharmacology & Biological Psychiatry* 29: 327–331.
Kobayashi, M., B. Devaraj, and H. Inaba. 1998. Observation of super-Poisson statistics of bacterial (*Photobacterium phosphoreum*) bioluminescence during the early stage of cell proliferation. *Physical Review E* 57: 2129–2133.
Lewis, C.D., G.L. Gebber, P.D. Larsen, and S.M. Barman. 2001. Long-term correlations in the spike trains of medullary sympathetic neurons. *Journal of Neurophysiology* 85: 1614–1622.
Löliger, S.A. 1990. *Relationship between subjective bliss, 5-hydroxy-3-indoleacetic acid and the collective practice of Maharishi's transcendental meditation and TM-Sidhi program*. Fairfield: Maharishi University of Management.
MacLean, C.R.K., K.G. Walton, S.R. Wenneberg, D.K. Levitsky, J.V. Mandarino, R. Waziri, et al. 1997. Effects of the transcendental meditation program on adaptive mechanisms: Changes in hormone levels and responses to stress after 4 months of practice. *Psychoneuroendocrinology* 22: 277–295.
Maharishi Mahesh Yogi. 1963. *The science of being and art of living*. New York: Penguin.
Maharishi Mahesh Yogi. 1975. Inauguration of the dawn of the age of enlightenment. Rheinweiler: Maharishi International University Press.
Maharishi Mahesh Yogi. 1986. *Life supported by natural law*. Washington, DC: Age of Enlightenment Press.
Mason, L.I., C.N. Alexander, F.T. Travis, G. Marsh, D.W. Orme-Johnson, J. Gackenbach, et al. 1997. Electrophysiological correlates of higher states of consciousness during sleep in long-term practitioners of the transcendental meditation program. *Sleep* 20: 102–110.
McElroy, W.D., and H.H. Seliger. 1962. Origin and evolution of bioluminescence. In *Horizons in biochemistry*, ed. M. Kasha and B. Pullman, 91–101. New York: Academic.
McElroy, W.D., and H.H. Seliger. 1963. The chemistry of light emission. *Advances in Enzymology and Related Areas of Molecular Biology* 25: 119–166.

McEwen, B.S. 1998. Protective and damaging effects of stress mediators. *New England Journal of Medicine* 338: 171–179.

Nakano, M., T. Noguchi, K. Sugioka, H. Fukuyama, and M. Sato. 1975. Spectroscopic evidence for the generation of singlet oxygen in the reduced nicotinamide adenine dinucleotide phosphate-dependent microsomal lipid peroxidation system. *Journal of Biological Chemistry* 250: 2404–2406.

Okuda, M., H.C. Lee, C. Kumar, and B. Chance. 1992. Oxygen radical generation during ischemia-reperfusion in the isolated perfused rat liver monitored by enhanced chemiluminescence. *Circulatory Shock* 38: 228–237.

Orme-Johnson, D.W. 2003. Preventing crime through the Maharishi Effect. *Journal of Offender Rehabilitation* 36: 257–281.

Orme-Johnson, D.W., and K.G. Walton. 1998. All approaches to preventing or reversing effects of stress are not the same. *American Journal of Health Promotion* 12: 297–299.

Orme-Johnson, D.W., C.N. Alexander, J.L. Davies, H.M. Chandler, and W.E. Larimore. 1988. International peace project in the Middle East: Effects of the Maharishi technology of the unified field. *Journal of Conflict Resolution* 32: 776–812.

Pugh, N.D.C., K.G. Walton, and K.L. Cavanaugh. 1988. Can time series analysis of serotonin turnover test the theory that consciousness is a field? *Society for Neuroscience Abstracts* 14: 372.

Rapgay, L., V.L. Rinpoche, and R. Jessum. 2000. Exploring the nature and functions of the mind: A Tibetan Buddhist meditative perspective. *Progress in Brain Research* 122: 507–515.

Riely, C.A., G. Cohen, and M. Lieberman. 1974. Ethane evolution: A new index of lipid peroxidation. *Science* 183: 208.

Sapolsky, R.M. 1992. *Stress, the aging brain, and the mechanisms of neuron death*. Cambridge: MIT Press.

Schneider, R.H., S.I. Nidisch, J.W. Salerno, H.M. Sharma, C.E. Robinson, R.J. Nidich, et al. 1998. Lower lipid peroxide levels in practitioners of the transcendental meditation program. *Psychosomatic Medicine* 60: 38–41.

Seeman, T.E., B.H. Singer, J.W. Rowe, R.I. Horwitz, and B.S. McEwen. 1997. Price of adaptation – Allostatic load and its health consequences. *Archives of Internal Medicine* 157: 2259–2268.

Seliger, H.H., and W.D. McElroy. 1960. Spectral emission and quantum yield of firefly bioluminescence. *Archives of Biochemistry and Biophysics* 88: 136–141.

Selye, H. 1936. A Syndrome produced by diverse nocuous agents. *Nature* 138: 32.

Selye, H. 1956. *The stress of life*. New York: McGrawHill.

Senders, A., H. Wahbeh, R. Spain, and L. Shinto. 2012. Mind-body medicine for multiple sclerosis: A systematic review. *Autoimmune Diseases*. doi:10.1155/2012/567324.

Shtilerman, M., G.H. Lorimer, and S.W. Englander. 1999. Chaperonin function: Folding by forced unfolding. *Science* 284: 822–825.

Soti, C., and P. Csermely. 2003. Aging and molecular chaperones. *Experimental Gerontology* 38: 1037–1040.

Sugioka, K., and M. Nakano. 1976. A possible mechanism of the generation of singlet molecular oxygen in NADPH-dependent microsomal lipid peroxidation. *Biochimica et Biophysica Acta* 423: 203–216.

Tarusov, B.N., N.A.I. Polidova, and A.I. Zhuravlev. 1961. Study on ultra-weak spontaneous luminescence of animal cells. *Biofizika* 6: 490–492.

Tarusov, B.N., N.A.I. Polidova, A.I. Zhuralev, and E.N. Sekamova. 1962. Ultraweak spontaneous luminescence in animal tissue. *Tsitologiya* 4: 696–699.

Teich, M.C. 1989. Fractal character of the auditory neural spike train. *IEEE Transactions on Biomedical Engineering* 36: 150–160.

Travis, F. 2001. Autonomic and EEG patterns distinguish transcending from other experiences during transcendental meditation practice. *International Journal of Psychophysiology* 42: 1–9.

Travis, F., and C. Pearson. 1999. Pure consciousness: Distinct phenomenological and physiological correlates of "consciousness itself". *International Journal of Neuroscience* 100: 77–89.

Travis, F., and R.K. Wallace. 1999. Autonomic and EEG patterns during eyes-closed rest and transcendental meditation (TM) practice: The basis for a neural model of TM practice. *Consciousness and Cognition* 8: 302–318.

Travis, F., J. Tecce, A. Arenander, and R.K. Wallace. 2002. Patterns of EEG coherence, power, and contingent negative variation characterize the integration of transcendental and waking states. *Biological Psychology* 61: 293–319.
Travis, F., A. Arenander, and D. DuBois. 2004. Psychological and physiological characteristics of a proposed object-referral/self-referral continuum of self-awareness. *Consciousness and Cognition* 13: 401–420.
Van Wijk, R., and E.P.A. Van Wijk. 2004. Human photon emission. *Recent Research Developments in Photochemistry & Photobiology* 7: 139–173.
Van Wijk, E.P.A., and R. Van Wijk. 2005a. Multi-site registration and spectral analysis of spontaneous emission from human body. *Forschende Komplementärmedizin/Research in Complementary Medicine* 12: 96–106.
Van Wijk, R., and E.P.A. Van Wijk. 2005b. An introduction to human biophoton emission. *Forschende Komplementärmedizin/Research in Complementary Medicine* 12: 77–83.
Van Wijk, E.P.A., and R. Van Wijk. 2006. Ultra weak photon emission of human body. In *Biophotonics – Optical science and engineering for the 21st century*, ed. X. Shen and R. Van Wijk, 171–184. New York: Kluwer.
Van Wijk, E.P.A., J.H.R. Maes, and J.M.H. Vossen. 1997a. Competition between a simultaneous and a serial feature in an appetitive feature-positive and feature-negative discrimination. *The Quarterly Journal of Experimental Psychology* 50: 17–39.
Van Wijk, E.P.A., J.H.R. Maes, and J.M.H. Vossen. 1997b. Interference with potential contextual occasion setting by a conditioned inhibitor in an appetitive discrimination procedure. *Learning and Motivation* 28: 382–403.
Van Wijk, E.P.A., J. Ackerman, and R. Van Wijk. 2005. Effect of meditation on ultraweak photon emission from hands and forehead. *Forschende Komplementärmedizin/Research in Complementary Medicine* 12: 107–112.
Van Wijk, E.P.A., M. Kobayashi, and R. Van Wijk. 2006a. Spatial characterization of human ultra-weak photon emission. In *Biophotonics and coherent systems in biology*, ed. L. Beloussov, V.L. Voeikov, and V.S. Martynyuk, 177–190. New York: Kluwer.
Van Wijk, E.P.A., H. Koch, S. Bosman, and R. Van Wijk. 2006b. Spatial characterization of human ultra-weak photon emission in TM practitioners and control subjects. *Journal of Alternative and Complementary Medicine* 12: 31–38.
Van Wijk, R., M. Kobayashi, and E.P.A. Van Wijk. 2006c. Spatial characterization of human ultra-weak photon emission. *Journal of Photochemistry & Photobiology B* 83: 69–76.
Van Wijk, E.P.A., R. Ludtke, and R. Van Wijk. 2008a. Differential effects of relaxation techniques on ultra-weak photon emission. *Journal of Alternative and Complementary Medicine* 14: 241–250.
Van Wijk, R., E.P.A. Van Wijk, F.A.C. Wiegant, and J. Ives. 2008b. Free radicals and low-level photon emission in human pathogenesis: State of the art. *Indian Journal of Experimental Biology* 46: 273–309.
Van Wijk, R., S. Bosman, J. Ackerman, and E.P.A. Van Wijk. 2008c. Correlation between fluctuations in human ultra-weak photon emission and EEG alpha rhythm. *Neuroquantology* 6: 452–463.
Van Wijk, E.P.A., R. Van Wijk, R. Bajpai, and J. Van der Greef. 2010. Statistical analysis of the spontaneously emitted photon signals from palm and dorsal sides of both hands in human subjects. *Journal of Photochemistry & Photobiology B* 99: 133–143.
Van Wijk, E.P.A., J. Van der Greef, and R. Van Wijk. 2011. Photon counts statistics in leukocyte cell dynamics. *Journal of Physics Conference Series* 329: 012021.
Vekaria, M. 2003. *Biophoton emission and intentionality*. Encinitas: Californian Institute for Human Science.
Walton, K.G., and D.K. Levitsky. 1994. A neuroendocrine mechanism for the reduction of drug use and addictions by transcendental meditation. *Alcoholism Treatment Quarterly* 11: 89–117.
Walton, K.G., and D.K. Levitsky. 2003. Effects of the transcendental meditation program on neuroendocrine abnormalities associated with aggression and crime. *Journal of Offender Rehabilitation* 36: 67–87.
Walton, K.G., and N.D.C. Pugh. 1995. Stress, steroids, and “ojas”: Neuroendocrine mechanisms and current promise of ancient approaches to disease prevention. *Indian Journal of Physiological Pharmacology* 39: 3–36.

Walton, K.G., N.D.C. Pugh, P. Gelderloos, and P. Macrae. 1995. Stress reduction and preventing hypertension: Preliminary support for a psychoneuroendocrine mechanism. *Journal of Alternative and Complementary Medicine* 1: 263–283.
Werner, G. 2010. Fractals in the nervous system: Conceptual implications for theoretical neuroscience. *Frontiers in Physiology* 6: 1–15.
Westerheide, S.D., and R.I. Morimoto. 2005. Heat shock response modulators as therapeutic tools for diseases of protein conformation. *Journal of Biological Chemistry* 280: 33097–33100.
Yadav, R.K., R.B. Ray, R. Vempati, and R.L. Bijlani. 2005. Effect of comprehensive yoga-based life style modification program on lipid peroxidation. *Indian Journal of Physiological Pharmacology* 49: 358–362.

# Mindfulness in German Schools (MISCHO): A Specifically Tailored Training Program: Concept, Implementation and Empirical Results

**Vera Kaltwasser, Sebastian Sauer, and Niko Kohls**

**Abstract** Research on mindfulness and its relationship with health, distress and well-being has blossomed over the last decades. Although there are many studies highlighting the importance of mindfulness for health-related parameters, empirical research has also started to look into the not primarily health related aspects of mindfulness such as cognitive attention, self-regulation, empathy, social and ethical competence and creative thinking. As many of these aspects are not only related to health but also to social, moral and professional behavior, mindfulness has also become a focus of interest in the educational sciences and pedagogy. In this chapter, we provide some background relevant for understanding the discourses about mindfulness in the context of education and discuss how mindfulness programs may be implemented in school settings and additionally describe the prerequisites for school teachers who want to teach mindfulness. Finally, the development and pilot-evaluation of the Mindfulness in Schools Program (MISCHO, the German program name is AISCHU® – Achtsamkeit in der Schule) is described.

V. Kaltwasser (✉)
Elisabethen-Gymnasium Frankfurt, Hessen, Germany
e-mail: VKaltwasser@aol.com; kw@elisabethenschule.net

S. Sauer
GRP – Generation Research Program, Human Science Center, University of Munich, Munich, Germany
e-mail: sauer@grp.hwz.uni-muenchen.de

N. Kohls
University of Applied Sciences Coburg, Integrative Health Promotion, Friedrich-Streib-Str. 2, D-96450, Coburg, Germany

Brain, Mind and Healing Program, Samueli Institute, Alexandria, USA
e-mail: niko.kohls@hs-coburg.de

S. Schmidt and H. Walach (eds.), *Meditation – Neuroscientific Approaches and Philosophical Implications*, Studies in Neuroscience, Consciousness and Spirituality 2,
DOI 10.1007/978-3-319-01634-4_20, 

## Introduction

Research on mindfulness and its relationship with health, distress and well-being related variables has blossomed over the last decades. Mindfulness can be understood in secular terms as the mental ability to focus on the direct and immediate perception of the present moment with a state of non-judgmental awareness and the voluntarily suspension of evaluative cognitive feedback (Hayes and Shenk 2004). There are many studies highlighting the evidence that mindfulness is beneficial for psychological and physical well-being both in clinical and nonclinical populations (Baer 2003; Bohlmeijer et al. 2009; Eberth and Sedlmeier 2012; Grossman et al. 2004; Sedlmeier et al. 2012). In recent years, empirical research has also started to look into the not primarily health related aspects of mindfulness such as increase of cognitive attention, self-regulation, empathy, social and ethical competence and creative thinking that are not only of high relevance to educational, learning and managerial processes but also pivotal for social, professional and moral behavior (Dane 2011; Himelstein et al. 2012; Ostafin and Kassman 2012; Sauer et al. 2011a; Sauer and Kohls 2011; Zelazo and Lyons 2012). It is for that reason that mindfulness has also become a focus of attention in many branches of the educational sciences and pedagogy, for example in the context of school and university education as well as advanced vocational training (Albrecht et al. 2012; Huppert and Johnson 2010; Lynch et al. 2011; Walach et al. 2007). There is a growing consensus among neuroscientists, educators and teachers that it is pivotal for educational processes to not only learn to memorize, evaluate and contextualize factual information but also to attain self-regulation strategies as well as to acquire metacognitive problem-solving skills preferentially at an early stage in the educational trajectory. Practicing mindfulness may be seen as perfect venue for developing these virtues and competencies as it allows individuals to learn to intentionally monitor, assess and modulate cognitive, emotional and motivational processes associated with behavior (Farb et al. 2007). As a consequence, successive behavioral patterns may become less impulsive and ego-centered and actually more reflected and socially benevolent. In short, mindfulness does not only allow an individual to deal with the distress of everyday life and acquire generic problem solving skills, but also to develop self-reflection, empathy and social and moral behavior. It is obvious that this would offer advantages both for the individual as well as the natural and social habitat.

The first part of this contribution deals with pivotal questions associated with bringing mindfulness into application in school settings such as:

- Why is it important to introduce mindfulness to the German school system?
- How can mindfulness be implemented in everyday school lessons?
- What are the prerequisites for teachers who want to teach mindfulness?
- How should a curriculum of mindfulness be designed so that it is effective, applicable and scientifically grounded?

As our background is the German school system, we can only offer experiences and insights associated with this context, but we are confident that most statements may be generalized and transferred to another cultural and educational context. In the second part of this contribution, evaluation results from a pilot project bringing mindfulness to high schools are reported.

## The Comprehensive Conceptualization of Education as 'Bildung' in the Humboldtian Tradition

According to the Humboldtian tradition, the German term for "education" – *Bildung* – refers to the continuing process of extensively forming and consciously developing an individual's intellectual, emotional and social abilities and human virtues. This may be achieved by an active process of contemplating, evaluating, ordering and framing conscious sensations of the intérieur and extérieur milieu in a systematic, conscientious, prudent, empathic and compassionate way. *Bildung* refers not only to an active but actually interactive process. The ability to perceive, understand and appreciate other individuals is naturally dependent on the capability to sense and interpret one's own experiences, thoughts, emotions, motives, desires and expectancies as well as the aptitude to gauge the internal and external resources. Research concerning the effects of mindfulness has a special appeal here, as it indicates that both stereotypic thinking and autopilot-like behavioral patterns may be reduced by means of practicing mindfulness (Langer 1989).

## Autopilot Mode, Mindlessness and Mindfulness

Instinctual behavior in situations which are perceived as dangerous is life-saving. In addition to the evolutionarily shaped automatized behavioral patterns, there are also individual autopilot structures that have been developed in the course of an individual's development. For example, emotional behavior patterns and ways of communicating with others are deeply ingrained. These autopilot patterns, although they can be life-saving, can also become obstructive. Particularly if specific overlearned patterns of behavior are what allowed an individual to succeed in the past, human beings are prone to permitting the autopilot mode to take control over their minds and bodies. In sharp contrast, acting consciously in reflected and self-determined ways requires getting out of the autopilot mode. Only if we become aware of how severely ingrained autopilot-like behavioral structures are and how they may restrict the individual freedom of choice, we then may be able to acknowledge the necessity of cultivating and enhancing our introspective skills. Technically speaking, purely observing from moment to moment without evaluating the sensations that are present is exactly what characterizes mindfulness, and research has not only shown that everyone is capable of being mindful but also corroborated that the ability to be mindful can be trained (Brown and Ryan 2003; Davidson et al. 2003). This is what the inaugurator who brought mindfulness into a modern medical context, Jon Kabat-Zinn, has described as cultivating a certain mindset that is characterized by a certain form of benevolent curiosity nurturing the attitude of "a beginner's mind" (Kabat-Zinn 1991, 1994).

We believe that by cultivating and fostering this spirit of mindfulness in a classroom setting, teachers may be able to initiate an educational transformation process in their students that does not only help improve learning capacities and academic performance but actually fosters personality development also: If students are able to understand how their experiences are being shaped by expectancies, habitualized assumptions, belief-systems and implicit cognitive-emotional evaluations related to their biography, they may become gradually more aware of the component of relative constructivism inherent to every perceptional process. As a consequence, not only may the ability to deal with distress and hardship increase, but the actual way students perceive and appraise themselves, their peers and the environment they are embedded in may also profoundly change. For example self-imputation on the basis of consistent behavioral patterns ("I am just a shy person") may be interpreted as stereotyped and unreflected pattern of self-attributions which can be changed. This is important because teenagers and adolescents are at a higher risk of suffering from low self-esteem, distress and mental problems (Laufer and Laufer 2011). Becoming more mindful of these inner processes may lead to improved self-regulation and enhanced resilience resulting in a firmer and more benevolent self-model (Sauer et al. 2011b).

## The Impact of Neuroscientific Insights Upon Contemporary Learning Concepts

In German pedagogical concepts the role of the body has traditionally played only a minor role. This conceptual assumption has left an imprint on German teaching concepts, as the importance and dominance of abstract, rational thinking has customarily been emphasized in German school and university education. In brief, topics of education were organized in a threefold manner, as cognitive-rational, contextual-emotional and body-oriented. However, when books such as Daniel Goleman's treatise about emotional intelligence were published in the middle of the 1990s, popularized insights stemming from Howard Gardner's work about the close interdependency between body, mind and emotions were made available to a greater public, reaching therefore also educators and teachers (Goleman 1995). About the same time Damasio's theory on the deception of Cartesian thinking additionally fanned the assumption that the Brain ~ Mind ~ Body relationship has to be seen as mutually interdependent, with reason and emotion representing complementary aspects of the mind (Damasio 1994, 1999). In 2002, German neuroscientist Manfred Spitzer started to familiarize German teachers and educators with the latest discoveries of neuroscience and highlighted the importance of emotional processes and distress for learning, memorization and recollection processes (Spitzer 2002).

How have these debates affected pedagogy and school education? First of all, there was growing interest in the neuroscientifically grounded exploration of learning processes. Second, as a consequence of the "emotional shift" within neuroscience and philosophy of mind (Dalgleish 2004; Davidson 2012; LeDoux and Bemporad 1997), the pivotal role of emotional regulation for learning was increasingly reflected in the scientific literature related to didactics, pedagogy and education. The importance of emotional and motivational processes for successful conscious state-oriented learning was thereby "rediscovered", but the numerous publications and professional debates concerning the role of embodiment, emotion and intrinsic motivation have only rarely exhibited impact on the prevailing teaching curricula and didactical concepts. Concomitantly, educational demands and standards were raised and, as a consequence, the work load and amount of tests and exams has increased. New methods for testing school performance such as multiple-choice tests were also introduced. This has in turn potentially fanned distress levels in pupils and students.

## Factors Associated with Positive Learning Experiences and Favorable Outcomes

What we know about factors associated with a positive learning experience and favorable learning outcomes can be concisely summarized:

First of all, it is well known that students are able to improve the learning outcome when they are highly and intrinsically motivated, i.e. they have a personal and emotionally fanned interest in the topic that is congruent with the contextual setting. Many research studies have empirically proven that individuals learn best when they are aware of the relevance of the learning subject and therefore attribute meaning and importance to the learning process.

It is additionally helpful if individuals become actively engaged in the learning process, are given tangible feedback about their development and are given enough time and sufficient but not excessive degrees of freedom so that they may look for ways of how to solve a problem on their own without losing track of the problem (Spitzer 1996, 2002; Hattie and Timperley 2007).

It is also important that students are neither under-challenged nor overburdened with their learning tasks. Empirical studies have shown that the relationship between arousal and performance can be described as a curvilinear, inverted u-shaped function known as the Yerkes-Dodson law (Yerkes and Dodson 1908). This rule mandates that performance increases with arousal level, until a certain point is reached. After that point, the level of arousal and correspondingly the amount of allostatic load become too high and performance decreases.

## High Distress Levels Among German Students and Teachers

The overall workload and general pace of learning processes in German schools is increasing. A clear sign of this is the fact that recently the number of advanced secondary school years was reduced from 13 to 12 years, without a corresponding reduction of the quantity of the teaching curricula. With the introduction of the *Programme for International Student Assessment* (PISA) in 2000, the scholastic performance of 15-year-old school pupils on mathematics, science, and reading was compared among nations. As a consequence of intermediate results, various projects aimed at introducing and improving quality management were implemented in all school types across Germany. Though it is understandable that educational standards should be transparent and comparable, we opine that there is a tendency visible in PISA to overestimate factual knowledge, while the importance of knowledge mining, problem solving and skills of self-regulation capacities is not (fully) acknowledged. As the school curricula are closely aligned with academic performance parameter associated with PISA, students are correspondingly frequently confronted with informational overload on a quantitative and qualitative level without being taught the tools of collecting, analyzing, organizing and interpreting the provided information in a way that allows them to attribute personal meaning to the abundance of factual knowledge.

Distress levels among German pupils and students were found to be high: For example, a recent representative survey commanded in 2009 by one of Germany's largest governmental health insurance bodies revealed that 90 % of German pupils and university students complain about distress and 30 % report to be permanently or frequently distressed; 37 % of the pupils additionally reported to be afraid of their professional future (FAZ-Institut and Techniker-Krankenkasse 2009). Although the main factors associated with increased distress were exams and workload, empirical findings also suggest that addiction to new technologies, particularly to online social networking, in young people may become a new mental health challenge for society (Byun et al. 2009; Kuss 2012). There is increasing evidence suggesting that the overload stemming from external digital stimuli may not only lead to attentional, cognitive and emotional deficits but also result in a decline of social competencies (Spitzer 2012; Turkle 2011). As restricting access to social media in a free and open society is not a viable option, we believe that an educational system has the moral obligation to address upcoming issues such as rapid development of digital technology that may under some circumstances posit a potential threat to health and well-being. It is therefore a pivotal task for teachers to accompany students during this journey to a poignant understanding of their personalities, so that they can develop to be responsible agents for themselves and adequately prepared to be operative in the socio-technological context they are embedded in. Of course these demands put an extra burden on teachers, often exaggerated as they are frequently not appropriately prepared for properly dealing with such situations.

It is important to recall that distress levels among German teachers are already high, with empirical studies suggesting that up to 30 % of German teachers may be affected by severe mental health problems (Bauer et al. 2007; Unterbrink et al. 2007). It has been suggested that teacher's habitual judgmental patterns about their students' behavior may contribute to repeated experiences of unpleasant emotions such as tension, frustration, anxiety, anger, which may eventually lead to the development of a chronic distress syndrome (Chang 2009). Although the atmospheric climate in schools has probably changed as a consequence of the raised educational standards, several recent studies found that those teachers with training in gestalttherapy or gestaltpedagogy or mindfulness-based interventions are less prone to suffer from mental health issues (Dauber and Döring-Seipel 2009; Gold et al. 2010; Schaarschmidt 2007). These findings suggest that those teachers with additional introspective training may be able to not only improve their coping strategies with distress but that they are also prone to establish more satisfying and productive relationships with their students, thereby supporting the learning process.

## An Extended Educational Paradigm: Empowering Self-Regulation and Self-Learning Capacities in Addition to Teaching Factual Knowledge

A necessary prerequisite of successful and meaningful learning experiences that naturally result in personal growth is the insight of each individual that he or she is not a passive victim of learning procedures but actively responsible for their personal and academic success. But this insight must be met and nourished with appropriate teaching methods that focus on enhancing self-regulation capacities such as impulse-control, self-awareness, and self-efficacy. Developing self-regulation capacities may not only enhance resilience, coping with distress and cognitive abilities, but also increase empathy and pro-social behavior.

We will show in the remainder of this contribution that mindfulness based interventions, which are specifically tailored to the needs and demands of teachers and pupils, are a promising venue for achieving this goal. One may argue that these competencies should be primarily established in the natural habitat of young children, i.e. their families, and correspondingly that parents and relatives are responsible for this basic enculturation. This may be true in an ideal world, but there are many reasons why emotional and social education has gradually been transferred to schools. While this can be deplored as a sign of the demise of family responsibility, seen from a pragmatic point of view schools must deal with these changed circumstances because children and adolescents who are not capable of self-regulating, delaying gratification, or focusing attention cannot learn effectively.

## Mindfulness in Schools (MISCHO): A Concept for Bringing Mindfulness into the Classroom

Mindfulness in Schools (MISCHO, German: Achtsamkeit in der Schule [AISCHU®]) is a program in which specially trained teachers introduce and practice mindfulness during class with their students. It has been developed by Vera Kaltwasser as a 12-week mindfulness curriculum specifically tailored to the needs of students on the basis of her longstanding experience as a teacher, teacher trainer, MBSR trainer and Qiigong teacher. The program is designed mainly for the use in secondary education. So far this program has been only described in detail in German (Kaltwasser 2008, 2010). It has similarity with other mindfulness programs for education (Meiklejohn et al. 2012) but also some unique elements. It was designed for pupils from the age of 10 years onwards (up to the end of high school) and can be integrated into the everyday school lessons at school by teachers who have been especially trained and who have a regular personal meditation practice. MISCHO combines psychoeducative training with experiential learning components. Vera Kaltwasser's experience with MISCHO suggests that teachers with mindfulness training are more capable of establishing a suitable atmospheric climate for learning as well as dealing with daily demands in the classroom. However, systematic training in mindfulness-based interventions seems to be a prerequisite for learning how to teach mindfulness to students: Only teachers who have both a substantial amount of systematic personal experience with introspective/meditative practices and who are able to incorporate the attitude of mindfulness into their everyday teaching activities are able to systematically convey this attitude in a classroom setting in an authentic and therefore appropriate way. This is why we consider it absolutely indispensable that teachers are not only psychoeducationally trained but also practicing a form of mindfulness training such as yoga, qigong or related forms of mind-body-oriented practices on a regular basis (Walach et al. 2012). This combination of psychoeducational literacy with mind-body training will allow teachers to develop an intuitive understanding about which exercises are beneficial for the students and how they can evoke and uphold their motivation.

MISCHO therefore consists of three components that build upon each other and the teacher can implement the training in his everyday lessons in the following way:

1. This first part contains an experiential phase including several exercises that provide insights into the body-mind link from the first-person perspective. This experiential knowledge is then underpinned by psychoeducational components that impart basic neuroscientifically grounded knowledge about the mind-body relationship and explain underlying psychophysiological mechanisms so that students have the ability to understand the complementarity between neuronal bottom-up and top-down-processing. This theoretical input provides a conceptual context for reflecting the experiential parts of MISCHO. The introductory phase entails four 45-min lessons.
2. The second phase comprises various forms of formal practices and exercises in a classroom group setting. Students are trained at least three times a week for only 10 min. This small allotment of time is supposed to ease the implementation of MISCHO in a classroom setting without consuming too many time resources.
3. The third phase includes individual mindfulness practices that are practiced in the context of everyday life. While it is desirable to practice mindfulness as frequently and regularly as possible under real-life conditions, targets and practices are individually determined.

Table 1 provides an overview about the three phases of the MISCHO curriculum and describes the respective subcomponents:

**Table 1** Conceptual summary of the Mindfulness in Schools (MISCHO) Program Phase

| | Overall aim/description of component |
|---|---|
| 1. Introduction | Setting the stage, creating and enhancing student motivation and providing an experientially grounded conceptual knowledge platform (psycho-education) by encouraging students to experience the body-mind-inter-connectedness and autopilot/automatic responses. The "experiments" are done in a playful way so that pupils are encouraged to be curious and open to their experiences |
| Session 1 (45 min) | "Raisin-Experiment – Becoming a phenomenological researcher": Psychophysiological reactions associated with exploring the raisin such as saliva being produced while putting the raisin to the lips/the smell of the raisins triggering pleasant or unpleasant memories that can be experienced and observed in the present |
| | "Stroop-Effect – the power of automatized patters": demonstrates the influence of interference as a result of inconsistent stimuli in the reaction time of a task. In the classical Stroop-Test the name of a certain color (e.g., "blue," "green," or "red") is presented in another color (e.g., the word "blue" printed in yellow ink). Over 700 studies show that naming the color of the word takes longer and is more prone to error than when the color of the ink fits with the name of the color represented by the word |
| | "The invisible Gorilla" – allocation of attentional resources: Examples such as the famous invisible Gorilla (focusing attention at the cost of ignoring details) are used to demonstrate that human minds do not work the way human beings think they do (Chabris and Simons 2011) |
| | "The attentional blink – how sensory input is processed": Attentional blink is a perceptional phenomenon that occurs, when a sequence of visual stimuli is presented in rapid succession at the same spatial location on a computer screen. A participant will often not be able to detect a second stimulus presented in succession if it is presented between 200–500 ms after the first one. (recommended for older students, age 16 onwards) |
| Session 2 (45 min) | "Ill-tempered mood and shabbily walking go hand in hand":<br>The impact of posture on mood and emotional behavior is systematically explored |
| Session 3 (45 min) | "The conscious mind is not a master in its own house, because it has been sublet to unconscious processes": Personal experiences with the "experiments" in session 1 & 2 (becoming aware of the power of unconscious impulses) are reflected and students are encouraged to reflect and share biographic experiences when they were not in control of their mind and behavior (e.g. unable to focus and concentrate, habit of binge drinking/eating, craving, falling prey to hostile and aggressive thoughts and emotions) |
| Session 4 Psychoeducation | "Basic Neuroscience for Kids and Teenagers": Basics of neuroscience relevant for understanding involuntary and voluntary psychophysiological reactions such as the three part brain, autopilot patterns, overlearned reactions, mental representations and stereotypes, neuronal bottom up/ top down processing, (patho)physiology of distress and the role of the prefrontal cortex for gratification delay and impulse control are introduced |
| 2. Formal practice in the classroom (3 × 10 min/ week) | Setting the stage for ongoing training in the classroom, and enhancing motivation for training – getting to know oneself better, enhancing self-awareness and emotional regulation capacities, enabling students to switch from "mode of doing" to "mode of being" (Segal et al. 2002) |

(continued)

**Table 1** (continued)

| | Overall aim/description of component |
|---|---|
| Mind-body exercises trained in a sitting position | |
| 1. Basic Breathing Exercise | Introducing ritualized exercise patterns in a distinct bodily posture, sitting upright, eyes closed, palms up on thighs, rubbing palms together at the end of the exercise<br>*Instruction:* "Watching your breath without changing it, if your attention is wandering somewhere else, bring it back to the breath" |
| 2. Abdominal breathing/deep breathing exercise | Deep diaphragmatic breathing is characterized by contracting the diaphragm (a muscle located in a horizontal position in the middle of the thoracic and stomach cavity) so that an expansion of the abdomen rather than the chest is occurring |
| 3. Basic Body Scan exercise | Body scan is an interoceptive and introspective exercise where the individual body is mentally scanned for physical sensations. Body scan exercises start with the hands and/or feet (awareness of the fingers, one by one or awareness of every toe) |
| Mind-body exercises trained in a standing position | Several introspective exercises practiced in a standing posture derived from Qigong are trained |
| Guided Imagination Exercises | Guided imagination exercises are utilized for enhancing relaxation, self-awareness and self-calming by harnessing the abilities of creative imagination. Exercises include "the still and magical place – the sanctuary" as well as "scenes of joy" |
| Mental exercises/ exploring cognitive patterns and resulting emotional behavior | Mental exercises and the exploration of cognitive patterns and their resulting emotional behavior are trained in order to detect habitual patterns such as inner dialogue dynamics, discover and reduce habituated judgments as well as identify personal triggers for emotional behavior |
| Interpersonal and interactive exercises<br>The mindful dialogue | The 'mindful dialogue' has been specifically designed to build and refine the interpersonal and group dynamics skills of students (more details are provided later on in this chapter). The 'mindful dialogue' should only be introduced when the students feel at ease with the silent sitting exercise. It is frequently introduced after the 12-week-training. The practice is initiated in class at times that the teacher finds appropriate (e.g. there is a so-called "class-lesson" in which personal and social issues are being discussed) |
| 3. Individual mindfulness practice in everyday life enhances self-compassion and acceptance for others | Everyday practice allows individuals to become more aware of stressful situations, understand personal cues for stress, practice shift of mode, notice stereotyped personal concepts in interaction, practice new ways of behavior in a playful way (e.g. "mindful eating", "mindful talking"), and become more benevolent to self and others |

## A Detailed Description of the Components of the MISCHO Program

Having introduced the reader to the basic concepts of MISCHO as well as the prerequisites, we will now describe several aspects necessary for its successful implementation. We also discuss certain aspects that need be taken into account when MISCHO is taught to students of different age cohorts.

### *Introduction*

#### Kindling Motivation

The most important goal of the initial phase of the MISCHO program is setting the stage for the manifestation of mindfulness in a classroom setting. In other words, to kindle a motivational spark for embracing the spirit of mindfulness, an atmosphere of acceptance and non-judgment is essential. At the same time it is important to stimulate students' curiosity so that they are keen to take on the role of a researcher scrutinizing their mind in a systematic way. It is certainly beneficial that mind-body practices have become quite well-known and even popular in recent years. This augmented acceptance of mindfulness and curiosity for "mindfulness tools" has certainly paved the way for bringing mindfulness into the realm of education. However, it is still challenging for teachers to introduce exercises of mindfulness to students, as children and adolescents can be wary of unfamiliar "consciousness techniques" and somewhat afraid of introspective processes. Teens are particularly sensitive to engage in things that they assess as childish and "weird".

As such the first steps of curriculum implementation in the classroom setting have been carefully designed so that initial resistance of students – if there is any – can be reflected and become part of the learning process. The teacher's function as a role model for embodying mindfulness is very important because (s)he can set a real-life example by carefully listening and not immediately judging the students' comments but rather appreciating them even if they seem obstructive or inapt. By welcoming all that comes up in the process of teaching mindfulness, the teacher is able to embody the core attitude of mindfulness – being totally present in the moment, putting trust in the emerging group process without judging it. If a teacher has been able to incorporate such an attitude, (s)he can systematically improve the social atmosphere in the classroom. (S)he then is no longer a teacher who just instructs "mind-body-training", but rather embodies the essence of mindfulness. A mindful teacher conveys this attitude by delivering an implicit but atmospherically tangible message that can be both soothing and inspiring to students: "You are appreciated exactly the way you are and you do not have to strive to do/be better, be more intelligent and more head-and-shoulders above average to be acknowledged."

The relief students may start to feel in such an appreciative atmosphere cannot be overestimated, as mindfulness can become contagious in a positive way both for students and teachers.

### Phenomenological Experiments and First Person Oriented Mind-Body Research

Practicing mindfulness by means of systematic and guided introspection and self-awareness trainings opens an experiential space that allows individuals to investigate and analyze their mental processes in a profound way, i.e. getting to know emotions, cognitions, likes and dislikes, their bodily sensations and impulses in a private and existential manner even while embedded in a group setting. However, students are not only enabled to become phenomenological researchers of their own mental realm but also encouraged to reflect and share "phenomenological research results" with their fellow students.

Particularly the close interlink of the mind-body relationship that has frequently been described in scientific essays, can be explored by means of first person oriented mind-body research (Maturana and Varela 1975, 1980; Varelaet al. 1974). The students are encouraged to experientially acquire the insight that the interpretation of cognition, emotion, language, and consciousness are observer-dependent. However, many inner mental states are rather opaque to the observer, leading to automatic and involuntary reactions and behavioral patterns.

One of the phenomenological experiments that will be described in more detail is an adoption of the well-known "raisin-exercise" which is part of the standard MBSR-program as it was devised by Jon Kabat-Zinn. The goal of the raisin-exercise is apparently simple: phenomenologically exploring – i.e. observing, touching, smelling, tasting, eating – a single raisin as mindfully as possible – as if one has never eaten a raisin before. However, the students are – in contrast to Kabat-Zinn's original instruction – asked to close their eyes so that they are not distracted by visual percepts so that they can pay full attention to their mental landscape, i.e. they can sense the various impulses arising when an unknown small object – a raisin – is put in their palms. Before the "experiment" the teacher will enhance curiosity by asking the students what will be put into their hands, while their eyes are closed. By meticulously paying attention to these inner processes the students become aware of the subtle interdependence between thoughts and emotions and body sensations. Students are invited to surrender to upcoming impulses and associated mental events but they are encouraged to withstand as long as possible. Of course, they are free to decide if they want to follow the instructions and they are invited to reflect upon the inner processes associated with obedience and/or defiance. Thus, students are able to become aware of concomitant cognitions and emotions such as curiosity, boredom or fear. "I hoped it would not be a dead fly!" was a statement made by one of the students. When the students are asked to discover the raisin by smelling – they detect how the olfactory sensory input may evoke positive or negative memories. When the students are asked to touch the "object" with their lips, they are able to experience how saliva production in their mouth is instantly increased. They are then asked to take the raisin into their mouth without allowing them to bite or swallow for a considerable time. When they are finally permitted to bite and chew, they are asked to voluntary delay swallowing and to observe the impulse of wanting to swallow and savor the taste of the raisin.

The described experiment is supposed to experientially teach students how complex automatized behavioral patterns such as eating may be split up in minute sequences such as smelling, chewing, tasting, and swallowing that are consciously perceived. Having finished the experiment, students are invited to mentally review and recall the process, again with their eyes closed. Finally students are invited to share their observations and comments about their mental experiment with each other, thereby revealing the variety of subjective cognitions and emotions associated with the experiment. For example one student told how an instant feeling of nausea came about when he smelled the raisin. He explicated that he was forced in his childhood days by his grandmother to eat cake with raisins although he could not stand the taste of them. Although it is obvious that this phenomenological experiment may be helpful for preventing eating disorders, the primary aim of this "research" is enhancing student motivation. Students are usually baffled by involuntary reactions associated with "the experiments". Experimental setups utilizing well documented mechanisms such as the Stroop effect show in a very impressive manner that mental processes cannot be easily controlled under certain circumstances (see Table 1). There are of course many more striking experiments of similar kinds that can be used as pragmatic tools for phenomenologically exploring and discovering the mechanism of human perception and awareness.

## *Formal Meditative/Introspective Practice*

After having stirred motivation and introduced the students to the concept of mindfully observing the mind and its advantages, it is necessary to introduce some regular and systematic ritualized training. Students are additionally informed about contemporary neuroscientific findings concerning mechanism and effects of meditation/mind-body practices so that they are able to contextualize their experiences. Although a variety of body-oriented exercises in a sitting or a standing position is available, we will shortly describe four types that we deem to be useful and viable in a classroom setting (see also Table 1).

### Sitting Meditation

This intervention is characterized by sitting in an upright but relaxed position for minutes. The legs are slightly apart – shoulder width – and the feet are flat on the ground with the hands palm up on the upper legs. This posture constitutes the formal commencement of the practice time, and at the end of the session the students are asked to rub their hands together as a sign of having completed this phase of formal meditation. The basic exercise is twofold: (1) Becoming aware of the breath and gently observing it without voluntarily changing it, (2) and redirecting attention back to observing the breathe when the focus of attention is shifted away. In this context Mark Williams terminology of "mode of doing" (our thoughts, our plans, our expectations etc.) versus "mode of being" (sensory awareness of the breath or/ and the body) is being conveyed to the students. The first meditation phase will only last 30 s. In the following sessions, training time will gradually be extended until the students are able to sit for 10 min without experiencing difficulties concentrating on their breath.

## Abdominal Breathing, Body Scan, and Other Practices Derived from Qi-Gong

Other forms of mind-body practices such as abdominal breathing and variations of the body scan are taught as well. Qi-gong (pronounced "chee-gung," translated as cultivation of body-mind) is an established ancient Chinese mind-body system comprised of postures, exercises, and breathing techniques as well contemplative, meditative and introspective procedures intended to improve and intensify the body's *qi*. According to traditional Chinese philosophy, *qi* is the fundamental life energy responsible for health and vitality. Qi-gong combines slow movements, awareness of the breath and imagination e.g. of agreeable images. As muscle tension is often an indicator of inner tensions the regulation of this hypertension by practicing these soft movements is beneficial.

## Guided Imaginations

Guided imaginations are introspective techniques that are used to mindfully "tone" and modulate cognitions and emotions by means of visualization techniques. The capacity to create positive mental images with affirmative emotional connotations can be taught, trained and refined. For example, students are guided to create or conjure up a "magical location" in their mind – a place such as the top of a mountain with a great view where they are enjoying the landscape and feel well and happy. Another exercise directs students' imagination towards a positive, peaceful, agreeable and joyful incident in their past. The students are instructed to explore the visualized image by observing their joyful memory as a mental film from different perspectives. They are first invited to watch the scenery from an external camera-like perspective and then instructed to switch to the first-person perspective thereby reminded to pay attention to concomitant cognitions, emotions and bodily sensations. The ability to evoke positive emotions such as joy and gratitude is an effective antidote against stress, anxiety, anger or fear.

### Work with Cognitive Patterns

When the formal mind-body exercises have been well established and practiced for a while, students have begun to understand how close the connection between body and mind is and how they themselves can make a difference in the way they communicate with their mental realm. At this point they are adequately prepared to start working on their cognitive patterns, because the formal practice has naturally exhibited impact on the way the students perceive themselves and the outer world. The most important insight they may have gained by practicing mind-body practices is the ability to increase their degrees of experiential and behavioral freedom. In short, if students become more aware of their automatic impulses and patterns they have the freedom of choice. They are able to postpone a demand for the sake of a voluntarily chosen alternative aim that can only be reached by delaying gratification. Another vital insight for the students is an understanding of the way we attribute meaning to situations in the outside world that is naturally associated with prior learning experience and therefore also a function of individual biography. This insight is important for dealing with stressful events, as we are only able to identify, understand and withhold automatic reactions if we are able to identify their triggers. This occurs naturally with increased awareness so that the freedom of choice is enhanced – choosing one's reactions voluntarily is an achievement on the road to personal independence.

### Mindful Dialogue

At this stage of the MISCHO curriculum the "mindful dialogue" is introduced to the students. It is a method originally developed by Gregory Kramer (Kramer 2007) that has been adopted to fit the demands of a classroom environment. In short, applying the concept of mindfulness to communication and interaction processes allows both the speaker and listener to direct their full attention to the conversation with an appreciative attitude.

The exercise of the "mindful dialogue" is structured in such a way that the complexity of the communication process can be closely scrutinized and investigated by the students.

Given that impersonal communication such as e-mail, texting have steadily increased, the potential power of mindful communication cannot be underestimated. The title of Sherry Turkle's book "Alone Together" underlines how electronic media have established virtual communication paths that can be detrimental (Turkle 2011). Children and adolescents who prefer impersonal communication may be deprived of the ability to develop social skills and competencies that are vital for mutual understanding. Acquiring the competencies to decipher emotions, communicative signs, behavioral clues etc. and developing the ability to listen carefully are the functional and communicational building blocks of human society.

How is the "mindful dialogue" implemented in the MISCHO curriculum? Before the communicative dialogue is established, students are instructed to evoke a joyful experience in the past by means of guided imagination techniques (as described above). The students are then asked to choose a partner with whom they feel comfortable sharing their evoked memories. The apparently "simple" act of voluntarily choosing a suitable partner is dealt with in a conscious way: The students naturally pay attention to their motives, fears, hopes and expectations of finding a partner they deem suitable or inappropriate. While the corresponding mental processes are addressed by the teacher, they can now be experientially reflected by the students. In contrast, if the teacher selects students' dyads, students may be able to experience the difference when decisions are externally imposed at the cost of limiting the freedom of subjective choice. The teacher guides this process by drawing the students' attention to their immediate impulses and asking them to decide if they want to surrender to them or not.

For example, at the very beginning the students – having their eyes closed – are asked to internally decide if they want to begin to tell their joyful experience. What is their habitual role in conversations? Are they the person who takes the initiative, who talks a lot, maybe interrupts the counterpart, or are they prone to let the other one do the talking? Do they have difficulties finding something to say? Are they afraid of the reactions of their respective partners? These subtle internal processes can thereby be scrutinized and personal cognitive patterns can be detected; with this newfound awareness the student can playfully start to change habits. The narrator may decide to let the partner begin to talk and simultaneously the cognitions and emotions evoked by this new behavior can be experienced. "Trust emergence" is one of Gregory Kramer's guidelines suggesting that one has to put trust in oneself and the listener. The great challenge for the teachers teaching the "mindful dialogue" is creating an atmosphere of trust and acceptance.

The "mindful dialogue" is structured in a way that enables the student – having the eyes closed – to become aware of the complex internal realm and then – opening the eyes – either as a listener or a speaker to become aware of the complex situation of personal – face to face – communication.

## *Mindfulness Transferred to Everyday Life Creates Self-Compassion and Acceptance for Others*

Having obtained a basic understanding of the complex relationships between body and mind in the artificial context of the experiments, the next step is to bring mindfulness into everyday life. This entails the ability to monitor distress triggers such as the onset of resenting mental dialogues and the capacity to stop these procedures before they fan negative emotions and produce dysfunctional behavioral responses. It is therefore also important to introduce methods that enhance self-compassion (Neff 2011).

### Teaching MISCHO to Different Age Groups

The MISCHO curriculum as described above is best suited for pupils of the age of around 10 years.[1] Around the age of 10 neuronal self-regulation capacities have been developed to the extent that allow children to voluntarily focus their attention for a longer period of time (Siegel 2007). Pre-teens who have been trained in mindfulness may be able to benefit from this training during the phase of puberty as "they are being given the gift of prefrontal reinforcements for challenging times in the future" (Siegel 2007, pp. 274–275). Having acquired mindfulness competencies may be particularly fruitful because teens are frequently reluctant to share their inner feelings or partake in phenomenological experiments as described above. However, mindfulness training may bestow the ability to share inner states with less resistance. Once students arrive at an advanced stage in their high school career ("Oberstufe"), around ages 16–18, they frequently start to welcome mindfulness interventions because they realize that they are benefitting from them. For these age groups MISCHO contains more elaborate psychoeducative information components concerning the body-mind-relationship. Useful collaborations with biology teachers scrutinizing the physiology of the brain may additionally be established.

## A Pilot Evaluation Study of MISCHO in 10 Year Old High School Students

### *Aim and Design of the Study*

This pilot study aimed at investigating the effects of the MISCHO intervention in 10- to 12-year-old high school pupils on stress, well-being, attentional and emotional regulation capacity. Students attended the fifth class of a high school located in the Frankfurt region in Germany.

The class receiving the MISCHO intervention for a period of 12 weeks (September 2010–February 2011, vacation excluded) for 10–15 min three times a week consisted of N=29 students. Two other 5th grades consisting of N=25 and N=30 students were used as active and passive control groups, respectively. The active control group was provided with an active reading training with the same time intensity over the same period of time as the mindfulness intervention group, whereas the passive control group was not receiving any intervention. Table 2 provides sample characteristics:

A psychophysical test assessing distinct aspects of attentional capacities by means of gauging participants time discrimination ability (for the conceptual background,

[1] Teaching mindfulness to children at an earlier stage, for example in primary schools and even in kindergarten is nevertheless possible, although the curriculum has to be adapted to the developmental stage (Altner 2009). For example, vivid stories and definite, concrete mental images were found to be most adequate for this age group.

**Table 2** Sample demographics

| Class/group | Number of students | Age | Female gender |
|---|---|---|---|
| Mindfulness | 29 | 10.29 (SD=.64) | 36 % |
| Active reading | 25 | 10.46 (SD=.78) | 71 % |
| No intervention | 30 | 10.32 (SD=.78) | 36 % |

**Table 3** Measurements

| Measurement | Description |
|---|---|
| Duration Discrimination Ability | For the duration discrimination test, students were presented a standard experimental paradigm consisting of a pair of auditory stimuli and asked to assess which acoustic cue was presented for a longer period of time, the first or the second. One stimulus was always a standardized auditory sound signal at 440 Hz (standard stimulus). The target stimulus was an auditory stimulus of either positive ("someone laughing"), neutral ("a cow mooing") or negative emotional ("someone weeping") association. The standard stimulus was always presented either for 2 ("short sound") or 4 s ("long sound"), while the temporal length of the target stimulus was varied from +40 % to −40 % in steps of 6 differences. In total, 36 trials consisting of 12 positive, 12 neutral and 12 negative stimulus were conducted. 18 trials had the shorter and 18 other trials the longer stimulus |
| SSKJ 3-8 questionnaire | The SSKJ questionnaire (Questionnaire for the assessment of stress and stress coping in children and teenagers) is a psychometrically validated instrument that assesses vulnerability to distress, coping with stress as well as symptoms of distress during childhood and adolescence (Eschenbeck et al. 2006; Saile 2007). In this study, for the sake of brevity, apart from the vulnerability to distress and somatic and psychological symptoms of distress scales, only the constructive-palliative emotional regulation capacity was assessed due to its conceptual proximity to mindfulness |
| KINDL questionnaire | The KINDL questionnaire (a generic children's quality of life questionnaire) is a frequently used questionnaire instrument for assessing quality of life (QoL) as a function of psychological, physical, psychosocial well-being and daily functioning in children and adolescents (Bullinger et al. 1994; Ravens-Sieberer and Bullinger, 1998a, b). The instrument is based on a well-validated measurement model based on six subscales evaluating pivotal dimensions of children and adolescents QoL and well-being |

see the chapter by Wittmann & Schmidt in this volume) and self-attributed distress and well-being as well as emotional self-regulation capacity was collected before September 2010 (t1) and after the intervention in February 2011 (t2). An additional post-measurement was taken in June 2011 (t3) (Table 3).

## Results

Due to the small sample size of the study, differences in means between first and second (t1–t2) and first and third (t1–t3) measurement point were computed as effect sizes on the basis of Cohen's *d*. According to the convention, effect sizes $d<0.2$ indicate a small, $d\sim0.5$ a medium and $d>0.8$ a large effect.

## Duration Discrimination Ability

Figure 1 depicts the group outcome for the duration discrimination task for all three groups for all three measurement points.

## Questionnaires

1. SSKJ: Sum scores of the stress vulnerability as well as the (somatic and psychological) stress symptom (sub)scale(s) indicate high experienced distress levels, while high scores on the constructive-palliative emotional regulation capacity point to increased utilization of the respective coping strategy. No relevant differences between the three groups were found for vulnerability to distress (MISCHO: $d_{t1-t2}=.15$; $d_{t1-t3}=-.06$; active control: $d_{t1-t2}=.19$; $d_{t1-t3}=-.17$; passive control: $d_{t1-t2}=.34$; $d_{t1-t3}=.03$). In a similar vein, no relevant change in the mindfulness class was found for the somatic distress symptoms scale, while a sizeable difference was observed in the passive control group (MISCHO: $d_{t1-t2}=.25$; $d_{t1-t3}=.35$; active control: $d_{t1-t2}=.34$; $d_{t1-t3}=.30$; passive control: $d_{t1-t2}=.85$; $d_{t1-t3}=.18$). With regard to the psychological distress subscale, in comparison to the two other classes, the smallest increase in distress was observed in the mindfulness class (MISCHO: $d_{t1-t2}=.13$; $d_{t1-t3}=.26$; active control: $d_{t1-t2}=.40$; $d_{t1-t3}=.42$; passive control: $d_{t1-t2}=.60$; $d_{t1-t3}=.30$). Concerning the constructive-palliative emotional regulation capacity, the mindfulness class exhibited an increase, while a decrease respectively no change was observed in the active and

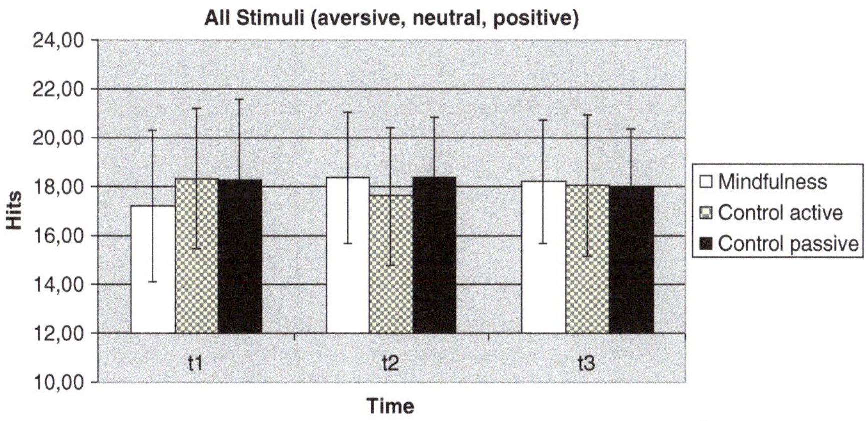


**Fig. 1** Total amount of correct hits only improved in the mindfulness class as indicated by the respective effect sizes (MISCHO: $d_{t1-t2}=.37$; $d_{t1-t3}=.32$; active control: $d_{t1-t2}=-.25$ und $d_{t1-t3}=-.10$; passive control: $d_{t1-t2}=.02$ und $d_{t1-t3}=-.09$). Subsequent analyses show that the effects are larger for negative (MISCHO: $d_{t1-t2}=.53$ und $d_{t1-t3}=.54$) and for longer acoustical stimuli (MISCHO: $d_{t1-t2}=.53$ und $d_{t1-t3}=.54$). Error bars indicate standard deviation

passive control groups (MISCHO: $d_{t1-t2}=.33$; $d_{T1-T3}=.20$; active control: $d_{t1-t2}=-.29$; $d_{t1-t3}=-.16$; passive control: $d_{t1-t2}=.02$; $d_{t1-t3}=-.05$).

2. KINDL: Higher sum scores can be interpreted as an indicator for a higher degree of self-reported health-related quality (QoL) of life of the respective child. While no substantial change in QoL was observed in the mindfulness class, QoL has decreased in the two other classes (MISCHO: $d_{t1-t2}=-.08$; $d_{t1-t3}=-.02$; active control: $d_{t1-t2}=-.57$; $d_{t1-t3}=-.68$; passive control: $d_{t1-t2}=-.38$; $d_{t1-t3}=-.11$). A similar result was obtained for somatic (MISCHO: $d_{t1-t2}=-.32$; $d_{t1-t3}=-.22$; active control: $d_{t1-t2}=-.76$; $d_{t1-t3}=-.50$; passive control: $d_{t1-t2}=-.71$; $d_{t1-t3}=-.16$) and psychological wellbeing (MISCHO: $d_{t1-t2}=.11$; $d_{t1-tT3}=.06$; active control: $d_{t1-t2}=-.50$; $d_{t1-t3}=-.54$; passive control: $d_{t1-t2}=-.54$; $d_{t1-t3}=-.19$). A possible explanation for the decreasing QoL over the course of the year may be an increased allostatic load due to performance tests. Family-related QoL has additionally increased in the mindfulness class, whereas it has decreased in the active and passive control class (MISCHO: $d_{t1-t2}=.18$; $d_{t1-t3}=.15$; active control: $d_{t1-t2}=-.46$; $d_{t1-t3}=-.58$; passive control: $d_{t1-t2}=-.27$; $d_{t1-t3}=-.24$). In contrast, no prae-post differences between groups were found for self-appreciation as well as friend- and school-related QoL.

## *Discussion*

The results of the duration discrimination test suggest that certain attentional aspects may have improved in the mindfulness class. However, from a statistical point of view no significant changes could be established, possibly due to differences in baseline, small sample size and random variation. It is thereby of interest to recall that particular acoustic cues associated with negative emotions as well as those cues displayed over a period of four seconds were more accurately differentiated by the students in the mindfulness class. This result may be interpreted as a "hard neuropsychological" effect suggesting that executive functions of attention and working memory improved in the mindfulness class. These cognitive functions have shown to be related to duration discrimination abilities (Pütz et al. 2012). Particularly the longer 4-s cues are more difficult to discriminate as their temporal length (two 4-s intervals with a pause in between last longer than 8 s) exceeds the temporal integration capacity of the "present moment" of around 3 s (Wittmann 2013). The question if the mindfulness class seems to have improved in certain cognitive functions as a result of the mindfulness intervention– as indicated by the effect sizes in comparison to the active and passive control group despite the lack of statistical significance – cannot be answered on the basis of the extant data and further research is warranted.

The impact of the mindfulness intervention can also be found in the questionnaire data, as the self-assessed constructive-palliative emotional regulation capacity was found to be improved only in the mindfulness class. Decrease in quality of life has been observed in other longitudinal data sets and may partially be interpreted as a shift of the internal reference point due to repeated measurement (Güthlin 2004).

The observed effects may cautiously be interpreted as practically relevant, although contextual effects may have occurred as a consequence of the non-randomly assigned group allocation. Nevertheless, if mindfulness interventions are supposed to be tested under real classroom situations, it is difficult to apply random group allocation procedures. However, despite the fact that the sample size was small and sample attrition occurred, we opine that the observed results are promising. From a pragmatic point of view, taken the feasibility aspect into account, our studies shows that mindfulness can be investigated in the classroom under real life conditions. While the effects sizes are small and we failed to obtain statistically significant group differences, it should also be taken into account that that the mindfulness intervention was only conducted three times a week for 10–15 min summing up to a total amount of 12 h of training over a period of 4 months. Therefore, assuming a linear trend and "dose–response" relationship of the impact of mindfulness on health, well-being and attentional capacities, it may be extrapolated that the systematic integration of mindfulness in every day school life could produce substantive effects on an array of variables relevant for health, well-being, attentional and cognitive performance, self-regulation capacities as well as social competencies.

## Conclusion

As we have shown in this contribution, the cultivation of mindfulness allows for investigating, understanding and reflecting oneself in a systematic way and thereby enhancing intellectual, attentional, emotional, and behavioral self-regulation capacities. We think that getting to know oneself by understanding the inner milieu is a prerequisite for understanding the outer world and other human beings. Being able to control one's impulses is the prerequisite not only for delaying gratification and enlarging self-efficacy, but also in the development of emotional and social competence. Moreover, being mindful can prevent an individual from seeking a "quick fix" of their feelings of unease by smoking, drinking alcohol, taking drugs or over-eating. Additionally, an increased awareness concerning the relativity of all individual perception processes and experientially grounded knowledge about the importance of implicit and automatized processes on thinking, feeling and behaving may even lead to individual transformation and development. Finally, being mindful allows the attitude of taking time and of experimenting with ideas and behavior "off the beaten track" as mindfulness may enhance creativity and out-of-the-box thinking. We believe on the basis of our pilot study that mindfulness can be easily and pragmatically implemented in the classroom and that both students and teachers may tremendously profit from a systematic integration of mindfulness in everyday school life. Further research on "mindfulness in education" is called for as well as the education of teachers to both learn and practice mindfulness and subsequently teach these skills to students.

**Acknowledgments** The pilot evaluation study was kindly supported by the German Qigong Society (http://www.qigong-gesellschaft.de/). Niko Kohls work is made possible through the long standing and generous support of the Samueli Institute, Alexandria USA. We are grateful for the support of Marc Wittmann, who has helped us with designing and interpreting the duration discrimination test and provided helpful comments on an earlier version of the manuscript. Jennifer Oyaga, Marcel Dickmann and Alexander Thölke helped with data management.

## References

Albrecht, N., P. Albrecht, and M. Cohen. 2012. Mindfully teaching in the classroom: A literature review. *Australian Journal of Teacher Education* 37(12): 1–14.

Altner, N. 2009. *Achtsam mit Kindern leben*. München: Kösel.

Baer, R.A. 2003. Mindfulness training as a clinical intervention: A conceptual and empirical review. *Clinical Psychology: Science and Practice* 10: 125–143.

Bauer, J., T. Unterbrink, A. Hack, R. Pfeifer, V. Buhl-Grießhaber, U. Müller, et al. 2007. Working conditions, adverse events and mental health problems in a sample of 949 German teachers. *International Archives of Occupational and Environmental Health* 80(5): 442–449.

Bohlmeijer, E., R. Prenger, E. Taal, and P. Cuijpers. 2009. The effects of mindfulness-based stress reduction therapy on mental health of adults with a chronic medical disease: A meta-analysis. *Journal of Psychosomatic Research* 68(6): 539–544.

Brown, K.W., and R.M. Ryan. 2003. The benefits of being present: Mindfulness and its role in psychological well-being. *Journal of Personality and Social Psychology* 84: 822–848.

Bullinger, M., S. von Mackensen, and I. Kirchberger. 1994. KINDL – ein Fragebogen zur Erfassung der gesundheitsbezogenen Lebensqualität von Kindern. *Zeitschrift für Gesundheitspsychologie* 2: 64–77.

Byun, S., C. Ruffini, J.E. Mills, A.C. Douglas, M. Niang, S. Stepchenkova, et al. 2009. Internet addiction: Metasynthesis of 1996–2006 quantitative research. *CyberPsychology & Behavior* 12(2): 203–207.

Chabris, C., and D. Simons. 2011. *The invisible gorilla: And other ways our intuitions deceive us*. New York: Broadway Books.

Chang, M.L. 2009. An appraisal perspective of teacher burnout: Examining the emotional work of teachers. *Educational Psychology Review* 21(3): 193–218.

Chiesa, A., and A. Serretti. 2009. Mindfulness-based stress reduction for stress management in healthy people: A review and meta-analysis. *The Journal of Alternative and Complementary Medicine* 15(5): 593–600. doi:10.1089/acm.2008.0495.

Dalgleish, T. 2004. The emotional brain. *Nature Reviews Neuroscience* 5(7): 583–589.

Damasio, A. 1994. *Descarte's error: Emotion, reason, and the human brain*. New York: Grosset/Putnam.

Damasio, A. 1999. *The feeling of what happens: Body and emotion in the making of consciousness*. New York: Harcourt.

Dane, E. 2011. Paying attention to mindfulness and its effects on task performance in the workplace. *Journal of Management* 37(4): 997–1018.

Dauber, H. 1997. *Grundlagen Humanistischer Pädagogik*. Bad Heilbrunn: Klinkhardt.

Dauber, H., and E. Döring-Seipel. 2009. Sind gestaltpädagogisch arbeitende Lehrerinnen und Lehrer gesünder? Ergebnisse einer empirischen Untersuchung zur Salutogenese. *Zeitschrift für Gestaltpädagogik* 20(2): 49–54.

Davidson, R. 2012. *The emotional brain*. London: Penguin.

Davidson, R.J., J. Kabat-Zinn, J. Schumacher, M. Rosenkranz, D. Muller, S.F. Santorelli, et al. 2003. Alterations in brain and immune function produced by mindfulness meditation. *Psychosomatic Medicine* 65: 564–570.

Eberth, J., and P. Sedlmeier. 2012. The effects of mindfulness meditation: A meta-analysis. *Mindfulness* 3(3): 174–189.

Eschenbeck, H., C. Kohlmann, A. Lohaus, and J. Klein-Heßling. 2006. Die Diagnostik von Stressbewältigung mit dem "Fragebogen zur Erhebung von Stress und Stressbewältigung im Kindes-und Jugendalter "(SSKJ 3-8). *Diagnostica* 52(3): 131–142.

Farb, N.A.S., Z.V. Segal, H. Mayberg, J. Bean, D. McKeon, Z. Fatima, et al. 2007. Attending to the present: Mindfulness meditation reveals distinct neural modes of self-reference. *Social Cognitive and Affective Neuroscience* 2(4): 313–322. doi:10.1093/scan/nsm030.

FAZ-Institut, & Techniker-Krankenkasse. 2009. F.A.Z.-Institut & Techniker Krankenkasse. *Kundenkompass Stress – Aktuelle Bevölkerung-sbefragung: Ausmaß, Ursachen und Auswirkungen von Stress in Deutschland*. Frankfurt am Main/Hamburg

Gold, E., A. Smith, I. Hopper, D. Herne, G. Tansey, and C. Hulland. 2010. Mindfulness-based stress reduction (MBSR) for primary school teachers. *Journal of Child and Family Studies* 19(2): 184–189.

Goleman, D. 1995. *Emotional intelligence*. New York: Bantham.

Grossman, P., S. Schmidt, L. Niemann, and H. Walach. 2004. Mindfulness based stress reduction and health: A meta-analysis. *Journal of Psychosomatic Research* 37: 35–43.

Güthlin, C. 2004. Response shift: Alte Probleme der Veränderungsmessung, neu angewendet auf gesundheitsbezogene Lebensqualität. *Zeitschrift für Medizinische Psychologie* 13(4): 165–174.

Hattie, J., and H. Timperley. 2007. The power of feedback. *Review of Educational Research* 77(1): 81–112.

Hayes, S.C., and C. Shenk. 2004. Operationalizing mindfulness without unnecessary attachments. *Clinical Psychology: Science and Practice* 11(3): 249–254.

Himelstein, S., A. Hastings, S. Shapiro, and M. Heery. 2012. Mindfulness training for self-regulation and stress with incarcerated youth A pilot study. *Probation Journal* 59(2): 151–165.

Huppert, F.A., and D.M. Johnson. 2010. A controlled trial of mindfulness training in schools: The importance of practice for an impact on well-being. *The Journal of Positive Psychology* 5(4): 264–274.

Kabat-Zinn, J. 1991. *Full catastrophe living: Using the wisdom of your body and mind to face stress, pain, and illness*. New York: Delacorte.

Kabat-Zinn, J. 1994. *Wherever you go, there you are*. New York: Hyperion Books.

Kaltwasser, V. 2008. *Achtsamkeit in der Schule: Stille-Inseln im Unterricht: Entspannung und Konzentration*. Weinheim: Beltz.

Kaltwasser, V. 2010. *Persönlichkeit und Präsenz. Achtsamkeit im Lehrerberuf*. Weinheim: Beltz.

Kramer, G. 2007. *Insight dialogue: The interpersonal path to freedom*. Boston: Shambhala.

Kuss, D. 2012. AS32-01-internet addiction or else? A way out of conceptual crisis and a call for nosological precision. *European Psychiatry* 27: 1.

Langer, E.J. 1989. *Mindfulness*. Reading: Addison-Wesley/Addison Wesley Longman.

Laufer, M.E., and M. Laufer. 2011. *Adolescence and developmental breakdown: A psychoanalytic view*. London: Karnac Books.

LeDoux, J., and J.R. Bemporad. 1997. The emotional brain. *Journal of the American Academy of Psychoanalysis* 25(3): 525–528.

Lynch, S., M. Gander, N. Kohls, B. Kudielka, and H. Walach. 2011. Mindfulness-based coping with university life: A non-randomized wait-list-controlled pilot evaluation. *Stress and Health* 27(5): 365–375. doi:10.1002/smi.1382.

Maturana, H.R., and F.J. Varela. 1975. *Autopoietic systems*. Biological Computer Laboratory BCL – Report 94. Urbana: University of Illinois.

Maturana, H.R., and F.J. Varela. 1980. *Autopoiesis and cognition: The realization of the living*, Boston studies in the philophy of science, vol. 42. Boston: Reidel.

Meiklejohn, J., C. Phillips, M.L. Freedman, M. Griffin, G. Biegel, A. Roach, et al. 2012. Integrating mindfulness training into K-12 education: Fostering the resilience of teachers and students. *Mindfulness* 3(4): 291–307.

Neff, K. 2011. *Self-compassion*. New York: Harper-Collins.

Ostafin, B.D., and K.T. Kassman. 2012. Stepping out of history: Mindfulness improves insight problem solving. *Consciousness and Cognition* 21: 1031–1036.

Pütz, P., P. Ulbrich, J. Churan, M. Fink, and M. Wittmann. 2012. Duration discrimination in the context of age, sex, and cognition. *Journal of Cognitive Psychology* 24(8): 893–900.

Ravens-Sieberer, U., and M. Bullinger. 1998a. Assessing health-realted quality of life in chronically ill children with the German KINDL: First psychometric and content anaytical results. *Quality of Life Research* 7: 399–407.

Ravens-Sieberer, U., and M. Bullinger. 1998b. News from the KINDL-questionnaire – A new version for adolescents. *Quality of Life Research* 7: 653.

Saile, H. 2007. Fragebogen zur Erhebung von Stress und Stressbewältigung im Kindes- und Jugendalter (SSKJ 3-8). *Zeitschrift für Klinische Psychologie und Psychotherapie* 36(2): 142–143. doi:10.1026/1616-3443.36.2.142.

Sauer, S., and N. Kohls. 2011. Mindfulness in leadership: Does being mindful enhance leaders' business success? In *Culture and neural frames of cognition and communication*, vol. 3, ed. S. Han and E. Pöppel, 287–308. Heidelberg: Springer.

Sauer, S., K. Andert, N. Kohls, and F. Müller. 2011a. Mindful leadership: Sind achtsame Führungskräfte leistungsfähigere Führungskräfte? *Gruppendynamik und Organisationsberatung* 42(4): 339–349. doi:10.1007/s11612-011-0164-5.

Sauer, S., S. Lynch, H. Walach, and N. Kohls. 2011b. Dialectics of mindfulness: Implications for western medicine. *Philosophy, Ethics, and Humanities in Medicine* 6(1): 10.

Schaarschmidt, U. 2007. *Gerüstet für den Schulalltag: Psychologische Unterstützungsangebote für Lehrerinnen und Lehrer*. Weinheim: Beltz.

Sedlmeier, P., J. Eberth, M. Schwarz, D. Zimmermann, F. Haarig, S. Jaeger, et al. 2012. The psychological effects of meditation: A meta-analysis. *Psychological Bulletin* 138(6): 1139–1171.

Segal, Z.V., J.M.G. Williams, and J.D. Teasdale. 2002. *Mindfulness-based cognitive therapy for depression: A new approach to preventive relapse*. New York: Guilford Press.

Siegel, D. 2007. *The mindful brain: Reflection and attunement in the cultivation of well-being*. New York: WW Norton.

Spitzer, M. 1996. *Geist im Netz. Modelle für Lernen, Denken und Handeln*. Heidelberg: Spektrum Akademischer Verlag.

Spitzer, M. 2002. *Lernen. Gehirnforschung und die Schule des Lebens*. Oldenburg: Spektrum Verlag.

Spitzer, M. 2012. *Digitale demenz*. München: Droemer.

Turkle, S. 2011. *Alone together: Why we expect more from technology and less from each other*. Cambridge, MA: Perseus Books.

Unterbrink, T., A. Hack, R. Pfeifer, V. Buhl-Grießhaber, U. Müller, H. Wesche, et al. 2007. Burnout and effort–reward-imbalance in a sample of 949 German teachers. *International Archives of Occupational and Environmental Health* 80(5): 433–441.

Varela, F.J., H.R. Maturana, and R.B. Uribe. 1974. Autopoiesis: The organization of living systems, its characterization and a model. *Biosystems* 5: 187–196.

Walach, H., E. Nord, C. Zier, B. Dietz-Waschkowski, S. Kersig, and H. Schüpbach. 2007. Mindfulness-based stress reduction as a method for personnel development: A pilot evaluation. *International Journal of Stress Management* 14(2): 188–198.

Walach, H., M. Gander Ferrari, S. Sauer, and N. Kohls. 2012. Mind-body-practices in integrative medicine. *Religions* 3: 50–81.

Wittmann, M. 2013. The inner sense of time: How the brain creates a representation of duration. *Nature Reviews Neuroscience* 14(3): 217–223.

Yerkes, R., and J. Dodson. 1908. The relation of strength of stimulus to rapidity of habit-formation. *Journal of Comparative Neurology and Psychology* 18: 459–482.

Zelazo, P.D., and K.E. Lyons. 2012. The potential benefits of mindfulness training in early childhood: A developmental social cognitive neuroscience perspective. *Child Development Perspectives* 6(2): 154–160. doi:10.1111/j.1750-8606.2012.00241.x.

# Index



S. Schmidt and H. Walach (eds.), *Meditation – Neuroscientific Approaches and Philosophical Implications*, Studies in Neuroscience, Consciousness and Spirituality 2, DOI 10.1007/978-3-319-01634-4, 

CPSIA information can be obtained
at www.ICGtesting.com
Printed in the USA
LVHW02*1918080718
583084LV00002B/138/P

9 783319 016337